EXPERIMENTAL
FOOD
SCIENCE

FOOD SCIENCE AND TECHNOLOGY
A Series of Monographs

EXPERIMENTAL
FOOD
SCIENCE 3RD EDITION

Marjorie P. Penfield
Ada Marie Campbell
The University of Tennessee, Knoxville

Academic Press, Inc.
Harcourt Brace Jovanovich, Publishers
San Diego   New York   Boston
London   Sydney   Tokyo   Toronto

Copyright © 1990, 1979, 1962 by Academic Press, Inc.
All Rights Reserved.
No part of this publication may be reproduced or transmitted in any form
or by any means, electronic or mechanical, including photocopy,
recording, or any information storage and retrieval system, without
permission in writing from the publisher.

Academic Press, Inc.
San Diego, California 92101

United Kingdom Edition published by
Academic Press Limited
24–28 Oval Road, London NW1 7DX

Library of Congress Cataloging-in-Publication Data

Penfield, Marjorie Porter.
    Experimental food science / Marjorie P. Penfield, Ada Marie
  Campbell. -- 3rd ed.
      p.   cm. -- (Food science and technology)
    Includes bibliographical references.
    ISBN   0-12-157920-4   (alk. paper)
    1. Food. 2. Food--Laboratory manuals.   I. Campbell, Ada Marie,
  Date. II. Title. III. Series.
  TP370.P37   1990
  664--dc20   90-294
                                                                  CIP

Printed in the United States of America
90 91 92 93   9 8 7 6 5 4 3 2 1

CONTENTS

PREFACE

As in the previous editions, this edition of *Experimental Food Science* has two main purposes: to present the scientific basis for understanding the nature of food, and to promote the principles of experimental methodology as applied to food. The book is intended for use in a first course in experimental foods.

The experimental approach will be the emphasis of most courses for which this book will be used. This new edition has been reorganized with an early presentation of basic information on methods, planning, and evaluation for those who will be doing individual work and for those who will only use the methods in class experiments. Although objective measurements should be made on products whenever possible, the lack of all or some of the equipment for making such measurements need not eliminate experiments from the class schedule. Some experiments require only the simplest equipment; improvised tests are described in Appendix D. Sensory evaluation is nearly always feasible. The chapter on sensory methods has been expanded to provide more specific direction for students. Scorecards for several tests as well as statistical charts for analysis of the results are included.

The format is slightly changed. Each chapter begins with an outline to present an overview and facilitate the locating of sections. Cross-references have been increased throughout the book. Celsius temperatures are used throughout, but a temperature conversion table is included in Appendix B. Expanded suggested exercises have replaced detailed experiments at the ends of chapters in order to provide needed experience in planning experiments and to permit the relating of experiments to the interests and resources of specific groups and individuals. Basic formulas and procedures are included in Appendix A.

As with earlier editions, incorporation of recent research findings and recognition of specific technological advances related to food have been major goals of the revision process. Several chapters have been expanded. The meat chapter has been divided into separate chapters on meat, poultry, and fish in recognition of increased interest in poultry and fish. Meat, poultry, and fish cuts now are shown, with scientific names of muscles, to facilitate understanding of the research literature. The chapter on flour includes a much expanded section on nonwheat flours. Technological developments and increased public interest are recognized in sections such as those on surimi, modified starch, alternative sweeteners, synthetic fats, extruded foods, pasta, and flat breads. The subject of microwave heating has been expanded, and some discussion of water activity has been incorporated. A separate chapter on microbiology expands the treatment of basic principles, with emphasis on foodborne illnesses. Emerging pathogens are covered as their importance continues to be recognized.

Emphasis on relationships between chemical and physical properties is continued. Some background in organic chemistry is assumed, as is some fundamental knowledge about food. The subject matter presentation draws heavily not only on the principles of related sciences, but also on published food research, both early and recent. The reference lists, though far from complete, are meant to be representative. It is hoped that the student will gain an appreciation for the tremendous contribution of early workers in the field, as well as for the advances resulting from more recent work and will recognize the dynamic nature of the field. The perceptive student will observe that new information results in new questions to be answered, and that new approaches to food study become possible as technological advances occur.

The student is encouraged to read both textbook material and research literature critically, to examine evidence presented, and to evaluate statements made and conclusions drawn. In this way the student will realize that reading should be an active, evaluating process, not a passive, accepting one.

Suggestions made by users of the earlier editions have been helpful during the revision process and are gratefully acknowledged. Students in many classes have provided inspiration through their questions, comments, needs, and interests. Their contributions are immeasurable.

PART I

FOOD
EXPERIMENTATION

FOOD EXPERIMENTATION

I. INTRODUCTION

A scientific approach to the experimental study of food is presented in this book. The experimental study of food is concerned with why foods are handled, processed, and prepared as they are, how and why variations in ingredients or treatments influence the quality of food, and how this knowledge can be used to improve the quality of food products. The scientific approach includes three basic steps: defining the problem and arriving at the hypothesis; testing the hypothesis in a carefully designed and controlled experiment; and accepting or rejecting the hypothesis in a report of the results. The scientific study of food is an exciting field of investigation. Adequate answers to some food-related problems have been found, often by applying sciences, such as chemistry, physics, biology, and microbiology, that are basic to the study of foods. Answers to other problems are only partial or not yet known. Incomplete answers and the unknown offer challenges to the food scientist. Students can find challenging questions to study in an experimental foods class.

This book is divided into two parts. Part I explores methods used in food science and should serve as a reference for laboratory work and for understanding the basic food science principles discussed in Part II. Understanding of the methods described in Chapters 3 and 4 will facilitate an understanding of the literature in food science. Part I can serve as a guide to the student who is starting an independent problem in experimental foods, who wants to expand the suggested exercises outlined at the ends of the chapters, or who wants to develop experiments using the formulas in the Appendix as a guide.

This textbook contributes to several sections of a research report. Part II is essentially a review of literature, both recent and older. The most current literature must always be reviewed, because information in food science changes rapidly; older literature provides the foundation for our understanding of food and its functional properties. Suggestions for reviewing the literature are included in Chapter 2. Appendix A contains experimental formulas. The procedures in Appendix A, although written as instructions, provide information for writing procedures for a report.

The other essential parts of the research report—the data, the discussion of the results, the conclusions, and the reference list—should be incorporated into the student's notebook or laboratory report. The reference list should give credit to the sources of information used in preparation of the report. Those sources should be cited in the text. For reference lists in this text and citations within the text, the Institute of Food Technologists' (IFT) style with some modification has been used. Details of this style are found in the style guide published by the IFT (1988). The style has been modified for this book to provide chapter numbers and inclusive page numbers. This will facilitate the process of obtaining copies of papers via interlibrary loan if not available to the reader. In addition, states are spelled out to facilitate information retrieval by international readers. For class reports, either this style or one of the styles used in other professional journals may be selected. It is important, however, that the same style be used throughout a report.

Suggested exercises at the end of each chapter include suggested experimental variations in treatment or ingredients. Basic formulas and procedures for some basic products are given in the Appendix. For other suggested exercises literature references may be given for formulas and procedure. The product from the basic formula serves as a control, in other words, a basis for comparison of the experimental products. Variations not described in the suggested exercise may better serve the needs and interests of a class. For example, related problems of current concern and interest may serve as a basis for class or individual projects—interesting projects may be suggested by class members. It is assumed that students using this book have some background in food science and thus can be involved in planning the details of experiments based on the suggested exercises.

Carefully controlled experiments with appropriate replication are necessary if the results are to be meaningful and, in the case of more extensive research studies, worthy of publication. Experiments should be controlled experiments. In other words, in a simple controlled experiment one factor (independent variable) is varied while all other conditions that might affect the results are controlled as much as possible. The effects of variation of that factor on selected quality attributes of the products (dependent variables) are measured. In a study of the effect of level of oat bran on the volume of muffins, level of oat bran is the independent variable and volume of the muffins is the dependent variable.

If more than one factor is altered (there is more than one independent variable), it is difficult without more than simple statistical analysis to determine the cause of any changes that might be observed in the quality of the products. More than one factor may be varied in a more complex experiment, which is called a factorial experiment. For example, to study the effects of both sugar and salt level on the quality of yeast bread a study could be designed to look at all possible combinations of sugar and salt levels. If three sugar levels and three salt levels are of interest, nine combinations would be studied. Statistical analysis of the data would make it possible to determine the effect of sugar level alone, the effect of salt level alone, and the effects of the various combinations. It is beyond the scope of this book to give a detailed description of the principles of experimental design and statistical analysis that must be applied in studies of this type. A general overview of experimental design is presented by Joglekar and May (1987). For the advanced student, several books on experimental design and/or statistical analysis might be consulted (Bender et al., 1982; Gacula and Singh, 1984; O'Mahony, 1986).

Temperature and humidity in the laboratory are difficult to control but may influence the results of an experiment. Faulty measurements, misinterpretation of the instructions, and variation in individual techniques are other possible unexplained variables that should be considered. A record of any unplanned variables should be made in the laboratory notebook so that they may be considered when the results are evaluated and discussed. An unplanned variable, if its presence is known, may be most helpful in explaining the results of an experiment. Research workers have sometimes changed the direction of work because of accidental occurrence of a condition that gave a clue to the solution of a problem. The procedures used to control the experimental conditions are discussed in this chapter. These procedures should be studied carefully and followed in individual or class experiments.

II. USING AND PRESENTING FORMULAS _____

Formulas (recipes) for products may be presented in several different forms. When using formulas it is important to note the method of presentation. Misinterpretation might result in faulty measurement of ingredients. The first method is common to consumer recipes and involves volume measurements of ingredients in household measuring devices. The second is the presentation of ingredients in units of weight. This might be metric units in scientific publications or customary units of pounds and ounces in quantity food production recipes or food production formulations. Additionally, ingredients in formulas may be presented as percentages. Two methods are used for this. The first is the for-

TABLE I
Four Methods of Formula Presentation

Ingredients	Customary units	Weight (g)	Formula (%)	Bakers (%)
Cake flour	2 c	192.0	23.7	100.0
Sugar	1¼ c	250.0	30.9	130.2
Baking powder	2½ tsp	7.5	0.9	3.9
Salt	1 tsp	5.5	0.7	2.9
Shortening	⅓ c	62.0	7.7	32.3
Milk	1 c	240.0	29.6	125.0
Vanilla	1 tsp	3.0	0.4	1.6
Egg	1	50.0	6.2	26.0
Total:	—	810.0	100.0	421.9

mula percent method, in which each ingredient is expressed as a percentage of the total weight of ingredients. The second is Bakers percent, which is sometimes referred to as flour weight basis. Obviously, this system is used for bakery products. In this method, ingredients are expressed as a percentage of the weight of the flour(s). Flour is listed as singular or plural because for a formula with several types of flour the total weight of flour is used as the basis for calculating the percentages of other ingredients. A single formula is expressed using each of the four methods in Table I. Note that in the Bakers percent system the flour(s) will always be 100% and the total percentage will be greater than 100. In the formula percent system the total percentage will always be 100. Obviously, when publishing or reporting a formula, the units (volume or weight) or method of expression (formula percent or Bakers percent) must be noted to enable readers, if they choose, to use the formula successfully.

III. CONTROLLING EXPERIMENTS _____

A. Uniformity of Ingredients

The uniformity of each ingredient must be carefully controlled. If possible, sufficient ingredients for the entire experiment should be procured at one time and should be well mixed prior to sampling. Such ingredients as flour, sugar, and shortening are easily mixed and stored in quantity, but perishable ingredients such as milk and eggs usually are obtained daily. Greater uniformity can be obtained by the use of reconstituted nonfat dry milk or buttermilk in place of their fresh counterparts. Reconstituted dried egg whites may be used in place of fresh whites but may require formula adjustments because processing influences their functional properties, as discussed in Chapter 7. A few foods, such as meat, present special sampling problems that must be considered carefully in planning the experiment. Further information on sampling is found in Chapter 4.

B. Temperature Control

Temperature control and measurement usually are important in an experiment. In this book, temperatures are all given in degrees Celsius (centrigrade), the system normally used in scientific work. Conversion of Fahrenheit values to Celsius values may be accomplished by use of the equation $°C = (°F - 32) (5/9)$. Fahrenheit equivalents for the Celsius oven temperatures used in this book are given in Appendix B.

Ingredients should be at the same temperature for each replication of an experiment. For baking, most ingredients should be at room temperature. The initial temperature of meat to be heated should be uniform in a controlled study because it affects the heating rate (Heldman, 1975).

C. Measurements of Quantity

Quantity of each ingredient is controlled through careful measuring and/or weighing. At the consumer level, volume measuring devices are used. Household measuring devices were designed for quick and easy use with enough accuracy and precision for home use. Accuracy refers to whether the results agree with the true result. Thus, an accurate 100-ml pipet delivers 100 ml. Precision refers to agreement among repeated determinations and is not necessarily accompanied by accuracy. Household measuring utensils such as a 1-cup liquid measure lack precision because their diameters at the point of measurement are larger than the diameters of more precise measuring devices, such as 100-ml pipets, 100-ml burets, or 100-ml cylinders. A difference of a few drops in the volume of the liquid measure shows very clearly in a pipet or buret, somewhat less clearly in a graduated cylinder, and far less clearly in a household measuring device. Therefore, a measuring utensil no larger than required is selected when possible. For example, a 10-ml graduated cylinder will measure 9 ml more precisely than will a 100-ml cylinder. Similarly, a 250-ml measure is used instead of a 500- or 1000-ml measure for amounts of more than 100 ml but less than 250 ml.

Errors in measuring food by volume may be caused by the manner in which measuring devices are used. Liquid measurements should be read at eye level and the position of the bottom of the meniscus noted.

The manner in which food is handled for measurement will influence the precision of the measurement. For example, if the measuring cup is dipped into the flour, the cup of flour will weigh from 130 to 158 g; if the flour is spooned into the cup without sifting, the weight will range from 121 to 144 g; and if the flour is sifted and then spooned, the weight will range from 92 to 120 g (Arlin et al., 1964). The factors that affect the weight of a cup of flour were studied by Grewe (1932a,b,c). Shortening is packed into the cup to avoid enclosing air bubbles. The weight of a measured cup of food may reflect variations in the food itself as well as in the way in which the food is measured.

For experimental work, the amount of each ingredient in a formula usually is controlled by weighing, except that liquids sometimes are measured in a

TABLE II
Weight and Volume Measurements of
Selected Food Materials[a]

Food material	Weight of 1 c (g)
Corn syrup, dark or light	328
Cornstarch	125
Cornmeal	
White, degerminated	138
White, self-rising, wheat flour added	141
Yellow	
Degerminated	151
Stone ground	132
Cream	
Half-and-half	242
Whipping	232
Fats and oils	
Butter	227[b]
Lard	205
Margarine, regular	225
Margarine, soft	208
Oil, cooking	209
Shortening, hydrogenated	187
Flours	
Barley, unsifted, spooned	102
Buckwheat	98[b]
Oat, sifted, spooned	
Coarse grind	120
Fine grind	96
Potato, unsifted, spooned	179
Rice, unsifted, spooned	
Brown	158
White	149
Rye	
Dark, stirred, spooned	127
Light	
Unsifted, spooned	101
Sifted, spooned	88
Soy, full-fat, unsifted, unspooned	96
Triticale, unsifted, spooned	119[c]
Wheat	
All-purpose, hard wheat	
Sifted, spooned	115[c]
Unsifted, spooned	125[c]
All-purpose, soft wheat	
Sifted, spooned	98[c]
Unsifted, spooned	116[c]
Bread	
Sifted, spooned	117
Unsifted, spooned	123
Unsifted, dipped	136
Cake	
Sifted, spooned	96
Unsifted, spooned	111

TABLE II (*continued*)

Food material	Weight of 1 c (g)
Gluten	
Sifted, spooned	136
Unsifted, spooned	135
Self-rising	
Sifted, spooned	106
Unsifted, spooned	125
Whole wheat, stirred, spooned	120
Honey, strained	325
Milk	
Whole, fresh, fluid	241
Skim or buttermilk	244[b]
Nonfat dry, unreconstituted	
Regular	134
Instant	74
Molasses	309
Sugar	
Brown, packed	211
Brownulated	152
Confectioner's	
Sifted	95
Unsifted	113
Granulated	196

	Weight of 1 tsp (g)
Baking powder	
Phosphate	3.8[b]
SAS–phosphate[e]	3.0[b]
Tartrate	2.8[b]
Baking soda	4.1[d]
Cream of tartar	3.1[d]
Salt, free flowing	5.5[b]

[a] Source: Except as noted, the data are from Fulton *et al.* (1977).
[b] From Adams (1975).
[c] From class data, The University of Tennessee, Knoxville, Tennessee.
[d] From AHEA (1980).
[e] SAS, Sodium aluminum sulfate.

graduated cylinder. The volume and weight measurements shown in Table II and in more extensive tables published by the American Home Economics Association (AHEA, 1980) and the United States Department of Agriculture (Adams, 1975; Fulton *et al.*, 1977) make it possible to convert one type of measurement to the other. Nutrient labels for products may be an additional source of information on the weight of a household measure of the ingredient.

For laboratory work, top-loading electronic balances are convenient. When food is to be weighed into a container, the container can first be weighed, but more often the tare mechanism of the balance is used to negate the weight of the container. If trip balances are used, the procedure of counterbalancing may be used to negate the weight of the container. Counterbalancing involves placing the empty container on the left side of the balance, which has been zeroed. A lighter container is placed on the right pan. Sufficient shot is poured into the right container to balance the two sides. Then, to weigh the sample, the desired weight is placed on the right by moving the riders and adding additional weights if needed. Food is added to the container on the left until the two sides balance.

With any type of balance, time can be saved by weighing all the foods to be used for a series of experiments at one time. Rigid freezer containers are convenient for weighing because they are light and can be covered if the food is to be stored before use. Plastic freezer bags may take less room during storage; therefore, they may be convenient. To facilitate weighing they can be placed inside a beaker on the balance. To weigh small amounts of food, small papers can be used. Small plastic portion control containers with lids such as those used in food service for condiments also are convenient. It is often possible for experienced workers to weigh foods more rapidly than they could measure them by household methods.

The suitability of a balance for weighing small quantities depends on its sensitivity. Sensitivity can be defined as the weight necessary to shift the scale by one division. Thus if the sensitivity of a top-loading balance is 0.01 g with a load of 10 g the potential error is 1 part in 1000 (0.1%). This very small error can be contrasted with the error of about 10% that might occur if an ingredient is measured with household methods. For trip balances, which might be used for class work because they are less expensive than top-loading electronic balances, the sensitivity may be 0.1 g. For the load of 10 g the error would be 1 part in 100 or 1%. For a load of 1 g the error would be 1 part in 10. Thus balances with such sensitivities are not appropriate for small quantities. More sensitive analytical balances may therefore be more appropriate for use with small quantities in research. The most accurate method should be selected for measuring or weighing the ingredients in an experiment to ensure that the same quantity is used for each treatment and for each replication.

IV. CONTROLLING TECHNIQUES

Variations in the techniques used to prepare samples for experimental work are often more difficult to control than variations in the quality and amount of food used, especially for batters and doughs. For food research, mixing is usually controlled by use of electric mixers connected to an electronic timer or timed with a stop watch. The same mixer, or identical ones, should be used throughout an experiment. If mixers are not available, some other method of control may be substituted. Timing hand mixing or counting the number of strokes is an alternative. Counting may be more reliable than timing since mixing speed varies with the worker. When strokes are counted, however, they should be as nearly the same type and strength as possible. Standardization of techniques may take practice.

The effects of individual differences in technique can be reduced in class work if a team of students does an experiment that includes several variations and the basic formula. The work is divided as it would be for an assembly line, so that one student does the same step on each product. In making biscuits, for example, one student might weigh the ingredients and bake the biscuits; a second student might mix; and a third might knead, roll, and cut. If some steps are time consuming or tiring, they can be divided. For example, if a product is to be kneaded 200 strokes, 1 student can knead the first 100 strokes, and another student the second 100.

Identical equipment should be used for each variation in an experiment. If instrumental methods are used for evaluation of the food, the same instrument should be used each time in order to avoid variability attributable to differences between instruments. Conditions under which instruments are used also should be constant. Obviously, each product should be labeled at all times during preparation and evaluation.

V. EVALUATING RESULTS OF CLASS EXPERIMENTS

Both objective and sensory evaluation may be used to determine the effects of treatments on the quality of products. These two types of evaluation techniques are described in more detail in Chapters 3 and 4.

Objective tests are less subject to human variability or error than are sensory tests and are of value if related to the sensory characteristics of the product. Objective tests should be used for class experiments if the specialized equipment for appropriate testing is available. Results of such tests are recorded on the chalkboard or duplicated so that all class members may use the results in writing laboratory reports.

Each class member should have the opportunity to evaluate the samples for each experiment. The principles of sensory evaluation of food, as discussed in Chapter 4, should be used in conducting sensory tests in class. Before evaluation begins, it is helpful to discuss the qualities of a desirable product. All samples should be coded with random numbers. The students record their observations on scorecards prepared for the experiment or devised by the class according to one of the techniques described in Chapter 4. Individual evaluation, followed by group discussion as the identity of the samples is revealed, is particularly useful in allowing students to understand the effects of the treatments studied on the characteristics of the food. It is appropriate to summarize and discuss the responses of the total class to get a broader idea of the quality of the product.

During the actual judging, it is important to avoid talking because judges are easily distracted and biased. Care should be exercised so that the number of samples is not too large for careful judging. If freezer space is available, it is sometimes convenient to freeze samples like bread, which require a long time for preparation, for testing at a later date when more time is available.

VI. REPORTING THE RESULTS ───────────────

A. Recording Data

Experimental work is not valuable unless a written record of the results is made. This record is best made with ink in a bound notebook as the experiment is being done. Copying from pieces of paper should be avoided because of possible errors, wasted time, and the danger of losing the papers. Tables for recording data should be prepared in advance of the laboratory session, allowing space for descriptive terms and remarks, as well as for sensory scores. If data that have been written in a notebook are to be omitted for any reason a single line is drawn through them, with a note about the reason for the omission. Results must not be rejected only because they are not expected.

The necessary parts of an experiment report are as follow: objectives or purpose, methods, results (data), discussion of results, conclusions, and reference list. If the methods used in an experiment have been published a reference to them is sufficient unless modifications have been made. Modifications of a published experiment must be described fully so that the reader of the report can repeat the experiment if desired. Data are easier to locate and interpret if they are summarized in tabular form rather than text form. A carefully worded title for each table is desirable so that the reader of the report can understand how the data were collected and what they mean.

B. Analyzing and Interpreting Data

Before drawing conclusions, it is necessary to interpret the data. A study of the purpose and plan of the experiment and of the tabulated results will show the comparisons that can be made. As this comparison is being made, one of the first questions is whether any differences that may be observed in the data are real differences resulting from the treatments, or whether they are due to chance variation. In research this question is appropriately answered by the use of statistics. For classroom experiments, one practical standard is that a real difference between or among treatments is greater than the difference between replications of the same experiment. Common sense also indicates that a result is more likely to be significant if results of all tests made on the product point in the same direction and if a gradual change in properties accompanies a gradual change in an ingredient or a procedure.

As more curricula include a basic course in statistics it will become appropriate to give the opportunity for students to apply the knowledge gained in the class to projects in experimental foods. Basic techniques that might be used are means and standard deviations as a minimum. If two treatments are studied, a *t*-test can be used to determine if the differences are statistically significant. For more treatments, an analysis of variance (ANOVA) is used to determine if there is a significant difference among the means for the various treatments. If there is a significant difference, a mean separation test is used to determine which mean is different from which mean. An ANOVA basically indicates whether the variance between treatments is greater than the variance within treatments and enables one to say with confidence that there is or is not a difference between treatments. Sometimes no difference is as important as a difference. For example, if a new ingredient is being compared to a currently used ingredient in a product, no difference would mean that it is all right to substitute the new ingredient for the old.

Variability cannot be eliminated entirely from an experiment. Uncontrollable variability is referred to as experimental error and contributes to variability in the results of a study. Experimental error must be minimized if real differences are to be identified. Replication or repetition of an experiment is important in minimizing experimental error. Repeated determinations or measurements on a single sample also are important. These are called replicates or, specifically, duplicates or triplicates if the determination is done two or three times.

After evaluating any differences observed in the main portion of the experiment, it is wise to look for additional relationships and interpretations of the data in order to make the most of the experimental work that has been done. It is equally important, however, to avoid drawing conclusions and assuming relationships that are not warranted by the data and the scope of the experiment.

C. Drawing Conclusions

After the data are interpreted, conclusions can be drawn. These must be based on the observations made in the laboratory, not on the reading or any feeling of what the results ought to be. If the results of an experiment do not agree with published work, differences in experimental conditions often can be found. Exploration of inconsistencies is likely to lead to an explanation. The conclusions must be limited by the conditions of the experiment. Thus a method that proved best for cooking frozen broccoli may not be best for other frozen vegetables or even for fresh broccoli. It is good practice to use the phrase "under the conditions of this experiment" in interpreting results in order to avoid any temptation to generalize more than the experiment justifies.

Suggested Exercises

1. Select a recipe from a recipe book and express it using the appropriate methods as shown in Table I. Remember that Bakers percent is appropriate for products containing flour.
2. Find formulas in the literature expressed using the methods illustrated in Table I. Convert each to the other systems.
3. Evaluate the accuracy of household measuring devices. Accuracy may be checked by weighing on an appropriate balance or by measuring the volume of the water held by the measuring device in a 100- or 250-ml graduated cylinder. Repeat each determination at least three times. Compare the average values with the standard capacity for the utensils.

   For most class work, it can be assumed that the weight of water in grams equals its volume in milliliters. This is strictly true only at 3.98°C, the temperature of maximum density for pure water. At higher temperatures, 1 ml of water weighs less than 1 g. Thus 250 ml of water weigh 249.6 g at 20°C and 249.3 g at 25°C. For such precise work as calibrating burets or pipets, corrections should be made for variations in the density of water. Consult the "Handbook of Chemistry and Physics" (CRC, 1989–1990) for information on the density of water at various temperatures.

Do any of the utensils exceed the acceptable tolerances? How do the variations of your individual determinations from your averages compare with the difference between your average and the standard capacity?

4. Evaluate the accuracy of recommended household methods of measuring various foods. Before beginning, check the accuracy of the measuring utensil that you plan to use. Follow the procedure outlined in exercise 3. Measure the selected food as described in the listing below. Weigh the measured quantity. Repeat the measure at least five times. If possible, make each observation on a portion of food not measured previously. Average your observations. Determine the deviation of each observation from the average.

   a. Sift all-purpose flour once and spoon lightly into the measuring utensil. Avoid shaking. Level with the straight edge of a metal spatula.
   b. Stir whole-grain flours and meals lightly but do not sift, then measure like all-purpose flour.
   c. Sift white sugar only if it is lumpy. Without shaking, fill the measuring device until it overflows, then level with the straight edge of a spatula.
   d. Pack brown sugar firmly into the measuring utensil, then level with a spatula.
   e. Press solid fat firmly into the container until it is full, then level with a spatula.

   How does your average compare with the published weight? Are the deviations from the average important in household measurements? Were deviations greater for some foods than for others? Why should ingredients be weighed rather than measured by volume for experimental foods work?

5. Study the effects of the following variables on measurement of the weight of a cup of flour. Do each experiment 5 times (10 if time permits), average the results, and determine the deviation from the average. Compare both the averages and the variability for the experiments.

   a. Sift three times before measuring.
   b. Sift directly into a cup instead of using the standard method of sifting and then spooning into the cup.
   c. Shake the cup slightly to level the flour while spooning sifted flour into it.
   d. Spoon unsifted flour into the cup instead of using the standard method.
   e. Dip the measure into unsifted flour instead of using the standard method.
   f. Compare different types of flour.
   g. Compare 5 (10 if time permits) measurements taken by the same person with single measurements taken by 5 or 10 people.

   Why is it recommended that flour be sifted before measuring?

6. Study one of the suggested exercises in Part II for variations in amount and type of ingredient, equipment, procedures, techniques, and experimental

conditions. Notice what factors are suggested for study. Decide which variables must be controlled and which will be impossible to control if the experiment is conducted.

7. Calibrate the ovens in the laboratory with either thermocouples and temperature recorder or an oven thermometer. Check at 149, 177, 204, and 232°C.

8. Read an assigned food research paper in a current journal and be able to discuss the following questions:

    a. What was the purpose of the experiment? What were the independent variables? What were the dependent variables?

    b. What variables were controlled?

    c. What uncontrolled variables were taken into account in the interpretation of the results?

    d. What objective and sensory methods of analysis were used?

    e. Did the investigators achieve their objectives?

    f. What new problems became apparent as a result of this study?

    g. What suggestions do you have about the procedures used in the experiment?

References

Adams, C. F. 1975. Nutritive values of American foods. Agric. Handbook 456, Agricultural Research Service, USDA, Washington, DC.

AHEA. 1980. "Handbook of Food Preparation." American Home Economics Association, Washington, DC.

Arlin, M. L., Nielson, M. M., and Hall, F. T. 1964. The effect of different methods of flour measurement on the quality of plain two-egg cakes. *J. Home Econ.* **56**: 399–401.

Bender, F. E., Douglass, L. W., and Kramer, A. 1982. "Statistical Methods for Food and Agriculture." Avi Publ. Co., New York.

CRC. 1989–1990. "Handbook of Chemistry and Physics," 70th edn., Weast, R.C. (Ed.). CRC Press, Inc., Boca Raton, Florida.

Fulton, L., Matthews, E., and Davis, C. 1977. Average weight of a measured cup of various foods. Home Econ. Research Rpt. 41, USDA, Washington, DC.

Gacula, M. C., Jr. and Singh, J. 1984. "Statistical Methods in Food and Consumer Research." Academic Press, New York.

Grewe, E. 1932a. Variation in the weight of a given volume of different flours. I. Normal variations. *Cereal Chem.* **9**: 311–316.

Grewe, E. 1932b. Variation in the weight of a given volume of different flours. II. The result of the use of different wheat. *Cereal Chem.* **9**: 531–534.

Grewe, E. 1932c. Variation in the weight of a given volume of different flours. III. Causes for variation milling, blending, handling, and time of storage. *Cereal Chem.* **9**: 628–636.

Heldman, D. R. 1975. Heat transfer in meat. *Recip. Meat Conf. Proc.* **28**: 314–325.

IFT. 1988. Style guide for research papers. *J. Food Sci.* **53**: 1583–1586.

Joglekar, A. M. and May, A. 1987. Product excellence through design of experiments. *Cereal Chem.* **32**: 857–858, 860–862, 864–867, 868.

O'Mahony, M. 1986. "Sensory Evaluation of Food." Marcel Dekker, Inc., New York.

PLANNING THE EXPERIMENT

An independent problem is an integral part of many advanced courses in food science. The student, having become accustomed to following instructions in conducting experiments, ultimately is expected to select and define a problem, plan and conduct the study, and write a report. This chapter deals with the preparations for conducting an individual study. The principles involved apply to the gamut of studies, ranging from a problem constituting only a small portion of a course to a really extensive project.

I. SELECTING AND DEFINING THE PROBLEM

Ideally a student arrives at the stage of selecting a problem for independent study having been well exposed to the experimental approach and to the food science literature. These previous academic experiences, plus everyday encounters with practical food-related questions, may have resulted in a store of ideas awaiting just such an opportunity. More often the unanswered questions are submerged and the ideas for first problems are elusive.

The search for ideas should not be left until the last minute. A leisurely browsing through the shelves of a food market may produce ideas involving use of newly available ingredients or new uses and/or treatments for established products. Information concerning current developments, obtained through the news media and popular magazines, frequently leads to ideas. Articles, sections featuring new ingredients, and advertisements for new ingredients in trade journals, such as *Food Technology*, alert readers to new ingredient possibilities. Conversation with friends and family members may expose questions concerning ingredients and methodology that could be answered through experimentation. Product development for modified diets provides opportunities that are of special interest to future dietitians. Reading research reports and textbooks should make one aware of gaps in the available information and of apparent discrepancies or inconsistencies in accepted theory.

Not every independent problem makes an original contribution to knowledge, but every independent problem should be worth doing. Careful selection of a problem enhances its value as a learning experience and contributes to the personal satisfaction derived from conducting the study.

Successful selection of a problem involves consideration of the available resources, including time, and of the likelihood that the question that is posed really can be answered, or the problem solved, under the prevailing conditions. In a class situation it is not unusual for each student to be asked for more than one problem suggestion because the instructor must take a broader view than the students, considering the total space, equipment, and financial resources available, as well as the total learning experience. Each student learns from observing others as well as from conducting an individual study; therefore, the instructor is concerned with the range of experiences provided through all of the problems undertaken within a group.

Further definition of a problem usually is necessary after the initial selection. Suppose an increased level of dietary fiber is of interest. A first step in narrowing the topic might be a decision as to whether to study a single type of fiber at several levels or several types at a single level. Other decisions would involve the product to which the fiber will be added and the criteria of performance. The further definition might result in a decision to study volume and tenderness of biscuits containing oat bran at replacement levels of 0, 10, 20, and 40%. A broad topic now has been narrowed to one involving a specific experimental ingredient used at specified levels in a selected product; two criteria of performance—volume and tenderness—also have been identified.

II. REVIEWING THE LITERATURE

A review of pertinent literature frequently aids in problem definition and in development of the specific plan. Whether or not a need for such aid is felt, a literature review prior to undertaking the work is important and frequently prevents regrets later. If the problem has been selected and defined in advance of the literature search, the review should provide information as to methodology used in previous studies. Such information could be helpful in the development of a specific plan.

Early establishment of an orderly procedure for conducting a literature review is of lasting usefulness. Textbooks, with their usual lists of references, provide a convenient starting point for reading on a given subject. Review articles such as those found in *Advances in Food Research* and in *CRC Critical Reviews in Food Science and Nutrition* (formerly *CRC Critical Reviews in Food Technology*) have rather inclusive bibliographies. Research articles, however, constitute the primary source of information and should be used as extensively as possible. Abstract journals and periodical indexes are used for locating pertinent references. Most of such publications provide a list of the periodicals abstracted or indexed. Some of the most pertinent abstract journals and indexes are described briefly here.

Food Science and Technology Abstracts, published monthly since 1969, is the abstract journal that probably is most useful to students in experimental food science courses. Many journals are abstracted and the abstracts are divided into 20 sections; some sections are devoted to specific food groups, such as cereals and bakery products, and some to other food-related categories, such as food microbiology and basic food science. Each issue has an author index and a detailed subject index. In addition, there is an annual cumulative index.

In *Biological Abstracts*, food technology and nutrition are among the general headings. Subheadings under food technology include topics such as baking technology, cereal chemistry, and dairy products. Section 17 in *Chemical Abstracts* is food and feed technology. *Nutrition Abstracts and Reviews* might be helpful, particularly to future dietitians. It has both a general subject index and a specific subject index.

Indexes provide less information than do abstract journals but are helpful. The *Biological and Agricultural Index* is published monthly and the third issue of each quarter is cumulative. Food science and nutrition are among the subject fields indexed; listings fall under a single alphabet rather than in subject matter sections. The journals that are most used by food science students are included in the list of journals indexed.

Among the disciplines covered by *Science Citation Index* is food science and technology; most of the pertinent journals are indexed. Nutrition and dietetics is another discipline covered; the *Journal of Nutrition* is indexed but the *Journal of the American Dietetic Association* is not.

The *Applied Science and Technological Index*, as its name suggests, has a technological emphasis in its listings. For example, *Food Technology* and *Food Engineering* are indexed but the *Journal of Food Science* is not. There is a single alphabetical listing, but under Food there is a list of subcategories, such as "food, low caloric."

Current Contents is not exactly an index but neither is it an abstract journal. Published weekly, it consists entirely of tables of contents. Since its initiation in 1958, it has expanded into eight separate titles. The one that covers food science periodicals is titled *Current Contents: Agriculture, Biology, and Environmental Sciences.*

One of the many features of the *Journal of the American Dietetic Association* each month is a group of abstracts of pertinent articles from a variety of journals. This collection is particularly useful in a situation where library resources are limited.

The most recent issues of abstract journals and indexes cannot cover the most recent research literature. Therefore, it is necessary also to go through recent issues of the research journals.

Effective use of abstracting and indexing journals, as well as of the index of a single issue of a research journal, involves some skill in selecting subject entries to search. In studying tables of contents, recognition of pertinence in rather obscure titles also is a skill that may be developed with experience and increasing knowledge of the subject matter.

Szilard (1987) described different types of publications available in the field of food and nutrition. Included were specific sources of information in various areas, such as special diets and the food groups. Szilard also provided considerable information about the use of a library.

Roundy and Mair (1985, p. 42) detailed the process of gathering information in the library. They described the kinds of materials usually available and their use; even the use of the card catalog was discussed. Many libraries now have their holdings computerized and users obtain the information found in traditional card files from computers—another time saver.

Some students, especially graduate students, have access to on-line information retrieval systems by way of terminals that are linked with computers through which data bases are accessible. Data bases are machine-readable files representing various sources of information, including journal articles. Of the data bases that cover food science, FSTA (*Food Science and Technology Abstracts*) is especially useful. On-line searching may not be helpful if information dated prior to 1969 is needed because most computerized data bases do not cover earlier years. However, *Science Citation Index* has been available in machine-readable form, as well as in the printed form, since 1961 (Cheney and Williams, 1980). Brooks and Touliatos (1989) pointed out the existence of some data bases that do not have print counterparts; *Agricola*, *Foods Adlibra*, and *Medline* are examples. They also described a sample search.

On-line searches usually are made by professionals who are trained to conduct them rather than by the users themselves, but effective communication between the searchers and the users is important. The key words and phrases to be used determine the breadth of the search and must be agreed upon in advance.

In addition to comprehensiveness, the major advantage of on-line information searches is saving of time; a print-out of the information requested can be obtained in a short time. This may not be important in the preparation of a class report, but students should know of the existence of such services for possible future use.

Roundy and Mair (1985, pp. 53, 60) provided some information for students who do have access to on-line information retrieval and wish to use it. Chandler (1982) discussed on-line information services as well as other sources of information.

Once a list of references has been compiled, finding the call numbers of the journals needed can be done most quickly with a serials catalog, which lists alphabetically the journals available in the library. Most libraries have serials catalogs in the reference room or the periodical room or both. Journals also may be included in computerized catalogs.

The actual mechanics of taking notes from the research articles can be handled with varying degrees of efficiency. Separate index cards for individual references are convenient for taking notes because they can be arranged as needed when the review actually is being written. Roundy and Mair (1985, p. 64) suggested the use of 5 × 8-in. cards rather than smaller cards or pieces of paper. They recommended writing a topic heading in the upper right margin of each card. The format for a single article should begin with the complete reference. Consistently writing the reference according to the exact bibliographical form to be used in the final report helps one become accustomed to that form and also helps prevent a necessary return to the library to fill in an accidental gap. The reference cards can be organized eventually to follow the outline for the paper.

Effective use of the time spent in the library involves exercise of judgment as to when to read for detail and when to scan. Scanning ability is developed with practice and saves considerable time. If the reader, having scanned an article, is uncertain as to its pertinence to the study at hand, a notation on the reference card as to the type of information included will prevent loss of the reference, while requiring little time. An obviously pertinent article will be read carefully, and adequately detailed notes will be taken. It is advantageous to make notes in such a way as to prevent plagiarism later when the review is written. Direct quotations should be either avoided during note taking or identified clearly with quotation marks.

The total learning experience in reviewing the literature is enhanced if the habit of reading critically is developed early. Raising questions concerning experimental plans, specific procedures used, presentation and interpretation of data, and even writing style is mental exercise that can benefit one's own work.

III. WRITING THE PLAN

Before proceeding with a plan, the investigator should write the specific question that is to be answered if this has not been done previously. With the literature review completed, a final judgment now is made as to whether and how the question can be answered. A written plan is prepared.

The written plan includes first a meaningful title, and then a justification for the proposed study. The justification frequently is based on the related literature or on the lack thereof and leads into a statement of the objective(s). It is important that the objective(s) be attainable. The planned procedure for meeting the objective(s) follows and includes the experimental design, details of the procedures to be used, and copies of any forms to be used in collecting data.

The description of the experimental design specifies the variable(s) and its (their) levels, the extent of replication planned, and the amount of work to be done each day. The procedures include the formulas to be used, the sources of materials, the controls to be applied to the procurement and use of materials, the standardization and controls to be applied to any food preparation, the methods of product evaluation to be used, and the data analysis to be conducted.

Some preliminary work usually is necessary before all decisions pertaining to the plan can be made. The preliminary work not only provides information as to whether the plan is realistic from the standpoint of time, but it also makes the worker aware of decisions that must be made in advance of the actual study. Preweighing of staple supplies contributes to efficiency as well as to control of experimental conditions but requires a plan for storage of the preweighed supplies.

Conscious choices sometimes must be made between alternative control measures. For example, consider the hypothetical experiment discussed in section I. Presumably every factor should be controlled except the fiber level (and flour level because of the use of fiber as flour replacement). However, the fiber concentration probably will affect the dough consistency achieved with addition of a specified amount of milk, and dough consistency is important because biscuit dough usually is kneaded by hand. A choice might be made to control the milk not at a single specified amount, but at the amounts (determined in preliminary work) that produce doughs of similar handling characteristics with the different levels of fiber. The need for making such choices should be anticipated so that important decisions will not have to be made under stress after the study begins.

If any special forms are to be used for collecting data, as stated previously, they are included in the plan; Table III in Chapter 9 is an example. A scorecard for sensory evaluation (for example, Figs. 4.4–4.8 and 4.11) is another type of special form. On occasion, a diagram showing the sampling plan for sensory and/or objective testing (for example, Fig. 4.1) is pertinent. Codes for samples to be subjected to sensory evaluation can be worked out in advance. Treatment of the data, such as the statistical analysis, if any, should be predetermined and stated in the plan. At the end of the plan the references cited in the justification and/or procedure are listed.

In a class situation, a market order usually must be included with the plan. The order should clearly differentiate between those staple supplies that are to be obtained only in the beginning and the perishables that are needed each time. Equipment that is needed also should be ordered. It usually is not safe to make any assumptions as to availability, even of equipment that normally is in the laboratory.

Further detail concerning various aspects of planning experiments is found in Harrison's (1972) chapter in "Food Theory and Applications."

Suggested Exercises

1. List some questions about food that could be answered through experimental work. Work out a general plan for answering each experimentally.
2. List the major abstracting, indexing, and research journals that are in your field and available in your library. Annotate the list with information such as frequency of publication and coverage.
3. Compile a bibliography on a selected subject, using a serial index. Locate one of the articles and prepare a reference card according to the format suggested in this chapter.
4. As a class determine key words for a computer search for a class experiment. Have the librarian conduct the search. Evaluate the usefulness of each reference obtained.

References

Brooks, A. and Touliatos, J. 1989 Computer searches: A guide for practitioners and researchers. *J. Home Econ.* **81**(2): 23–26, 39–40.

Chandler, G. 1982. "How to Find Out," 5th edn. Pergamon Press, New York.

Cheney, F. N. and Williams, W. J. 1980. "Fundamental Reference Sources," 2nd edn. American Library Association, Chicago, Illinois.

Harrison, D. 1972. Planning and conducting experiments. *In* "Food Theory and Applications," Paul, P. C. and Palmer, H. H. (Eds.). John Wiley and Sons, New York.

Roundy, N. and Mair, D. 1985. "Strategies for Technical Communication." Little, Brown and Co., Boston, Massachusetts.

Szilard, P. 1987. "Food and Nutrition Information Guide." Libraries Unlimited, Inc., Littleton, Colorado.

EVALUATING FOOD BY OBJECTIVE METHODS

Methods used by food scientists to evaluate food quality include objective and sensory methods. Those that do not depend on the observations of an individual and can be repeated using an instrument or a standard procedure are described as objective methods (IFT, 1964). The advantages of these tests are many. They may offer a permanent record of results and invite confidence because they are reproducible and less subject to error than sensory methods of evaluation. However, if results of objective and sensory methods do not correlate, they may not be measuring the same component of quality and hence the chemical or physical method may not be useful for the study. For example, determination of the alcohol-insoluble solids content of vegetables gives an index of plant tissue maturity but may not be related to overall textural quality (Szczesniak, 1973, pp. 109–116). High correlation between an instrumental method and sensory testing is a necessary prerequisite to the use of an instrumental method in quality control (Szczesniak, 1987).

I. CATEGORIES OF OBJECTIVE METHODS ⸻

Objective methods may be classified into four categories. Instrumental or objective methods include a wide variety of tests.

A. Chemical Methods

Chemical methods include the determination of the nutritive value of foods before and after cooking, as well as constituents that affect the palatability of a food such as peroxides in fats and components responsible for flavor and color. Accurate methods for analysis of nutrient content of foods are critical to nutritional status studies, to accurate information on the effect of processing treatments on nutritive values, and to nutrient labeling. Chemical methods will not be included here except to refer the reader to publications of various professional organizations (AACC, 1983; AOAC, 1984, 1985; AOCS, 1981), which give the methods for determining various constituents in specific foods and are kept up to date by continuous study of analytical methods and by periodic revisions.

B. Physicochemical Methods

Certain physicochemical determinations are important in food analysis. Of these, probably the one most frequently used is hydrogen ion concentration, or pH, which is discussed in Chapter 6. Measurement with a refractometer of refractive index, the angle at which light is bent by certain substances, is useful in finding the sugar concentrations of syrups and the degree of hydrogenation of fats (Carlson, 1972). The refractive index as related to fat is discussed in Chapter 15.

C. Microscopic Examination

Some properties of foods depend on the structure or physical arrangement of their components. Microscopic examination of foods such as mayonnaise, whipped cream, fondant, and cake batter may yield valuable information. Painter (1981) described the differentiation among components of a batter viewed through a microscope. Fat distribution and air cell characteristics can be noted.

TABLE I
Studies Involving Scanning Electron Microscopy

Food	References
Baked products	Cloke *et al.* (1984a); Pomeranz *et al.* (1984)
Eggs	Woodward and Cotterill (1986, 1987)
Emulsions	Barbut (1988); Klemaszewski *et al.* (1989); Yang and Cotterill (1989)
Extruded products	Harper (1986); Stanley (1986)
Fish	Bremner and Hallett (1985)
Fruits	Mohr (1973); Bolin and Huxsoll (1987)
Meat	Cheng and Parrish (1976); Jones *et al.* (1976)
Milk and other food gels	Kalab and Harwalkar (1973)
Starch systems	Miller *et al.* (1973); Chinnaswamy *et al.* (1989)
Vegetables	Davis *et al.* (1976); Haard *et al.* (1976); Hung and Thompson (1989); Floros and Chinnan (1988)

Fiber diameters and sarcomere lengths of meat samples can be measured with the ocular measuring device of a microscope (Tuma *et al.*, 1962; Hegarty and Allen, 1975). Examination of meat, vegetable, and fruit tissue is more difficult because it involves making histological sections. Although such studies require training and experience, they are essential in some types of food research. Scanning electron microscopy (SEM), a technique that makes it possible to view a three-dimensional image of the structure of materials, also is used in food research. Several SEM photomicrographs are included in Chapters 7, 9, 16, 17, and 24 of this book. Study of SEM photomicrographs facilitates understanding of changes that occur in foods subjected to various treatments. Scanning electron microscopy has been used as a research tool in studies of many foods. References to such studies are summarized in Table I. The general principles of food microscopy (Vaughan, 1979), electron microscopy in food science (Pomeranz, 1976), and microscopy of cereal products (Bechtel, 1983) have been reviewed.

D. Physical Property Evaluation

The objective measurements discussed in this chapter concern the physical properties of foods. Some measurements, such as those of temperature or the amount of liquid drained from the food on standing, are simple and used frequently. Others, on which attention will be focused, are of special interest because they may be related to the results of sensory tests. These tests are made with special instruments or with improvised devices. Reviews of measurements that are especially applicable to fruits and vegetables (Kramer and Twigg, 1959; Mohsenin, 1986), meat (Szczesniak and Torgenson, 1965), and baked products

(Funk *et al.*, 1969) can be found in the literature. Methods for all foods have been discussed (deMan *et al.*, 1976; Bourne, 1982).

The advisability of substituting an improvised device for a manufactured instrument depends, of course, on the availability of the manufactured instrument, which in turn may depend on the amount of use that it will receive and the cost. Improvised devices such as those described in this chapter may be used to meet the needs for class projects. The creative worker may be able to improvise others, basing them on the principles of operation of instruments described in the literature.

Commercially available equipment discussed in this chapter is listed in Appendix C. Sources for the purchase also are listed. A price range is included to give a general idea of costs. The ranges are wide to cover price fluctuations but may not be wide enough to cover all possible variations in models of instrument. References describing the apparatus and its use are included in the discussion. Directions for use of improvised equipment are included in Appendix D.

II. APPEARANCE

The appearance of foods can be recorded by means of photography or, in some cases, photocopying. A photograph provides a record of size only if scales are included in the picture. The grain of baked products is visible in a photograph when lighting is controlled carefully so that an optimum amount of light falls on the object. An example of a photograph is shown in Fig. 3.1. A record of the appearance of baked products that is somewhat less clear than a photograph, but is satisfactory for many purposes, is obtained with a photocopy machine. Photocopies provide a record of the actual size and shape of a sliced or halved baked product and give some record of the grain. An example of how one might describe the grain of baked products is given by Painter (1981). The process is

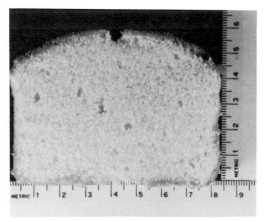

FIGURE 3.1
Photograph of a cake as a record of size, shape, and grain. (Courtesy of Rogers Penfield.)

more successful with light-colored products than with products like chocolate cake. A photocopy can be made by placing a sheet of clear plastic film on the glass plate of the machine and placing the sample(s) on the film before making the copy. Quality is improved if a piece of white paper is placed on top of the samples. Notations of significant characteristics of each sample may be made on the copy. A set of photocopies of this type is shown in Fig. 3.2.

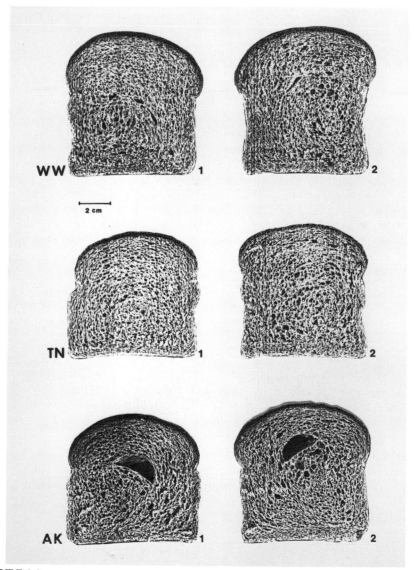

FIGURE 3.2
Photocopies of bread as a record of shape and grain. (From Swanson and Penfield, 1988; Reprinted from *Journal of Food Science*, **53**, No. 3, pp. 896–901. Copyright © 1988 by Institute of Food Technologists.)

III. COLOR

Preliminary acceptance or rejection of a food usually is based on the visual appearance, including the color. Therefore, color, an important quality attribute of food, is studied in many ways. Only a brief introduction to the study of food color can be given here. The discussion is designed to facilitate a basic understanding of the various color values that may be found in the literature or are provided by instruments used for color analysis in food research and quality control. Basic information on color theory is presented by Billmeyer and Saltzman (1981, Chapters 1–2) and Francis and Clydesdale (1975, Chapters 1–8) and in several review articles (Clydesdale, 1972, 1976; Noble, 1975; Setser, 1984).

A rapid method of measuring color that is satisfactory for some purposes is to match the color of the food to a sample of color or a color chip found in a color atlas. A book of color samples should be available in most college and university libraries (Maerz and Paul, 1950; Villalobos-Dominguez and Villalobos, 1947; Kornerup and Wanscher, 1962). Although this procedure furnishes a permanent record of the color of a food, it is not entirely satisfactory because it is difficult to match the food with one small block of color on a chart containing many such blocks. Availability of a standard color sample for a product facilitates visual comparison. Billmeyer and Saltzman (1981, p. 72) suggested that comparison of a sample with a standard color sample under standardized lighting is a good technique for determining that two samples are not identical.

Specification of color may be based on one of several three-dimensional color solids. For example, in the Munsell system a color may be assigned the descriptor $H V/C$. H defines the hue or location on the 100-hue scale on the circumference of the solid. Value (V) indicates the position that the color occupies on the vertical plane of the solid. The bottom of the plane is absolute black (0/) and the top is absolutely white (10/). C designates chroma and refers to the distance from the center of the color space. The value /0 at the center represents neutral gray and values of /14 or even /16 represent increasing strength of the color (Clydesdale, 1984, pp. 97–98).

Disk colorimetry was developed to facilitate the use of the Munsell color system in evaluating the color of agricultural products in United States Department of Agriculture (USDA) laboratories (Nickerson, 1946). Two to four disks that have been cut along the radius are interlocked. Selection of the colored disks depends on the color to be matched. The color of the disks is blended by spinning them on a shaft of a small motor so that one color is seen with no flicker. The proportions of the four disks are adjusted until the color of the sample is matched. A method of controlling lighting is necessary and is provided in commercially available disk colorimeters. The color is described in terms of the amount of each disk exposed (Clydesdale, 1984, p. 116).

There are two types of instrumental methods of color analysis. Transmission spectrophotometry is used to measure the intensity of light that passes through a clear or transparent solution. Measurements of this type frequently are used in quantitative analysis of specific compounds. Because most foods are opaque in nature, reflectance spectrophotometry, or more commonly, tristimulus colorimetry, is used in food research. In a tristimulus system, color is specified by three attributes: hue, value, and chroma in the Munsell system, or dominant wavelength, lightness (%Y), and purity in the CIE (Commission Internationale de L'Éclairage) system. The three attributes, respectively, refer to the actual color, the luminosity, and the strength of the color. Several colorimeters are available for determination of the data necessary for specification of these attributes.

The Hunter color difference meter is an instrument that is used frequently in food research. Measurements from this instrument are based on the Hunter color solid. Spaces within the solid may be located from the values L, a, and b. L is value and ranges from white to black (100 to 0). Red is represented by $+a$, green by $-a$, yellow by $+b$, and blue by $-b$. From these values, chroma or saturation index is calculated with the formula $(a^2 + b^2)^{1/2}$. Chroma is equivalent to purity in the CIE system. Hue is expressed as hue angle (θ_s), $\tan^{-1} a/b$ (Clydesdale, 1984, p. 125). For convenience, Hunter data frequently are expressed in terms of L, a, and b in the literature (Clydesdale, 1984, pp. 120–136).

Measurement of difference in color of two products or a product and a standard may be of interest. The instrumentation previously described may be used for such measurements. CIE values (XYZ) or Hunter values may be used to mathematically express color differences.

Color is a sensory attribute. Instruments are used to describe the sensation that is perceived by the human. They may facilitate color evaluation in research and quality control. Review papers that illustrate use of the principles discussed here include those by Clydesdale and Francis (1969) and Setser (1984).

IV. GEOMETRICAL CHARACTERISTICS ──────────
A. Size and Shape

Measurement of the size of food material particles or structural components may be of interest in research and quality control. Traditional methods involve use of sieves or other sorting devices for separation and sizing of particles. Standards of identity for flour and meals include specification of the percentage of the product that will pass through each of a series of sieves (FDA, 1988a). Because particle size of fiber ingredients will influence mouth feel of baked products, these ingredients may be sieved prior to inclusion.

Computer techniques may be used to measure size and shape parameters of various food materials and components. Russ *et al.* (1988) described several applications in which area, length, width, perimeter, convexity, and other parameters were measured. Size of rice grains can be analyzed to determine if the rice can be sold as short or long grain. Image analysis of bread cells can be used to measure cell size distribution within a sample. An ink print (Fig. 3.3) or other representation of the crumb is analyzed. Additional data presented by Russ *et al.* (1988) showed how cookie size related to the number of chocolate chips included in the cookies. Klemaszewski *et al.* (1989) described the use of image

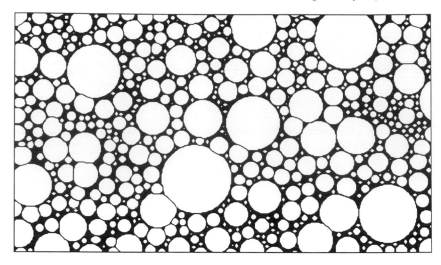

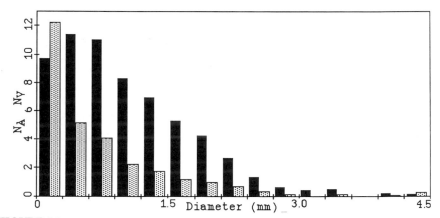

FIGURE 3.3

Top: Discriminated image of cells in bread, obtained by cutting bread and applying an ink pad to the surface. *Bottom:* Calculated spherical cell size distribution (black bars) compared to measured size distribution of circular sections of cells (gray bars) shown above. (From Russ *et al.*, 1988; Reprinted from *Food Technology*, **42**, No. 2, pp. 94, 96–102. Copyright © 1988 by Institute of Food Technologists.)

analysis for studying emulsion structure and breakdown. Image analysis may be used in conjunction with SEM to study size of microstructures (Bolin and Huxsoll, 1987). Image analysis depends on the use of a digitizer attached to a computer system. Such systems can determine frequencies, measure size, and calculate various size ratios.

Shape of cakes is an important characteristic. The American Association of Cereal Chemists (AACC) provides a method for evaluating the symmetry, uniformity, and shrinkage of layer cakes. A template is provided for these measurements and is shown in Fig. 3.4. Another dimensional measurement is the cookie

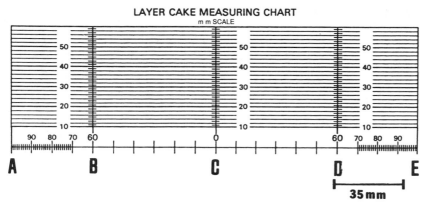

FIGURE 3.4
Template for measuring indices of cake quality. Volume index = B + C + D; symmetry index = (2C − B − D); uniformity index = B − D. (From AACC, 1983, method 10-91.)

spread test (AACC, 1983, method 10-50D). For pastry, the height of a stack of four pieces is measured to give an indication of the flakiness of the product.

B. Volume

An important indicator of quality of a baked product is volume. The effects of changes in formulation and/or handling on volume frequently are studied.

1. Displacement Method

In the displacement method, the volume of a baked product is found by difference after measuring the volume of low-density seeds, such as rapeseeds, held in a container with and without the product. The container can be a straight-sided box, a can, a crystallizing dish, or a volumeter, an apparatus built like an hour glass (Cathcart and Cole, 1938). If a cake is baked in a pan with sides that are higher than the product, its seed displacement can be found by pouring seeds on top of the cake while it is still in the pan. The volume of the seeds required to fill the pan then is subtracted from the volume of the empty pan (Brown and Zabik, 1967). After the volume of a baked product has been determined, it may be divided by the weight of the product and reported as specific volume, in units of cubic centimeters per gram. This calculation facilitates comparison of products having different weights. The comparison also is facilitated if the same weight of batter is baked for each product. When the displacement method is used, sensory evaluation of the baked product must be delayed or done on a duplicate product because seed displacement must be done on the entire product at a time when it is not soft, often after 24 hr. Other problems associated with direct measurement of volume with rapeseeds is collapse of the structure under the weight of the seeds and incomplete removal of the cake from the pan (Cloke *et al.*, 1984b). Thus alternative methods of studying volume are used.

2. Index to Volume

An index to volume can be found by measuring the area of a slice of food with a planimeter or a digitizer system attached to a computer. The edges of the slice are traced or a photocopy is used, and the area is found. It is important to use a slice that is representative of the product, such as a center slice, or to use several slices from the same positions in each product and average the areas. Index to volume sometimes is more convenient than seed displacement because a slice can be traced quickly and the area found at any convenient time, making it possible to complete sensory evaluation of the product immediately. An alternative system for measuring an index to volume is to measure the height of several slices of the product. Height is measured at the outer edges, the center and the points halfway between the center and the outer edges. The five values are averaged and reported as an index to volume called standing height (Tinklin and Vail, 1946). To ensure validity of index to volume values, equal weights of batter must be baked in identical pans. The AACC template which was previously described can also be used for determining the volume of layer cakes. The method is similar to the measurement of standing height previously described (AACC, 1983, method 10-90).

Cloke *et al.* (1984b) studied alternatives to rapeseed displacement and concluded that any of several indices will give information on general trends. Variation in indices is seen with variation in cake contour. However, variation in volume measurement with rapeseed displacement may also be seen. It can be concluded that no method is completely satisfactory. However, it also can be assumed that it is appropriate to use one method consistently within a study.

V. TEXTURE

The objective measurement of texture is complex because it must reflect the action of the mouth in removing food from an eating utensil, the action of the tongue and jaws in moving the food, and the action of the teeth in cutting, tearing, shearing, grinding, and squeezing food. Textural characteristics include the mechanical properties of hardness, cohesiveness, adhesiveness, fracturability, viscosity, gumminess, springiness, and chewiness, as well as geometrical parameters, which are discussed in Chapter 4, and fat and moisture content (Szczesniak, 1963, 1986). Therefore, it is not surprising that there are many ways to evaluate the texture of various foods. Most of these tests involve subjecting the food material to stress or applying force to it. The effect of that stress is seen in the strain produced or a change in the physical properties of the food. In some tests, the strain or deformation produced is measured as a function of the stress. Instruments in this group include cone penetrometers and the Adams and Bostwick consistometers. In other tests, the stress required to effect a specific strain is measured. Many tests of this type are used in food research, including compression and shearing tests. Many of the tests that will be described in this section may be done by attaching a test cell to a universal testing machine (UTM); UTMs are multipurpose instruments that are used to measure the amount of force required to complete a test of the mechanical properties of the test material as well as other parameters. A more complete description of the use of UTMs for texture profile analysis is given at the end of this section.

A. Texture of Meat

Some of the early measurements of texture of meat were made with a pene-trometer (Noble *et al.*, 1934). McCrae and Paul (1974) used a cone penetrome-ter, similar to the one shown in Fig. 3.5, in meat studies, suggesting that the action of the cone was similar to the action of the teeth biting into a piece of

FIGURE 3.5
Universal penetrometer (model 4101). (Courtesy of Lab-Line Instruments, Inc., Melrose Park, Illinois.)

meat. A flat plunger attached to a UTM may be used in meat studies as described by Bouton *et al.* (1971). Hardness (kilograms) is the force required to penetrate the sample to a depth of 80% of its thickness. The probe is allowed to penetrate the sample two times to simulate two "bites." The ratio of the work done during the second bite to the work done during the first bite is termed cohesiveness. The closer the ratio is to one, the more cohesive the sample. A third parameter, chewiness, is expressed in kilograms of force and is the product of hardness × cohesiveness.

The most commonly used device for the evaluation of meat tenderness is the Warner–Bratzler shear, shown in Fig. 3.6. The force necessary to shear a cylindrical sample of meat 1.27 or 2.54 cm in diameter against the dull edge of a triangular opening is measured (Bratzler, 1932). The shearing portion of this instrument may be adapted for attachment to a UTM. If used in this manner, two parameters may be measured as described by Larmond and Petrasovits (1972). Firmness in kilograms per minute is calculated as the slope of the line

FIGURE 3.6
Warner–Bratzler shear. (Courtesy of G-R Electric Mfg. Company, Manhattan, Kansas.)

drawn from the origin of the curve to the peak and shear cohesiveness is defined as the peak force in kilograms. Sensory scores usually are highly correlated with Warner–Bratzler shear values, possibly because the dull edge against which the meat is torn simulates the grinding surfaces of the teeth, which also are dull. However, some investigators have shown weak correlation between Warner–Bratzler shear values and sensory scores for tenderness. Care must be exercised when cylindrical samples or cores are removed from the meat. Meat should be at room or refrigerator temperature to avoid irregularly shaped cores. Williams *et al.* (1983) indicated that shear value varies with position in the muscle, thus reinforcing the need to be consistent in position of taking cores for Warner–Bratzler shear testing. Differences were found between longissimus samples stored for 2 hr before coring and for longer times. Thus time of holding before coring is also a source of variation in values. Kastner and Henrickson (1969) suggested that a mechanical device be used to ensure cores of uniform diameter. Cores usually are taken so that shearing occurs at a 90° angle to the fiber. Applying force at that angle takes more force than applying force parallel to the fibers (Murray and Martin, 1980). Szczesniak (1986) reviewed factors that affect shear values.

Tensile properties of meat samples also have been investigated. The amount of force required to pull apart a small sample of meat with the muscle fibers oriented parallel to the force is suggested to be a measure of the strength of the muscle fibers, whereas the amount of force required to pull apart a sample with fibers perpendicular to the force is suggested to be a measure of the strength of the connective tissue holding the muscle fibers together (Bouton and Harris, 1972). Tests of this nature are made with a UTM fitted with grips to hold the meat sample while it is stretched until it breaks.

As the consumption of processed meats increases, questions should be raised about the use of these "traditional" methods of evaluating meat tenderness for these products. Processed meats including frankfurters and other emulsion-based products are more homogeneous than intact muscle foods. Restructured products also are more homogeneous but may present sampling problems if bind is weak. Little information can be found on measurement of texture of these products. Quinn *et al.* (1979) reported that the universal cone penetrometer and universal food rheometer (a universal testing machine) readings can be used to monitor texture of wieners because values obtained correlate highly with sensory scores for textural parameters. Prusa *et al.* (1982) described simulated incisor and molar attachments for the Instron and used them to evaluate the tenderness of poultry meat frankfurters. They concluded that for the products tested, the relationships between sensory scores for firmness and Instron values derived with a Warner–Bratzler shear, puncture probe, and simulated incisor were strong. This indicated that the Warner–Bratzler shear and a puncture probe, tests used with intact muscle studies, were suitable for studies with frankfurters. Lee *et al.* (1987) evaluated frankfurters using 11 different tests with an Instron. They found that three measurements, compressive force at failure, compressive force at 50% compression, and maximum shear force, correlated well with sensory scores for elasticity, firmness, and chewiness.

For restructured meats, cohesiveness or strength of bind is an important quality attribute. MacNeil and Mast (1989) described an attachment for a texture press system that pulled the steaks apart on a horizontal plane, giving a measure of "resistance to tear." Values obtained varied with changes in formulations, suggesting that the method is feasible for the study of cohesiveness in restructured products. Some investigators do, however, use traditional methods of evaluation for restructured products, including penetration and Warner–Bratzler shear (Costello *et al.*, 1985), straight-blade shear test (Berry *et al.*, 1987), and Kramer shear (Strange and Whiting, 1988).

B. Texture of Fruits and Vegetables

Puncture testing most frequently is used to evaluate the firmness of fruit and vegetable tissue. The Magness–Taylor puncture tester is a commonly used apparatus that measures the amount of force required to penetrate the sample to a specific depth (Bourne, 1965). The amount of force required varies as the probe passes through the skin and then through the flesh. The single pea puncture maturometer is used to measure skin properties (Hung and Thompson, 1989). Hung and Thompson used an Ottawa texture measuring system back extrusion cell to evaluate packability and chewiness. The Kramer shear press, which consists of a rectangular box with evenly spaced slits in the bottom, frequently is used to study the tenderness of fruits and vegetables. A series of 10 blades is moved through a sample of food in the box. As the blades move, the food is compressed, sheared, and extruded through the opening in the box. Many foods have been tested with this system (Szczesniak *et al.*, 1970). Many of the test devices for fruits and vegetables may be attached to an Instron universal testing machine.

C. Texture of Liquids and Viscoelastic Foods

Rheology is the science of flow and deformation of materials, both liquid and solid; therefore, an understanding of its principles is important to the study of food texture. Application of rheology to food research was reviewed by Scott Blair (1958) and Finney (1972).

1. Viscosity and Consistency

All liquids flow; some do not flow readily. Their resistance to flow, or viscosity, is caused by attraction between molecules of the liquid and/or larger particles. In a pure liquid, this attraction or internal friction is greater between large and well-hydrated molecules than between smaller molecules. Temperature must be controlled closely in measurements of viscosity because as the temperature of a pure liquid increases, its viscosity decreases. Absolute viscosity is measured in terms of the amount of work required to maintain a certain rate of flow. Its unit is the poise, which is defined as a force of 1 dyn/cm^2, which produces a 1 cm/sec difference in velocity of two planes separated by 1 cm of liquid. Absolute viscosity is a characteristic of Newtonian fluids or homogeneous liquids such as sugar syrups, oils, very dilute fruit juices, and solutions of some gums. Processed whole, 2%, 1%, and nonfat milks also have been shown to be Newtonian (Wayne and Shoemaker, 1988). The term apparent viscosity should be used to refer to the flow characteristics of non-Newtonian fluids. Relative viscosity is found by comparing the rate of flow of a liquid with that of a reference liquid, usually water. The rate of flow through a tube in comparison to the rate of flow of water through the same tube is a simple measurement of relative viscosity. An ordinary pipet with a part of the tip cut off, if necessary, for more rapid flow can be used. Pipets especially designed for viscosity measurements, like the Ostwald pipet, are available.

Apparent viscosity or consistency can be evaluated with several tests. The time required for cake batter to flow between two marks on the stem of a funnel is a simple measurement of this type (Tinklin and Vail, 1946). Grawemeyer and Pfund (1943) described the line-spread test, which is suitable for foods like white sauce, starch puddings, applesauce, and cake batters. For this test, food is placed in a hollow cylinder (Appendix D). The cylinder is lifted when the food is at the desired temperature and the product is allowed to spread for a specified period of time (30 sec to 2 min). Consistency is reported as distance (in centimeters) spread in the designated time period.

The Adams consistometer (Adams and Birdsall, 1946) is a commercially available apparatus, which was first used with creamed corn. It consists of a funnel-like reservoir and a metal plate with concentric rings and works on the same principle as the line-spread test. The Bostwick consistometer measures the distance that a food flows under its own weight down a slanted trough in a given period of time. Its use is designated in the standard of identity for tomato catsup (FDA, 1988b). Foehse and Hoseney (1988) used the Bostwick to study fluidity of corn flour/water batters. Use of the Bostwick in place of a rotational viscometer minimized the problem associated with sedimentation of batter components during testing. Elling (1988) suggested that the consistometer can be used in a rapid test of flour quality. Thus it would seem that it could be used to study the influence of varying ingredients on batter quality.

Several other methods can be used for viscosity and consistency measurements. For example, the length of time required for a steel or glass ball to fall through a column of test material is measured in an improvised device (Morse *et al.*, 1950) or a falling ball viscometer. Several rotational viscometers are available. A spindle attached to the Brookfield viscometer rotates in the test material and the amount of drag on the spindle is measured. The Brookfield viscometer (Fig. 3.7) has been used with many foods, including syrups (Collins and Dincer, 1973), protein slurries (Fleming *et al.*, 1975), and gums (Balmaceda *et al.*, 1973). For very viscous materials, the Brookfield can be mounted on a helipath stand that gradually lowers the spindle through the test material for testing of undisturbed material. In the MacMichael viscometer, a horizontal disk is suspended in the liquid by means of a steel wire, and the force required to hold the disk stationary is measured while the outside cup is rotated at a constant speed. In the Stormer viscometer, measurements are based on the rate of rotation of a

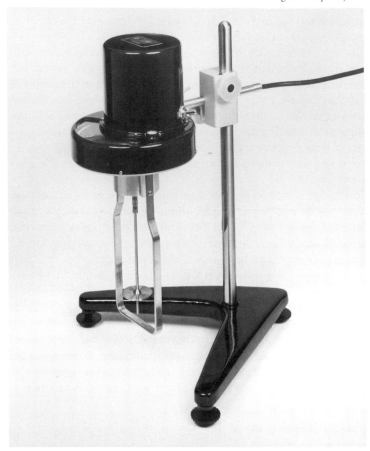

FIGURE 3.7
Brookfield viscometer. (Courtesy of Brookfield Engineering Laboratories, Stoughton, Massachusetts; photograph by Hutchins Photography, Inc.)

cylinder immersed in a sample and impelled by a uniform force. For some problems, change in consistency must be followed over a period of time, as in heating and cooking of a starch paste.

Consistency measurements that are important in the flour and starch industries can be made with a Brabender amylograph. A starch and water suspension is rotated and stirred in a sample cup as it is heated or cooled. Changes in consistency are recorded in graphic form in an arbitrary unit of consistency called Brabender units (BU). Interpretation of curves from this instrument is explained by Shuey and Tipples (1980) and Mazurs and co-workers (1957) and discussed in Chapter 16.

2. Elasticity

The science of rheology includes study of elastic solids as well as viscous liquids. These solids cannot be said to flow, but they can be deformed by force and recover when the force is removed. Elastic solids include gels such as pectin gels, baked custards, rennet curds, gelatin gels, and starch gels. The firmness of gels is indicated by the extent to which they retain their height when turned from a container or by their resistance to penetration by a variety of instruments.

The percentage sag of a gel can be measured simply by determining the height of the gel before and after it is removed from its container. This can be done as described in Appendix D. Sag then is expressed as a percentage of the height before unmolding. The Exchange ridgelimeter is a simple instrument used for direct measurement of percentage of sag. It was designed specifically for evaluation of pectin gels that will sag in the range of 10 to 40% (Ehrlich, 1968).

Various methods are used to measure the resistance of a gel to penetration. The penetrometer (Fig. 3.5) can be adapted for measurements on baked custards, starch gels, and pectin gels by substituting a light-weight cone for the heavier cones available from the manufacturer. A device improvised from common laboratory equipment and operating on the same principle as the penetrometer was described by Hanning and co-workers (1955). Johnson and Breene (1988) compared several methods for pectin gel strength measurements. Three instruments (Pektinometer, Chatillon, and Instron) were used to either push or pull a breaking figure into or out of the sample. All three instruments seemed to provide feasible methods for gel evaluation. Daget and Collyer (1984) described seven types of tests for gel strength (double compression, compression to rupture, tension, adhesiveness, cone penetration, extrusion, and puncture). All test devices were attached to a UTM. This work illustrates the many approaches that may be taken to testing of gel textural characteristics.

D. Texture of Spreadable Foods

Consumer acceptance of some products depends on the characteristics of the product when it is spread on another product. Thus the study of butter characteristics as evaluated with a cone penetrometer has been reported. Dixon and Parekh (1979) reported that an index based on cone angle and depth of penetration was related to perception of the spreadability of butter.

E. Texture of Doughs and Baked Products

Texture measurements frequently are made on batters and doughs as a way of predicting the quality of the final baked product. The consistency of batters has been measured with line-spread apparatus. Hunter and co-workers (1950) suggested that very viscous batters of a given specific gravity are indicative of fine dispersion of incorporated air, whereas a high line-spread reading in association with the same specific gravity is indicative of the dispersion of air in larger units. Other viscosity and consistency measuring devices also can be used to evaluate batters. The rheological properties of dough are related to the quality of the finished product. The consistency and stability of doughs can be measured with a farinograph, which is designed to measure the force required to turn mixer blades at a constant speed during mixing of the dough (Locken et al., 1972). The force increases as gluten is being developed and later decreases as gluten is broken down with continued mixing. Similar information can be obtained with the mixograph (Shuey, 1975).

Spread ratios may be measured as indicators of cookie dough quality. A method outlined by the AACC (1983, method 10-50D) involves measuring the width of six cookies laid side by side, rotating each cookie 90°, measuring the width (W) again, and averaging the two readings. The cookies also are stacked two times in different orders. The height is measured each time (T) and averaged. From these measurements, the W/T ratio is calculated and a correction factor applied. In addition cookie thickness may be measured (Arndt and Wehling, 1989).

The evaluation of the texture of baked products is approached in several ways. The breaking strength of pastries, cookies, and crackers can be determined with a shortometer, developed by Bailey (1934) and shown in Fig. 3.8. The pastry or other wafer (approximately 4.5 × 9.0 cm) is placed across two horizontal bars; a single upper bar is brought down by means of a motor until it breaks the wafer, and the force is recorded. Shortometer values were significantly correlated with sensory values for tenderness, and values obtained with a Kramer shear press also were significantly related to sensory scores for tenderness (Stinson and Huck, 1969). Similar breaking tests may be accomplished with appropriate attachments for a UTM (Arndt and Wehling, 1989).

FIGURE 3.8
Bailey shortometer. (Courtesy of Computer Controlled Machines, Northfield, Minnesota.)

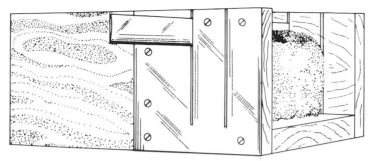

FIGURE 3.9
Cutting box used to obtain slices of uniform thicknesses.

Compression testing is used frequently to evaluate crumb firmness or softness of baked products. Softness was defined as the degree of compression under a constant load, whereas firmness was defined as the amount of force required to attain a specified compression (Babb, 1965). The Baker compressimeter can be used to measure both of these textural characteristics (Guy, 1982; Kamel, 1987) and is described in an AACC method for crumb firmness (AACC, 1983, method 74-10). A penetrometer fitted with a flat disk (Funk *et al.*, 1969; Kamel and Rasper, 1986) also may be used for measurements of softness. Gaines and Donelson (1985) described the use of a Struct-O-Graph for measurement of the distance a cake sample is compressed to a resistance of 450 and 225 cm·g for layer and angel food cakes, respectively. Both cake and bread firmness measurements may be made with the Instron UTM (Baker *et al.*, 1987; Walker *et al.*, 1987). Compression testing for cookie breaking strength also has been reported (Doescher *et al.*, 1987).

Sampling and conditions of testing must be carefully controlled in measurements of crumb firmness. Uniform slices for these and other tests are obtained with a cutting box like the one shown in Fig. 3.9. Position in the loaf or cake should be controlled as variation may be seen with position (Baker *et al.*, 1987). Values obtained become more variable as depth of compression increases, so Baker *et al.* (1988) suggested that for bread testing on a UTM 25% compression of a 25-mm slice be used as is recommended for the AACC method 74-09 (AACC, 1983).

F. Texture as Measured by Multipurpose Instruments

Texture as perceived by humans is not a simple characteristic but a composite of characteristics. Thus it is understandable that methods for evaluating more than one characteristic have been developed. Texture profile analysis by sensory panels is mentioned in Chapter 4. It also is possible to use instrumental methods to arrive at a texture profile for most foods. The General Foods texturometer was developed to give values to the parameters of the General Foods texture profile

(Friedman *et al.*, 1963). Parameters evaluated included hardness, cohesiveness, adhesiveness, fracturability, gumminess, springiness, and chewiness. Viscosity, the eighth profile parameter, usually is evaluated with a viscometer such as the Brookfield viscometer. The Instron UTM was first used by Bourne (1968) for the objective texture profile analysis of food. A picture of a recent model for food testing is shown in Fig. 3.10. The crosshead of the UTM moves up or down at a constant rate of speed. As it moves, the force that is required to compress a sample of food is continuously recorded. For texture profile analysis, the

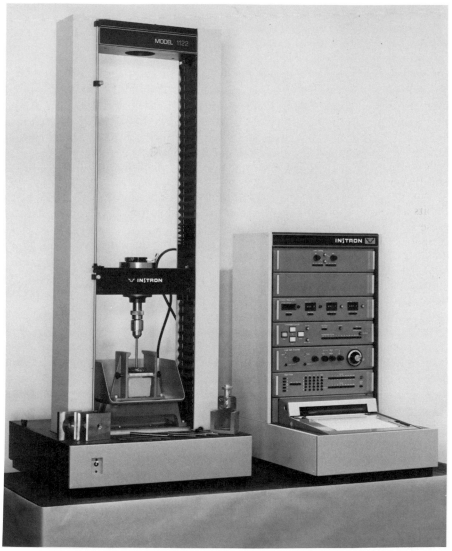

FIGURE 3.10
Universal testing machine. (Courtesy of Instron Corp., Canton, Massachusetts.)

instrument is allowed to compress the sample of food twice to represent two bites. From the two curves, seven parameters of texture may be measured. Sample curves representing two bites and showing six of the seven parameters are shown in Fig. 3.11. The figures are to be read from right to left. Hardness, a measure of force, is the peak height at maximum compression on bite 1 (H). If a break in the curve occurs before maximum compression the food exhibits the characteristic of fracturability. The food tested for Fig. 3.11 did not exhibit the characteristic of fracturability. The first peak on the apple curve (Fig. 3.12) is fracturability, which also is measured in kilograms of force. Adhesiveness is represented by the area of the curve under the baseline (A_3) and reflects the amount of work needed to remove the food from the compression device. Springiness (S_2) is indicated by the distance of compression during the second bite and represents the distance that the product sprang back after the compressive forces were removed. It also may be calculated as a ratio of S_2/S_1. Cohesiveness is calculated by dividing the area under the curve for the second bite (A_2) by that of the first bite (A_1). Because it is a ratio, it has no units. Gumminess is calculated by the formula hardness × cohesiveness and is a measure of force. Chewiness is equal to hardness × cohesiveness × springiness and is a measure of work. Not all foods will exhibit all of these characteristics. Sample curves from four very different foods are shown in Fig. 3.12 to illustrate this fact. Sample sizes were approximately the same for all products. Note that the scales are different for the top and bottom curves. One can in many cases identify a food by the characteristic shape of the texture profile. From these curves one can, for example, note that marshmallows are very springy, cheddar cheese exhibits adhesion, apples are fracturable, and bread is soft. Apples and cheddar cheese are not very cohesive.

Other UTMs are available for testing of this type and texture profile analyses of many foods have been reported in the literature. These reports include studies of gels (Daget and Collyer, 1984), rice (Okabe, 1979), frankfurters (Quinn *et al.*, 1979; Hargett *et al.*, 1980), baked products (Baker *et al.*, 1987; Walker *et al.*, 1987), and apples (McLellan *et al.*, 1984).

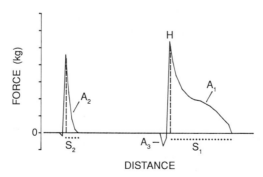

FIGURE 3.11
Instron curve showing measurements for determining textural parameters. Curve is read from right to left. H, Hardness; area A_2/area A_1 = cohesiveness; A_3, adhesiveness; S_2, springiness; $H \times A_2/A_1$ = gumminess; $H \times A_2/A_1 \times S_2$ = chewiness. See text for further explanation.

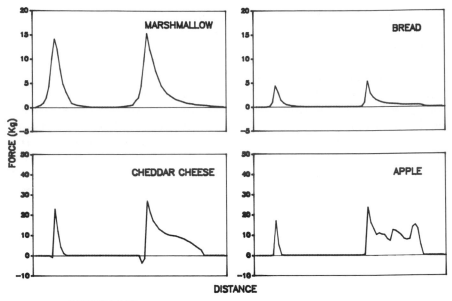

FIGURE 3.12
Instron texture profile curves for four foods of varying textural qualities.

In addition to the compression type of attachments used in texture profile analysis, other test cells may be attached to the UTM, including probes for puncture testing, extrusion cells, Warner–Bratzler shear attachments, and Kramer shear cells. Breene (1975) reviewed the use of instrumental methods for texture profile analysis.

G. Texture – Moisture and Fat Components

1. Press Fluid Measurements

Press fluid measurements involve the mechanical measurement of fluid from a sample. Measurements of this type are used in meat studies, although the amount of fluid that can be pressed from a vegetable such as corn is used as an index of maturity and succulence. Press fluid values for meat are indicators of water-binding capacity. Fluids may be pressed from a sample by packing it into a Carver cell, a device resembling a cylinder and piston, and exerting high force with a hydraulic press. An alternative method involves placing small samples of meat on filter paper and pressing them between acrylic sheets. The meat area is divided by the fluid area to give an expressible moisture index. Subtracting this ratio from 1 gives an index of water-binding capacity (Shaffer et al., 1973). Wismer–Pedersen (1987, pp. 141–159) indicated that this technique may be used with just Plexiglas plates and some means of screwing them together to apply force. This could be used in a situation where a hydraulic press is not available. An alternative method for objective evaluation of juiciness is outlined by Bouton and co-workers (1975). A significant relationship between sensory scores for juiciness and the amount of fluid pressed from a sample of meat centrifuged at 100,000 g was reported.

2. Moisture and Fat Determinations

The moisture content of food materials may be determined as an indicator of the quality of food. Traditionally methods involve a long, slow drying process in an oven or vacuum oven. Methods for specific foods are outlined by the Association of Official Analytical Chemists (AOAC, 1984). Karmas (1980) summarized methods of determining moisture content and taking samples for such measurements. Lee and Latham (1976) reported a procedure for using the microwave oven for more rapid moisture determinations.

Fat content also may be measured according to AOAC methods. In addition, Marriott *et al.* (1985) outlined rapid methods for fat and moisture determinations for meat. Such rapid methods would be useful for class experiments when time is limited.

VI. MISCELLANEOUS TESTS ————————————————

Specific gravity and the similar measurement, density, indicate the amount of air incorporated into products such as whipped cream, egg white foams, creamed shortening, and cake batter. The determination of specific gravity, described in Appendix D, involves dividing the weight of the food packed into a small, even-rimmed cylindrical container by the weight of the water held by the same container (Lee *et al.*, 1982). Density would be determined by dividing the weight of the food packed into the container by the volume of the container. Specific gravity might also be measured with a pycnometer (Pearce *et al.*, 1984). Low specific gravity or density values for batter, which indicate that a large amount of air is present, are associated with good volume of cakes in some reports (Jooste and Mackey, 1952).

References

AACC. 1983. "Approved Methods of the American Association of Cereal Chemists," 8th edn. AACC, St. Paul, MN.

Adams, M. C. and Birdsall, E. L. 1946. New consistometer measures corn consistency. *Food Ind.* **18:** 844–846, 992, 994.

AOAC. 1984. "Official Methods of Analysis," 14th edn. Association of Official Analytical Chemists, Washington, DC.

AOAC. 1985. "Changes in Official Methods of Analysis—1st Supplement," 14th edn. Association of Official Analytical Chemists, Washington, DC.

AOCS. 1981. "Official and Tentative Methods," 3rd edn. American Oil Chemists' Society, Champaign, Illinois.

Arndt, E. A. and Wehling, R. L. 1989. Evaluation of sweetener syrups derived from whey as replacements for sucrose in sugar-snap cookies. *Cereal Foods World* **34:** 423–424, 426–428.

Babb, A. T. S. 1965. A recording instrument for the rapid evaluation of the compressibility of bakery goods. *J. Sci. Food Agric.* **16:** 670–679.

Bailey, C. H. 1934. An automatic shortometer. *Cereal Chem.* **11:** 160–163.

Baker, A. E., Dibben, R. A., and Ponte, J. G. 1987. Comparison of bread firmness measurements by four instruments. *Cereal Foods World* **32**: 486–489.

Baker, A. E., Walker, C. E., and Kemp, K. 1988. An optimum compression depth for measuring bread crumb firmness. *Cereal Chem.* **65**: 302–307.

Balmaceda, E., Rha, C.-K., and Huang, F. 1973. Rheological properties of hydrocolloids. *J. Food Sci.* **38**: 1169–1173.

Barbut, S. 1988. Microstructure of reduced salt meat batters as affected by polyphosphates and chopping time. *J. Food Sci.* **53**: 1300–1304.

Bechtel, D. B. 1983. "New Frontiers in Food Microstructure." American Association of Cereal Chemists, St. Paul, Minnesota.

Berry, B. W., Smith, J. J., and Secrist, J. L. 1987. Effects of flake size on textural and cooking properties of restructured beef and pork steaks. *J. Food Sci.* **52**: 558–563.

Billmeyer, F. W. and Saltzman, M. 1981. "Principles of Color Technology," 2nd edn. John Wiley and Sons, New York.

Bolin, H. R. and Huxsoll, C. C. 1987. Scanning electron microscope/image analyzer determination of dimensional postharvest changes in fruit cells. *J. Food Sci.* **52**: 1649–1650, 1698.

Bourne, M. C. 1965. Studies on punch testing of apples. *Food Technol.* **19**: 413–415.

Bourne, M. C. 1968. Texture profiling of ripening pears. *J. Food Sci.* **33**: 223–226.

Bourne, M. C. 1982. "Food Texture and Viscosity." Academic Press, New York.

Bouton, P. E. and Harris, P. V. 1972. The effects of some post slaughter treatments on the mechanical properties of bovine and ovine muscle. *J. Food Sci.* **37**: 534–539.

Bouton, P. E., Harris, P. V., and Shorthose, W. R. 1971. Effects of ultimate pH upon water-holding capacity and tenderness of mutton. *J. Food Sci.* **36**: 435–439.

Bouton, P. E., Ford, A. L., Harris, P. V., and Ratcliff, D. 1975. Objective assessment of meat juiciness. *J. Food Sci.* **40**: 884–885.

Bratzler, L. J. 1932. Measuring the tenderness of meat by means of a mechanical shear. Master's thesis, Kansas State College, Manhattan.

Breene, W. M. 1975. Application of texture profile analysis to instrumental food texture evaluation. *J. Texture Studies* **6**: 53–82.

Bremner, H. A. and Hallett, I. C. 1985. Muscle fiber—connective tissue junctions in the fish blue grenadier (*Macruronus novaezelandiae*): A scanning electron miscroscope study. *J. Food Sci.* **50**: 975–984.

Brown, S. L. and Zabik, M. E. 1967. Effect of heat treatments on the physical and functional properties of liquid and spray-dried egg albumen. *Food Technol.* **21**: 87–92.

Carlson, D. 1972. Critical angle refractometer. *Food Technol.* **26**(5): 84–88, 89.

Cathcart, W. H. and Cole, L. C. 1938. Wide-range volume-measuring apparatus for bread. *Cereal Chem.* **15**: 69–79.

Cheng, C. S. and Parrish, F. C., Jr. 1976. Scanning electron microscopy of bovine muscle: Effect of heating on ultrastructure. *J. Food Sci.* **41**: 1449–1454.

Chinnaswamy, R., Hanna, M. A., and Zoebel, H. F. 1989. Microstructural, physiochemical, and macromolecular changes in extrusion-cooked and retrograded corn starch. *Cereal Foods World* **34**: 415, 417–422.

Cloke, J. D., Davis, E. A., and Gordon, J. 1984a. Relationship of heat transfer and water-loss rates to crumb-structure development as influenced by monoglycerides. *Cereal Chem.* **61**: 363–370.

Cloke, J. D., Davis, E. A., and Gordon, J. 1984b. Volume measurements calculated by several methods using cross-sectional tracings of cake. *Cereal Chem.* **61**: 375–378.

Clydesdale, F. M. 1972. Measuring the color of foods. *Food Technol.* **26**(7): 45–51.

Clydesdale, F. M. 1976. Instrumental techniques for color measurement of foods. *Food Technol.* **30**(10): 52–54, 58–59.

Clydesdale, F. M. 1984. Color measurement. *In* "Food Analysis: Principles and Techniques. Vol. I. Physical Characterization," Gruenwedel, D. W. and Whitaker, J. (Eds.). Marcel Dekker, Inc., New York.

Clydesdale, F.M. and Francis, F. J. 1969. Colorimetry of foods. I. Correlation of raw, transformed and reduced data with visual rankings for spinach puree. *J. Food Sci.* **34**: 349–352.

Collins, J. L. and Dincer, B. 1973. Rheological properties of syrups containing gums. *J. Food Sci.* **38**: 489–492.

Costello, C. A., Penfield, M. P., and Riemann, J. R. 1985. Quality of restructured steaks: Effects of days on feed, fat level, and cooking method. *J. Food Sci.* **49**: 685–689.

Daget, N. and Collyer, S. 1984. Comparison between quantitative descriptive analysis and physical measurements of gel systems and evaluation of the sensorial method. *J. Texture Studies* **15**: 227–245.

Davis, E. S., Gordon, J., and Hutchinson, T. E. 1976. Scanning electron microscope studies on carrots: Effects of cooking on the xylem and phloem. *Home Econ. Research J.* **4**: 214–224.

deMan, J. M., Voisey, P. W., Rasper, V. F., and Stanley, D.W. 1976. "Rheology and Texture in Food Quality." Avi Publ. Co., Westport, Connecticut.

Dixon, B. D. and Parekh, J. V. 1979. Use of the cone penetrometer for testing the firmness of butter. *J. Texture Studies* **10**: 421–434.

Doescher, L. C., Hoseney, R. C., and Milliken, G. A. 1987. A mechanism for cookie dough setting. *Cereal Chem.* **64**: 158–163.

Ehrlich, R. M. 1968. Controlling gel quality by choice and proper use of pectin. *Food Prod. Dev.* **2**(1): 74–75, 40–41.

Elling, H. 1988. Rapid flour quality tests. *Cereal Foods World* **33**: 495–497.

FDA. 1988a. Cereal flours and related products. *In* "Code of Federal Regulations," Title 21, Section 136. U.S. Govt. Printing Office, Washington, DC.

FDA. 1988b. Catsup. *In* "Code of Federal Regulations," Title 21, Section 155.194. U.S. Govt. Printing Office, Washington, DC.

Finney, E. E., Jr. 1972. Elementary concepts of rheology relevant to food texture studies. *Food Technol.* **26**(2): 68–77.

Fleming. S. E., Sosulski, F. W., and Hamon, N. W. 1975. Gelation and thickening phenomena of vegetable proteins. *J. Food Sci.* **40**: 805–807.

Floros, J. D. and Chinnan, M. S. 1988. Microstructural changes during steam peeling of fruits and vegetables. *J. Food Sci.* **53**: 849–853.

Foehse, K. B. and Hoseney, R. C. 1988. Factors affecting the Bostwick fluidity of corn flour/water systems. *Cereal Chem.* **65**: 497–500.

Francis, F. J. and Clydesdale, F. M. 1975. "Food Colorimetry: Theory and Applications." Avi Publ. Co., Westport, Connecticut.

Friedman, H. H., Whitney, J. E., and Szczesniak, A. S. 1963. The texturometer—a new instrument for objective texture measurement. *J. Food Sci.* **28**: 390–396.

Funk, K., Zabik, M. E., and El-Gidaily, D. A. 1969. Objective measurements for baked products. *J. Home Econ.* **61**: 119–123.

Gaines, C. S. and Donelson, J. R. 1985. Effect of varying flour protein content on angel food and high-ratio white layer cake size and tenderness. *Cereal Chem.* **62**: 63–66.

Grawemeyer, E. A. and Pfund, M. C. 1943. Line-spread as an objective test for consistency. *Food Research* **8**: 105–108.

Guy, E. J. 1982. Evaluation of sweet whey solids in yellow layer cakes with special emphasis on fragility. *Bakers Digest* **56**(4): 8, 10–12.

Haard, N. F., Medina, M. B., and Greenhut, V. A. 1976. Scanning electron microscopy of sweet potato root tissue exhibiting "hardcore." *J. Food Sci.* **41**: 1378–1382.

Hanning, F., Bloch, J. deG., and Siemers, L. L. 1955. The quality of starch puddings containing whey and/or non-fat milk solids. *J. Home Econ.* **47**: 107–109.

Hargett, S. M., Blumer, T. N., Hamann, D. D., Keeton, J. T., and Monroe, R. J. 1980. Effect of sodium acid pyrophosphate on sensory, chemical, and physical properties of frankfurters. *J. Food Sci.* **45**: 905–911.

Harper, J. M. 1986. Extrusion texturization of foods. *Food Technol.* **40**(3): 70, 72–76.

Hegarty, P. V. J. and Allen, C. E. 1975. Thermal effects on the length of sarcomeres in muscles held at different tensions. *J. Food Sci.* **40**: 24–27.

Hung, Y.-C. and Thompson, D. R. 1989. Changes in texture of green peas during freezing and frozen storage. *J. Food Sci.* **54**: 96–101.

Hunter, M. B., Briant, A. M., and Personius, C. J. 1950. Cake quality and batter structure. Bull. 860, Cornell Univ. Agric. Exp. Stn., Ithaca, New York.

IFT. 1964. Sensory testing guide for panel evaluation of foods and beverages. *Food Technol.* **18**: 1135–1141.

Johnson, R. M. and Breene, W. M. 1988. Pectin gel strength measurements. *Food Technol.* **42**(2): 87–93.

Jones, S. B., Carroll, R. J., and Cavanaugh, J. R. 1976. Muscle samples for scanning electron microscopy: Preparative techniques and general morphology. *J. Food Sci.* **41**: 867–873.

Jooste, M. E. and Mackey, A. D. 1952. Cake structure and palatability as affected by emulsifying agents and baking temperatures. *Food Research* **17**: 185–196.

Kalab, M. and Harwalkar, V. R. 1973. Milk gel structure. I. Application of scanning electron microscopy to milk and other foods. *J. Dairy Sci.* **56**: 835–842.

Kamel, B. 1987. Bread firmness measurement with emphasis on Baker compressimeter. *Cereal Foods World* **32**: 472–473, 475–476.

Kamel, B. and Rasper, V. F. 1986. Comparison of Precision penetrometer and Baker compressimeter in testing bread crumb firmness. *Cereal Foods World* **31**: 269–270, 272–274.

Karmas, E. 1980. Techniques for measurement of moisture content of foods. *Food Technol.* **34**(5): 52–59.

Kastner, C. L. and Henrickson, R. L. 1969. Providing uniform meat cores for mechanical stress force measurement. *J. Food Sci.* **34**: 603–605.

Klemaszewski, J. L., Hague, Z., and Kinsella, J. E. 1989. An electronic system for determining droplet size and dynamic breakdown of protein stabilized emulsions. *J. Food Sci.* **54**: 440–445.

Kornerup, A. and Wanscher, J. H. 1962. "Reinhold Color Atlas." Reinhold Publishing Co., New York.

Kramer, A. and Twigg, B. A. 1959. Principles and instrumentation for the physical measurement of food quality with special reference to fruit and vegetable products. *Adv. Food Research* **9**: 153–220.

Larmond, E. and Petrasovits, A. 1972. Relationship between Warner-Bratzler and sensory determinations of beef tenderness by the method of paired comparisons. *Can. Inst. Food Sci. Technol. J.* **5**(3): 138–144.

Lee, C. C., Hoseney, R. C., and Varriano-Marston, E. 1982. Development of a laboratory-scale single-stage cake mix. *Cereal Chem.* **59**: 389–392.

Lee, C.M., Whiting, R. C., and Jenkins, R. K. 1987. Texture and sensory evaluation of frankfurters with different formulations and processes. *J. Food Sci.* **52**: 896–900.

Lee, J. W. S. and Latham, S. D. 1976. Rapid moisture determination by a commercial-type microwave oven technique. *J. Food Sci.* **41**: 1487–1488.

Locken, L., Loska, S., and Shuey, W. 1972. "The Farinograph Handbook." American Association of Cereal Chemists, St. Paul, Minnesota.

MacNeil, J. H. and Mast, M. G. 1989. Resistance to tear: An instrumental measure of the cohesive properties of muscle food products, *J. Food Sci.* **54**: 750–751.

Maerz, A. J. and Paul, M. R. 1950. "A Dictionary of Color," 2nd edn. McGraw-Hill Book Co., New York.

Marriott, N. G., Smith, G. C., Carpenter, Z. L., and Dutson, T. R. 1985. Rapid analytical methods for ground beef. *J. Food Qual.* **8**: 153–160.

Mazurs, E. G., Schoch, T. J., and Kite, F. E. 1957. Graphic analysis of the Brabender viscosity curves of various starches. *Cereal Chem.* **34**: 141–152.

McCrae, S. E. and Paul, P. C. 1974. Rate of heating as it affects the solubilization of beef muscle collagen. *J. Food Sci.* **39**: 18–21.

McLellan, M. R., Kime, R. W., and Lind, L. R. 1984. Relationship of objective measurements to sensory components of canned applesauce and apple slices. *J. Food Sci.* **49**: 756–758.

Miller, B. S., Derby, R. I., and Trimbo, H. B. 1973. A pictorial explanation for the increase in viscosity of heated wheat starch-water suspension. *Cereal Chem.* **50**: 271–280.

Mohr, W. P. 1973. Applesauce grain. *J. Texture Studies* **4**: 263–268.

Mohsenin, N. M. 1986. "Physical Properties of Plant and Animal Materials," 2nd edn. Gordon and Breach Science Publishers, New York.

Morse, L. M., Davis, D. S., and Jack, E. L. 1950. Use and properties of non-fat milk solids in food preparation. I. Effect of viscosity and gel strength. II. Use in typical foods. *Food Research* **15**: 200–222.

Murray, A. C. and Martin, A. H. 1980. Effect of muscle fiber angle on Warner-Bratzler shear values. *J. Food Sci.* **45**: 1428–1429.

Nickerson, D. 1946. Color measurement and its application to the grading of agricultural products. A handbook on the method of disk colorimetry. Misc. Publ. 580, USDA, Washington, DC.

Noble, A. C. 1975. Instrumental analysis of the sensory properties of food. *Food Technol.* **29**(12): 56–60.

Noble, I. T., Halliday, E. G., and Klaas, H. K. 1934. Studies on tenderness and juiciness of cooked meat. *J. Home Econ.* **26**: 238–242.

Okabe, M. 1979. Texture measurement of cooked rice and its relationship to the eating quality. *J. Texture Studies* **10**: 131–152.

Painter, K. A. 1981. Functions and requirements of fats and emulsifiers in prepared cake mixes. *J. Amer. Oil Chemists' Soc.* **58**: 92–95.

Pearce, L. E., Davis, E. A., and Gordon, J. 1984. Thermal properties and structural characteristics of model cake batters containing nonfat dry milk. *Cereal Chem.* **61**: 549–554.

Pomeranz, Y. 1976. Scanning electron microscopy in food science and technology. *Adv. Food Research* **22**: 205–307.

Pomeranz, Y., Meyer, D., and Seibel, W. 1984. Wheat, wheat-rye, and rye dough and bread studied by scanning electron microscopy. *Cereal Chem.* **61**: 53–59.

Prusa, K., Bowers, J., and Chambers, E. IV. 1982. Instron measurements and sensory scores for texture of poultry meat and frankfurters. *J. Food Sci.* **47**: 653–654.

Quinn, J. R., Raymond, D. P., and Larmond, E. 1979. Instrumental measurement of wiener texture. *Can. Inst. Food Sci. Technol. J.* **12**: 154–156.

Russ, J. C., Stewart, W. D., and Russ, J. C. 1988. The measurement of macroscopic images. *Food Technol.* **42**(2): 94, 96–102.

Scott Blair, G. W. 1958. Rheology in food research. *Adv. Food Research* **8**: 1–61.

Setser, C. S. 1984. Color: Reflection and transmission. *J. Food Qual.* **6**: 183–197.

Shaffer, T. A., Harrison, D. L., and Anderson, L. L. 1973. Effects of end point and oven temperatures on beef roasts cooked in oven film bags and open pans. *J. Food Sci.* **38**: 1205–1210.

Shuey, W. C. 1975. Practical instruments for rheological measurements on wheat products. *Cereal Chem.* **52**(3, Part II): 42r–81r.

Shuey, W. C. and Tipples, K. H. 1980. "The Amylograph Handbook." American Association of Cereal Chemists, St. Paul, Minnesota.

Stanley, D. W. 1986. Chemical and structural determinants of texture of fabricated foods. *Food Technol.* **40**(3): 65–68, 76.

Stinson, C. G. and Huck, M. B. 1969. A comparison of four methods for pastry tenderness evaluation. *J. Food Sci.* **34**: 537–538.

Strange, E. D. and Whiting, R. C. 1988. Effect of temperature and collagen extractability and Kramer shear force on restructured beef. *J. Food Sci.* **53**: 1224–1225, 1233.

Swanson, R. B. and Penfield, M. P. 1988. Barley flour and salt level selection for a whole-grain bread formula. *J. Food Sci.* **53**: 896–901.

Szczesniak, A. S. 1963. Classification of textural characteristics. *J. Food Sci.* **28**: 385–389.

Szczesniak, A. S. 1973. Indirect methods of objective texture measurements. *In* "Texture Measurements of Food," Kramer, A. and Szczesniak, A. S. (Eds.). Reidel Publ. Co., Boston, Massachusetts.

Szczesniak, A. S. 1986. Sensory texture evaluation methodology. *Recip. Meat Conf. Proc.* **39**: 86–95.

Szczesniak, A. S. 1987. Correlating sensory with instrumental texture measurements—an overview of recent developments. *J. Texture Studies* **18**: 1–15.

Szczesniak, A. S. and Torgenson, K. W. 1965. Methods of meat texture measurement viewed from the background of factors affecting tenderness. *Adv. Food Research* **14**: 33–165.

Szczesniak, A. S., Humbaugh, P. R., and Block, H. W. 1970. Behavior of different foods in the standard shear compression cell of the shear press and the effect of sample weight on peak area and maximum force. *J. Texture Studies* **1**: 356–378.

Tinklin, L. and Vail, G. E. 1946. Effect of method of combining the ingredients upon the quality of the finished cake. *Cereal Chem.* **23**: 155–165.

Tuma, H. J., Venable, J. H., Wuthier, P. R., and Henrickson, R. L. 1962. Relationships of fiber diameter to tenderness and meatiness as influenced by bovine age. *J. Animal Sci.* **21**: 33–36.

Vaughan, J. G. (Ed.). 1979. "Food Microscopy." Academic Press, New York.

Villalobos-Dominguez, C. and Villalobos, J. 1947. "Villalobos Color Atlas." Stechert-Hafner, New York.

Walker, C. E., West, D.I., Pierce, M. M., and Buck, J. S. 1987. Cake firmness measurement by the Universal Testing Machine. *Cereal Foods World* **32**: 477, 478–479.

Wayne, J. E. B. and Shoemaker, C. F. 1988. Rheological characterization of commercially processed fluid milks. *J. Texture Studies* **19**: 143–152.

Williams, J. C., Field, R. A., and Riley, M. L. 1983. Influence of storage times after cooking on Warner-Bratzler shear values of beef roasts. *J. Food Sci.* **48**: 309–310, 312.

Wismer-Pedersen, J. 1987. Chemistry of animal tissues. Part 5—water. *In* "The Science of Meat and Meat Products," 3rd edn., Price, J. F. and Schweigert, B. S. (Eds.). Food and Nutrition Press, Inc., Westport, Connecticut.

Woodward, S. A. and Cotterill, O. J. 1986. Texture and microstructure of heat-formed egg white gels. *J. Food Sci.* **51**: 333–339.

Woodward, S. A. and Cotterill, O. J. 1987. Texture and microstructure of cooked whole egg yolks and heat-formed gels of stirred egg yolk. *J. Food Sci.* **52**: 63–67.

Yang, S.-S. and Cotterill, O. J. 1989. Physical and functional properties of 10% salted egg yolk in mayonnaise. *J. Food Sci.* **54**: 210–213.

EVALUATING FOOD BY SENSORY METHODS

Sensory evaluation of food or evaluation of food quality by a panel of judges is essential to many food studies because it answers the important questions of how a food looks, smells, feels, and tastes. Some might include sounds in this list; crisp foods may not seem to be crisp if you cannot hear them when you bite and chew. Whether a food product will be accepted or not depends on the integration of the consumer's perception of color, texture, and flavor into an overall impression of quality. Chemical and physical tests of food quality will not provide this type of information. Therefore, sensory evaluation is an essential component of a food research project or product development.

I. SENSORY EVALUATION DEFINED

The Sensory Division of the Institute of Food Technologists (IFT, 1981b) defines sensory evaluation as "a scientific discipline used to evoke, measure, analyze, and interpret reactions to those characteristics of foods and materials as they are perceived by the senses of sight, smell, taste, touch, and hearing."[1]

Sensory evaluation tests may be used in product development (Civille, 1978; Erhardt, 1978; Moskowitz, 1983), research, quality control (Nakayama and Wessman, 1979; Reece, 1979), and shelf-life studies (Dethmers, 1979; Labuza and Schmidl, 1988). In each of these applications, sensory evaluation data may be used as the basis for decision making (Meilgaard *et al.*, 1987a, pp. 3–4).

The answers to the questions posed in sensory evaluation may seem easy, but they are actually difficult because the answers depend on human judgment, which is individual and not always consistent. Many types of sensory tests are used in food science studies. In analytical sensory testing, panelists may be asked to discriminate between or among samples (difference tests) or describe or score the quality of a product (descriptive tests). In affective testing panelists are asked to rate the acceptability of a product and/or describe their preference for a product. The test selected will depend on the objectives of the investigation. Clear statement of the objectives of the project and of the sensory test is essential to the successful use of sensory test methods.

In this chapter the basics of sensory testing are introduced. More complete treatments of the subject, including the physiological and psychological aspects, are available in several books on sensory evaluation of food (Amerine *et al.*, 1965; Jellinek, 1985; Meilgaard *et al.*, 1987a,b; Piggott, 1988; Stone and Sidel, 1985). Basic procedures for sensory evaluation also are outlined in these books as well as in "The Manual of Sensory Testing Methods" of the American Society for Testing and Materials (ASTM, 1968). Guidelines for sensory evaluation of specific foods are provided in several sources (Cross *et al.*, 1978; AMSA, 1983a,b; Szczesniak, 1986; Meilgaard *et al.*, 1987b; Bodyfelt *et al.*, 1988).

---

[1] From IFT, 1981b; Reprinted from *Food Technology* **35**, No. 11, p. 50. Copyright © 1981 Institute of Food Technologists.

II. PREPARATION AND SAMPLING _____

Only food that is prepared by a method that can be duplicated is worth the effort of evaluating. When, and only when, the conditions of preparation are controlled carefully and defined can differences in quality attributes (dependent variables) be attributed to known variables (independent variables). Careful sampling of the food also is necessary. A well-mixed homogeneous sample is ideal, but is possible with only a few foods such as applesauce, milk, yogurt, fruit juices, and mashed potatoes. If canned or frozen foods are being sampled, it is desirable to mix the contents of several lots of food before sampling for preparation.

Special care is necessary to obtain reliable results with nonhomogeneous materials. For example, one muscle is selected from a slice of meat for evaluation. If possible, a judge is given a sample from the same area of the same numbered slice from each roast in a series. Paired roasts (representing the same location from the left and right side of the animal) are used if possible to minimize animal variation. A sampling plan for both sensory and physical testing from a study of fish (Madeira and Penfield, 1985) is shown in Fig. 4.1. Similar plans can be found in the literature for other food products. A judge should be given a

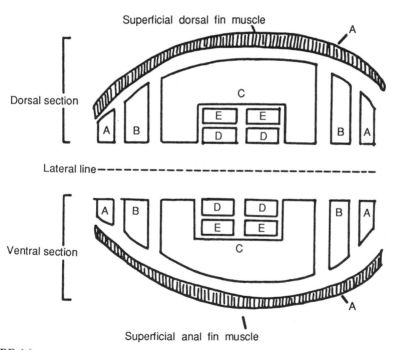

FIGURE 4.1
Sampling plan for objective measurements and trained sensory panel evaluation of turbot fillet cooked in conventional and microwave ovens. A, Discarded; B, used for raw pH and raw moisture content; C, used for Kramer shear test and cooked moisture content; D, samples for one trained panelist; E, samples for second trained panelist. (From Madeira and Penfield, 1985; Reprinted from *Journal of Food Science*, Volume **50**, No. 1, pp. 172–177. Copyright © 1985 by Institute of Food Technologists.)

sample from the same position in each cake or loaf of bread since these products vary from end to end. Outside slices of meat or baked products should not be included as samples.

All samples should be equally fresh when judged. It sometimes is possible to keep products baked at different times in satisfactory condition by freezing for short periods of times. Such a practice should be used only if absolutely necessary, as in a class situation when limited laboratory time prohibits preparation and evaluation of products in the same period. Care must be taken in selection of wrapping materials and temperature of storage to minimize changes attributable to freezing. Differences in length of storage times should be minimized.

III. PRESENTATION OF SAMPLES

Several factors must be controlled in conducting a sensory evaluation test in order to minimize experimental error in the data. Some specific guidelines for the factors that are discussed in this section are made in the ASTM Standard Practice E480-84, "Establishing Conditions for Laboratory Sensory Evaluation of Foods and Beverages" (ASTM, 1989).

A. Temperature of Samples

It is important that all samples in a series be at the same temperature when judged. In order to obtain meaningful results in affective testing, it is logical that the food should be tasted at the customary serving temperature. Cardello and Maller (1982) reported that acceptability scores for several foods including beef stew, creamed corn, pork sausage, scrambled eggs, meat loaf, and green beans were highest when the foods were tasted at the normal serving temperature. However, three foods frequently served at ambient temperatures (apple pie, ham, and biscuits) were most acceptable when tested at warmer temperatures.

If the usual temperature is very hot or very cold a more moderate temperature is appropriate, since the sense of taste is rendered less acute at extreme temperatures. Temperatures of not lower than 7°C or higher than 77°C should be used (ASTM, 1968, p. 14). The most appropriate temperature for analytical testing may depend on the product and test objective. Olson *et al.* (1980) reported that panelists rated 50°C steak samples to be more flavorful and juicy than 22°C samples. Tenderness scores did not vary with sample temperature.

If samples are to be served warm, several techniques may be used to maintain the temperature. Caporaso (1978) suggested double boilers for holding meat samples for testing. Care must be exercised to ensure that samples being held at warm temperatures do not lose moisture and become dry.

B. Serving Utensils

Serving containers for all samples in a series should be identical in size, color, and shape. White or clear containers are preferred so that the color of the food will show clearly and not be altered by its surroundings. Glass is preferable. However, for large panels glass may not be practical. Plastic or other disposable containers may be used if they do not impart odor or taste to the sample. If samples are placed in containers and stored prior to testing, care must be taken to ensure that flavors are not imparted to the sample from the container during the storage period. Styrofoam cups may be useful for their insulating properties (ASTM, 1989, Designation 480-84). Forks, spoons, and knives, if required, should also be carefully selected to ensure that they do not impart flavors to the sample. Care must be exercised in the washing and rinsing of nondisposable utensils. Detergent residues may affect the flavor of samples served in the containers.

C. Sample Size

Sufficient food is required to provide each judge with enough sample to enable him/her to make the decisions required by the test. In general, enough sample for at least two bites or sips of each sample should be provided. For some experiments a larger sample may be needed. For affective testing larger samples usually are provided. However, normal serving-sized samples are not necessary. Specific sample size recommendations are given in Designation E480-84 by the ASTM (1989).

D. Sample Coding

Samples are best identified by a code rather than a descriptive name that may bias the perception of the panelists. Codes such as A, B, and C or 1, 2, and 3 are undesirable because the first of a series suggests first choice or best quality to the judges. Randomly selected letters, three-digit randomly selected numbers, geometric shapes, colors, or symbols may be used as codes. Color codes may be particularly useful for consumer testing (Beckman et al., 1984). Three-digit random numbers are used most commonly and can be selected from a random number table in a statistics book or sensory evaluation reference book (Larmond, 1977, pp. 72–73; O'Mahony, 1986, pp. 473–474; Meilgaard et al., 1987b, pp. 413–414), generated with a computer program for random numbers, or selected with three successive rolls of a 10-sided die. When one series of samples is to be evaluated several times, judging will be less biased if the codes are altered each time.

E. Order of Presentation

In some experiments, judges tend to score the first samples presented higher than others. Kim and Setser (1980) noted that panelists tended to prefer the second of two cake samples presented in a paired preference test. These effects of position can be minimized by having the judges evaluate an additional or "warm-up" sample before the experimental series. In other experiments judges may base their evaluations on previously presented samples. A good sample preceding a poor sample may result in a lower than normal score for the poor sample, whereas the opposite order of presentation results in a higher than normal score for the good sample (Amerine *et al.*, 1965). Effects such as this, known as the contrast effect, can be minimized in experiments with a small number of samples by using all possible orders of sample presentation an equal number of times. For example, if 50 judges are asked to tell if two samples (A and B) are different or which of two samples (A and B) they prefer, 25 should receive sample A first and 25 should receive sample B first. This technique is known as balancing the order of presentation. If three samples (A, B, and C) are to be evaluated in a ranking or descriptive test, they can be presented in six possible orders (ABC, ACB, BAC, BCA, CAB, and CBA). A multiple of six judges should be used to achieve a balanced order of presentation. The possible orders should be used in random sequence as determined with a random number table or "drawing from a hat." Obviously, as the number of samples to be tested by each judge increases, balancing becomes more complex and sometimes impractical. For example, with 4 samples, there are 24 (4 × 3 × 2 × 1) possible orders of presentation. If 24 judges were used the orders could be balanced. In experiments with large numbers of samples, randomizing the order of presentation will help to minimize psychological effects (Larmond, 1973, pp. 17–19). Random orders may be determined by following orders in a random number table in a statistics book, generated with a computer program for random numbers, or determined by drawing from a hat.

F. Number of Samples

The number of samples that can be judged efficiently in one session is limited. Recommendations for the exact number differ as the type of food and type of test vary (ASTM, 1968, pp. 15–16; IFT, 1981b). More samples can be scored in color evaluation than in texture or flavor evaluations which involve consumption of the product. More bland than strong-flavored samples can be evaluated in one session. Fewer samples can be evaluated when complex score cards are used than when simple evaluation systems are used. Judges should not be asked to score so many samples that they will become fatigued and therefore inefficient. As a general guideline, six single samples or pairs of samples or four triangles constitute the maximum number that should be presented to a trained panel for evaluation. O'Mahony *et al.* (1988) presented more samples than this to panelists in tests involving 10 triangles. In addition, they suggested that giving warm-up samples before difference tests may in fact improve performance. Thus, there is some disagreement as to the number of samples that a panelist can effectively test in a session. However, it should be noted that inexperienced consumer panels should be given fewer samples than trained or experienced panelists (ASTM, 1968, pp. 15–16).

G. Time of Testing

The best time for judging is midmorning or midafternoon, when the judges are neither too well fed nor too hungry. If a study is to be replicated, panels should meet at the same time each day.

H. Rinsing

Generally it is recommended that each panelist be provided with a glass of neutral, room-temperature water for rinsing between samples. Unsalted crackers, apple wedges, or very dilute lemon juice may be used to remove flavors and food residues from the mouth. Cold water is avoided because it may dull the sense of taste. Good quality water should be used.

IV. ENVIRONMENT FOR TESTING

The environment is an important factor in sensory evaluation of food and should be without distractions so that the judges can concentrate on testing of the food. Judgments should be made independently; therefore, some form of separation of judges is desirable. The best way to satisfy this factor is to seat each judge in an individual compartment, which is properly lighted, painted a neutral color such as white, beige, or gray, and temperature controlled. Such an arrangement, which may include sliding doors or a hatch for sample presentation as shown in Fig. 4.2, usually is available only in large food research facilities or

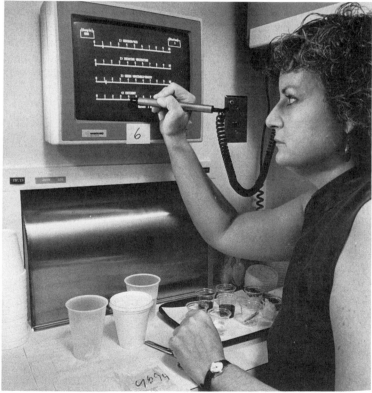

FIGURE 4.2
Individual sensory evaluation booths with controlled lighting and pass through for sample presentation. (Courtesy of USDA.)

FIGURE 4.3
Portable sensory evaluation booth. (Photo—University of Tennessee, Knoxville.)

large university research laboratories. A similar effect is accomplished by placing tables against a wall and separating them with panels about 1.2 m high, or by installing portable carrels on laboratory tables or counters. The carrels should be 45 to 50 cm high. Such an arrangement is diagrammed in Fig. 4.3. A room other than the preparation room should be used for testing if available. The room should be well ventilated and free from foreign odors. Panelists should not have to walk through the preparation area to get to the testing area. Convenient, comfortable seating for the judges is essential. Detailed information on physical requirements for a sensory evaluation laboratory may be found in the American Society of Testing and Materials publication, "Physical Requirements for Sensory Evaluation Laboratories" (Eggert and Zook, 1986).

Talking during judging is avoided because judges may be influenced by the opinions of others. However, discussion after sessions for scoring is desirable because it may lead to better performances by the judges and help to maintain their enthusiasm and interest.

A trend in construction of new or updating of older laboratories for sensory evaluation is the inclusion of a computer terminal or data-recording device in each booth (Anon., 1988; Brady *et al.*, 1985; McLellan and Cash, 1983; Winn, 1988). This allows for direct entry of data into a personal computer with potential for transfer of data to a mainframe computer. King and Morzenti (1988) in preliminary studies found less variation in panel responses when a computer rather than paper ballots was used for recording responses.

V. SELECTION AND TRAINING OF PANELISTS

A. Selection of a Panel

Procedures used to select a panel for analytical testing will depend on the type of sensory testing to be done as well as on the circumstances. If few persons are available, it may be necessary to accept anyone who will serve. It is, of course, better to select the most able judges from a larger group. In any case, it is important that the panelists be available for the entire period of the experiment, interested in the problem, and willing to serve. They should be in good health. People with colds are unable to evaluate foods accurately. Potential panelists should be surveyed to determine if they have food allergies or sensitivities.

For discrimination and descriptive testing, some investigators have selected panelists on the basis of their thresholds for basic tastes. However, the results of these tests have not been especially useful in predicting sensitivity to the taste of a food (Boggs and Hanson, 1949). It has been suggested that the ability to recognize the four basic tastes—sweet, sour, salty, and bitter—should be a minimum requirement in flavor work (Martin, 1973). Jellinek (1985, pp. 40–58) provides instruction for testing of this ability. The ability to recognize duplicate samples in a triangle test is considered important (Martin, 1973) and is described by Cross and co-workers (1978, pp. 9–10) as a basis for selection of panelists for meat evaluation. In studies on the effects of reheating on turkey breasts and rolls, Cremer and Richman (1987) and Cremer and Hartley (1988) asked potential panelists to participate in triangle tests to differentiate between freshly cooked and reheated turkey and fresh and rancid oils.

Another practical method of selecting panelists is to present actual test samples to the potential judges, replicating the samples presented to each judge several times. The consistency of each panelist in scoring the samples then is studied with the use of analysis of variance. Those who perform consistently and are able to differentiate among samples are retained for the panel. Since panelists vary in their ability to evaluate different foods, it is appropriate to select a panel for each food investigated.

For flavor and texture profiling techniques specific tests are used for selection of panelists. For example, for flavor profile analysis panelists are asked to identify the four basic tastes in water solutions, rank solutions of the basic tastes in order of increasing intensity, and identify at least 11 of 20 odorants including acetic acid, ammonia, amyl acetate, benzaldehyde, captax, d-fenchyl-alcohol, geraniol, methyl salicylate, phenol, vanillin, aniseed, cade, cassia, clary sage, cloves, eucalyptus, lavender, lime, linseed, and peppermint (Spencer, 1971).

Panelists for affective testing, frequently called consumer panelists, are selected to represent some segment of the population. Thus, socioeconomic, cultural, and geographic factors rather than ability are used as a basis for selection of the panelists (ASTM, 1979, Chapter 4). In a food company, consumer panelists may be recruited from the community as described by Nally (1987). Hess and Setser (1986) showed that individuals with diabetes mellitus or hypoglycemia rated aspartame- and fructose-sweetened cakes differently than did healthy panelists. The importance of employing users rather than nonusers for sensory testing was evident.

B. Size of Panel

The size of the panel is a question important to most experimental foods students and to most research workers. Availability of qualified panelists and differences in those panelists as well as the inherent variability in the product should influence the size of the panel. It should, of course, be as large as possible in order to reduce the experimental error, thus improving the reliability of the results. Several extra judges usually are included in a panel because of the inevitability of some absences. A minimum of 5 panelists has been suggested for descriptive and discrimination testing, whereas at least 50 to 100 should be used for affective testing with consumer panels (ASTM, 1968, pp. 3–7). The Sensory Evaluation Division of the IFT suggests that 24 consumer panelists are adequate for "rough product screening" (IFT, 1981b). The lack of experience of consumer panelists increases experimental error and therefore necessitates a large number of panelists.

C. Training the Panelists

Panelists for discrimination and descriptive testing may be trained before the experiment is started. Important functions of the training period are to show the judges that effort and concentration are essential in evaluation of foods and to develop a common understanding of terminology and general as well as specific procedures among the panelists. For example, as a general guideline, panelists should be asked to refrain from eating a meal for at least 60 min and from smoking, snacking, or chewing gum for 20 min before test sessions (ASTM, 1968, p. 9). The training period may be started by explaining enough of the problem to arouse the interest of the judges. With some problems, however, such as those involving off-flavors, it may be necessary to withhold some information to avoid causing bias. Interest also may be stimulated by asking the panelists to help design the score card, but, in any case, clear instructions on the scoring method are essential. Discussion among the panelists and the investigator concerning standards for the food to be tested is desirable.

Another helpful device in training judges is use of reference standards. Standards may help panelists to define terms and understand the range of a scale, and may actually reduce the times required for training (Rainey, 1986). Examples of reference standards are given by Szczesniak *et al.* (1963) and Muñoz (1986). A reference can be labeled to serve as a standard, or coded and presented with unknown samples. If coded, its identity can be revealed later and the findings discussed. The availability of such a sample depends, of course, on the nature of the problem. If it is possible to provide a valid standard throughout the experiment as well as in the training period, judging will be improved. However, if the quality of the standard is not constant, a standard will be misleading. Judges can be helped by scoring duplicate samples and later learning their identities. Samples of varying quality should be provided during the training period to give experience in using the extremes of the score card. It can be seen from this discussion that some steps in selecting and training a panel can be combined. Certainly some members of a panel can be omitted on the basis of their performance during the training period if necessary. Both selection and training of judges are described by several authors (ASTM, 1981; Meilgaard *et al.*, 1987b, Chapter 10; Stone and Sidel, 1985, pp. 4–47).

VI. TYPES OF TESTS

Sensory tests may be divided into three groups on the basis of the type of information that they provide. The three types are discrimination, descriptive, and affective (acceptance–preference) (IFT, 1981b). Selection of the appropriate test should be based on clearly defined objectives for the project.

A. Discrimination Tests

Discrimination tests are divided into two groups, difference and sensitivity. Difference tests are used to determine if there is a difference between or among samples. The methods of discrimination testing commonly used are the paired comparison, triangle, duo–trio, and ranking tests. Tests of this type are especially useful in selecting and training judges, in product development, and in quality control. They are relatively easy for the judges as they require only a short memory of food quality. One disadvantage of some of these methods is that only two samples are compared at a time. Thus it may be necessary to pair every sample in a series with every other sample. The number of samples that must be evaluated by each judge may be as high as or higher than in other methods. A greater quantity of each sample must be prepared. Another disadvantage is that, although data obtained with these methods indicate whether there is a difference between two samples and may give an indication of the direction of that difference, they do not reveal the degree of difference unless additional questions are asked. O'Mahony *et al.* (1988) suggested that "warm-up" samples may improve the performance of judges in discrimination tests.

For a paired comparison text, two samples are presented together and the judge is asked whether there is a difference in the samples with respect to a specific characteristic. A simple paired comparison test scorecard is shown in Fig. 4.4. In a directional paired comparison test (Fig. 4.5), the panelists may be asked to identify the sweeter piece of cake or the more tender sample of meat. The judges may be asked to indicate a choice even though they feel that there is no difference. In a paired comparison test, the judge has a 50% chance of selecting either sample as being different. The statistical significance of the test results can be determined by statistical analysis or by the use of special tables based on the use of appropriate statistical tests in Appendix E. Tables may also be found in other sources (Amerine *et al.*, 1965, p. 525; ASTM, 1968, pp. 64–67; Lar-

```
┌─────────────────────────────────────────────┐
│                                               │
│         Paired Comparison Questionnaire       │
│      Evaluate each of these two samples. Are they │
│      alike?                                    │
│                                               │
│                                               │
│              _____ yes _____ no             │
│                                               │
│                                               │
│      Comments:                                │
│                                               │
│                                               │
│                                               │
│                                               │
└─────────────────────────────────────────────┘
```

FIGURE 4.4
Questionnaire for paired comparison test.

```
┌─────────────────────────────────────────────┐
│                                               │
│         Directional Paired Comparison         │
│                 Questionnaire                 │
│                                               │
│                                               │
│      Evaluate each of these two samples for   │
│      sweetness.  Which is sweeter?            │
│                                               │
│          Sample code   Check sweeter sample   │
│                                               │
│             _____         _____             │
│             _____         _____             │
│                                               │
│      Comments:                                │
│                                               │
│                                               │
└─────────────────────────────────────────────┘
```

FIGURE 4.5
Questionnaire for directional paired comparison test.

mond, 1977, p. 64; Meilgaard *et al.*, 1987b, pp. 132–137; O'Mahony, 1986, pp. 411–412; Roessler *et al.*, 1978; Stone and Sidel, 1985, pp. 165–166, 168–169).

In the triangle test, three samples are presented, two of which are duplicates. The panelists are asked to identify the "odd" sample as shown in the sample scorecard (Fig. 4.6). The order of sample presentation should be varied because judges tend to choose the central sample of the three as the odd sample. The triangle test is useful when differences between samples are small. It is not of value when differences are so large that they are obvious. Since the judge is asked to select one sample from a set of three, the chance of guessing correctly is 33.3%. Tables for determining the statistical significance of the results are available in Appendix E or several books (Amerine *et al.*, 1965, pp. 526–527; ASTM, 1968, p. 68; Larmond, 1977, p. 63; O'Mahony, 1986, p. 413; Stone and Sidel, 1985, pp. 167–168). If the critical number of judges correctly identifies the odd sample, it may be concluded that the two samples are different. For example, if 32 judges from a panel of 70 correctly identify the odd sample, the samples are

---

Triangle Test Questionnaire
There are three samples on the tray. Two of the samples are alike.   Taste the samples in the order shown below. Indicate the odd sample by placing a check beside its code. Please describe the difference in the space provided.

       Sample code   Odd sample

        ‾‾‾‾‾   ‾‾‾‾‾
        ‾‾‾‾‾   ‾‾‾‾‾

Difference observed:_____

_____

FIGURE 4.6
Questionnaire for triangle test.

Duo–Trio Questionnaire
There are three samples on the tray. One is marked "R" for reference and the other two are coded. One of the coded samples is identical to R. Indicate which sample is different from R.

       Sample code   Check odd sample

        ‾‾‾‾   ‾‾‾‾‾‾
        ‾‾‾‾   ‾‾‾‾‾‾

Comments:

FIGURE 4.7
Questionnaire for duo–trio test.

assumed to be different at the 5% level. If, however, only 30 correctly identify the odd sample, the samples do not differ statistically.

The duo–trio test also involves three samples, but the judge is informed that the first sample presented is a control (Fig. 4.7). The panelist is asked to indicate which of the other two samples differs from the control. Therefore, the chance of guessing is 50%. Tables used for determining the statistical significance of the results are available in Appendix E or in several texts (Amerine *et al.*, 1965, p. 525; ASTM, 1968, pp. 64, 67; Larmond, 1977, p. 64; O'Mahony, 1986, p. 411; Stone and Sidel, 1985, pp. 165–166).

An additional approach to difference testing is the use of signal detection. O'Mahony (1979, 1986, pp. 389–397) describes application of this technique to sensory evaluation of foods. With the procedure the probability that the difference between two samples will be detected is determined by calculating the *R* index. This value is a probability value and thus can range from 0.00 to 1.00. The closer it is to 1.0 the greater the probability that the difference between two samples will be detected. For testing, panelists are given several pairs of the samples (A and B) in the test as warm-up samples. The samples are tasted until the panelist is sure that he/she can identify each of the samples. They are then given a series of the two samples in random order. The number of samples is 20 or more. As each sample is tasted, the panelist indicates if he/she is sure it is A, thinks it is A but is not sure, thinks it is B but is not sure, or is sure it is B. Results are tallied and the *R* index is calculated.

Judges participating in ranking tests for more than two samples are asked to rank a series of samples with respect to one or more specific characteristics such as a flavor, an odor, a color, or a textural parameter (Fig. 4.8). Numerical values are assigned to the results, with the sample representing the highest degree of the attribute receiving the highest value. Results can be evaluated statistically by consulting tables presented by Kahan *et al.* (1973) or Basker (1988a,b). This test is especially useful when an entire series of samples is available at one time and is to be ranked for one characteristic. It is not generally satisfactory when products available at different times are to be compared, when there is little difference among samples, or when information is desired on several characteristics at the same time. Ranking can be used to determine if there is a difference among a group of samples, which sample differs from which samples, and in what direction the differences occur. Ranking does not give a measure of the degree of difference among the samples.

Taste threshold tests are also discrimination tests, belonging to the group called sensitivity tests. These tests are conducted to determine the lowest concentration of substance that can be detected (absolute or detection threshold) or the lowest concentration of a substance required for identification of the substance (recognition or identification threshold). Threshold tests may involve the evaluation of acuity for the four basic tastes, odor notes, or variations in concentration of some constituent of a food. Tests of taste threshold have been made by adding small amounts of evaporated milk to whole milk; salt, sugar, or lemon juice to canned foods; or a sample that has deteriorated to a fresh sample. A method of determining odor and taste thresholds is outlined in Standard Practice E679-79 of the ASTM (1989). In this method a series of samples of

Ranking Questionnaire

Rank the samples for saltiness. Test the samples in the order given (____ ____ ____ ____). Assign the rank of 4 to the saltiest sample and a rank of 1 to the least salty sample. Place the code beside the appropriate rank.

Code	Rank	
____	1	Least salty
____	2	
____	3	
____	4	Most salty

Comments:

FIGURE 4.8
Questionnaire for ranking test.

ascending concentration of the test material is presented to the panelist. Each sample is the odd sample in a triangle test in which water samples are the like samples. Thresholds are the points at which the panelists begin to correctly identify the odd sample (detection) and correctly identify (recognition) the taste or odor sensation.

B. Descriptive Tests

The objective of descriptive testing is to characterize and/or compare samples with respect to one or more specific characteristics. Tests in this category include attribute rating, texture or flavor profiling, and quantitative descriptive analysis as well as some modifications of these methods.

For attribute rating, scaling or scoring techniques are used in the evaluation of food quality. Tests vary with the food and with the purpose of the experiment. Judges are asked to rate products on graphic scales (line scales anchored by bipolar terms or terms reflecting extremes in intensity), on scales consisting of a series of numbers representing a range of low to high intensity of a characteristic, or on category scales consisting of a series of descriptive words or statements representing successive levels of a characteristic. Examples of these types of scales are shown in Fig. 4.9. Category scales consisting of numbers may require more intensive training of judges to develop an understanding of the scale. The use of understandable adjectives may make judging relatively easy and precise. For example, juiciness, a textural attribute of meat, might be evaluated on a scale consisting of the terms extremely juicy, moderately juicy, slightly juicy, slightly dry, moderately dry, and extremely dry. To facilitate tabulation and averaging of the scores, a series of descriptive terms is assigned numerical values. The

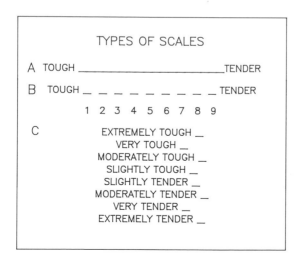

FIGURE 4.9
Types of scales used for descriptive testing: (A) line scale; (B) number scale; (C) word or descriptor scale.

TABLE I

Sample Terms to Anchor Ends of Scales for
Descriptive Testing

Thin—thick	Light—dark
Tough—tender	Not at all salty—very salty
Dry—wet	Not at all bitter—very bitter
Dry—moist	Not at all sweet—very sweet
Dry—juicy	Not at all sour—very sour
Fine—coarse	Not at all greasy—very greasy
Smooth—gritty	Very loose mass—compact mass
Dull—shiny	Open—very compact
Airy—compact	Does not spring back—springy
Soft—hard	

number 1 is assigned to the least desirable and/or least intense term and, in the case of the juiciness scale, 6 would be assigned to the term extremely juicy. These numbers are not included on the score card so that the judges will think in terms of the descriptive adjectives rather than the numerical scores. Similar scales can be devised for other quality attributes. Terms used in descriptive testing should be clearly defined so that the judges will understand them in order to allow for differentiation among the products that are being evaluated (Civille and Lawless, 1986). Sample sets of descriptive terms are included in Table I.

Another scaling technique that has been adapted for use in food evaluation is magnitude estimation. In this technique, which is a type of ratio scaling, panelists are asked to develop the numerical scale that they will use in their own evaluations. For example, if asked to evaluate the sweetness of orange juice, a judge may assign a score of 50 to the first sample. If the judge feels that the second sample is twice as sweet, a score of 100 is assigned; if it is half as sweet, a score of 25 is assigned. The ratio of the scores assigned, not the scores, is the important result. This method obviously gives information that cannot be obtained with more traditional scoring techniques. It is useful in evaluating the feasibility of substituting one ingredient such as a new sweetener for the presently used ingredient (Moskowitz, 1974).

Shand *et al.* (1985) compared several types of scales for evaluation of meat and found that differences among samples were likely to be found with a category scale and least likely to be found with a line scale. Panelists like magnitude estimation the least and category scaling the most. Choice of scale should be based on ability of panelists to differentiate among samples.

Extensive preliminary work on the part of the research worker and the judging panel may be necessary for the development of a score card such as the one in Table II. Various characteristics are listed and points assigned depending on the investigator's estimation of the relative importance of various attributes and defects in the quality of the product. Training of the judges using such a score card would be necessary so that the panel would have common understanding of the terminology. A similar scoring system for cakes (AACC method 10-90) was presented by the American Association of Cereal Chemists (AACC, 1983).

TABLE II
Scoring System for Cakes[a]

Qualities	Values used for statistical analysis[b]	Qualities	Values used for statistical analysis[b]
Size of cells		Tenderness	
Large	3	Crumbly	3
Medium	10	Very tender	8
Small	10	Tender	10
Very fine	3	Slightly tender	8
Compact	1	Tough	3
Distribution of cells		Moisture	
Uniform	6	Moist	10
Irregular	4	Dry	5
Tunneled	2	Wet	2
Crumb characteristics		Flavor	
Velvety	10	Well balanced	10
Slightly harsh	8	Sweet	5
Very harsh	2	Salt	2
		Bitter	2

[a]From Hunter et al. (1950). (Reprinted with permission of the New York State College of Human Ecology, a statutory college of the State University, Cornell University, Ithaca, New York.)

[b]These values did not appear on the judges' scorecards.

In order to obtain useful information about the quality of test products, scales should not reflect just the opinions of the judges. Terms such as desirable, acceptable, and undesirable do only that. Those terms are appropriate for affective testing but not for descriptive sensory evaluation.

Another descriptive method for testing the quality of food is the flavor profile method. This method is used to define and analyze flavor (Caul, 1957). Flavor factors that are perceived are called character notes. A list of those notes is made by each panel member during preliminary work on the food being investigated. The lists then are compared, and agreement is reached on which notes are to be used in further work. The intensity of each character note and the amplitude of the overall aroma and taste are rated. For example, the character notes listed for one sample of catsup were as follows: sweet, salt, molasses, sour, cooked tomato, and spice complex. A classic application of this method may be found in a paper by Berry et al. (1980). Flavor profiles of cooked beef loin steaks from carcasses of varying maturity group are reported. A modified application of the method is illustrated in a paper on fish by Prell and Sawyer (1988). Aroma notes including briny, sweet, fresh fish, old fish, stale fish, sour, shellfish, gamey fish, fish oil, earthy, nutty-buttery, musty, and scorched were identified by the panelists. In addition, 16 flavor notes were identified. Results of the study enabled the investigators to group the fish into groups of similar flavor characteristics. For example, haddock, wolffish, tilefish, pollack, cod (market), flounder, and cusk were grouped together. Bluefish and mackerel were found to be similar; both were high in flavor intensity.

The textural characteristics of food may be evaluated with a similar technique (Brandt *et al.*, 1963; Civille and Liska, 1975, Muñoz, 1986). A texture profile panel evaluates the mechanical, geometrical, fat, and moisture properties of food. Mechanical properties evaluated may include hardness, fracturability, springiness, cohesiveness, chewiness, gumminess, adhesiveness, and viscosity (Szczesniak, 1963, 1975; Szczesniak *et al.*, 1963). The definitions for six of these characteristics are given in Table III. Foods representing varying degrees of the mechanical properties other than springiness and cohesiveness are described by Szczesniak and co-workers (1963). Muñoz (1986) presented some modifications of the definitions and texture scales developed by Szczesniak and co-workers. In addition, Muñoz presented definitions and reference scales for adhesiveness to palate, teeth, and lips, cohesiveness, denseness, wetness, self-adhesiveness, roughness, cohesiveness of mass, moisture absorption, and manual adhesiveness. The definitions of these terms and the scales described by both of these individuals are useful in designing scoring systems for texture evaluations and training judges to evaluate the textural characteristics of food. Fish species also have been evaluated by a texture profile panel as they were by a flavor profile panel (Cardello *et al.*, 1982). Characteristics evaluated included hardness, flakiness, chewiness, fibrousness, moistness, and oily mouth coating. Groupings of fish with similar textures were made. Textural groupings differ from the flavor groupings previously discussed (Prell and Sawyer, 1988).

The profiling techniques have been used commercially for comparing competing products, for quality control, and for product development. Used in their most comprehensive form, these techniques may not be as generally useful in

TABLE III

Definitions for Texture Profile Analysis[a]

Term	Definition
Hardness	Force required to penetrate a substance with the molar teeth
Adhesiveness	Force required to remove the material that adheres to the mouth (generally to the palate)
Fracturability[b]	Ease with which a sample crumbles, cracks, or shatters
Chewiness	Length of time in seconds required to masticate a sample at the rate of one chew per second in order to reduce it to the consistency satisfactory for swallowing
Gumminess	Denseness that persists throughout chewing (refers to semisolid foods)
Viscosity	Force required to draw a liquid from a spoon over the tongue

[a] From Szczesniak *et al.* (1963). (Adapted from *Journal of Food Science*, **28**, No. 4, pp. 397–403. Copyright © 1963 by Institute of Food Technologists.)
[b] Originally termed brittleness.

class or research in a university setting as some of the other methods because they require a highly trained panel and because the results may, depending on the type of scales used, not lend themselves to statistical analysis and may be difficult to interpret. However, a modification of the texture profile analysis technique by Szczesniak and co-workers (1975) minimizes some of these limitations. In consumer texture profile analysis untrained panelists evaluate the textural characteristics of products on six-point scales ranging from "not at all" to "very much so." In addition, the panelists are asked to score an "ideal" product using their mental image of the degree to which the product should exhibit the characteristics. Comparison of scores for the ideal with scores for the test products is useful in product development. Variation among panel scores may be higher for this technique than for trained panels but consumer or untrained panelists can accomplish the task of providing information about the characteristics of the products. Although originally suggested for texture evaluation, this method also may be used for evaluation of appearance and flavor characteristics (Swanson and Penfield, 1988). Scores for the test products may be plotted as deviations from scores for the ideal as shown in Fig. 4.10. This technique may be particularly useful in class projects when time for training of panelists is not available but a reasonable number of panelists (approximately 25) can be recruited.

Quantitative descriptive analysis (QDA) involves the identification and quantification of the sensory characteristics of a product. The panel under the guidance of a panel leader identifies and defines these characteristics. Each identified characteristic usually is rated on a linear scale anchored by the terms, weak and intense. Data can be analyzed statistically and presented graphically to illustrate

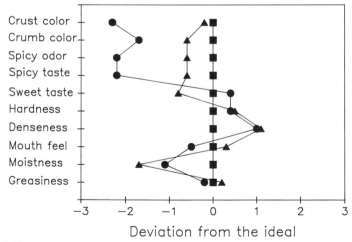

FIGURE 4.10

Results of consumer profile test in which panelists provided scores for the "ideal" spiced doughnut and evaluated two samples of commercially available doughnuts. ■, Ideal; ●, doughnut A; ▲, doughnut B.

TABLE IV
Sensory Profile Attributes Used for Snack Foods[a]

Sensory attribute	Scale								
Hardness	Soft	1	2	3	4	5	6	7	Hard
Chewiness	Tender	1	2	3	4	5	6	7	Tough
Moisture	Moist	1	2	3	4	5	6	7	Dry
Flavor balance	Blended	1	2	3	4	5	6	7	Disjointed
Fullness	Full	1	2	3	4	5	6	7	Thin
Aromatics	Slight	1	2	3	4	5	6	7	Strong
Sweet taste	Slight	1	2	3	4	5	6	7	Strong
Mouthfeel	Slight	1	2	3	4	5	6	7	Strong
Aftertaste	Short	1	2	3	4	5	6	7	Long
"Others"	None	1	2	3	4	5	6	7	Many

[a] From Mook (1984).

differences. Quantitative descriptive analysis is useful in product development (Stone *et al.*, 1974). A similar technique described by Mook (1984) is an outgrowth of the flavor profile analysis method. Attributes to be evaluated are determined by the panel and evaluated on a seven-point scale. References may be established for points on the scales. Terms used for the evaluation of snack products (cookies and crackers) are shown in Table IV. Terms such as these could be used for any of the descriptive techniques.

C. Affective Tests

Affective testing is used to determine if panelists like a product, if they prefer one product to another, and/or if they intend to use a product (acceptance). These types of evaluation may appear to be identical. However, it is possible for a judge to show a strong preference for a sample, but not use it or accept it for reasons other than its likeability. Affective tests may be called acceptance, preference, or consumer tests. Large panels (50–100) are used in this type of sensory evaluation and are often called consumer panels because untrained, inexperienced (naive) judges are used.

Several types of tests may be used to answer the questions posed in this type of testing. The most commonly used evaluation technique for measuring acceptability and/or preference is the hedonic scale. The word hedonic is defined as "pertaining to, or consisting in, pleasure." Hedonic scales may have five to nine points. The phrases used in a nine-point hedonic scale are shown in Fig. 4.11. In addition to determining the acceptability of a single product, differences in responses to foods can be determined with the hedonic scale. It was originated and has been used extensively by the Quartermaster Food and Container Institute of the Armed Forces.

```
┌─────────────────────────────────────────────────────┐
│              Hedonic Scorecard                        │
│  Taste each sample and mark how much you like         │
│  or dislike it.                        Code           │
│                                                       │
│  Like extremely              ____  ____  ____         │
│  Like very much              ____  ____  ____         │
│  Like moderately             ____  ____  ____         │
│  Like slightly               ____  ____  ____         │
│  Neither like nor dislike    ____  ____  ____         │
│  Dislike slightly            ____  ____  ____         │
│  Dislike moderately          ____  ____  ____         │
│  Dislike very much           ____  ____  ____         │
│  Dislike extremely           ____  ____  ____         │
│                                                       │
│  Comments:                                            │
│                                                       │
└─────────────────────────────────────────────────────┘
```

FIGURE 4.11

Scorecard for hedonic test.

"Smiley" questionnaire

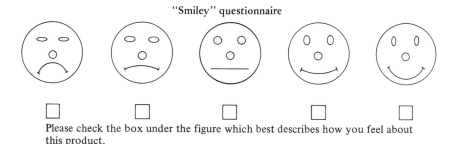

Please check the box under the figure which best describes how you feel about this product.

FIGURE 4.12

Facial hedonic scale developed by the Continental Can Company. (From Ellis, 1964, p. 187).

Although verbal hedonic scales like the one just described have been used frequently, Jones and Thurstone (1955) found that misinterpretations of the term "dislike moderately" often occurred. A modification of the hedonic scale was introduced to minimize misinterpretations of terms. The modification, the facial hedonic scale, is useful with young children and others with limited reading ability. If consists of five, seven, or nine faces depicitng varying degrees of pleasure or displeasure. One example of a facial hedonic is shown in Fig. 4.12. Many variations of this basic idea have also been used (Ellis, 1968). Moskowitz (1985, p. 150) described the use of a "Snoopy" hedonic scale.

Schutz (1965) developed a nine-point food action or attitude rating scale to determine what actions an individual expects to take with regard to a food product. The scale, which is a measure of overall food acceptance, is shown in Table V. The FACT test, as the food action rating scale is often called, is more sensitive than the hedonic scale. This probably is because panelists tend to be more realistic when evaluating or predicting actions (Schutz, 1965). However, this method should not be used to replace the hedonic scale but should be used to supplement it.

Several discrimination and descriptive tests can be adapted for affective test-

TABLE V
Food Attitude Rating Scale for FACT Method[a]

I would eat this every opportunity that I had.
I would eat this very often.
I would frequently eat this.
I like this and would eat it now and then.
I would eat this if available but would not go out of my way.
I don't like this but would eat it on an occasion.
I would hardly ever eat this.
I would eat this only if there were no other food choices.
I would eat this only if forced to.

[a] From Schutz (1965). (Reprinted from *Journal of Food Science* **30**, No. 2, pp. 365–374. Copyright © 1965 by Institute of Food Technologists.) FACT, Food action rating scale.

ing. Paired comparisons, ranking, and magnitude estimation tests can be used for this type of testing by modification of the score card and instruction to the panelists.

VII. INTERPRETATION OF RESULTS

The results of many of the descriptive and affective sensory methods described can be studied by tabulating the data, including the score of each judge for each sample, the means, the ranges, and the deviations from the mean. Some of the variability in results is attributable to the samples themselves, and may be a combination of differences in the raw materials and in the method of preparation. Sources of error in the judging include variability in the performance of one judge on duplicate samples as well as variability among several judges on the same sample.

After the data have been tabulated and averaged, the answer to the question posed by the experimenter may be obvious, and further analysis unnecessary. Certainly common sense is a useful aid in the interpretation of experimental results. Its application suggests the examination of means in light of the number of samples in each treatment and the variation among the samples. When there is much variation among replicate samples, differences among averages must be large to be meaningful unless an unusually large number of samples was used.

A study of the data such as described above is not adequate when the investigator wishes to state with confidence that the results obtained are statistically significant. In this case, a statistical analysis of the results is necessary. If an analysis is to be made, the original experiment should have been planned with statistical analysis in mind, because it is difficult, and sometimes impossible, to apply statistics to a completed experiment not appropriately planned. References have been given to tables based on statistical analysis for several discrimination and descriptive tests. The analysis of data from other tests may be more complex. Various methods of testing the significance of differences among means may be used depending on the experimental plan. Analysis of variance, with a multiple range test when appropriate, is very useful in many cases. Correlations, or an indication of the relationship between two variables, can be calculated between descriptive sensory data and objective data. Suggested sources of information on statistical analysis include several of the references at the end of the chapter (ASTM, 1968; Larmond, 1977; Stone and Sidel, 1985; O'Mahony, 1986; Amerine *et al.*, 1965).

Access to data handling and statistical analysis software for personal computers or to mainframe computers may facilitate data handling and analysis (Aust, 1984; Pecore, 1984; Savoca, 1984). Computer analysis is essential for large, comprehensive sensory projects such as the ones on fish texture and flavor previously discussed (Cardello *et al.*, 1982; Prell and Sawyer, 1988). The development of software for direct computer entry of data by panelists and the subsequent analysis by personal or mainframe computer will facilitate this process.

VIII. PRESENTATION OF RESULTS

Presentation of results from experimental food studies is discussed in Chapter 5. Sensory Evaluation Division guidelines (IFT) are available for help in presenting information on sensory studies (IFT, 1981a,c). In a food company setting, results may be presented in a more abbreviated format to the individual who requested the test (Stone and Sidel, 1985, pp. 50–55).

Suggested Exercises

1. Conduct a paired comparison or a triangle test. For samples you might use products prepared in the laboratory that have differences in formulas that may result in small differences in some quality attribute of the product or you might choose commercially available products that may have small differences such as a product that contains a nonsugar sweetener and the sugar-containing counterpart. Adapt the scorecard in Fig. 4.4 or 4.5. Analyze the results using the appropriate table (Appendix E). More extensive tables are in several sources (Amerine *et al.*, 1965, pp. 525–527; ASTM, 1968, p. 64; Larmond, 1977, pp. 63–64; O'Mahony, 1986, pp. 411–413; Roessler *et al.*, 1978; Stone and Sidel, 1985, pp. 165–169).
2. Add varying amounts of lemon juice to tomato juice (0, 10, 20, and 30% lemon juice on a volume basis). Ask panelists to rank the samples according to the increasing degree of acidity. Analyze the results as suggested by Kahan *et al.* (1973) or Larmond (1977, pp. 38–41). The experiment could be modified by adding varying amounts of salt to tomato juice or varying amounts of sugar to apple juice. The scorecard in Fig. 4.8 may be adapted for this experiment.
3. Develop a descriptive analysis system for a food product with the help of several class members. Establish references for points on the scales. Use the

scorecard several times on samples of varying quality. Discuss the adequacy of the scoring system and make appropriate revisions in the system.

4. Determine the thresholds of the class for NaCl or sucrose by following the ASTM Standard Procedure E679-79 (ASTM, 1989).

5. Select four similar products from the supermarket (i.e., four brownie mixes). Prepare as directed. Recruit a consumer panel to rank the products in order of preference. Analyze the results as suggested by Basker (1988a,b).

6. Determine if panelists can detect a difference between an artificial flavoring and a natural extract using a triangle or duo–trio test. Sugar cubes may be used as a carrier. Using a dropper or pipet, place three drops of flavoring agent on each sugar cube. Instruct panelists to smell the cube and suck the liquid from it for testing for differences. It is not essential to consume the sugar cube. Analyze the results using tables listed in exercise 1.

References

AACC. 1983. "Approved Methods of the American Association of Cereal Chemists," 8th edn. AACC, St. Paul, Minnesota.

Amerine, M. A., Pangborn, R. M., and Roessler, E. B. 1965. "Principles of Sensory Evaluation of Food." Academic Press, New York.

AMSA. 1983a. Guidelines for sensory, physical, and chemical measurements of ham. *Recip. Meat Conf. Proc.* **36**: 213–220.

AMSA. 1983b. Guidelines for sensory, physical, and chemical measurements in ground beef. *Recip. Meat Conf. Proc.* **36**: 221–228.

Anon. 1988. Computers tell "how sweet it is." *Food Technol.* **42**(11): 98, 100.

ASTM. 1968. Manual on sensory testing methods. ASTM Spec. Tech. Publ. 434, Amer. Soc. Testing Materials, Philadelphia, Pennsylvania.

ASTM. 1979. Manual on consumer sensory evaluation. ASTM Spec. Tech. Publ. 682, Amer. Soc. Testing Materials, Philadelphia, Pennsylvania.

ASTM. 1981. Guidelines for the selection and training of panelists. ASTM Spec. Tech. Publ. 758, Amer. Soc. Testing Materials, Philadelphia, Pennsylvania.

ASTM. 1989. "1989 Annual Book of ASTM Standards. Vol. 15.07. End Use Products." Amer. Soc. Testing Materials, Philadelphia, Pennsylvania.

Aust, L. B. 1984. Computers as an aid in discrimination testing. *Food Technol.* **38**(9): 71–73.

Basker, D. 1988a. Critical values for differences among rank sums for multiple comparisons. *Food Technol.* **42**(2): 79–84.

Basker, D. 1988b. Critical values for differences among rank sums for multiple comparisons by small taste panels. *Food Technol.* **42**(11): 88–89.

Beckman, K. J., Chambers, E. IV, and Gnagi, M. M. 1984. Color codes for paired preference and hedonic testing. *J. Food Sci.* **49**: 115–116.

Berry, B. W., Maga, J. A., Calkins, C. R., Wells, L. H., Carpenter, Z. L., and Cross, H. R. 1980. Flavor profile analysis of cooked beef loin steaks. *J. Food Sci.* **45**: 1113–1115, 1121.

Bodyfelt, F. W., Tobias, J., and Trout, G. M. 1988. "The Sensory Evaluation of Dairy Products." Van Nostrand Reinhold, New York.

Boggs, M. M. and Hanson, H. L. 1949. Analysis of foods by sensory difference tests. *Adv. Food Research* **2**: 219–258.

Brady, P. L., Ketelsen, S. M., and Ketelsen, J. P. 1985. Computerized system for collection and analysis of sensory data. *Food Technol.* **39**(5): 82, 84, 86, 88.

Brandt, M. A., Skinner, E. Z., and Coleman, J. A. 1963. Texture profile method. *J. Food Sci.* **28**: 404–409.

Caporaso, F. 1978. A simple technique for maintaining temperature of cooked meat samples prior to sensory evaluation. *J. Food Sci.* **43**: 1041–1042.

Cardello, A. V. and Maller, O. 1982. Acceptability of water, selected beverages and foods as a function of serving temperature. *J. Food Sci.* **47**: 1549–1552.

Cardello, A. V., Sawyer, F. M., Maller, O., and Digman, L. 1982. Sensory evaluation of the texture and appearance of North Atlantic fish. *J. Food Sci.* **47**: 1818–1823.

Caul, J. F. 1957. The profile method of flavor analysis. *Adv. Food Research* **7**: 1–40.

Civille, G. V. 1978. Case studies demonstrating the role of sensory evaluation in product development. *Food Technol.* **32**(11): 59–60.

Civille, G. V. and Lawless, H. T. 1986. The importance of language in describing perceptions. *J. Sensory Studies* **1**: 203–215.

Civille, G. V. and Liska, I. H. 1975. Modifications and applications to foods of the General Foods sensory texture profile technique. *J. Texture Studies* **6**: 19–32.

Cremer, M. L. and Hartley, S. K. 1988. Sensory quality of turkey rolls roasted at two temperatures and reheated with varying sauces in three types of institutional ovens. *J. Food Sci.* **53**: 1605–1609.

Cremer, M. L. and Richman, D. K. 1987. Sensory quality of turkey breasts and energy consumption for roasting in a convection oven and reheating in infrared, microwave, and convection ovens. *J. Food Sci.* **52**: 846–850.

Cross, H. R., Bernholdt, H. F., Dikeman, M. E., Greene, B. E., Moody, W. G., Staggs, R., and West, R. L. 1978. Guidelines for cookery and sensory evaluation of meat. Amer. Meat Sci. Assoc. and Nat'l. Livestock and Meat Board, Chicago, Illinois.

Dethmers, A. E. 1979. Utilizing sensory evaluation to determine product shelf life. *Food Technol.* **33**(9): 40–42.

Eggert, J. and Zook, K. 1986. Physical requirements guidelines for sensory evaluation laboratories. Publ. 913, Amer. Soc. Testing Materials, Philadelphia, Pennsylvania.

Ellis, B. H. 1964. Flavor evaluation as a means of product evaluation. Proceedings 11th Ann. Meeting, Society Soft Drink Technologists, Washington, DC.

Ellis, B. H. 1968. Preference testing methodology. *Food Technol.* **22**: 583–588, 590.

Erhardt, J. P. 1978. The role of the sensory analyst in product development. *Food Technol.* **32**(11): 57–58, 66.

Hess, D. A. and Setser, C. S. 1986. Comparison of aspartame- and fructose-sweetened cakes: importance of panels of users for evaluation of alternative sweeteners. *J. Amer. Dietet. Assoc.* **86**: 919–923.

Hunter, M. B., Briant, A. M., and Personius, C. J. 1950. Cake quality and batter structure. Bull. 860, Cornell Univ. Agric. Exp. Stn., Ithaca, NY.

IFT. 1981a. *Guidelines for the preparation and review of papers reporting sensory evaluation data.* Food Technol. **35**(4): 16–17.

IFT. 1981b. Sensory evaluation guides for testing food and beverage products. *Food Technol.* **35**(11): 50–59.

IFT. 1981c. Guidelines for the preparation and review of papers reporting sensory evaluation data. *J. Food Sci.* **46**: 1294–1295.

Jellinek, G. 1985. "Sensory Evaluation of Food: Theory and Practice." VCH Publishers, Deerfield Beach, Florida.

Jones, L. V. and Thurstone, L. L. 1955. The psychophysics of semantics: An experimental investigation. *J. Appl. Psychol.* **39**: 31–36.

Kahan, G., Cooper, D., Papavasiliou, A., and Kramer, A. 1973. Expanded tables for determining significance of differences for ranked data. *Food Technol.* **27**(5): 61–69.

Kim, K. and Setser, C. S. 1980. Presentation order bias in consumer preference studies on sponge cakes. *J. Food Sci.* **45**: 1073–1074.

King, A. J. and Morzenti, A. 1988. Response freedom in computerized and mechanical modes of sensory scoring. *Food Technol.* **42**(10): 150, 152, 159–160.

Labuza, T. P. and Schmidl, M. K. 1988. Use of sensory data in the shelf life testing of foods: Principles and graphical methods for evaluation. *Cereal Foods World* **33**: 193–194, 196–198, 200–205.

Larmond, E. 1973. Physical requirements for sensory testing. *Food Technol.* **27**(11): 28–30, 32.

Larmond, E. 1977. Laboratory methods for sensory evaluation of food. Publ. 1637, Canada Dept. Agric., Ottawa, Ontario.

Madeira, K. and Penfield, M. P. 1985. Turbot fillet sections cooked by microwave and conventional heating methods: Objective and sensory evaluation. *J. Food Sci.* **50**: 172–177.

Martin, S. L. 1973. Selection and training of sensory judges. *Food Technol.* **27**(11): 22, 24, 26.

McLellan, M. R. and Cash, J. N. 1983. Computerized sensory evaluation: A prototype data-collection system. *Food Technol.* **37**: 97–99.

Meilgaard, M. M., Civille, G. V., and Carr, B. T. 1987a. "Sensory Evaluation Techniques," Vol. I. CRC Press, Inc., Boca Raton, Florida.

Meilgaard, M. M., Civille, G. V., and Carr, B. T. 1987b. "Sensory Evaluation Techniques," Vol. II. CRC Press, Inc., Boca Raton, Florida.

Mook, J. H. 1984. Correlation of consumer and professional descriptions. *Cereal Foods World* **29**: 403–405.

Moskowitz, H. R. 1974. Sensory evaluation by magnitude estimation. *Food Technol.* **28**(11): 16, 18, 20–21.

Moskowitz, H. R. 1983. "Product Testing and Sensory Evaluation of Foods." Food and Nutrition Press, Inc., Westport, Connecticut.

Moskowitz, H. R. 1985. "New Directions for Product Testing and Sensory Analysis of Foods." Food and Nutrition Press, Inc., Westport, Connecticut.

Muñoz, A. M. 1986. Development and application of texture reference scales. *J. Sensory Studies* **1**: 55–84.

Nakayama, M. and Wessman, C. 1979. Application of sensory evaluation to the routine maintenance of product quality. *Food Technol.* **33**(9): 38–39.

Nally, C. L. 1987. Implementation of consumer taste panels. *J. Sensory Studies* **2**: 77–83.

Olson, D. G., Caporaso, F., and Mandigo, R. W. 1980. Effects of serving temperature on sensory evaluation of beef steaks from different muscles and carcass maturities. *J. Food Sci.* **45**: 627–628, 631.

O'Mahony, M. 1979. Short-cut signal detection measures for sensory analysis. *J. Food Sci.* **44**: 302–303.

O'Mahony, M. 1986. "Sensory Evaluation of Food. Statistical Methods and Procedures." Marcel Dekker, Inc., New York.

O'Mahony, M., Thieme, U., and Goldstein, L. R. 1988. The warm-up effect as a means of increasing the discriminability of sensory difference tests. *J. Food Sci.* **53**: 1848–1850.

Pecore, S. D. 1984. Computer-assisted consumer testing. *Food Technol.* **38**(9): 78–80.

Piggott, J. R. 1988. "Sensory Analysis of Foods," 2nd edn. Elsevier Applied Science Publishers, New York.

Prell, P. A. and Sawyer, F.M. 1988. Flavor profiles of 17 species of North Atlantic fish. *J. Food Sci.* **53**: 1036–1042.

Rainey, B. 1986. Importance of reference standards in training panelists. *J. Sensory Studies* **1**: 149–154.

Reece, R. 1979. A quality assurance perspective of sensory evaluation. *Food Technol.* **33**(9): 37.

Roessler, E. B., Pangborn, R. M., Sidel, J. L., and Stone, H. 1978. Expanded statistical tables for estimating significance in paired-preference, paired-difference, duo-trio and triangle tests. *J. Food Sci.* **43**: 940–943, 947.

Savoca, M. R. 1984. Computer applications in descriptive testing. *Food Technol.* **38**(9): 74–77.

Schutz, H. G. 1965. A food action rating scale for measuring food acceptance. *J. Food Sci.* **30**: 365–374.

Shand, P. J., Hawrysh, Z. J., Hardin, R. T., and Jeremiah, L. E. 1985. Descriptive sensory assessment of beef steaks by category scaling, line scaling, and magnitude estimation. *J. Food Sci.* **50**: 495–499.

Spencer, H. W. 1971. Techniques in the sensory analysis of flavours. *The Flavour Industry* May: 293–302.

Stone, H. and Sidel, J. L. 1985. "Sensory Evaluation Practices." Academic Press, Orlando, Florida.

Stone, H., Sidel, J., Oliver, S., Woolsey, A., and Singleton, R. C. 1974. Sensory evaluation by quantitative descriptive analysis. *Food Technol.* **28**(11): 24, 26, 28–29, 32–33.

Swanson, R. B. and Penfield, M. P. 1988. Barley flour level and salt level for whole-grain bread formulas. *J. Food Sci.* **53**: 896–901.

Szczesniak, A. S. 1963. Classification of textural characteristics. *J. Food Sci.* **28**: 385–389.

Szczesniak, A. S. 1975. General Foods texture profile revisited—ten year perspective. *J. Texture Studies* **6**: 5–17.

Szczesniak, A. S. 1986. Sensory texture evaluation methodology. *Recip. Meat Conf. Proc.* **39**: 86–89.

Szczesniak, A. S., Brandt, M. A., and Friedman, H. H. 1963. Development of standard rating scales for mechanical parameters of texture and correlation between the objective and sensory methods of texture evaluation. *J. Food Sci.* **28**: 397–403.

Szczesniak, A. S., Loew, B. J., and Skinner, E. Z. 1975. Consumer texture profile technique. *J. Food Sci.* **40**: 1253–1256.

Winn, R. L. 1988. Touch screen system for sensory evaluation. *Food Technol.* **42**(11): 68–70.

PREPARING THE REPORT

Reporting experimental work after its completion usually is required, and writing the report often is dreaded. The task need not be overwhelming if it is approached systematically, with some prior thought and attention to a few principles. Leaving the preparation until the evening before the due date is almost certain to result in a case of panic rather than a satisfying learning

A word processor, if available, simplifies the writing task. Isolated sections of the report—particularly the introduction, the review of literature, and the list of references—can be put on a computer disk in advance and revised at intervals; the result of several revisions should be a relatively polished manuscript. The principle also applies to handwriting or typing, but use of a word processor simplifies revision and saves time. The revised portions are available for quick printing. Further saving of time results because repeated proofreading of unchanged sections is unnecessary when the word processor is used. In addition, elimination of the need for retyping when changes are made results in more nearly error-free pages; a common problem with typewritten corrections is the creeping in of new errors.

Whether a word processor or a typewriter is used, the importance of proofreading cannot be overemphasized. Typographical errors are far too common.

Programs that check spelling facilitate but do not eliminate the need to proof carefully. The writer must check words in context. For example, because "k" and "l" are side by side on the keyboard, cook easily becomes cool or vice versa. A spell-check program will find that either is a correct spelling, but the meaning of the sentence certainly differs! Cooling meat to 65°C is certainly different from cooking meat to 65°C.

The main sections of the report are the following: introduction, review of the literature, materials and methods (procedures), results and discussion, summary and conclusions, and references. The headings are subject to some variation and the sections are subject to some combining.

I. THE ORGANIZATION OF THE REPORT

A. Introduction, Review of Literature

The introduction includes a statement of purpose and provides justification for the work. The literature review frequently is combined with the introduction, providing the background for the stated purpose of the study and the justification. In this case, the review probably will have been written during the planning phase. If the collection of data has covered a period of even a month, some new related information may have appeared in the research literature; therefore, current issues of research journals should be consulted and pertinent new material incorporated.

The review of literature may be a separate section if it is long and if the introduction does not depend on it. A long review should not consist merely of separate summaries but should be organized into a cohesive exposition of the subject. Preparation of an outline is helpful in achieving organization and in excluding irrelevant and repetitious material.

B. Materials and Methods

The purpose of this section is to state clearly what was done. Decisions concerning the materials and methods will of necessity have been made during the planning stage; therefore, much of the writing of this section could have been completed before the data were collected. Possibly some revision and/or elaboration will be necessary because of the course of events during the study. For example, the actual number of replications might have differed from the number planned. Unforeseen difficulties might have necessitated changes in materials or methods. Early results might even have brought about a change in direction of the study.

Judgment is needed as to the detail in which methods are described. Woodford (1968) agreed with the frequently stated rule of thumb that the detail should be sufficient to permit the reader to repeat the experiment. He went on to caution, however, that "who the reader is" should be defined. Unless some background on the part of the reader is assumed, this section might be intolerably long. Woodford suggested a constant self-quetsioning: "Are these details essential to the success of the experiment?"

C. Results and Discussion

The results and discussion may be either combined or presented as separate sections. In a short report they are likely to be combined. Tables and figures are helpful in the presentation of results. In many books on technical writing, including those of Barnett (1987), Carosso (1986), Lefferts (1981), and Roundy and Mair (1985), sections are devoted to visual aids. Additional information concerning tables and figures is found in the publication manual of the American Psychological Association (1983) and in the style guide of the Council of Biological Editors (CBE, 1978).

1. Tables

Tables should be prepared before any statements of results are written. Eventually the tables will be integrated into the text, but the text should be fitted to the tables rather than the tables to the text. Tables and text should supplement, not duplicate, each other. The text should emphasize the "big picture," not the details presented in the table.

The content of the tables requires some thought. A decision that must be made is the extent to which the raw data are to be included along with the averages and other calculated values. Consider a study that involves sensory evaluation. The complete data would show the extent of variation among the judges but would add considerable bulk. Another decision might involve inconclusive data: Should they be presented in tabular form or should they be covered by a statement in the text? Judgment must be applied when answering such questions in individual cases.

The amount of information to be included in a single table is another important consideration. There is an advantage in presenting the results of more than one type of evaluation in the same table because doing so makes certain relationships apparent. For example, presenting the results of the objective measurements and the sensory evaluation side by side would show a relationship between a physical characteristic and acceptance of a product.

Tabular form varies somewhat with publications, but there are some principles to be observed. Tables are placed after their first mention in the text. They are numbered consecutively through the report. Each title, placed above the table, should be as concise as possible while giving the reader a comprehensive idea of the content. Side headings and column headings are used as needed to identify the treatments and the nature of the values. Units of measurement must be given for all values but not individually. They can be at the top of the columns or in the left-hand column (stub), depending on what is convenient in a given situation. However, it can be advantageous to arrange the tabular data so that comparisons are made vertically rather than horizontally; listing levels of independent (controlled) variables in the left-hand column and the dependent (experimental) variables across the top of the table permits vertical comparisons and involves placement of units of measurement at the tops of columns. A table with that arrangement is found in the sample report later in the chapter (Table (II). Table III in the report is arranged with the dependent variables in the left column and the independent variables across the top. Sometimes all of the values in a single table have the same unit, such as percentage, which can be conveniently specified in the title. Standard abbreviations should be used for common units; unusual abbreviations should be explained, usually in table footnotes.

Accuracy in entering figures in a table obviously is important. Decimal points should be aligned. A zero should precede a decimal fraction. Consideration should be given to significant digits; in general, the calculations should be carried to as many digits as practicality permits, and then the values should be rounded. The need for rounding applies to values obtained via computer analysis, as well as to other calculated values. For a given measurement, the number of significant digits reported should reflect the accuracy of the measurement.

2. Figures

Figures such as instrument tracings and photographs represent primary results. Graphs, such as plotted curves, histograms, and bar graphs, are figures that can be derived from the numerical data. Figures contribute, along with tables, to the description of the experiment. Their uniqueness lies in their ability to record natural appearances, as in the case of light and electron micrographs, or to show trends and relationships, as with graphs.

Figures, like tables, are numbered consecutively through the report. Like tables, they should be self-explanatory. Legends, normally placed below the figures, should indicate the exact nature of the information presented without unnecessary details.

Some general rules exist for constructing graphs. Usually there should not be more than three curves on a graph. Clear labeling and differentiation between lines are important; different types of lines (solid, dotted, dashed) and different geometric shapes at data points are helpful. Legends should be complete, and axes should be clearly and accurately labeled.

Levels of an independent (controlled) variable—for example, time in a storage study, temperature in a storage or baking study, concentration of an ingredient—are marked on the x (horizontal) axis. Measurement values for a dependent variable are plotted on the y (vertical) axis; thus the height of a bar graph or the height of a curve at a designated point reflects the magnitude of a specific value or average. The scale should originate with zero unless the measurement values are clustered at high levels, in which case the y axis should be broken. Lefferts (1981) illustrated different types of graphs, indicated the usefulness of each, and gave step-by-step directions for drawing them.

With appropriate software and a printer that has graphics capability, figures can be computer drawn. Data can be plotted to produce curves; such figures are suitable for publication. Other types of graphs, such as bar graphs and pie charts, can be drawn also. Even pictorial applications are possible; the double helix shown in Chapter 16 is an example. For class purposes, a dot matrix printer

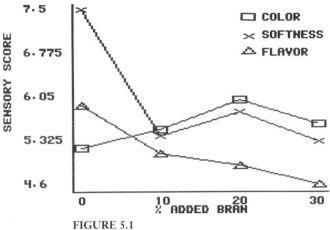

FIGURE 5.1
Line graph drawn with dot matrix printer.

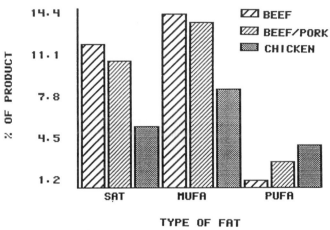

FIGURE 5.2
Bar graph drawn with dot matrix printer.

program can provide acceptable figures; Figs. 5.1, 5.2, and 5.3 are examples. Among the many books on computer graphics are those of Conklin (1983) and Jefimenko (1987). They provide specific instructions for production of different types of graphics.

3. Conclusions

The Results and Discussion section might end with conclusions, or the section might be followed by a separate summary and statement of conclusions. When conclusions are drawn, they should be stated in relation to the specific experimental conditions; it cannot be assumed that the same results would be obtained under different experimental conditions. Sometimes suggestions are made for future work. If this is done, the suggestions should be realistic.

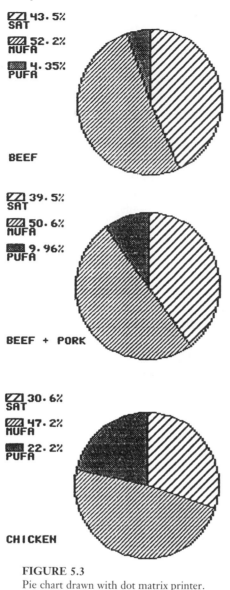

FIGURE 5.3
Pie chart drawn with dot matrix printer.

D. References

The list of references at the end of the report should follow a definite form consistently, as discussed in Chapters 1 and 2. Again, frequent use of published research reports as a guide is encouraged. Because of compilation of a reference list during the preliminary review of literature (Chapter 2), only insertion of later references should be necessary when the final report is written. The list should include all of the references that have been cited and none that has not.

II. THE LANGUAGE OF THE REPORT _____

The language of the report begins with the title, which should be concise and yet indicative of the content. Beginning a title with a key word, as opposed to "A Comparison of. . . ," "A Study of. . . ," or "The Effect of. . . ," is recommended. For example, "The Effect of Triticale Flour on the Volume of Bread" might better be written "Bread Volume as Affected by Triticale Flour" or "Triticale Flour in Bread: Effect on Loaf Volume," depending on whether bread volume or triticale flour is the desired emphasis.

Plagiarism should be avoided throughout the report. The review of literature probably is the section in which it is most likely to occur unless care is taken. As suggested in Chapter 2, putting direct quotes in quotation marks while taking reading notes is helpful in avoiding plagiarism during the later writing.

The third person passive voice, which is used (probably overused) commonly in scientific writing, predominates in the methods and materials section. However, there is a growing trend toward increased use of the first person active voice. A hazard of the passive voice is the dangling participle, which occurs all too commonly in scientific writing. A participle functions as an adjective, but when the passive voice is used, the noun or pronoun to be modified is likely to be absent. This is not an argument against the passive voice but against dangling participles. An example of incorrect usage is the following: "Viscosity was measured using a Brookfield viscometer." Viscosity did not use a viscometer; the worker did. In this case the problem is easily avoided by writing, "Viscosity was measured with a Brookfield viscometer."

Subject–verb agreement requires constant attention; a common writing (and speaking) error is failure of the verb to agree with its subject. The error often occurs when a phrase intervenes and the last word of the phrase differs in number from the subject. Incorrect: "The volume of each of the cakes were measured. . ."; the number of the verb is dictated by its subject, *volume*, not by the noun immediately preceding it. Correct: "The volume of each of the cakes was measured. . .". Note that *data*, *media*, and *phenomena*, contrary to popular usage, are plural nouns.

Either/or and *neither/nor* combinations can be troublesome. The verb takes the number of the nouns (or pronouns); no problem is presented in the following: "Neither the appearance nor the volume was affected by the substitution" or in "Neither the sensory characteristics nor the physical properties were. . .". A problem may occur when the nouns (or pronouns) differ in number. In this case, the verb agrees with the closer noun (or pronoun) as follows: "Neither the volume nor the sensory characteristics were. . .".

Restrictive and nonrestrictive clauses present hazards of improper use of the relative pronouns *that* and *which* and of improper punctuation. *That* introduces a restrictive clause, which is essential to identification of the modified noun or pronoun; the restrictive clause is not set off by punctuation. *Which* introduces a nonrestrictive clause, which is not essential to identification of the modified noun or pronoun and is set off by commas. The following sentences contain restrictive and nonrestrictive clauses, respectively: "The flavor of bread that contained 25% triticale flour and 75% wheat flour was preferred over the flavor of bread that contained 100% wheat flour." "The triticale bread, which had a lower volume than did the wheat bread, had a somewhat compact crumb."

A common misuse of terms involves words that frequently appear in the results and discussion section: *affect* and *effect*. *Affect* is used only as a verb; for example, "Oven temperature affects optimal baking time." *Effect* most frequently is used as a noun; for example, "The effect of oven temperature on optimal baking time is inverse and approximately linear." The other use of *effect* is infrequent; *effect* is used as a verb only in the sense of "to bring about." "An increase in oven temperature effects a decrease in optimal baking time" is correct, although that use of *effect* is uncommon. The introductory paragraph of Section V in Chapter 3 contains a correct use of *effect* as a verb. Other examples are in Sections I,A,1,a and VI,B of Chapter 9. In oral usage, the pronunciation of *effect* depends on its use; the first "e" is short when *effect* is a noun and long when *effect* is a verb.

Elimination of writing frills throughout the report contributes to control of length and to readability. Examples in the CBE style manual (CBE, 1978) include the following: "conducted inoculation experiments on" versus the more concise "inoculated"; "due to the fact that" or "for the reason that" versus "because"; the treatment having been performed" versus "after treatment"; "a number of" versus "a few" or "several"; "during the time that" versus "while"; "in order to" versus "to"; and "it would thus appear that" versus "apparently." In other cases, superfluous words represent redundancy; the terms "initial preparation," "first and foremost," "end result," "full and complete," and "each and every" are examples of redundancy.

Frequent use of the classic treatise by Strunk and White (1979) on English usage, principles of composition, and writing style can benefit anyone, student or not. Most freshman English textbooks also are useful references. Many books on technical writing have sections on usage and style; among them are those of Alley (1987), Barnett (1987), Carosso (1986), and Roundy and Mair (1985).

The data presented in tables and figures should be discussed but not belabored. The writer should briefly state and, if possible, explain the results and compare them with those obtained by other workers if references are available; an attempt should be made to explain apparent discrepancies. As mentioned previously, limitations of the study should be recognized; broad generalizations based on a class experiment are unwarranted.

Caution is needed in comparing treatment means when statistical tests of significance are not conducted. Small differences may or may not be real differences, as is discussed in Chapter 1. Statistical analysis often is feasible for an individual class problem. The discussion in Chapter 1 of data analysis in relation to various experimental plans is applicable to problems undertaken in a class.

III. A SAMPLE REPORT _____

A sample report[1] is presented below. While reading it, look for illustrations of the points that have been made in the preceding sections.

---

[1] Adapted from a class report by Betsy Bohannon (1985) and used with her permission.

PUMPKIN BREAD QUALITY AS
AFFECTED BY ADDED WHEAT BRAN ▰▰▰▰▰▰▰

In recent years, increased dietary fiber has been promoted extensively because of a variety of health benefits that have been recognized. An inexpen-
sive and easy way of increasing dietary fiber is the use of commercial wheat
bran, available in most grocery stores. This coarse bran, which resembles
sawdust, is difficult to swallow and is most acceptable when added to other
foods (Hui, 1983).

Acceptable layer cakes have been produced with up to 30% substitution of
wheat bran for flour (Jeltema, 1979; Shafer and Zabik, 1978). However, although products with bran at the 30% level were acceptable, their color was
significantly affected (Shafer and Zabik, 1978). The flavor of chiffon cakes
containing wheat bran at a higher level of substitution than 30% was not acceptable (Springsteen et al., 1977).

The naturally dark color of pumpkin bread should mask the effects of
wheat bran on appearance, and its pronounced flavor should tend to mask
the bran's effects on flavor. Dryer et al. (1982) found pumpkin bread to be a
good vehicle for substitution of navy bean flour. The purpose of this study
was to observe the effects of added wheat bran on characteristics and acceptability of pumpkin bread.

Procedure

Commercial wheat bran was added to pumpkin bread in amounts of 0, 20,
30, and 40% of the flour weight. The four breads were made in each of three
replications with the use of the following formula (Dryer et al., 1982):

Ingredients	Amounts (g)
All-purpose flour	100.65
Sulfate–phosphate baking powder	0.45
Baking soda	2.00
Salt	3.00
Cinnamon	0.50
Ground cloves	0.25
Brown sugar	135.35
Shortening	31.35
Whole egg	50.00
Canned pumpkin	123.00
Whole fluid milk	60.50
Vanilla	1.25

The amounts of bran added and the calculated crude fiber contents of the
four breads are shown in Table I.

TABLE I

Bran Additions and Crude Fiber Contents of
Pumpkin Breads

Flour (g)	Bran (Hodgson Mill)		Crude fiber[a] (g)	
	(%)	(g)	In loaf	Per slice[b]
100.65	—	—	—	—
100.65	20	20.13	2.2	0.32
100.65	30	30.20	3.3	0.47
100.65	40	40.26	4.4	0.63

[a] Based on 11% crude fiber in bran.
[b] Based on seven slices per loaf.

All nonperishable ingredients were weighed at the same time from the same source. Perishable ingredients were weighed before each replication.

The ingredients were mixed by a modified version of the AACC method 10-90 (1976). Flour, baking powder, baking soda, salt, cinnamon, and cloves were sifted together into a mixing bowl; the brown sugar, shortening, and appropriate amount of bran were added. The egg, pumpkin, milk, and vanilla were combined in a separate bowl, and 60% of this mixture was added to the dry ingredients. This combination was mixed with a Kitchen-Aid® mixer at low speed (setting 2) for 30 sec, scraped down, and mixed at medium speed (setting 4) for 2 min. One-half of the remaining pumpkin mixture was added, and mixing was continued at low speed for 15 min; the batter was scraped down, and mixing was continued at medium speed for 2 min. The remaining pumpkin mixture was added and the batter was mixed on low speed for 30 sec, scraped down, and mixed at medium speed for 2 min.

Batters (400 g of each) were weighed into greased, waxed paper lined 18.5 × 9-cm pans. The four breads were baked together in a preheated oven at 177°C until an inserted toothpick came out clean. The baked breads were cooled in the pans for 5 min, removed, and allowed to cool completely. The loaves in each replication were frozen when cool and tested 2 days later after being thawed at room temperature.

Evaluation

Three 2.5-cm slices were cut from one end of each loaf after removal of the crust. The remainder of the loaf was cut into 1.25-cm slices. A cutting box was used for obtaining uniform slices.

Index to Volume

The standing height of the first and middle 2.5-cm slices in each loaf was measured in centimeters with a vernier caliper. Each slice was measured in the center, at the ends, and halfway between each end and the center; the values were averaged.

```
┌─────────────────────────────────────────────────────────────────┐
│               Scorecard for Pumpkin Bread                         │
│                                                                   │
│   Evaluate the sample of pumpkin bread, placing a mark across each│
│   line to indicate the degree to which the specified characteristic is│
│   present.  Please rinse before going on to the next sample.      │
│                                                                   │
│   SAMPLE _____                                                  │
│                                                                   │
│   Color:              desirable _____ undesirable        │
│   Softness:               soft _____ firm                 │
│   Cell structure:         fine _____ coarse               │
│   Moistness:             moist _____ dry                  │
│   Flavor:                 good _____ poor                 │
│   Overall acceptability:  superior _____ unacceptable     │
│                                                                   │
│   Comments:                                                       │
│                                                                   │
└─────────────────────────────────────────────────────────────────┘
```

FIGURE 1
Scorecard for sensory evaluation of pumpkin bread. (Each line was 9 cm long on the actual scorecard and panelists received one scorecard for each sample.)

Compressibility

A 2.5-cm meat cylinder was used for cutting a sample from the center of each of the 2.5-cm slices used for standing height measurements. Compressibility was measured in the Instron universal testing machine at range 5, cross-head speed 50 mm/min, and chart speed 100 mm/min. Samples were compressed to 40% of their original height.

Sensory Evaluation

Six panelists evaluated the four bread samples in each of the three replications. Samples, all of the same size and shape, were cut from the 1.25-cm slices; they were randomly coded and their placement on each plate was random. Water was provided for rinsing between evaluations. Samples were evaluated in a testing room in which temperature and lighting were controlled. A scorecard (Fig. 1) with unstructured 9-cm scales was used; after the scales were marked, the marks were converted to values by division of each scale into nine 1-cm spaces and numbering the dividing lines so that nine indicated the most desirable quality.

The data were subjected to analysis of variance and Tukey's test (Larmond, 1977).

Results and Discussion

The four pumpkin breads did not differ significantly in either index to volume or compressibility (Table II). Smith and Hawrysh (1978) reported decreased height of chiffon cakes with the use of bran. They substituted bran for flour, whereas bran was added to a constant weight of flour in this study; the extra volume of the added bran possibly counteracted any decrease of volume that

TABLE II

Index to Volume and Compressibility of
Pumpkin Bread with Different Levels of
Added Bran[a]

Bran addition (% of flour wt.)	Measurements	
	Index to volume[b] (cm)	Compressibility (kg)
0	3.8	1.5
20	4.3	1.2
30	3.2	1.7
40	4.9	1.1
F value:	0.77 NS	0.44 NS

[a] Each value is an average for six determinations, two for each of three replications.

[b] Determined as average standing height.

TABLE III

Sensory Scores for Pumpkin Bread with Different Levels of Added Bran[a]

Characteristic	Percentage added bran				F value
	0	20	30	40	
Color[b]	5.2a	5.5a	6.0a	5.6a	0.80 NS
Softness[c]	7.5a	5.4b	5.8b	5.3b	10.85[d]
Cell structure[e]	7.4a	4.2b	3.5b	3.1b	18.60[d]
Moistness[f]	7.6a	5.8b	5.6b	4.7c	35.15[d]
Flavor[g]	5.9a	5.1ab	4.9ab	4.6b	3.40[d]
Acceptability[h]	5.9a	5.3a	5.1a	4.8a	1.29 NS

[a] Each value is an average for 18 judgments, 1 for each of 6 panelists in each of 3 replications; values are based on a 9-point unstructured scale.

[b] 9 = desirable, 1 = undesirable.

[c] 9 = soft, 1 = firm.

[d] $p < 0.05$. Values in a row followed by the different letters are significantly different.

[e] 9 = fine, 1 = coarse.

[f] 9 = moist, 1 = dry.

[g] 9 = good, 1 = poor.

[h] 9 = superior, 1 = unacceptable.

it might have caused. Or the decreased height observed by Smith and Hawrysh might have reflected the relatively delicate foam structure of chiffon cake; the bran could perhaps disrupt the cell walls and thus cause coalescence of air cells and decreased cake volume.

Sensory evaluation (Table III) showed no effect of added bran on the judges' evaluation of the desirability of the color of pumpkin bread. The naturally dark color of the bread successfully masked the dark color of the bran.

All of the breads containing bran were considered firmer than the control bread, probably because of their higher solids content. However, the level of added bran made no significant difference.

Cell structure was coarser in the breads containing bran than in the control bread. Again, the level of added bran was not important.

All of the breads containing added bran were rated less moist than the control breads; the bread with the most bran was judged significantly less moist than the breads with the other levels of added bran. Although the use of bran might be expected to increase the moisture in bread because of its water-binding capacity (Smith and Hawrysh, 1978), the high solids content with added bran, as opposed to bran substituted for flour, might have overcome possible effects due to water binding.

Bran at the highest level of addition had a negative effect on flavor. The other two levels of added bran did not significantly affect the flavor.

Although the panelists perceived effects on characteristics other than color, especially with addition of bran at the level of 40% of the flour weight, the differences among the breads did not affect the panelists' acceptance of them; overall acceptability did not differ significantly among the breads containing 0, 10, 20, and 40% of added bran. Thus pumpkin bread apparently is a satisfactory vehicle for added dietary bran. Further study might involve various levels of substitution of bran for flour rather than addition of various amounts of bran to a constant amount of flour.

References

AACC. 1976. "Approved Methods," 7th edn. American Association of Cereal Chemists, St. Paul.

Bohannon, B. 1989. Personal communication, University of Tennessee Hospital, Knoxville.

Dryer, S. B., Phillips, S. G., Powell, T. S., Uebersax, M. A., and Zabik, M. E. 1982. Dry roasted navy bean flour incorporation in a quick bread. *Cereal Chem.* **59:** 319.

Hul, Y. H. 1983. "Human Nutrition and Diet Therapy." Wadsworth, Inc., Belmont, CA.

Jeltema, M. A. 1979. Fiber components—quantitation and relationship to cake quality. *Dissertation Abstracts International, Sect. B:* 1614. University Microfilms International, Ann Arbor, MI.

Larmond, E. 1977. Laboratory methods for sensory evaluation of food. Publ. 1637, Canada Dept. Agric., Ottawa, Ontario.

Shafer, M. A. M. and Zabik, M. E. 1978. Dietary fiber sources for baked products: Comparison of wheat brans and other cereal brans in layer cakes. *J. Food Sci.* **43:** 375.

Smith, D. A. and Hawrysh, Z. J. 1978. Quality characteristics of wheat-bran chiffon cakes. *J. Amer. Dietet. Assoc.* **72:** 599.

Springsteen, E., Zabik, M. E., and Shafer, M. A. M. 1977. Note on layer cakes containing 30 to 70% wheat bran. *Cereal Chem.* **54:** 193.

IV. ORAL REPORTS

Oral reports of experimental work frequently are made in classes. Such reports are deadly if read and may be stilted if presented from memory. Talking from an outline, from brief notes, or from well-prepared slides or overhead transparencies requires careful preparation but eliminates the disadvantages of either reading or memorizing the report. Speaking loudly enough to be heard (and without dropping the voice at the ends of words or sentences), varying the pitch of the voice, varying the rate of speaking, and making eye contact with listeners help hold the audience's attention (Carosso, 1986, p. 522). Mannerisms distract an audience and should be avoided (Barnett, 1987, p. 174).

An oral report of an experimental project is organized similarly to the written report but includes much less detail. This is true especially with respect to methods, unless the study happens to have dealt with methodology. Visual aids are helpful in presenting a statement of purpose, a brief description of the experimental plan, and the results.

A major difference between oral and written reports involves the tables and figures. For oral reports, visual aids must be relatively simple to enable the listeners to digest the information. An overhead projector usually is available for showing transparencies. If one is not available, and if the group is small, a large flip pad of paper on an easel can serve the purpose. The entire group must be able to read any visual material shown; adequate time must be allowed and the material must be easily read. A rule of thumb for a visual on an easel is an inch of letter or numeral height for every 10 ft of distance between the easel and the audience (Carosso, 1986, p. 527).

The oral report should be not only well prepared but also carefully timed. A speaker who exceeds the time limit risks boring the audience and, in a classroom, falling into disfavor with the instructor. If the reporting schedule specifies time for questions and answers, adherence to that schedule is important; using the discussion time for finishing the report is poor practice.

The language of the oral report is as important as that of the written report, though usually less formal. Care is needed to present the important information and to avoid wordiness. Pronunciations should be checked in advance if there is any question regarding correctness.

Practicing an oral report with a listener who is not totally familiar with the work can be helpful in locating problem areas. That listener also can help the presenter anticipate questions that might be asked following the classroom presentation. All of the elements of preparation of the report contribute to the ease and effectiveness of the presentation.

Suggested Exercises

1. Select one or more research articles that are of interest to you. Without reading any text, select some tables and/or figures and for each write a summary of the results. Now read the pertinent portions of text and compare your statements with those of the author(s). Are your interpretations of the information in the tables and/or figures in agreement? If not, are you at fault or are the authors? To what extent did the design of the visual aids make this exercise unnecessarily difficult?
2. Prepare visual aids for presenting the following data obtained in a class experiment in which ground beef patties with and without filler (bread crumbs) were oven cooked in custard cups for 18, 22, and 26 min. Total cooking loss was determined as the difference between original weight and

cooked weight; drip loss was determined as the difference between weight of cup with drippings and weight of empty cup; and evaporative loss was calculated as the difference between total loss and drip loss. Losses in grams were converted to percentages of original weight. Each value is the average for four trials. Total losses with no filler: 20, 25, and 32% for 18, 22, and 26 min. respectively; with filler: 11, 18, and 23% for 18, 22, and 26 min, respectively. Drip losses with no filler: 13, 14, and 14% for 18, 22, and 26 min, respectively; with filler: 6, 7, and 8% for 18, 22, and 26 min, respectively. Evaporative losses with no filler: 7, 11, and 18% for 18, 22, and 26 min, respectively; with filler: 5, 11, and 15% for 18, 22, and 26 min, respectively.

3. Read the following paragraphs and identify the correct and incorrect usages listed after the paragraphs:

Triticale is a wheat–rye hybrid which plant breeders developed for the purpose of trying to improve on the breadmaking properties of rye flour and on the nutritional value of wheat flour. Because of the large number of wheat and rye varieties available, the end result of cross-breeding is subject to variation. In any case, triticale flour has breadmaking and nutritional properties that are intermediate between those of flours from the parent grains.

Comparing wheat, rye, and triticale flours, each of them are used in a standard bread formula. During mixing, water absorption, which is relatively high for a wheat bread flour, is lower for triticale flour and lowest for rye flour. Substituting either rye or triticale flour for wheat flour in bread negatively effects loaf volume and crumb grain; the effects are greater with rye flour than with triticale flour.

The nutritional affect, that is of particular interest, involves lysine, the limiting amino acid in wheat; the lysine concentration in triticale flour is higher than that of wheat flour but lower than that of rye flour.

Identify:
a. both correct and incorrect word use and/or punctuation involving restrictive and nonrestrictive clauses
b. dangling participle(s)
c lack of agreement in subject and verb number
d. redundancy
e. other superfluous words
f. both correct and incorrect use of *affect* and *effect*

4. Study the sample report and identify specific applications of points made earlier in this chapter. This exercise might include some suggestions for improvement.

References

Alley, M. 1987. "The Craft of Scientific Writing." Prentice-Hall, Inc., Englewood Cliffs, New Jersey.

American Psychological Association. 1983. "Publication Manual of the American Psychological Association." APA, Washington, DC.

Barnett, M. T. 1987. "Writing for Technicians." Delmar Publishers, Inc., Albany, New York.

Carosso, R. B. 1986. "Technical Communication." Wadsworth Publishing Co., Belmont, California.

CBE Style Manual Committee. 1978. "Council of Biology Editors Style Manual," 4th edn. American Institute of Biological Sciences, Arlington, Virginia.

Conklin, D. 1983. "PC Graphics: Charts, Graphs, Games, and Art on the IBM PC Computer." John Wiley and Sons, New York.

Jefimenko, O. D. 1987. "Scientific Graphics with Lotus 1-2-3™." Electret Scientific Co., Star City, West Virginia.

Lefferts, R. 1981. "Elements of Graphics." Harper and Row, New York.

Roundy, N. and Mair, D. 1985. "Strategies for Technical Communication." Little, Brown and Co., Boston, Massachusetts.

Strunk, W. and White, E. B. 1979. "The Elements of Style," 3rd edn. Macmillan Co., New York.

Woodford, F. P. 1968. "Scientific Writing for Graduate Students." Rockefeller University Press, New York.

PART II

FOOD SCIENCE TODAY

INTRODUCTION TO FOOD SCIENCE

Throughout most of this book, the chemical and physical nature of food is studied, along with the chemical and physical changes that occur when foods are exposed to various conditions. The relationships between chemical and physical properties are emphasized. A brief orientation to some underlying scientific principles is presented in this chapter. It provides needed background for some students and a review for others.

I. WATER

Water, the universal chemical substance so important to the properties of food, is considered first. Not only is water a component of all foods, but it contributes significantly to the physical differences among foods and to the changes that foods undergo. Food structure depends on water; try to imagine making a food emulsion, a crystalline candy, or a frozen dessert without water. Water is important chemically as a reactant in hydrolysis, which has tremendous effects on food properties. Chemical reactions not involving water directly also are important to food properties, both positively and negatively, and water plays a role in the occurrence of the reactions. It dissolves other chemicals and provides the mobility required to bring reactants together. On the other hand, water may interfere with chemical reactions by diluting reactants so that interaction becomes difficult. Some compounds do not react without first being ionized in water. Substances that are not truly soluble in water are dispersed as larger particles in many food dispersions.

Water can exist in three physical states: gas (vapor), liquid, or solid. Heat transfer, either from or into water, is necessary for its conversion from one state to another. Conversion from water vapor to liquid or from liquid to solid (ice) is exothermic; heat must be removed. Conversion from solid to liquid or from liquid to vapor is endothermic; heat must be supplied. The amount of heat that must be transferred into a given amount of water to produce a change in physical state or temperature equals the amount of heat that must be supplied or removed in order to reverse the change. The amounts of energy that must be supplied or removed to change only the physical state of water at the solid–liquid level and at the liquid–gas level are relatively large: 80 cal/g for the heat of fusion at 0°C and 540 cal/g for the heat of vaporization at 100°C. The amount of energy that must be supplied or removed to change the temperature without affecting the physical state also is large; the heat capacity (specific heat) is about 1 cal/g/°C and varies slightly with the temperature of the water.

The properties of water are closely related to its chemical nature. The covalent bonds between oxygen and the two hydrogen atoms form an angle of 104.5°. Although the positive charges equal the negative charges in the water molecule, the asymmetric arrangement of the atoms represents an imbalance of the charges. The oxygen atom with its two negative charges is on one side of the molecule and the two hydrogens, each with its one positive charge, are on the other side. This polar nature of the water molecule is responsible for its hydrogen-bonding ability. Each hydrogen atom is attracted electrostatically to the oxygen of another water molecule, and each oxygen atom can attract two hydrogen atoms. Thus each water molecule may be connected, through hydrogen bonding, to four other water molecules, and these in turn may be hydrogen bonded to other water molecules, forming a three-dimensional structure. The actual extent of the clustering depends on the temperature. There is so much hydrogen bonding in ice that an extremely ordered, rigid structure exists. When the ice is melted and the water heated, molecular motion results in breaking of hydrogen bonds until, when the vapor state is reached, there are none.

Even at a constant temperature, continual forming and breaking of hydrogen bonds among adjacent water molecules occurs. Individual hydrogen bonds are weak compared with covalent bonds, but their net effect can be great. Hydrogen bonding is responsible for the relatively large amounts of energy involved in changing the physical state and the temperature of water. Not only are the boiling point, freezing point, heat capacity, heat of fusion, and heat of vaporization relatively high, but so also is surface tension. Hydrogen bonding also is responsible for the relatively low vapor pressure of water. Although this discussion emphasizes hydrogen bonding between water molecules, other types of compounds, such as those having hydroxyl, carboxyl, or amino groups, also may participate.

The rather large energy requirements mentioned above are significant with regard to the function of water as a medium for cooling and heating. Water in the form of ice is particularly effective as a cooling medium because, in order to melt, it must absorb a large amount of energy. If it obtains that energy as heat from food, the food is cooled. For example, a home ice cream freezer has an inner container made of metal and an outer container made of wood or another poor conductor; thus, the melting of ice in the ice–salt mixture between these containers removes heat mostly from the mixture to be frozen, rather than from the surrounding air. Water is effective as a heating medium because to become hot it must absorb a large amount of energy from the source of heat. The energy then is available for transfer to food.

The physical constants of water, in addition to those mentioned above, include the following: freezing point (melting point), 0°C; boiling point, 100°C at sea level and normal atmospheric pressure; and density, 1g/cm^3 at 4°C. The density of water decreases below and above 4°C; it is 0.958 g/cm^3 at 100°C (CRC, 1989–1990).

Natural waters are not pure but are actually very dilute solutions of salts. The mineral content of water differs both quantitatively and qualitatively with source and can affect the performance of water in food. Pyler (1988, pp. 551–560) discussed various salts and their properties.

Much of the water in food is physically trapped in capillaries. Although it does not flow freely from the food, it can freeze and in other ways behave like pure water. The ability of food to hold entrapped water is water-holding capacity.

Some of the water in food is bound. Bound water has had many definitions over the years. Fennema proposed that it be defined as the water that "exists in the vicinity of solutes and other nonaqueous constituents, exhibits reduced molecular mobility and other significantly altered properties as compared with bulk water in the same system, and does not freeze at −40°C."[1] The most tightly bound water is that which interacts with ions. Some water is bound to nonionizing solutes that have hydrogen-bonding capacity, and some water interacts with hydrophilic groups of macromolecules. The greater the degree of binding, the

---

[1] Reprinted from Fennema, (1985, p. 37).

less is the activity with respect to crystallization, freezing, evaporation, support of chemical and microbial activity, and release from the food when pressure is applied. The extent of water binding affects the response of a food to various treatments and is itself affected by treatments.

Although perishability of food is related to its water content, two foods of the same water content can have quite different shelf lives because of a difference in water activity. Water activity (a_w) is defined as the ratio of p, the vapor pressure of water in a product (or the partial pressure of water in the atmosphere above the food or solution) at a specified temperature, to p_0, the vapor pressure of pure water at the same temperature. The water activity of a sample is equal to 0.01 × the relative humidity of an atmosphere in equilibrium with the sample (Troller and Christian, 1978, p. 3), a relationship that can be used in determination of water activity. Values for water activity are lower than 1.0 because the presence of nonaqueous nonvolatile substances lowers vapor pressure (p). Some foods are grouped according to their approximate a_w in Table I, and the relationship of water activity to various deteriorative reactions is shown in Fig. 6.1.

Reduction of the a_w of foods can be achieved not only by dehydration but also by freezing; formation of ice crystals removes water from solution, thus concen-

TABLE I

Approximate Water Activity Values of Some Foods and of
Sodium Chloride and Sucrose Solutions[a]

a_w	NaCl (%)	Sucrose (%)	Foods
1.00–0.95	0–8	0–44	Fresh meat, fruit, vegetables, canned fruit in syrup, canned vegetables in brine, frankfurters, liver sausage, margarine, butter, low-salt bacon
0.95–0.90	8–14	44–59	Processed cheese, bakery goods, high-moisture prunes, raw ham, dry sausage, high-salt bacon, orange juice concentrate
0.90–0.80	14–19	59–saturation (0.86 a_w)	Aged cheddar cheese, sweetened condensed milk, Hungarian salami, jams, candied peel, margarine
0.80–0.70	19–saturation (0.75 a_w)	—	Molasses, soft dried figs, heavily salted fish
0.70–0.60	—	—	Parmesan cheese, dried fruit, corn syrup, licorice
0.60–0.50	—	—	Chocolate, confectionary, honey, noodles
0.40	—	—	Dried egg, cocoa
0.30	—	—	Dried potato flakes, potato crisps, crackers, cake mixes, pecan halves
0.20	—	—	Dried milk, dried vegetables, chopped walnuts

[a] From Troller and Christian (1978).

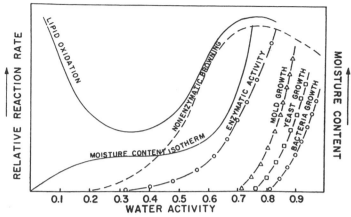

FIGURE 6.1

Relative reaction rates of deteriorative reactions as a function of the water activity of foods. (From Labuza, 1971b; Reprinted from "Proceedings, SOS/70, Third International Congress of Food Science and Technology," p. 633. Copyright © 1971 by Institute of Food Technologists.)

trating the solutes. Addition of solutes such as sucrose, sodium chloride, and glycerol is another means of lowering a_w. Intermediate moisture foods (IMF, normally 20–50% moisture) are soft enough to be eaten without hydration and yet can have low enough a_w levels (normally 0.7–0.9) to not be susceptible to bacterial growth (Troller and Christian, 1978, p. 186). They include both pet foods and human foods. Some foods, dried fruits, for example, naturally contain enough sugar for adequate control of a_w and do not need added humectants. With other products, humectants such as glycerol are added to lower a_w.

Among the many reports of analytical values for water activity and studies of the effects of water activity on specific foods and food phenomena are those involving crisp snack foods (Katz and Labuza, 1981), canned foods (Alzamora and Chirife, 1983), dried prunes and raisins (Bolin, 1980), intermediate moisture foods (Singh *et al.*, 1983), layer cakes (Sych *et al.*, 1987), processed cheese (Magrini *et al.*, 1983), meat products (Vallejo-Cordoba *et al.*, 1986), apples (Bourne, 1986), food powders (Moreyra and Peleg, 1981), enzyme activity (Acker, 1969), microbial activity (Beuchat, 1981), browning reactions (Tiribio *et al.*, 1984), and lipid oxidation (Labuza, 1971a).

Moisture sorption isotherms are useful in predicting food stability and in establishing conditions for adjusting water content. A sorption isotherm is a curve that expresses the relationship between the moisture content of a given food and its water activity during water uptake (adsorption or resorption) or loss (desorption). As indicated by the terminology, the temperature must be constant; water activity increases with increasing temperature. For most foods the curve is S shaped. In the lower part of the isotherm (A), the water is particularly tightly bound, consisting of a monomolecular layer. In the middle region (B), the water is bound but less firmly than in region A; additional layers are built up on the first monolayer. In region C, water is loosely held, primarily by physical entrapment. If sorption isotherms are obtained for a sample that is dried and then

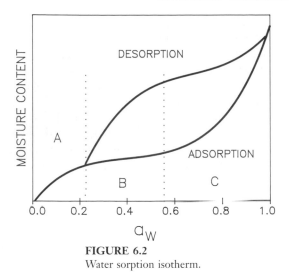

FIGURE 6.2
Water sorption isotherm.

rehydrated, the resorption isotherm usually does not superimpose on the desorption isotherm (Fig. 6.2), a situation that is called hysteresis and has not been fully explained. Hysteresis is significant because whether a given moisture content is achieved by desorption or by adsorption influences the water activity of a food system and thus the potential for various deteriorative changes.

Among the many discussions of water activity and sorption isotherms are those of deMan (1980, pp. 7–15), Fennema (1985, pp. 46–60), and Labuza (1968, 1971b). Labuza (1984, pp. 64–76) described methods of determining water activity and sorption isotherms. Troller and Christian (1978) discussed, in addition to the basic concepts and methods of measurements, the effects of a_w on enzymatic reactions, nonenzymatic browning, lipid oxidation, microbial growth, and food properties.

II. FOOD DISPERSIONS

Water is the usual dispersion medium in foods, though certain food constituents, notably fat-soluble vitamins and some flavor compounds, are dissolved in fat. Fat is also the dispersion medium in water-in-oil emulsions. Solutions are discussed first, then colloidal dispersions.

A. Solutions

A solution is a homogeneous mixture of the molecules and/or ions of different substances, one of which usually is water. The particles of the dissolved substance are so small and have so much kinetic energy that they can make room for themselves between water molecules. In a sugar solution, the particles that are distributed uniformly among water molecules are sugar molecules. In a solution of sodium chloride, which ionizes into Na^+ and Cl^-, the particles that are uniformly mixed with water molecules are ions. The dissolved substance, whether a nonionizing compound or an ionizing compound, is the solute and the medium in which it is dissolved is the solvent.

Solubility is the amount of solute that can be dissolved in a given amount of solvent at a given temperature. Solubilities of inorganic and organic compounds are given in tables of physical constants (CRC, 1989–1990) as g/100 g solvent at specified temperatures. The effect of temperature on solubility varies with solutes. Increasing temperature greatly increases the solubility of many solutes and has less effect on the solubility of others. The extent of the solubility of a water-soluble compound at a given temperature depends on the attraction of its molecules or ions to water as compared with the attraction of its molecules or ions to one another.

The concentration of a solution is the amount of the solute dissolved in a specified amount of the solvent or of the solution and is expressed in several ways, including the following: percentage of solute by weight; percentage of solute by volume if the solute is a liquid; molarity, which is the number of moles (gram molecular weights) of solute dissolved in a liter of solution; and normality, which is the number of gram equivalent weights of solute dissolved in a liter of solution. The details of these methods of expressing concentration can be reviewed in an elementary chemistry textbook. The important point here is the need to be specific in identifying the concentration of a solution used in experimental work. In the case of a specified percentage concentration, it is essential to indicate whether the amount of solute is g/100 g solution (percentage by weight), ml/100 ml solution (percentage by volume, applicable when the solute is a liquid), g/100 ml solution, g/100 g solvent, ml/100 ml solvent, or g/100 ml solvent. The latter four possibilities are not true percentages but are sometimes used as a matter of convenience and loosely designated as percentages. Note that an excess of solute lying undissolved in a solution does not contribute to the concentration.

A saturated solution contains as much solute as can be dissolved at a given temperature; further addition of solute at that temperature results in no change in concentration. An unsaturated solution is less concentrated and a supersaturated solution is more concentrated; supersaturation is achieved by preparation of a saturated solution at an elevated temperature and cooling without agitation or seeding. A supersaturated solution is unstable; it is difficult to keep the excess solute from crystallizing out.

Several properties of solutions are particularly important in food preparation. Among these are the colligative properties. Colligative properties are those that depend on the number of solute particles (molecules and/or ions) in a given amount of solvent and not on the nature of the solute (aside from the effect of its molecular size on the maximum number of particles that can be in a given amount of solvent). The colligative properties discussed here are vapor pressure, boiling point, freezing point, and osmotic pressure.

1. Vapor Pressure

It is obvious to anyone who has boiled water that a liquid vaporizes at its boiling point. What is less obvious, but still apparent if one leaves a container of water standing uncovered for a long time at room temperature, is that vaporization also occurs at lower temperatures, though more slowly. If the container is closed, vaporization occurs, but condensation of the vapor back into the water also takes place. Vapor pressure is the pressure exerted by the gas above the liquid when equilibrium exists—that is, when evaporation and condensation occur at the same rate. Vapor pressure increases with increasing temperature. It decreases with the addition of a nonvolatile solute because the solvent constitutes a smaller proportion of the liquid, and the nonvolatile solute occupies space without contributing to vapor pressure. The greater the concentration of solute, the greater is the reduction in vapor pressure.

2. Boiling Point

The boiling point of a liquid is the temperature at which the vapor pressure equals the external pressure, which is atmospheric pressure under ordinary conditions. The greater the reduction in vapor pressure resulting from the presence of a nonvolatile solute, the higher is the temperature required to raise the vapor pressure sufficiently to permit boiling. The extent of the effect is directly proportional to the number of solute particles in a given quantity of solvent. One mole of a nonvolatile nonionizing solute raises the boiling point of a liter of water 0.52°C. The effect of nonvolatile nonionizing solutes on the boiling point of water is applied in the use of a thermometer for achieving the desired extent of concentration of a jelly or candy mixture.

One mole of a nonvolatile ionizing solute raises the boiling point of a liter of water 0.52°C per ion formed. Sodium chloride forms two ions and calcium chloride forms three ions. A mole of sodium chloride, therefore, raises the boiling point of a liter of water 1.04°C, and a mole of calcium chloride raises the boiling point of a liter of water 1.56°C. The three preceding sentences assume complete ionization; this is not necessarily the case, but the maximum effect is approached.

Volatile solutes actually lower the boiling point because of their contribution to, rather than hindrance of, vaporization. With the increased vapor pressure, a lower temperature than 100°C permits water to boil.

References to 100°C as the boiling point of water are based on standard atmospheric pressure, 760 mm Hg. At reduced atmospheric pressures, such as at high altitudes, pure water boils at a lower temperature than at standard atmospheric pressure, and a given solution boils at a correspondingly lower temperature than does the same solution at standard atmosphere. The boiling point of water is decreased by 1°C with each increase in altitude of 960 ft. Conversely, in a pressure cooker, in which steam is retained, a greater external pressure must be overcome and boiling point is elevated.

3. Freezing Point

Solutes depress the freezing point of water because they disrupt structure. The extent of the effect on a liter of water is 1.86°C/mol of a nonionizing solute. Ice crystals will begin to form at −1.86°C in a sucrose solution containing 342 g sucrose/liter. The corresponding effect is 1.86°C/mol of ions formed in the case of an ionizing solute. Therefore, a mole of sodium chloride lowers the freezing point of a liter of water 3.72°C, and a mole of calcium chloride lowers the freezing point of a liter of water 5.58°C. The effects do not differ for volatile and nonvolatile solutes, and freezing point is essentially unaffected by atmospheric pressure.

4. Osmotic Pressure

Osmosis is the passage of water molecules through a semipermeable membrane that separates either pure water and a solution or two solutions that differ as to concentration of dissolved particles. If pure water and an aqueous solution are on opposite sides of a membrane through which water molecules can pass, water molecules pass through the membrane in both directions, but much more rapidly from pure water to the solution than vice versa. If the membrane separates solutions of two concentrations, the net flow of water will be from the more dilute solution to the more concentrated one. The more rapid passage is in the direction in which equalization of concentrations can be approached. The rate at which osmosis occurs depends on the difference in solute concentration on the two sides of the membrane. Osmosis is particularly rapid when the difference in concentration is large; it is retarded as equilibrium is approached.

The occurrence of osmosis causes a change in the relative volumes of the two liquids. The volume of the solution that becomes more dilute increases. Osmotic pressure is measured as the pressure required to prevent that increase in volume. Considerable osmosis occurs in living tissues. Membranes are changed after death, but selective permeability is still evident in some fruits that have an intact firm skin, such as plums. Plums cooked in syrup having a sugar concentration greater than that of the plums are likely to shrivel because of the passage of water through the skin into the syrup.

Any general chemistry textbook provides further information on the subject of solutions.

B. Colloidal Dispersions

The dispersed particles in a colloidal dispersion are sufficiently large that their numbers in a given volume of dispersion medium cannot be great enough to permit effects on colligative properties comparable to those observed in a solution. On the other hand, the particles are not large enough to be observed with an optical microscope, as they are in suspensions. The particles in colloidal dispersions are huge molecules or aggregates of molecules and their size falls into a rather wide range, specified as 1–200 nm by Keenan *et al.* (1980, p. 360) and as large as 1–1000 nm by Zapsalis and Beck (1985, p. 507).

Many substances can exist, depending on the conditions, in true solution, in colloidal dispersion, or in suspension. Sugar is an example. When the syrup for a crystalline candy has been concentrated, the sucrose still is in solution in the form of molecules. In the crystallization process, crystal growth involves increasing particle size first to colloidal dimensions and then to suspended crystals. Some of the sucrose remains in solution, as is discussed in Chapter 24.

Foams and emulsions, because of their properties, frequently are considered as a special class of colloidal dispersion, even though their dispersed particles are large enough to be observed with an optical microscope. Gels also are frequently discussed with colloidal systems.

1. Terminology

The dispersion medium is the continuous phase, sometimes called the external phase. The dispersed substance (the colloid) is the dispersed, discontinuous, or internal phase. It consists of colloidal particles. A sol is a solid-in-liquid colloidal dispersion in the liquid state. When gelatin has been hydrated with cold water and then dispersed with hot water, it is a sol. Sols in which water is the continuous phase are hydrosols. Most food sols are hydrosols. Sols often are changed to gels. A gel is a liquid-in-solid dispersion. A foam such as whipped cream is a gas-in-liquid dispersion; a foam such as a marshmallow is a gas-in-solid dispersion. An emulsion is a liquid-in-liquid dispersion. All combinations of physical states are possible except gas in gas. Those combinations mentioned above are most characteristic of food dispersions.

Several terms apply to the colloid itself rather than to the colloid system. A lyophilic colloid has an affinity for the dispersion medium. When the dispersion medium is water, the lyophilic colloid more specifically is hydrophilic. A lyophobic (or hydrophobic) colloid does not have an affinity for the dispersion medium. Therefore, it can be dispersed only through special treatment, as with a protective colloid. A protective colloid is a material that is lyophilic and can be absorbed on the surface of the lyophobic colloidal particle, thus conferring lyophilic dispersal properties. Colloidal particles frequently are called micelles, particularly those formed by condensation.

2. Preparation

Colloids can be prepared by reducing the size of large particles (dispersal) or by increasing the size of small particles until they are of colloidal size (condensation). Certain substances, such as gelatin and dried albumen, form colloidal dispersions in water without any special treatment to change their particle size. Reduction of particle size of other substances may be brought about chemically, as by hydrolysis, or mechanically, as by grinding, mixing at high speeds, or forcing a liquid dispersion through a very small opening. The growth of crystals from molecules is a condensation process.

3. Stabilization

Frequently it is desirable to maintain a colloidal dispersion as unchanged as possible. Stability necessitates preventing aggregation of particles. There are several mechanisms of stabilization.

a. Solvation. This mechanism, called hydration when water is the dispersion medium, is characteristic of hydrophilic colloids. Hydrophilic colloids have groups that bind water. The effect is a layer of water that surrounds each colloidal particle and moves with the particle. Such particles do not aggregate unless they are destabilized. Most food colloids are hydrophilic, and thus their dispersions are stabilized by solvation. Caseins are hydrophobic individually, but they occur in milk combined into micelles that are structured in such a way that hydrophilic groups are concentrated at the particle surface.

b. Adsorption of Surface-Active Substances. A protective colloid is an example of a surface-active substance. Though a substance contributing this type of stabilizing effect does not have to be a colloid, it does have to be able to orient itself on the particle surface in such a way that a protective film is formed to prevent aggregation. Such behavior is characteristic of foam stabilizers and emulsifying agents. Surfactant properties are discussed further in Chapter 15.

c. Presence of Electric Charge on Particle Surfaces.

Carboxyl, sulfate, phosphate, and amino groups may be present and may be charged. When all the particles carry the same net charge, they repel one another. The effect is stabilization of the dispersion. A substance that does not have charged groups might adsorb either positive or negative ions preferentially. Whether the adsorbed ions are positive or negative, their like charges cause the particles to repel one another. The charged surfaces of the particles attract counterions (ions of the opposite charge) from the dispersion medium, resulting in an electrical double layer.

Lyophilic colloids may be stabilized simultaneously by hydration and electric charge. Pectin is an example of such a colloid.

4. Destabilization

Destabilization is removal of the stabilizing factor and consequent aggregation into particles of larger than colloidal dimensions. Colloidal dispersions stabilized by hydration are destabilized by competitive removal of the water layer. In the making of pectin jelly, the sugar contributes to gel structure in that it dehydrates the pectin particles. The effectiveness of many surface-active substances can be reduced by heating. The stabilizing role of an electric charge can be eliminated by neutralization of the charge. For example, the acid in pectin jelly neutralizes the charge on the pectin particles. Salts also are destabilizing substances. With stabilizing forces overcome, interactions between particles become possible and gels may form, flocculation may occur, or aggregates may precipitate, depending on the specific substances and the prevailing conditions. In the case of gel formation, bonding occurs between dispersed particles to such an extent that a solid rigid structure is formed, with water trapped in the meshes of the solid network. Note that there is a reversal of phases in gelation. A solid-in-liquid dispersion is changed to a liquid-in-solid dispersion.

Many sources of information on the general principles of colloid chemistry are available, including the discussions by Whitney (1977) and Zapsalis and Beck (1985).

5. Some Specific Colloidal Substances

Proteins represent a major group of colloids in food and are discussed below. Hydrocolloids, which have come into prominence in food applications in recent years, also are discussed briefly.

a. Proteins.

Specific proteins are discussed throughout the book, but some general information is provided here. Amino acids are the structural units of proteins. The characteristic feature of the amino acids that differentiates them from fatty acids is the amino group on the α carbon:

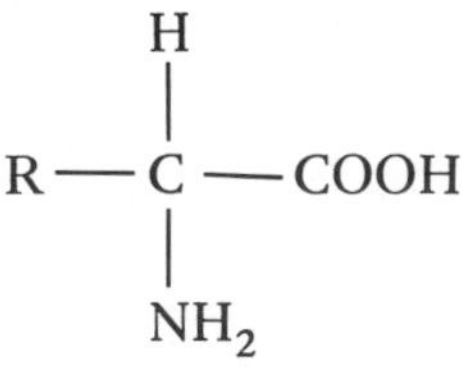

$$H$$
$$R — C — COOH$$
$$NH_2$$

Amino acid

Amino acids are joined into long chains through peptide linkages. A peptide linkage is a carbon–nitrogen bond joining a carboxyl carbon of one amino acid and an α-amino nitrogen of another amino acid with the splitting off of a molecule of water:

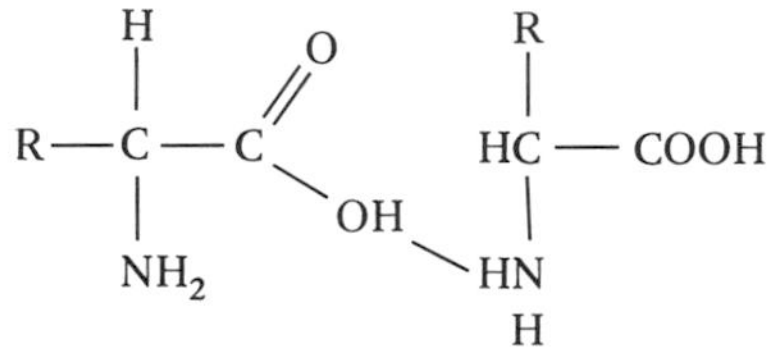

Peptide linkage

Note that the amino group of one amino acid and the carboxyl group of the other are free to form peptide linkages with other amino acids. About 20 amino acids predominate in proteins, though many more exist. The large number of different structural units and the great length of protein chains make it easy to recognize that the potential for variation is large.

A protein molecule is represented diagrammatically as a backbone in which —C—C—N— is a repeating unit and the distinctive portions of the individual amino acid molecules, represented by R, extend as side chains:

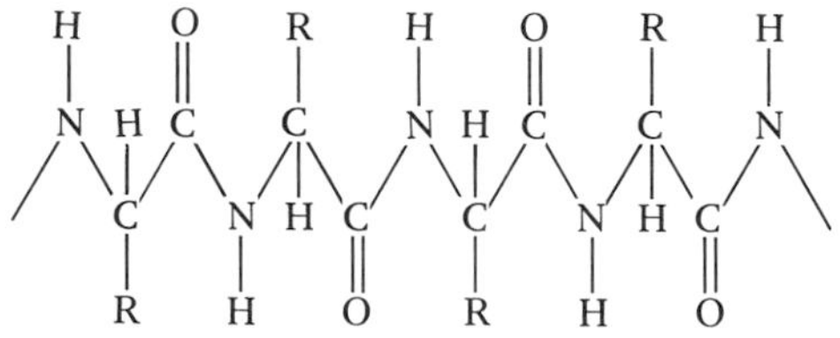

Backbone structure

The repeating —C—C—N— segment of a polypeptide backbone consists of an α carbon, a carboxyl carbon, and an α-amino nitrogen. The specific amino acids and their sequence in the polypeptide chain constitute the primary structure of a protein. Only rather recently have protein chemists been able to establish the primary structure of many proteins. The magnitude of their task is apparent if one considers that κ-casein, not one of the larger food protein molecules, has 169 amino acid residues.

If there were no structure in proteins beyond the primary structure, all proteins would be monomeric extended chains. However, proteins differ in their physical and biological properties to a greater extent than can be explained by differences in primary structure. Individual proteins in their native state have characteristic conformations based on secondary, tertiary, and quaternary structure. Secondary structure results from interactions between amino acid residues that are rather close to one another in the sequence, resulting in regular helical or pleated structures. The classical example is the α helix with hydrogen bonding between carboxyl oxygen atoms and amino nitrogen atoms of peptide bonds that are four residues apart:

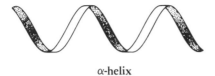

α-helix

The helical conformation usually does not exist continuously over the entire chain:

Chain with helices

Not all amino acids can be involved in helical structures. Steric hindrance and electrostatic repulsion of some side chains can prevent helix formation. The presence of proline or hydroxyproline in a polypeptide chain introduces a bend. A random coil is typical of certain proteins having constraints to helix formation. The β structure is a secondary structure in which hydrogen bonding holds chains in zigzag shapes, and joins them to each other to form pleated sheets.

Further structure is provided at the tertiary level. Tertiary structure results from interaction between amino acid residues that are farther apart in the polypeptide chain than those involved in secondary structure:

Tertiary structure

Tertiary interactions, resulting in folding of chains, are less regular than secondary but are not random. They are genetically determined and are important to properties of specific proteins. The interactions are varied and include hydrogen bonding, electrostatic interaction, hydrophobic interaction, and disulfide bonds. Quaternary structure results from interaction between polypeptide chains:

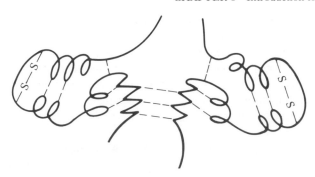

Quaternary structure

Conformation is the net result of the above levels of structure. For each protein there is a three-dimensional shape, or conformation, that is most stable under normal conditions of temperature, ionic strength, and pH; this is the "native conformation." The particle of native protein represents a minimum level of free energy.

The nature and sequence of the amino acid residues in a protein molecule determine the kinds of intrachain and interchain bonding that are possible contributors to secondary, tertiary, and quaternary structure. Some of the amino acids (for example, glycine, alanine, valine, leucine, and isoleucine) have unsubstituted hydrocarbon chains; their R groups are hydrophobic. Others have chains with substituents such as hydroxyl (—OH) groups (for example, serine and threonine); "extra" carboxyl (—COOH) groups (aspartic and glutamic acids and asparagine); "extra" amino (—NH$_2$) groups (lysine, arginine, histidine); sulfur groups (cysteine, cystine, and methionine); or ring structures (phenylalanine, tyrosine, and tryptophan).

There is much potential for hydrogen bonding in proteins because of the —OH, —NH$_2$, —COOH, and some other groups. Ionic bonding is possible because the exposed —COOH groups can be negatively charged and the exposed —NH$_2$ groups can be positively charged. Covalent bonding can occur when cysteine, with its —SH group, is present; when two —SH groups get close enough together under oxidizing conditions, a disulfide (—S—S—) bond forms between them. Disulfide bonding can occur between portions of a single chain and also between chains. Hydrophobic bonding, or more accurately hydrophobic interaction, involves the hydrophobic (nonpolar) unsubstituted hydrocarbon chains or rings of certain amino acids. Such groups do not have an inherent attraction for one another but share a dislike for water. They come into intimate contact with one another in their attempt to avoid water. Not surprisingly, these groups tend to be oriented toward the interior of a protein molecule because proteins occur in aqueous environments. Hydrophobic interaction probably has a large influence on the folding pattern. The amounts of energy associated with different types of bonds differ greatly, ranging from 30 to 100 kcal/mol for covalent to 10–20 for ionic, 2–10 for hydrogen, and 1–3 for hydrophobic bonding (Pomeranz, 1985, p. 153). Even weak bonding can have large overall effects because of a large extent of interaction.

The native conformation of a protein molecule is stabilized, as well as formed, , by the various types of interactions between amino acid residues. With many proteins, native structure does not extend beyond the tertiary level. With others, association between monomers is characteristic.

Protein denaturation is a common occurrence in food. Sometimes it is desired and sometimes it is not. Denaturation consists of nonproteolytic changes in the three-dimensional native protein structure; it does not include breaking of peptide bonds. Some protein chemists also exclude the breaking of disulfide bonds from the definition of denaturation. In any case, the process results in the unfolding of molecules.

Denaturation is not an all-or-nothing phenomenon. It may involve conformational change that is so slight as to be practically undetectable by most techniques. When denaturation is sufficient to cause detectable effects, those effects can be quite variable, but they always include decreased solubility. Viscosity, another property that is important to food properties, usually is increased. Heat is the primary factor contributing to protein denaturation in food. Salts, pH, and mechanical shear also can have important effects.

Renaturation is reversal of denaturation. It is more likely to be partial than complete. The heat resistance of the enzymes catalase and peroxidase is a reflection of protein renaturation.

Coagulation follows denaturation. Once extensive denaturation has occurred, new cross-bonding between chains can occur extensively because the chains are more extended and interaction is easier. More groups are exposed and molecules can approach each other closely over larger portions of their chains. With increased cross-bonding, flocculation occurs, and finally a visible coagulum appears. If the coagulating mixture is being stirred, the coagulum is manifest as increased viscosity caused by small curds. If the mixture is undisturbed during coagulation and conditions are otherwise favorable, the coagulum is continuous, resulting in a gel. A gel such as an egg gel, formed through protein coagulation, is not a reversible structure.

Gelation of a gelatin sol occurs through hydrogen bonding after the gelatin chains have been dispersed and the sol has been cooled without agitation. This solidification, which is not an example of coagulation even though gelatin is a protein, is reversible. The hydrogen bonds in a gelatin gel can be broken by heat and the sol reestablished. Cooling then can result in gelation again.

Further information concerning proteins is available from many sources, including Cheftel *et al.* (1985) and Zapsalis and Beck (1985). Finley (1989) presented an overview of the effects of food processing on proteins.

b. Hydrocolloids. Hydrocolloids, or gums, are natural or synthetic in origin. In both cases, they consist of complex polysaccharide derivatives. Many natural hydrocolloids are obtained from sea plants. Alginates, carrageenan, and agar are examples. Other natural hydrocolloids are obtained from land plants. Examples are tree exudates such as gum arabic, gum karaya, and gum tragacanth, as well as seed gums such as guar gum and locust bean gum. Synthetic hydrocolloids, such as carboxymethyl cellulose (CMC), are derivatives of cellulose. Microcrystalline cellulose is a nonfibrous form of cellulose. Xanthan gum is a biosynthetic hydrocolloid. It is synthesized by the bacterial species *Xanthomonas campestris* in a medium containing glucose and certain salts.

Hydrocolloids differ as to whether they have charged groups, and for those that do, the extent of charge. Charged groups in various hydrocolloids include sulfate and carboxyl groups. The presence of such groups affects the response to ions, such as Ca^{2+}, in food mixtures. Hydrocolloids also differ in their sensitivity to acid and to heat, in their solubility, and in their gel-forming ability.

The basic physical property that hydrocolloids have in common is their ability to absorb relatively large amounts of water. This absorption results in high viscosity, which sometimes is an end in itself. Sometimes the increased viscosity contributes to other desired effects of hydrocolloids in food formulation, such as control of crystallization and prevention of the settling out of solids. Other functions in food include reduction of syneresis (separation of liquid from a gel or very viscous mixture), reduction of evaporation rate, protection of emulsions (possibly through an effect on viscosity), formation of gel structure, and contributions to special textural effects, such as gumminess. A study of the labels on a large variety of processed and formulated foods gives insight into the widespread use of hydrocolloids.

The subject of hydrocolloids was discussed further by Glicksman (1982), Igoe (1982), Whistler and Daniel (1985, pp. 121–134), and Zapsalis and Beck (1985, pp. 374–389).

III. ACIDITY AND HYDROGEN ION CONCENTRATION (pH) ————

The acidity of foods or of the medium in which they are cooked frequently has an important effect on certain qualities of the product, such as the color of a cooked green vegetable or the thiamine retention of quick breads. Acidity often is expressed as pH.

A. Acidity

Acids are substances that donate the hydrogen ion, H^+, and bases are substances that furnish the hydroxide ion, OH^-. Acids and bases are proton donors and receptors, respectively. The concentration of acids and bases often is expressed as molarity or normality. Normality is somewhat more convenient than molarity in considering solutions that react with one another, because it takes into account the number of replaceable hydrogen atoms or hydroxyl groups. A given volume of a normal solution of any acid will react with the same volume of a normal solution of any base.

Although it is convenient for many purposes, normality does not always give all of the necessary information about the acidity of a solution. For example, 0.1 N solutions of hydrochloric and acetic acids are quite different. The 0.1 N acetic acid solution, which is equivalent to highly diluted vinegar, tastes much less sour than the hydrochloric acid solution and reacts less strongly with indicators and other substances. The reason for these differences is that sensation of sourness and certain other effects depend on the concentration of hydrogen ions in the solution. Far more hydrogen ions are present in the hydrochloric acid solution than in the acetic acid solution because the hydrochloric acid is almost completely ionized and the acetic acid is only about 1% ionized. Since both are 0.1 N, however, they react to the same extent with a base in titration. Titration gives a measure of total acidity rather than of hydrogen ion concentration.

B. Hydrogen Ion Concentration

Hydrogen ion concentration (pH), sometimes called active acidity to distinguish it from total acidity, is important in a study of food. It usually is expressed as pH so that the values obtained will be simple numbers. If the hydrogen ion concentration is expressed as moles of hydrogen ion per liter, an awkward number results, such as 0.0000001 mol for pure water. This decimal also can be written $1/10^7$ or 10^{-7}. For convenience, the exponent in the expression $1 \times 1/10^7$ is used to designate pH. The p stands for power and the H for hydrogen ion.

The corresponding values for pH and hydrogen ion concentration in Table II show that very acidic solutions have a high concentration of hydrogen ions and, therefore, a low pH. Note that the most acidic solution contains some hydroxyl ions and the most basic solution contains some hydrogen ions. This is explained by the slight ionization of water to form both hydrogen and hydroxyl ions. Pure water, with equal concentrations of hydrogen and hydroxyl ions, is neutral (pH

TABLE II
Concentration of Hydrogen and Hydroxyl Ions in
Solutions of Different pH[a]

Reaction	pH	Hydrogen ion concentration (mol/liter)	Hydroxyl ion concentration (mol/liter)
	0	10^0	10^{-14}
	1	10^{-1}	10^{-13}
	2	10^{-2}	10^{-12}
Acidic	3	10^{-3}	10^{-11}
	4	10^{-4}	10^{-10}
	5	10^{-5}	10^{-9}
	6	10^{-6}	10^{-8}
Neutral	7	10^{-7}	10^{-7}
	8	10^{-8}	10^{-6}
	9	10^{-9}	10^{-5}
	10	10^{-10}	10^{-4}
Basic	11	10^{-11}	10^{-3}
	12	10^{-12}	10^{-2}
	13	10^{-13}	10^{-1}
	14	10^{-14}	10^0

[a] The product of the hydrogen ion concentration and the hydroxyl ion concentration in water and in dilute aqueous solutions is a constant, 1.00×10^{-14} at 25°C. This fact and the knowledge that numbers can be multiplied by adding their exponents explain why the sum of the exponents in these two columns for the same pH is always -14. It will be recalled that $10^0 = 1$.

TABLE III
pH Values of 0.1 N Solutions of
Several Acids and Bases

Acid or base	pH
Hydrochloric acid	1.07
Sulfuric acid	1.23
Acetic acid	2.87
Ammonium hydroxide	11.27
Sodium hydroxide	13.07

7). A solution at any pH contains 10 times as many hydrogen ions as an equal volume of a solution of the next higher pH. Hydrogen ion concentration is 0.01 mol/liter at pH 2 and 0.001 mol/liter at pH 3.

It has been mentioned that although acid solutions of the same normality give the same value on titration, they may be quite different in hydrogen ion concentration. This is illustrated in Table III, in which the pH values of 0.1 N solutions of several acids and bases are shown. Hydrochloric and sulfuric acids, because they are almost completely ionized, have pH values approaching 1 at a normality of 0.1, whereas acetic acid, which is only slightly ionized, has a pH of 2.87 at a concentration of 0.1 N. Similarly, the 0.1 N solution of the highly ionized base sodium hydroxide has a much higher pH than that of a 0.1 N solution of the slightly ionized ammonium hydroxide.

1. Change of pH in Solutions

The pH of distilled water usually is below 7 because water absorbs carbon dioxide from the atmosphere. The unstable acid, carbonic acid, is formed when carbon dioxide dissolves in water. The pH of a solution or of water can be changed by the addition of acidic or basic substances. Often the acidic or basic compounds that alter the hydrogen ion concentration of water and of aqueous solutions are salts. Many salts ionize almost completely. A salt of a strong base and a weak acid raises the pH; a salt of a weak base and a strong acid lowers the pH; and a salt of a strong base and a strong acid does not affect pH.

Changes in pH resulting from the addition of acids, bases, or salts are much more pronounced in water than in foods because foods are buffered. Buffered solutions resist changes in hydrogen ion concentration on the addition of acid or base or on dilution. They do this by taking up hydrogen or hydroxyl ions to form slightly ionized substances. Buffered solutions prepared in the laboratory usually contain mixtures of weak acids, such as acetic acid or phosphoric acid, and their salts, although weak bases and their salts also can be used. The addition of hydrochloric acid to a solution containing equal parts of acetic acid and sodium acetate illustrates buffer action. As hydrochloric acid is added, the proportion of acetic acid increases while that of sodium acetate decreases, and sodium chloride is formed. Because of the slight ionization of acetic acid, there is no great change in pH until the sodium acetate in the original mixture is nearly exhausted. Similarly, the addition of sodium hydroxide to such a buffer solution increases the proportion of sodium acetate, but it changes the pH little until the acetic acid is almost exhausted. Foods are buffered by proteins and by salts.

2. Determination of pH

Indicator solutions and papers treated with indicator solutions can be used for colorimetric estimation of pH. An acid–base indicator changes color at a known pH and is a useful way to obtain information quickly without the use of special equipment. However, the accuracy of indicators is limited and the use of indicators with colored materials is difficult; therefore, pH is measured electrometrically in research. Electrometric determination of pH is possible because hydrogen ions carry a charge. When the reference electrode and a measuring electrode of a pH meter are put into a solution, an electromotive force between the electrodes is proportional to the hydrogen ion concentration of the solution. Measured as voltage, the electromotive force is converted into a pH reading by the meter. Buffers of known pH are used for calibration of the instrument prior to use. The standard buffer solution and the unknown solution should be at the same temperature.

IV. ENZYMES _____

Enzymes are biological catalysts. The reactions they catalyze would occur very slowly in their absence. Enzyme activity in foods is important from the standpoint of the enzymes naturally present in plant and animal tissues used as food and the enzymes added during food processing. Many of the chemical changes in living organisms are catalyzed by enzymes. In plant and animal foods, enzyme activity continues after death unless the enzymes are inactivated. Enzymes are used in the food industry to produce specific effects. Lactase treatment of milk (Chapter 8), enzymatic tenderization of meat (Chapter 9), and glucose isomerase treatment of corn syrups (Chapter 23) are among the numerous examples mentioned in this book.

A. Mechanism of Action

Enzyme activity involves the formation of a complex between the enzyme and the substrate. Formation of the complex is possible because the active site on the enzyme molecule complements a site on the substrate geometrically. The complex formation changes the conformation of a substrate molecule and expedites the reaction by putting a strain on the bond to be broken. This step, activation, greatly reduces the amount of energy required to cause the reaction. The reaction then proceeds with the formation of an enzyme–product complex. Finally the enzyme–product complex is split, with release of the product and regeneration of the enzyme, which then is available to complex with another molecule of the substrate. This reuse is the reason that enzymes, like inorganic catalysts, are effective in minute quantities.

Certain enzyme-catalyzed reactions in food do not occur in the intact tissue because of the physical separation of the enzyme and the substrate. Mechanical damage to the tissue can bring the enzyme and the substrate into contact with each other and permit reactions to occur.

B. Nomenclature and Nature of Enzymes

Nomenclature has not been very systematic. The trivial names of enzymes frequently identify enzymes according to the substrates on which they act. Examples are sucrase, lipase, and chlorophyllase. Indication of the type of reaction catalyzed also is common, as with dehydrogenase and invertase. The names of some enzymes identify both the substrate and the reaction catalyzed. Pectinmethylesterase and ascorbic acid oxidase are examples. Other trivial names, such as papain and pepsin, provide no information concerning the substrate or the reaction. An attempt has been made to establish a systematic and meaningful nomenclature system in which an enzyme name includes first the substrate and then the type of reaction. Reed (1975a, p. 27) gave the enzyme having the trivial name glucoamylase as an example. Its systematic name is $\alpha-1,4$-glucan glucohydrolase. Because systematic names often are cumbersome, enzymes have been assigned numerical codes, each consisting of four numerals separated by periods. The first number indicates the class (based on general function, such as hydrolase), and the succeeding numerals represent increasing descriptive specificity (Richardson and Hyslop, 1985, pp. 374–375).

Specificity of enzyme activity is variable and perhaps depends on the rigidity of the active site. Some enzymes are specific with regard to the type of bond that can be split but not the substrate. Somewhat greater specificity may be for a certain type of chemical bond but with specificity regarding locations of the bonds in substrate molecules (for example, adjacent to a particular group). Or an enzyme might be so specific that it catalyzes only one reaction involving one substrate. Sucrase (invertase) is an example.

Enzymes are proteins. Inactivation of enzymes by heat is an example of heat denaturation of protein. As is the case with other proteins, the primary structures of many enzymes have been determined. Leadlay (1978, pp. 12–17) summarized the process of determining primary structure. The specific three-dimensional conformation of any enzyme molecule is determined by the primary structure and is important to the enzyme's activity; an enzyme's active site must be available to the reactive portion of the substrate. Stability of enzyme structure depends on large numbers of weak bonds. Many enzymes function only in the presence of nonprotein cofactors. Some cofactors are metals; ascorbic acid oxidase, for example, requires copper. Other types of cofactors include some vitamin derivatives and some inorganic or organic molecules that are associated with the enzymes. Some cofactors, prosthetic groups, are tightly bound to protein moieties (apoenzymes), and others are only loosely bound. In addition to enzyme activators, there are enzyme inhibitors, some of which act by competing with substrate for active sites on enzyme molecules. Others have different mechanisms of action. For example, some inhibitors can act by binding at a site other than the active site and, in so doing, distort the active site so that it does not "fit" the substrate's binding site (Leadlay, 1978, p. 42).

C. Factors Affecting Activity

Enzyme activity is affected by several factors, one of the most important being temperature. For each enzyme there is an optimum temperature or temperature range. It is around 35–40°C for most enzymes, though there are notable exceptions and the range is broad for some enzymes (papain, for example). Enzyme activity is low at low temperatures and increases with increasing temperature, approximately doubling with every 10°C increase in temperature up to the optimum temperature. Further increases in temperature result in increasing denaturation until finally inactivation is complete. The optimum temperature depends on the pH of the medium. Wasserman (1984) discussed the production of enzymes having increased thermostability.

Enzyme activity is greatly affected by pH at a given temperature. Activity tends to be greatest at physiological pH levels. Changes in pH cause changes in ionization that can affect activity. Each enzyme shows activity within a pH range that is very narrow for some enzymes and broader for others. In the case of a narrow pH range for activity, buffering is important to continued enzyme activity. Irreversible enzyme inactivation because of denaturation occurs at high and low pH extremes; at pH levels that are outside the optimum range but not extreme, inactivation resulting from ionization may be reversible. A given enzyme can have different pH optima for different substrates (Reed, 1975b, p. 40). The activity of a given enzyme in relation to pH may vary with the source of the enzyme. West (1988) stated that, with some exceptions, fungal enzymes generally are most active at acid pH, whereas bacterial, yeast, and mammalian enzymes are most active at neutral or slightly alkaline pH. Ionic strength also is important (Zapsalis and Beck, 1985, pp. 150–151). Enzymes are used extensively in food processing. Because it sometimes is not desirable to adjust the pH of a food to meet the needs of a certain enzyme, the selection of an enzyme for a given process is important and needs to take into account the interdependence of temperature and pH effects, as well as the constraints of processing conditions.

Substrate concentration affects enzyme activity. If all other factors are constant, the initial reaction rate increases with the initial substrate concentration. When the initial substrate concentration is increased beyond a certain level, the enzyme becomes the limiting factor. Increasing the substrate concentration further, therefore, has no effect.

Enzyme concentration affects reaction rate in a similar fashion. With increased enzyme concentration, the initial reaction rate increases until the substrate concentration becomes limiting.

With a given combination of substrate concentration and enzyme concentration, the reaction rate changes during the course of the reaction. The initial reaction rate is relatively rapid. Later, the rate slows down and ultimately levels off. Product accumulation is responsible for this effect; the reaction is slowed as equilibrium is approached. A rather recent technological development, enzyme immobilization, has the potential of nearly eliminating the retarding effect. The enzyme is immobilized on a solid support that can bind the enzyme without masking the active sites that are needed for complexing the substrate. If the substrate is then passed through a column or packed bed of the solid, the reaction proceeds continuously, catalyzed by the enzyme, with the product(s) collected beyond the column or packed bed. Thus the products have little or no effect on the reaction rate. Even if a batch method is used, immobilization of the enzyme on the solid support has an advantage over traditional methods in that it permits reuse of the enzyme; the solid is readily separated from the system. In previous enzyme applications it was necessary to destroy enzymes in order to stop the reactions they catalyze. The technology has some negative aspects; for example, enzyme activity is decreased somewhat because the enzyme's attachment to the solid support decreases its accessibility to the substrate (Boyce, 1986).

The history, methods, and applications of enzyme immobilization technology were described by Messing (1975). Leadlay (1978, pp. 67–70) discussed different methods of enzyme immobilization, along with their advantages and disadvantages. Zapsalis and Beck (1985, pp. 184–187) summarized the technology of enzyme immobilization and described the major applications. Hultin (1983) described and evaluated immobilized enzyme technology.

Water activity (a_w) also affects enzyme activity because the availability of water to dissolve substrate and to permit diffusion of substrate to enzyme decreases with increased tightness of water binding. Most enzymes are inactive at low a_w levels and yet, at near neutral pH, retain their ability to become active when the a_w is increased, as with hydration.

Several enzymes and the nature of the reactions they catalyze are mentioned throughout this book. References on the general subject of enzymes include Pyler (1988, pp. 132–144), Reed (1975a,b), Richardson and Hyslop (1985), Stauffer (1987), and Zapsalis and Beck (1985). Boyce (1986) presented an overview of the use of enzymes in the food industry. Löffler (1986) discussed the industrial use of proteolytic enzymes.

V. BROWNING

Browning is a common occurrence in food. It is undesirable in some cases and desirable in others. The goal in handling food is to inhibit undesirable browning reactions as much as possible and to control desirable browning reactions. Both enzymatically catalyzed and nonenzymatic browning reactions occur in food.

A. Enzymatic Browning

Enzymatic (oxidative) browning occurs on the cut surfaces of certain fruits, such as apples, peaches, bananas, and pears. This type of browning requires three factors: substrate, which consists of polyphenolic compounds, a polyphenol oxidase (phenolase), which is an enzyme that can catalyze the first step in the reaction, and oxygen, which is a reactant. Copper is present as a prosthetic group on the enzyme molecule (Richardson and Hyslop, 1985, p. 446). Both the substrate, which frequently is the amino acid tyrosine, and the enzyme are present in susceptible fruits, and cutting into the tissue gives them access to one another. Oxygen becomes available when the cut surface is exposed to air. The enzyme catalyzes the oxidation of the polyphenolic substrate to a quinone. Quinones are not dark in color but are readily polymerized to dark-colored compounds.

Phenolase-catalyzed browning can have a few desirable effects, such as color development in cider, tea, and cocoa. However, the browning on cut surfaces of fruits is in no way desirable. Coseteng and Lee (1987) studied browning, as well as concentrations of phenolase and polyphenolics, at intervals during cold storage of nine apple cultivars. The extent of browning was highly correlated with concentration of the enzyme for some cultivars and with concentration of the substrate for other cultivars.

Preventive steps include inactivation of the enzyme by heat denaturation, the use of acid to inhibit enzyme activity and chelate (tie up) the copper, exclusion of oxygen, and reduction of quinones (by ascorbic acid) back to polyphenols before they can polymerize. Bisulfites also interfere with browning, possibly through an effect on the enzyme and possibly through reduction of quinones. Although sulfites are effective, they are no longer allowed in fresh fruits and vegetables in commercial food service because some consumers are sensitive to sulfites. Alternative dips, therefore, have been studied. Sapers and Ziolkowski (1987) found solutions of ascorbic acid and erythorbic acid to be effective with the two varieties of apples that they studied. The occurrence and prevention of enzymatic browning are discussed further in Chapter 14.

B. Nonenzymatic Browning

1. Carbonyl-Amine (Maillard) Browning

Carbonyl-amine browning, like enzymatic browning, actually involves more than one reaction. The total process is more complex than the process of enzymatic browning. The initial step is the condensation of the carbonyl group of a reducing sugar and a free amino group of a protein or a free amino acid. The condensation product loses water to form a Schiff base, which undergoes cyclization to a nitrogen-substituted glycosylamine. The glycosylamine in turn undergoes Amadori rearrangement to form a ketosamine, if the original sugar is an aldose, or an aldosamine, if the original sugar is a ketose. The brown pigments may arise thereafter through formation of many carbonyl derivatives and other reactions, followed by a series of fragmentations and polymerizations. Different pathways are possible, each resulting in brown pigments, called melanoidins, and volatile compounds that affect flavor. The specific sugars and amino acids involved and other reaction conditions affect the flavor that develops, as well as the rate and extent of browning.

The factors affecting carbonyl-amine browning include pH, temperature, moisture content, and water activity, as well as the sugars and amino acids available. Carbonyl-amine browning increases with increasing pH, especially in the pH range 7.8–9.2 (Whistler and Daniel, 1985, p. 103); little, if any, occurs at pH 6 or below. Browning also increases as temperature increases. Browning rate generally is higher at low-to-intermediate moisture levels than at very high or very low levels and higher at water activity levels between 0.6 and 0.8 than at higher and lower levels. Pentoses are more reactive than hexoses. Sucrose, being a nonreducing sugar, does not participate. Lysine and the other basic amino acids are particularly reactive because of their side-chain amino groups. Foods differ as to the relative availability of reducing sugars and amino acids. Marquez and Añon (1986), in a study of color development during the frying of potatoes, found that both reducing sugars and amino acids were involved in the color development, but the concentration of reducing sugars was the limiting factor. With two apple juice varieties, Toribio and Lozano (1986) found amino acid concentration to be the critical factor.

Both desirable and undesirable cases of carbonyl-amine browning are mentioned in various chapters of this book. If browning is not desirable, certain control measures are helpful. These include pH reduction, use of sulfur dioxide or sulfites, temperature control if feasible, and, in products containing little reducing sugar, removal of that sugar.

2. Sugar Browning (Caramelization)
Sugar browning differs from carbonyl-amine browning in several respects. It does not require nitrogen or any other nonsugar reactant. It can involve any kind of sugar, including the nonreducing sugar sucrose, which usually is used for commercial production of caramel syrups. Whereas carbonyl-amine browning, although accelerated by high temperatures, can occur at room temperature and lower if other conditions are favorable, caramelization requires very high temperatures. When a sugar is heated to temperatures above its melting point (for example, above 186°C for sucrose), dehydration reactions occur, resulting in the formation of furfural derivatives, which undergo a series of reactions ending with polymerization to brown compounds. Small amounts of acids and certain salts contribute to this type of browning (Whistler and Daniel, 1985, p. 98).

3. Ascorbic Acid Browning
Ascorbic acid plays a role in the browning of citrus juices and concentrates during storage. Degradation of ascorbic acid occurs, following either an aerobic or an anaerobic pathway. Sugar apparently is not involved. Nitrogen is not involved in the early reactions but may be involved ultimately in the formation of brown pigments. Metallic ions have a catalytic effect. The effect of pH depends on whether conditions are aerobic or anaerobic and whether a catalyst is present. Robertson and Samaniego (1986) reported a study in which ascorbic acid degradation during browning of lemon juice during storage appeared to be mostly anaerobic. Many reactions may contribute to the development of browning. Wong and Stanton (1989) reported that both ascorbic acid browning and Maillard (carbonyl-amine) browning are involved in browning of kiwi juice concentrate, with ascorbic acid browning predominating.

4. Prevention of Nonenzymatic Browning
Because there probably is considerable overlap among nonenzymatic browning reactions, control of browning in a given food system is subject to some trial and error. Certainly, low-temperature storage is helpful in any case. Sulfites also seem to inhibit a variety of browning reactions. To the extent that variation in

moisture and pH levels is feasible, these factors are best controlled according to the experimentally established conditions for stability of a specific food.

Among recent discussions of browning reactions are those of Cheftel *et al.* (1985, pp. 342–346), Danehy (1986), Namiki (1988), Richardson and Hyslop (1985, pp. 445–447), and Whistler and Daniel (1985, pp. 96–105).

VI. ENERGY TRANSFER AND CONVERSION AND MASS TRANSFER

Food preparation and food processing involve energy in many ways. Energy in the form of heat is transferred into and out of food in heating and cooling, respectively. Other forms of energy participate in food production, and thus energy conversions occur.

Whereas energy transfer involves a given form of energy moving from one place to another, energy conversion is a change of one form of energy to another form. Examples include conversion of electrical energy in an electric surface burner or chemical energy in a gas surface burner to heat, conversion of electrical energy to microwave energy in the magnetron of a microwave oven, and conversion of microwave energy to heat in the food that absorbs the microwaves.

Energy transfer often is associated with mass transfer, in which particles of a material actually migrate. Consider a loaf of baking bread. Some water migrates from the interior to the surface, and some evaporation from the surface occurs. The escaped water vapor continues to diffuse through the air in the oven because of a gradient in moisture concentration. The absorption of fat by a food during frying is an example of mass transfer, as is the leaching of soluble substances in a food into a heating medium. Examples of mass transfer are mentioned in several chapters of this book, though they are not necessarily so labeled. Energy conversion and transfer and mass transfer are highly interrelated.

A. Heat and Heat Transfer

Heat is not the only type of energy of interest with regard to energy transfer involving foods. Transfer of radiant energy, such as microwaves, gamma rays, and X rays, also is pertinent. Heat, however, is the major type of energy used for preparing and processing food. Holmes and Woodburn (1981) reviewed heat transfer in relation to the thermal treatment of food. In addition to reviewing methods of determining the extent of thermal stress, they reviewed studies in which temperature/time relationships have been reported for a variety of foods cooked under various conditions.

The transfer of heat into or out of food results from temperature differences, and the direction of transfer is from the hotter region to the colder region. The rate of transfer depends on the extent of the temperature difference; the greater the difference, the more rapid is the rate. The three methods of heat transfer are conduction, convection, and radiation.

In conduction of heat, thermal energy is transferred directly from molecule to molecule. Consider a food that is cooked in water in a saucepan on an electric burner. Molecules of the saucepan material vibrate because of energy received from the vibrating metal molecules of the hot burner. The thermal energy is conducted from the saucepan to the water with which it is in direct contact, then from that water to colder water and from heated water to the food. Further transfers by conduction occur between steam and the food if the food is not completely covered by water and between the outer and inner regions of the food. Materials from which containers and heating surfaces are made differ in their thermal conductivities. The thermal conductivity of some materials differs as follows: copper > aluminum > iron > Pyroceram glass (White, 1988). The thermal conductivity of water is greater than that of steam and air but lower than that of metals. However, the heat capacity of water is relatively high; the significance of the high heat capacity of water in heating food is mentioned earlier in this chapter.

Transfer of heat by convection occurs only in liquids and gases and involves motion of the fluid beyond the vibrational motion of individual molecules; actual currents develop. In most cases, the fluid receives heat by conduction from a hot surface. The water in the example of conduction cited above receives heat directly from the container, but the heat is transferred within the water by convection as well as by conduction. When food is deep fried, the heat is transferred from the fat to the food by conduction, but heat transfer occurs within the fluid fat by convection. In a conventional oven, convection currents develop naturally in the air because of temperature differences in the air; warmer (less dense) portions rise and colder (denser) portions fall.

Convection ovens were developed because heat transfer by natural convection is not efficient. In a convection oven, convection is forced; the streaming effect is intensified by a fan or a pump, and heat transfer, therefore, is more rapid than in a conventional oven. Even greater efficiency, plus greater uniformity of heat transfer, resulted from the development for the bakery industry of the impingement oven. Air impingement involves the use of strategically placed air nozzles for sweeping away the insulating layer of air immediately surrounding the product and thus increasing the rate of heat transfer to the product. Because surface moisture also is swept away, humidity in the oven must be controlled to permit only the desired amount of surface drying. Walker (1987) summarized methods of energy transfer and described the development and the principles of air impingement technology. Sato *et al.* (1987) studied the relation of air temperature and/or air velocity to the characteristics of sponge cake baked in a forced convection oven. Browning, shape, firmness, and volume were affected by variations in air temperature and air velocity.

The radiation of heat is involved in broiling, though heat conduction also is involved. The use of infrared radiation to keep foods hot is another application of thermal radiation.

B. Ionizing Radiation and Microwave Energy

X Rays and gamma rays are ionizing radiation and may be used for food preservation. Gamma radiation from a cobalt-60 or cesium-137 source is used far more extensively than is radiation with X rays. The process of irradiating foods to preserve them has been slow in gaining widespread acceptance. The large doses required to sterilize foods can produce undesirable effects on food properties; therefore, applications of irradiation that involve lower doses show more promise. One of the first applications of gamma radiation approved by the FDA was for elimination of insect infestation of wheat and flour; irradiation of potatoes for inhibition of sprouting was approved soon thereafter (Giddings and Welt, 1982). Relatively low radiation doses are effective for these uses. Approval came later for some applications involving irradiation of fresh fruits and vegetables and certain meats with doses that are much lower than those required for sterilization but useful in increasing keeping quality. However, although food safety is not a problem with the uses that have been approved, there still is some opposition by consumers to the concept, in part because of a negative connotation of the word irradiation. Another possible limitation to application of the technology is its cost. Rogan and Glaros (1988) described the use of cobalt-60 radiation in food processing, its functions, its effect on food components, and its risks and benefits.

Microwave cookery represents the major recent application of radiant energy. Chapters 7, 9, 10,11, 14, 19, and 21 in this book include examples of microwave use. Microwave ovens have become common kitchen appliances, prized for the speed with which they heat foods. The major commercial impact of microwave technology has been on development of microwavable foods and on packaging rather than on industrial food production processes.

The magnetron of a microwave oven generates an electromagnetic field in which positive and negative poles constantly and rapidly reverse direction. Transmitted to the food as microwaves, the electromagnetic energy interacts with the dipolar water molecules and with positive and negative ions of salts. Rapid molecular rotation occurs in response to the changing field, and generation of heat by molecular friction occurs.

The penetration of microwaves and conversion of microwave energy to heat in the food occur instantaneously. The depth of penetration, 2–3 cm, depends in part on the power output of the oven. The maximum power available depends on the specific oven; most are 600- to 700-W ovens, though the power level in some is as low as 400 W. In addition, most of the currently available ovens provide for selection of lower power levels for defrosting or for cooking foods that would be cooked conventionally at low temperatures. The reduced power settings do not give continuous reduced power but intermittent full power. For example, operation at the 30% power setting involves full microwave power 30% of the time and no power 70% of the time. During a period of no energy output in the cycle, heat is conducted from warmer to cooler areas, reducing localized overheating. The on–off cycling, therefore, contributes to uniform cooking of sensitive foods and foods that need adequate time for certain changes, such as starch gelatinization, to occur. Protection of texture and development of flavor often require the use of reduced microwave power.

The cycling that occurs with reduced power can be particularly beneficial in defrosting because the response of ice to microwaves is different from that of water. Ice permits deeper microwave penetration but has a lower rate of microwave absorption and, therefore, of heat development; therefore, a partially thawed food can have both hot and icy areas. Reduced power requires more time for thawing than does full power but provides a means of preventing localized overheating. A potential problem with use of reduced microwave power is the variation among different microwave ovens with respect to the time base (duration of on–off cycle). If the on–off cycle is relatively long and cooking time of a given food very short, cooking instructions for reduced power may not give reproducible results. For foods requiring longer cooking times, the time base is less important (O'Meara, 1989). The consumer normally does not know the time base of a given microwave oven. Although the capability for reduced power is common, the specific procedures for heating specific foods require some experimental work not only during product development but also during preparation in a specific microwave oven.

The composition and properties of food also affect depth of microwave penetration. The extent of microwave absorption and depth of penetration are negatively related. Relatively high moisture and salt contents and low solids are generally associated with relatively shallow penetration of microwaves (IFT, 1989a). Although penetration is relatively shallow with high water content, absorption of microwave energy is high, and heating occurs readily. Fats have low specific heat and, therefore, can readily reach high temperatures; however, their ability to distribute heat in a system is poor because of their low degree of electrical polarity. Penetration is more rapid and absorption is lower in porous than in dense food.

The portion of the food that is not heated by microwaves is heated by conduction of heat inward; therefore, the time required for cooking depends in part on the thickness of the food. Cooking time can be quite long if a large part of the heating must occur by conduction. The shape of the food is a factor, also. Food in a rectangular shape does not heat evenly; a doughnut shape gives the most even heating (Best, 1987). The time depends not only on the size and shape of individual pieces but also on the total amount of food being cooked.

Browning does not occur in an ordinary microwave oven because moisture diffuses outward and evaporates from the surface. Evaporation is a cooling process. The surface does not get hot enough or dry enough to brown. Combination microwave/convection ovens provide a means of browning surfaces while cooking food rapidly.

Pei (1982) compared the energy and mass transfers involved in conventional and microwave baking of bread. The advantages of combining microwave radiation with conventional baking of bread were described. Some studies involving microwave baking of cakes are discussed in Chapter 21.

The use of microwaves in food preparation presents unique challenges. One involves foods containing ground spices or herbs. The full flavor that highly seasoned foods attain in conventional cooking does not develop in the microwave oven, apparently because the flavor components in ground spices and herbs do not have time to diffuse through cell walls and throughout the product. Compensation by increase of seasoning levels is not always satisfactory, and Hassel (1989) described some flavor products that have been developed as alternatives to ground seasonings for use in microwave applications, as well as in some other convenience foods. Alternative flavor products include spice emulsions; spray-dried spice products; extensions that consist of essential oils and solvent-extracted spice components, plated onto carriers such as salt, dextrose, dextrins, and starches; liquid dispersions; and dry agglomerates of spice components.

Another challenge is the stabilization of formulated foods that are developed for microwave cooking and often frozen before distribution. Unless modifications are made in conventional formulas, unpleasant textural changes can occur. Microcrystalline cellulose, a particulate rather than fibrous form of α-cellulose, is useful in stabilizing emulsions, controlling ice crystal growth in products that are frozen, and maintaining product consistency (IFT, 1989b).

For applications in which browning is desirable, special products for coating food surfaces are being developed. A mixture of Maillard (carbonyl-amine) browning reactants and substances that have electric dipoles and thus heat readily is an example (Best, 1987).

Some products, such as microwave pizza and popcorn, require the development of localized high temperatures. Therefore, it has been necessary to develop special packaging, susceptor packaging, to meet those needs. Susceptors, consisting of metallized paperboard (IMPI, 1987), cause localized heating by absorbing energy especially strongly. Although the metal itself does not absorb microwave energy, it readily absorbs heat from other substances and becomes very hot. Another type of packaging for microwaving is a special absorbent material developed for cooking bacon and other fatty products.

Best (1987) discussed factors that must be considered in the formulation of microwavable foods and of packaging for them. The International Microwave Power Institute (IMPI, 1987) developed a handbook that provides considerable information concerning microwave ovens, accessories, and applications.

Suggested Exercises

1. Find sucrose and sodium chloride in the organic and inorganic sections, respectively, of the tables of physical constants in the "Handbook of Chemistry and Physics" (CRC, 1989–1990). Note the magnitude of the values for solubility in water in each case. Which is more soluble? Comment on the effect of temperature on the solubility of sucrose and of sodium chloride. Note the molecular weights of the two compounds. What is the significance of their molecular weights with regard to their effects on the boiling point of water?

2. In separate pans of boiling distilled water, make stepwise additions of sucrose and of sodium chloride. (The increments need to be larger for sucrose than for sodium chloride.) After each addition, stir to dissolve the solute, bring the solution back to a boil, and record the boiling point. Continue making additions and recording boiling points until you have demonstrated the following: (1) a difference in the amounts of the solutes required to bring about a given increase in the boiling point of a given amount of water (What controls will be necessary?); (2) a difference in the solubility of sucrose and sodium chloride; and (3) the relationship between solubility and effect on boiling point.

3. Sometimes comparisons are made in laboratory classes of the crystallization of sucrose and of glucose in a simple mixture such as a fondant. The subject matter of this chapter should have made it clear that heating both solutions to the same boiling point will not produce solutions of equal concentration. (A mole of sucrose is 342 g and a mole of glucose is 180 g.) Calculate the boiling point of a glucose solution that is theoretically equal in concentration (on a weight percentage basis) to a sucrose solution that boils at 114°C.

4. Mix equal parts of a commercial liquid pectin and distilled water. To 10-ml portions in 50-ml beakers, make the following additions: (1) 1.5 ml 0.1 N HCl, (2) 10 g sucrose, (3) 1.5 ml 0.1 N HCl and 10 g sucrose, and (4) 3 ml 70% ethanol. Prepare a second series of samples and heat to the boiling point. Cool to room temperature. Observe and explain the results.

5. Study labels of processed and formulated foods and make a list of the hydrocolloids included and the foods containing them. State the probable function(s) of the gum(s) in each food.

6. Measure the pH of tap water and distilled water. Boil and cool each type of water and again measure pH. Explain the results.

7. Study the buffering action of milk by finding the effects of acid and base on pH as determined by a colorimetric or an electrometric method. Find the pH of boiled, cooled distilled water. Then add to 10 ml of the water enough 0.1 N HCl to lower the pH by two units. Measure the amount of acid required. Next find the pH of a mixture containing half milk and half boiled distilled water before and after adding to 10 ml of the mixture the same amount of acid as was used with the water. If the pH change is less in the presence of milk than of water, the buffering effects of salts and protein are responsible.

8. Repeat exercise 7, using enough 0.1 N NaOH to raise the pH by two units.

9. Heat solutions containing three sugars (sucrose, glucose, and any pentose) and three amino acids (glutamic, glycine, and lysine) in all nine combinations. Compare as to extent of browning.

10. The actual power output of microwave ovens often is lower than the nominal wattage. Determine the power output of at least one microwave oven as follows (Van Zante, 1973, p. 49): Measure the temperature (in °F) of 475 ml (2 c) of cold water in a pint container and heat in a microwave oven for 1 min. After again recording the temperature, multiply the temperature increase by 17.5 to obtain the power output in watts per hour. How does the resulting value compare with the wattage specified by the manufacturer?

11. Power is not distributed evenly in a microwave oven. Several methods of obtaining information concerning power distribution have been used (Van Zante, 1973, pp. 55–57). Apply a method using egg whites, as follows: Put 30 g of egg white into each of nine custard cups. Place the cups equidistantly in the oven to form a 3 × 3 matrix. Turn on the power until the first sample appears to be coagulated. Remove all of the cups from the oven, placing them in the same relative positions. Observe coagulated and

uncoagulated portions and devise a method of recording results, such as with a diagram showing cooked and uncooked areas. Can you think of a way of observing a three-dimensional pattern?

References

Acker, L. W. 1969. Water activity and enzyme activity. *Food Technol.* **23**: 1257–1270.

Alzamora, S. M. and Chirife, J. 1983. The water activity of canned foods. *J. Food Sci.* **48**: 1385–1387.

Best, D. 1987. Microwave formulation: A new wave of thinking. *Prepared Foods* **156**(10): 70–74.

Beuchat, L. R. 1981. Microbial stability as affected by water activity. *Cereal Foods World* **26**: 345–349.

Bolin, H. R. 1980. Relation of moisture to water activity in prunes and raisins. *J. Food Sci.* **45**: 1190–1192.

Bourne, M. C. 1986. Effect of water activity on texture profile parameters of apple flesh. *J. Texture Studies* **17**: 331–340.

Boyce, C. O. L. 1986. Enzyme basics: A product development overview. *Bakers Digest* **60**(1): 6–12.

Cheftel, J. C., Cuq, J.-L., and Lorient, D. 1985. Amino acids, peptides, and proteins. *In* "Food Chemistry," 2nd edn., Fennema, O. R. (Ed.). Marcel Dekker, Inc., New York.

Coseteng, M. Y. and Lee, C. Y. 1987. Changes in apple polyphenoloxidase and polyphenol concentrations in relation to degree of browning. *J. Food Sci.* **52**: 985–989.

CRC. 1989–1990. "Handbook of Chemistry and Physics," 70th edn., Weast, R. C. (Ed.). CRC Press, Inc., Boca Raton, Florida.

Danehy, J. P. 1986. Maillard reactions: Nonenzymatic browning in food systems with special reference to the development of flavor. *Adv. Food Research* **30**: 77–138.

deMan, J. M. 1980. "Principles of Food Chemistry." Avi Publ. Co., Westport, Connecticut.

Fennema, O. R. 1985. Water and ice. *In* "Food Chemistry," 2nd edn. Fennema, O. R. (Ed.). Marcel Dekker, Inc., New York.

Finley, J. W. 1989. Effects of processing on proteins: An overview. In "Protein Quality and the Effects of Processing," Phillips, R. D. and Finley, J. W. (Eds.). Marcel Dekker, Inc., New York.

Giddings, G. G. and Welt, M. A. 1982. Radiation preservation of food. *Cereal Foods World* **27**: 17–20.

Glicksman, M. 1982. Functional properties of hydrocolloids. *In* "Food Hydrocolloids," Vol. I, Glicksman, M. (Ed.). CRC Press, Inc., Boca Raton, Florida.

Hassel, J. 1989. Spice alternatives for microwave applications. *Cereal Foods World* **34**: 340–341.

Holmes, Z. A. and Woodburn, M. 1981. Heat transfer and temperature of foods during processing. *CRC Crit. Rev. Food Sci. Nutr.* **14**: 231–294.

Hultin, H. O. 1983. Current and potential uses of immobilized enzymes. *Food Technol.* **37**(10): 66–82, 176.

IFT. 1989a. Microwave food processing. A scientific status summary by the IFT expert panel on food safety and nutrition. *Food Technol.* **43**(1): 117–126.

IFT. 1989b. Hydrocolloid functions to improve stability of microwavable foods. *Food Technol.* **43**(6): 96–100.

Igoe, R. S. 1982. Hydrocolloid interactions useful in food systems. *Food Technol.* **36**(4): 72–74.

IMPI. 1987. "Microwave Cooking Handbook." International Microwave Power Institute, Clifton, Virginia.

Katz, E. E. and Labuza, T. P. 1981. Effect of water activity on the sensory crispness and mechanical deformation of snack food products. *J. Food Sci.* **46**: 403–409.

Keenan, C.W., Kleinfelter, D.C., and Wood, J. H. 1980. "General College Chemistry," 6th edn. Harper and Row, San Francisco, California.

Labuza, T. P. 1968. Sorption phenomena in foods. *Food Technol.* **22**: 263–265, 268, 270, 272.

Labuza, T. P. 1971a. Kinetics of lipid oxidation in foods. *CRC Crit. Rev. Food Technol.* **2**: 355–405.

Labuza, T. P. 1971b. Properties of water as related to the keeping quality of foods. *In* "Proceedings, SOS/70, Third International Congress of Food Science and Technology." Institute of Food Technologists, Chicago, Illinois.

Labuza, T. P. 1984. "Moisture Sorption: Practical Aspects of Isotherm Measurement and Use." American Association of Cereal Chemists, St. Paul, Minnesota.

Leadlay, P. F. 1978. "An Introduction to Enzyme Chemistry." The Chemical Society, London.

Löffler, A. 1986. Proteolytic enzymes: Sources and applications. *Food Technol.* **40**(1): 63–70.

Magrini, R. C., Chirife, J., and Parada, J. L. 1983. A study of *Staphylococcus aureus* growth in model systems and processed cheese. *J. Food Sci.* **48**: 882–885.

Marquez, G. and Añon, M. C. 1986. Influence of reducing sugars and amino acids in the color development of fried potatoes. *J. Food Sci.* **51**: 157–160.

Messing, R. A. (Ed.). "Immobilized Enzymes for Industrial Reactors." Academic Press, New York.

Moreyra, R. and Peleg, M. 1981. Effect of equilibrium water activity on the bulk properties of selected food powders. *J. Food Sci.* **46**: 1918–1922.

Namiki, M. 1988. Chemistry of Maillard reactions: Recent studies on the browning reaction mechanism and the development of antioxidants and mutagens. *Adv. Food Research* **32**: 115–184.

O'Meara, J. P. 1989. Variable power: A dilemma for the microwave oven user. *Microwave World* **10**(2): 12–16.

Pei, D. C. T. 1982. Microwave baking—new developments. *Bakers Digest* **56**(1): 8–10, 32.

Pomeranz, Y. 1985. "Functional Properties of Food Components." Academic Press, Orlando, Florida.

Pyler, E. J. 1988. "Baking Science and Technology," 3rd edn., vol. I. Sosland Publishing Co., Merriam, Kansas.

Reed, G. 1975a. General characteristics of enzymes. *In* "Enzymes in Food Processing," 2nd edn., Reed, G. (Ed.). Academic Press, New York.

Reed, G. 1975b. Effect of temperature and pH. *In* "Enzymes in Food Processing," 2nd edn., Reed, G. (Ed.). Academic Press, New York.

Richardson, T. and Hyslop, D. B. 1985. Enzymes. *In* "Food Chemistry," 2nd edn., Fennema, O. R. (Ed.). Marcel Dekker, Inc., New York.

Robertson, G. L. and Samaniego, C. M. L. 1986. Effect of initial dissolved oxygen levels on the degradation of ascorbic acid and the browning of lemon juice during storage. *J. Food Sci.* **51**: 184–187, 192.

Rogan, A. and Glaros, G. 1988. Food irradiation: The process and implications for dietitians. *J. Amer. Dietet. Assoc.* **88**: 833–838.

Sapers, G. M. and Ziolkowski, M.A. 1987. Comparison of erythorbic and ascorbic acids as inhibitors of enzymatic browning in apple. *J. Food Sci.* **52**: 1732–1733, 1747.

Sato, H., Matsumura, T., and Shibukawa, S. 1987. Apparent heat transfer in a forced convection oven and properties of baked food. *J. Food Sci.* **52**: 185–188, 193.

Singh, R. K., Lund, D. B., and Buelow, F. H. 1983. Storage stability of intermediate moisture apples: Kinetics of quality change. *J. Food Sci.* **48**: 939–944.

Stauffer, C. E. 1987. Some general aspects of enzyme chemistry. *In* "Enzymes and Their Role in Cereal Technology," Kruger, J. E., Lineback, D., and Stauffer, C. E. (Eds.). American Association of Cereal Chemists, St. Paul, Minnesota.

Sych, J., Castaigne, F., and Lacroix, C. 1987. Effect of initial moisture content and storage relative humidity on textural changes of layer cakes during storage. *J. Food Sci.* **52**: 1604–1610.

Toribio, J. L. and Lozano, J. E. 1986. Heat induced browning of clarified apple juice at high temperatures. *J. Food Sci.* **51**: 172–175, 179.

Toribio, J. L., Nunes, R. V., and Lozano, J. E. 1984. Influence of water activity on the non-enzymatic browning of apple juice concentrate during storage. *J. Food Sci.* **49**: 1630–1631.

Troller, J. A. and Christian, J. H. B. 1978. "Water Activity and Food." Academic Press, New York.

Vallejo-Cordoba, B., Nakai, S., Powrie, W. D., and Beveridge, T. 1986. Protein hydrolysates for reducing water activity in meat products. *J. Food Sci.* **51**: 1156–1161.

Van Zante, H. J. 1973. "The Microwave Oven." Houghton Mifflin Co., Boston, Massachusetts.

Walker, C. E. 1987. Impingement oven technology—Part I. Principles. Research Dept. Tech. Bull. 9(11): 1–7, American Institute of Baking, Manhattan, Kansas.

Wasserman, B. P. 1984. Thermostable enzyme production. *Food Technol.* **38**(2): 78–89, 98.

West, S. 1988. The enzyme maze. *Food Technol.* **42**(4): 98–102.

Whistler, R. L. and Daniel, J. R. 1985. Carbohydrates. *In* "Food Chemistry," 2nd edn., Fennema, O. R. (Ed.). Marcel Dekker, Inc., New York.

White, F. M. 1988. "Heat and Mass Transfer." Addison-Wesley Publishing Co., Reading, Massachusetts.

Whitney, R. McL. 1977. Chemistry of colloid substances: General principles. *In* "Food Colloids," Graham, H. D. (Ed.). Avi Publ. Co., Westport, Connecticut.

Wong, M. and Stanton, D. W. 1989. Nonenzymic browning in kiwifruit juice concentrate systems during storage. *J. Food Sci.* **54:** 669–673.

Zapsalis, C. and Beck, R. A. 1985. "Food Chemistry and Nutritional Biochemistry." John Wiley and Sons, New York.

EGGS

Eggs serve important roles in many food products because of their diverse functional properties. The structure and composition of eggs in relation to their functional properties are considered in this chapter. The three functional properties of eggs that are of importance with respect to food preparation are coagulation, emulsification, and foaming.

I. STRUCTURE AND COMPOSITION

A diagram of an egg is shown in Fig. 7.1. The yolk contains alternate light and dark layers of yolk material surrounded by a colorless two-layered vitelline membrane. The yolk is held in place within the egg by means of the chalazae, opaque fibrous structures that extend into the white on either end of the egg and are continuous with the chalaziferous layer immediately surrounding the yolk. Just outside the chalaziferous layer (inner thick white) is the inner layer of thin

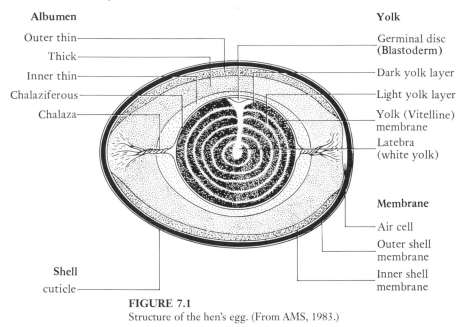

Albumen

Outer thin
Thick
Inner thin
Chalaziferous
Chalaza

Yolk

Germinal disc
(Blastoderm)
Dark yolk layer
Light yolk layer
Yolk (Vitelline)
membrane
Latebra
(white yolk)

Membrane

Air cell
Outer shell
membrane
Inner shell
membrane

Shell

cuticle

FIGURE 7.1
Structure of the hen's egg. (From AMS, 1983.)

white, surrounded by the thick (viscous) white and an outer layer of thin white. The thick white constitutes over half of the total fresh egg white. Within the shell are two membranes composed of protein–polysaccharide fibers. The membranes separate to form the air cell at the large end of the egg. The air cell forms as the egg cools and the contents contract following laying. The shell is granular in structure and sufficiently porous to permit respiration of a developing embryo. Microbial invasion is prevented by protein fibers in the pores and the shell is covered with a thin layer or cuticle, a protective protein layer (Powrie and Nakai, 1985, p. 830).

All the nutrients needed by the developing chick are supplied by the egg. The composition of the egg with respect to components important to its role in food processing and preparation is given in Table I. The white, frequently referred to as albumen, is composed largely of protein and water with only a trace of fat. Water (87–89%) and protein (9.7–10.6%) contents vary with the age of the hen (Powrie and Nakai, 1985, p. 832). Carbohydrate, though small in amount (about 1%), can cause problems in the drying of egg whites unless removed. Most of the free carbohydrate in egg white is glucose, a reducing sugar (Powrie and Nakai, 1985, pp. 832–833). Glucose will react with proteins in the carbonyl-amine reaction to produce undesirable brown discoloration in dried egg whites as well as the whites of hard-cooked eggs (Baker and Darfler, 1969).

The protein of albumen is a complex mixture. Powrie and Nakai (1986, p. 108) described the white ". . . as a protein system consisting of ovomucin fibers in an aqueous solution of numerous globular proteins." Most of the proteins are glycoproteins; some have lipids bound to them (Pomeranz, 1985, p. 196). Egg albumen proteins were discussed in detail by Vadehra and Nath

TABLE I

Relative Weights and Composition of Egg[a]

	Weight[b] (g)	Water (%)	Protein (%)	Fat (%)	Carbohydrate (%)
Whole	50	74.6	12.1	11.2	1.2
White	33	88.0	10.1	trace	0.8
Yolk	17	48.8	16.4	32.9	0.2

[a] From Adams (1975).
[b] Large egg.

(1973) and Powrie and Nakai (1985, pp. 833–837). Quantitatively, the three primary proteins are ovalbumin, ovotransferrin, and ovomucoid. Ovalbumin, a phosphoglycoprotein, constitutes 54% of the protein in egg white. Ovalbumin is resistant to denaturation by heat and becomes more so during storage via conversion to S-ovalbumin (Powrie and Nakai, 1985, p. 834). In its denatured state, ovalbumin is an important structural component of baked products. Ovotransferrin (conalbumin), a glycoprotein, constitutes 12% of the proteins and is characterized by its ability to bind divalent and trivalent metallic ions such as iron. It is more readily coagulated by heat and less readily coagulated by surface denaturation than is ovalbumin. Ovomucoid, a heat-resistant glycoprotein, constitutes about 11% of the protein portion of egg white and is an inhibitor of some proteases (Pomeranz, 1985, pp. 196–199). Other proteins are present in smaller quantities. Ovomucin, another glycoprotein, is thought to contribute to the thickness of the white. Thick white contains four times as much ovomucin as thin white (Powrie and Nakai, 1985, pp. 833–837). Ovomucin is not coagulated by heat (Johnson and Zabik, 1981a). Several globulins are present, including lysozyme, which exhibits bacteriolytic activity. The globulins, as a group, are important for the foaming properties of egg white. This has been shown in studies of the functional properties of duck egg white proteins. The addition of globulins to duck egg whites, which are inherently lacking in this group, results in improvement of the volume of angel food cakes made with the whites (MacDonnell *et al.*, 1955). The globulins represent about 8% of the proteins in egg white. Other proteins present in small amounts include ovoinhibitor and a protease inhibitor; ovoflavoprotein, an apoprotein that binds riboflavin as flavoprotein; ovomacroglobulin, a large glycoprotein; and avidin, which binds biotin, making it biologically unavailable (Powrie and Nakai, 1986). Avidin is rendered inactive by heating.

Yolk is more concentrated than white, containing less water, more protein, and a considerable portion of fat, as shown in Table I. The protein content is relatively constant whereas lipid varies from 32 to 35%. This variation is related to strain not diet (Powrie and Nakai, 1985, pp. 832–833). Diet does influence the fatty acid content of the yolk lipids. Changes are primarily in the levels of linoleic and oleic acids; levels of saturated fatty acids remain constant (Powrie and Nakai, 1986, pp. 105–107). Triacylglycerol (66%), phospholipid (28%), and cholesterol (5%) are the major components of the lipid fraction (Powrie and

Nakai, 1985, pp. 832–833). Concern for cholesterol has led to many attempts to reduce the cholesterol content in egg yolk as described by Mast and Clouser (1988, pp. 325–327).

Powrie and Nakai (1985, p. 377) described yolk as ". . . particles distributed in a protein (livetin) solution." The particles range in size and composition. Yolk may be separated into plasma, a clear supernatant, and a granular fraction by centrifugation. The plasma represents the major portion (78%) of the liquid yolk. Livetin, a globular protein fraction, and a low-density lipoprotein (LDL) fraction are the major components of the plasma. The LDL fraction of plasma (sometimes called very low-density lipoprotein fraction) has a lipid content of 84–89% and includes neutral lipids and phosphoplipids (Powrie and Nakai, 1985, p. 844). The granular fraction contains phosvitin (16%), lipovitellins (17%), and LDL (12%). Phosvitin serves as the carrier of iron in the yolk. Lipovitellin is a high-density lipoprotein fraction. The lipids in the LDL portion of the granular fraction include phospholipids, cholesterol, and triacylglycerols (Powrie and Nakai, 1985, p. 842).

The color of the yolk depends on the presence of carotenoids, primarily xanthophylls, with one or more hydroxyl groups that impart moderately polar characteristics to the pigment (Marusich and Bauernfeind, 1981). Lutein and zeaxanthin are the major xanthophylls in yolk and small amounts of cryptoxanthin and carotene also are found (Marusich and Bauernfeind, 1981). The color varies from a pale yellow to a bright red, depending on the amount and type of pigment present in the diet of the hen (Carlson and Halverson, 1964; Alam *et al.*, 1968). Alfalfa and corn are used in diets of hens to supply the xanthophylls which are rapidly transferred to the yolk of the egg (Marusich and Bauernfeind, 1981, pp. 323–327). Color of yolk may vary with time of the year. Janky (1986) found that pigmentation became less yellow from January to October and more yellow from October to December but found no evidence to indicate that consumers were concerned about this change in color.

In general, it can be concluded that yolk is a very complex system. The functionality of this complex system in food products is dependent on the physical and chemical characteristics of the various components.

II. EGG QUALITY

A. Evaluation of Egg Quality

1. Commercial Evaluation

Eggs are sold commercially according to weight and grade, factors that are independent of each other. The weight of eggs for the purpose of size classification is expressed in ounces per dozen: a minimum of 15 for pewee, 18 for small, 21 for medium, 24 for large, 27 for extra large, and 30 for jumbo eggs. This variation in size illustrates the desirability of weighing broken out egg when doing experimental work. When formulas specify a certain number of eggs it is assumed that medium or large eggs will be used. For commercial sorting of eggs into size categories, a range including the given weights is used. The average weights for a large whole egg, the white, and the yolk are given in Table I. For medium eggs, the values are 44, 29, and 15 g, respectively (Adams, 1975). These weights may be used to determine how much frozen bulk egg or pooled fresh egg should be used in a formula.

Commercial quality evaluation and grading of eggs involve procedures outlined by the United States Department of Agriculture (USDA, 1981; AMS, 1983). The quality of an egg is determined by evaluating a number of factors. The final score given to an egg can be no higher than the lowest score given to the egg for any one quality factor. Therefore, one factor alone may determine the quality score. Quality factors include both exterior and interior qualities. External factors include shell characteristics: shape and texture, soundness, cleanliness, and color. Composition and egg quality are not related to shell color in a practical sense (Curtis *et al.*, 1986). Shell color is used only for sorting eggs into groups of like color for sale and is not considered in USDA standards of quality or grades. Interior quality is evaluated commercially by candling, where the condition of the egg is judged as it is held up to a light in a dark room and rotated. The candling light reveals the condition of the shell, the size of the air cell, and the size, distinctness, color, and mobility of the yolk. Abnormalities indicative of inedible yolk also are evident. These include blood rings, meat spots, and mottled yolks. As the grade of the egg decreases, the size of the air cell increases, the white becomes less firm or thinner, and the yolk becomes more distinct in candling and appears flattened. The appearance of graded eggs that have been broken from the shell is shown in Fig. 7.2.

It cannot be assumed that every egg in a carton of a certain grade will meet all the requirements for that grade, because grade specifications allow individual eggs in a lot to be of lower quality than indicated. However, 72, 82, and 90% of Grade AA, A, and B eggs, respectively, must fall within the specified grade at the destination (USDA, 1981). This tolerance is intended to cover differences in the efficiency of graders, normal changes in the product under favorable conditions during reasonable periods between grading and subsequent inspection, and reasonable variations in interpretations by graders. Janky (1986) confirmed, in a study of eggs purchased on the retail market, that considerable variation in albumen quality may occur within one grade level within a dozen as well as over a period of a year.

The relationship of the candled quality of eggs to their performance in food preparation was studied by Dawson *et al.* (1956). Eggs of lower grade were obtained by holding fresh eggs under varying conditions. There was a direct relationship between the candled quality of the eggs and the volume, texture, tenderness, and acceptability of angel food cakes made from them. However, there was little or no relationship between the candled quality of the eggs and the firmness of baked custards as indicated by penetration values or standing height after unmolding. It was evident from this early study that the whites of AA or A quality are better for food preparation requiring whites than are those grading lower. The lower grades may be satisfactory for formulas requiring whole eggs. It is important to note that because egg-grading standards were revised in 1981 (USDA, 1981) grade C is no longer applicable. Individual eggs now are graded as AA, A, or B. Therefore, results of this early study (Dawson *et al.*, 1956) should be applied in terms of grades AA and A versus B. It also should be noted that eggs grading B are not readily available on the retail market.

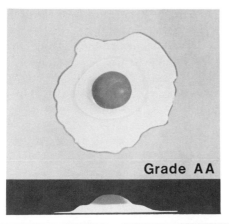

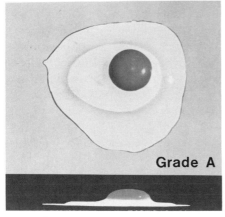
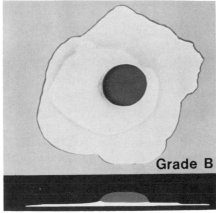

FIGURE 7.2
Broken out appearance (top and side views) of eggs of varying grades. Grade AA: Egg covers small area; much thick white surrounds yolk; has small amount of thin white; yolk round and upstanding. Grade A: Egg covers moderate area; has considerable thick white; medium amount of thin white; yolk round and upstanding. Grade B: Egg covers wide area; has small amount of thick white; much thin white; yolk somewhat flattened and enlarged. (From AMS, 1983.)

2. Evaluation for Research Purposes

Several evaluation methods depend on measurement of the height of the thick white, which can be measured with a commercially available micrometer attached to a tripod (AMS, 1983, pp. 26–27) or with the probe of a vernier caliper. The Haugh unit, which indicates the relationship of the height of the thick white to the weight of the egg, is the most widely used measure of albumen quality (AMS, 1983, p. 13). Tables for direct determination of Haugh units from height and weight measurements are available (AMS, 1983). As albumen quality decreases, Haugh units also decrease.

For experimental work it may be desirable to determine the relative proportions of thick, thin, and adhering albumen. The eggs are cracked and allowed to drain for 30 sec, after which the albumen adhering to the shell can be removed. The thin outer white will drain through a sieve and can be collected in a weighed container. The remaining portions of white can be transferred after separation from the yolk to a weighed container. After necessary weighing, the percentage of each fraction can be determined (Reinke *et al.*, 1973).

The condition of the yolk may be evaluated by determining the yolk index, which represents the height of the yolk divided by its width when the yolk is resting on a flat surface (Sauter *et al.*, 1951). As with Haugh units, deterioration of the egg causes a decrease in the yolk index.

B. Changes in Egg Quality during Storage

As soon as the egg is laid, changes begin to take place that lower its quality and eventually cause spoilage. These changes can be retarded by proper handling, but they cannot be prevented entirely. During aging, the size of the air cell increases, the yolk enlarges and its membrane weakens, the egg white becomes thinner, the egg becomes more alkaline, and the odor and flavor of the egg deteriorate. All of the changes are influenced by temperature. As temperature of storage increases, the rate of quality loss also increases. Thus, cold storage ($-1.1-0°C$ and relatively high humidity) sometimes is used to maintain quality for long periods of time. For general storage, temperatures of less than 10°C are recommended (Stadelman, 1986a, p. 67).

The increase in size of the air cell during storage is commercially important because it affects the appearance of eggs in candling. The air cell formed after the egg is laid continues to enlarge as water evaporates from the egg during storage. Moisture loss is retarded by storing the eggs at a relative humidity of 75–80% (Stadelman, 1986a, p. 69). Air cell size in itself is of less importance to the consumer even though it represents a loss of moisture, or a reduced yield and a changed appearance (Hale and Britton, 1974). The influence of air movement on weight loss was evaluated by Maurer (1975a) because refrigerators for home use have fans for air circulation and open containers for storage of eggs. Air speed did not influence weight loss. Eggs stored in open rather than closed containers lost weight to a greater but not statistically significant extent. Carton type had little influence on quality retention at refrigerator temperatures but was important at room temperature (Mellor *et al.*, 1975).

In addition to the loss of water through the shell, there is movement of water from the white to the yolk. The water content of yolk stored 50 days at 30°C increased from 49 to 50% (Vadehra and Nath, 1973). The transfer results in enlargement and decreased viscosity of the yolk, weakening of the vitelline membrane, and consequent flattening of the yolk when the egg is broken.

The thick egg white becomes thinner during storage. This is reflected in lowered albumen indices. Several theories have been proposed to explain the change. However, Powrie and Nakai (1986, pp. 115–116) implied that the mechanism described by Kato *et al.* (1979) seems most plausible. These investigators showed that *O*-glycosidically linked carbohydrate units were released as egg white became thinner. Release of the carbohydrate units causes dissociation of F-ovomucin, an important component in the gel structure of ovomucin.

The pH of the egg white rises with the loss of carbon dioxide from the egg from about 7.9–8.1 to about 9.1–9.3 in the first 5 days of storage. Changes in pH influence the functional properties of egg white. In contrast, yolk pH changes little from the 6.0 of a fresh yolk during storage.

Some deterioration in odor and flavor occurs during storage of eggs. Unpleasant flavors are absorbed by eggs if care is not taken to prevent odors in storage areas (MacLeod and Cave, 1977). In addition, characteristic stale odors and flavors develop in eggs during prolonged storage. If eggs are dipped in oil prior to storage, the off-flavors that develop may be more intense since volatile flavor components do not diffuse from the egg (Stadelman, 1986a, p. 70). The influence of storage temperature on flavor changes in products containing eggs of varying ages was studied by Dawson and co-workers (1956). Results from the study, which are illustrated in Fig. 7.3, suggest that changes occurring during storage at refrigerator temperatures are not rapid. Eggs evaluated when soft cooked maintained flavor equivalent to that of grade A eggs for 15 wk in cold storage, 13 wk in the refrigerator, or 1 wk at room temperature. As used for baked custards or angel food cake, however, eggs maintained quality equivalent to grade A for at least 24 wk in cold storage or a refrigerator. In cold storage, grade A quality can be maintained for as long as 6 months. Storage of eggs at room temperature may cause a decrease from grade A to B in 1 week. The importance of yolk changes in flavor deterioration is evident in Fig. 7.3. Scores for

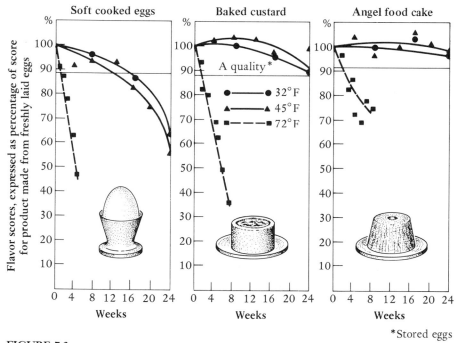

FIGURE 7.3

Flavor of products made from eggs held at three temperatures for varying lengths of time. (From Dawson et al., 1956.)

angel food cakes remained higher than scores for baked custards and soft-cooked eggs.

In addition to the inevitable changes that occur during the aging of eggs, microbial spoilage sometimes takes place. When the egg is laid, its contents are usually sterile. As the egg cools, however, microorganisms may be drawn in through the porous shell. The chances for microbial invasion are increased if the shell is dirty and is washed, even if a bacteriocide is present in the water (Board and Tranter, 1986, p. 92). Even after the microorganisms penetrate the shell, they encounter the natural defenses of the egg, which include the shell membrane, an alkaline pH, the bacteria-dissolving protein lysozyme, and the iron- and biotin-binding proteins, conalbumin and avidin. Usually heavy contamination overcomes the defensive mechanisms of the egg and causes spoilage during storage.

III. FUNCTIONALITY OF EGGS AND THEIR ROLES IN PRODUCTS

Eggs are used in many food products because of their functional properties, coagulation, emulsification, and foaming. They also may add color and flavor to products. Eggs may serve more than one function in a product and are therefore difficult to replace.

A. Coagulation

As with other high-protein foods, the denaturation and coagulation of egg proteins are important to the functional performance of eggs in a food system. Denaturation is a rearrangement of the orderly alignment of the molecules in a native protein. Peptide chains of a protein usually are coiled into a globular compact shape. This arrangement is stabilized with weak bonds between amino acid residues along the peptide chains. During denaturation, some of the forces that hold the chains in coiled and folded positions are disrupted, allowing the peptide chains to uncoil and unfold. New bonds are formed to give a new arrangement to the protein molecules. Ma and Holme (1982) showed that gel formation involves hydrophobic interactions and suggested that interchange of sulfhydryl-disulfide linkages also may contribute. The solubility of a protein is decreased by denaturation.

Coagulation is a term used to describe that entire process that results in a loss of solubility or a change from a fluid to a more solid state. The term gelation, meaning the formation of a gel, also is used to designate the loss of fluidity of egg white and yolk. The coagulation of eggs is responsible for the thickening effect that eggs have in products such as custards. Egg white begins to thicken as the temperature reaches 62°C, and at 65°C it will not flow. At 70°C the mass is fairly firm. Yolk coagulates at a somewhat higher temperature than white. It begins to thicken at 65°C and loses its fluidity at about 70°C. Coagulation does not occur instantaneously but gradually over a period of time. The reaction proceeds more rapidly as the temperature of heating is increased. Since heat is absorbed during coagulation of egg proteins, the reaction is endothermic. The firmness of an egg white coagulum or whole egg coagulum is dependent on the time and temperature of heating. Albumen coagulum increased in toughness as temperature of heating increased from 77 to 90°C and as time of heating increased from 7 to 60 min (Beveridge *et al.*, 1980), whereas whole egg coagulum toughened as much at 80°C as at higher temperatures (Beveridge and Ko, 1984).

Using scanning electron microscopy, Woodward and Cotterill (1986, 1987) showed the changes that occurred in the microstructure of heated albumen and yolk. The microstructure of albumen gels heated at various temperatures for varying periods of time is shown in Fig. 7.4. Increased cross-linking of the structure occurred as the gel was heated at 80°C for increasing lengths of time (A, C, and D). Heating at higher temperatures of 85 and 90°C for 30 and 50 min, respectively, resulted in a similar structure to that of the gels heated at 80°C for 15 min, even though the gels heated at the higher temperatures were harder than those heated at 80°C. Woodward and Cotterill (1986) indicated that these micrographs illustrate the transition from the sol state (soft) to the gel state (harder and cohesive) which results as the protein ovalbumin is denatured. It is important to note that the hardness of the gel formed was dependent on the pH of the albumen and the concentration of protein. As pH and concentration increased, gel hardness also increased.

Woodward and Cotterill (1987) also studied the microstructure of egg yolk. The microstructures of intact heated egg yolk and egg yolk stirred before gelation were both studied and are illustrated in Fig. 7.5 (A and B). A granular appearance was obvious for the intact yolk and reflects the mealy, crumbly texture associated with the yolk of a hard-cooked egg. In contrast, the structure of the stirred yolk was described as a protein matrix with spheres embedded. This structure obviously is responsible for the more cohesive, rubbery texture of yolk cooked after beating, which was shown with Instron measurements of texture. Hardness of yolk increased with heating at 85–90°C for up to 30 min. Instron measurements indicated that cross-linking was limited in yolk gels as all samples fractured when compressed.

1. Hard-Cooked Eggs

The length of time required for coagulation of eggs cooked in the shell depends on the heating temperature. When eggs were cooked at unusually low temperatures (85°C for 30 min) the yolk was not firm enough for slicing and the white did not coagulate firmly after 90 min at 72°C (Andross, 1940). The temperature at which eggs are cooked in the shell is varied by starting the eggs in cold water and bringing to a boil, or by starting them in boiling water and turning off the heat. In either method, the proportion of water to eggs and the size of the pan must be controlled for uniform results. Irmiter and co-workers (1970) compared eggs cooked by starting in cold water with eggs placed in boiling water and simmered for 18 min. Both methods produced satisfactory products, but the boiling water start was preferred because there were fewer cracked eggs and the eggs were easier to peel. In contrast, Hale and Holleran (1978) reported more cracks when a boiling water start was used. In addition, drilling or piercing the shell before cooking increased cracking during cooking.

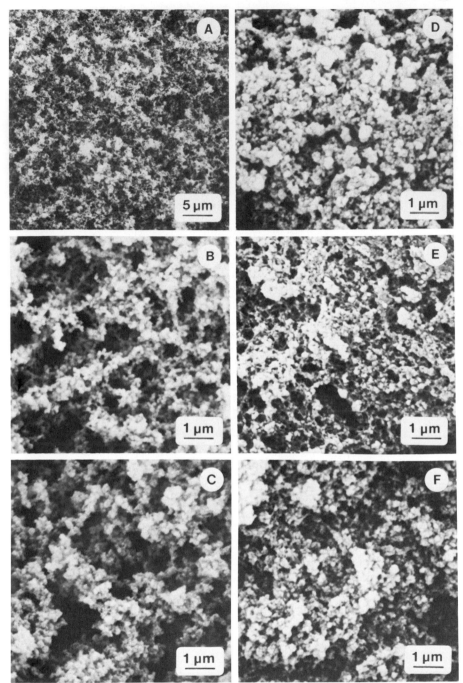

FIGURE 7.4
Microstructure of egg white gels (pH 9.0) prepared at various temperatures and times of heating.
(A) Heated 10 min at 80°C; (B) higher magnification of (A); (C) heated 12 min at 80°C; (D) heated
15 min at 80°C; (E) heated 30 min at 85°C; (F) heated 50 min at 90°C. (From Woodward and
Cotterill, 1986; Reprinted from *Journal of Food Science.* **51**, No. 2, pp. 333–339. Copyright © 1986
by Institute of Food Technologists.)

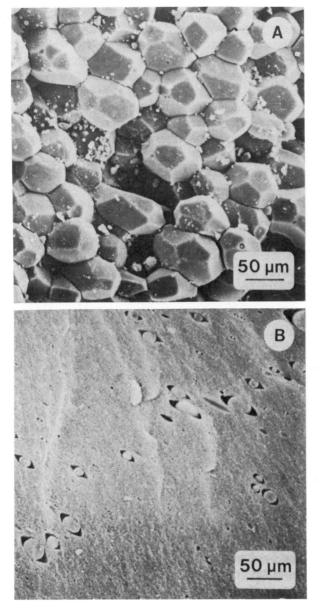

FIGURE 7.5
Microstructure of (A) shell-cooked egg yolk and (B) stirred cooked yolk. (From Woodward and Cotterill, 1987; Reprinted from *Journal of Food Science.* **52**, No. 1, pp. 63–67. Copyright © 1987 by Institute of Food Technologists.)

Difficulty in peeling is one problem associated with hard-cooked eggs. Eggs cooked after 7–10 days of storage peel easily with minimum loss of egg white and surface damage (Hale and Britton, 1974; Spencer and Tryhnew, 1973). Ease of peeling is related to the rise in pH that occurs with aging (Reinke *et al.*, 1973). Undesirable changes are associated with the change in pH. Therefore, other methods of increasing ease of peeling fresh eggs have been sought. Hale and Britton (1974) reported that eggs cooked within 24 hr after laying were easier to peel if they were cooled rapidly in liquid nitrogen (30 sec), ice water (1 min), or cool air (25°C for 10 min) and reheated immediately in boiling water for 10 sec.

The aging of the egg influences the appearance of hard-cooked eggs in another way. Yolk position is important in food service operations. The yolks of eggs cooked when only 4 days old were more ideally positioned than the yolks of eggs cooked after 11 days in an upright position (Grunden *et al.*, 1975). Storage position of eggs influences centering of the yolk. Cardetti and co-workers (1979) reported that storing eggs on their sides was best, followed by small end up.

Sometimes the surface of the yolks of hard-cooked eggs becomes greenish as ferrous sulfide is formed during cooking. The iron for the compound comes from the yolk. As the pH of the yolk increased, more ferrous sulfide was formed (Baker *et al.*, 1967). Formation of ferrous sulfide thus is more likely to occur in older eggs. As with other chemical reactions, the formation of the green color is favored by heat, whether due to overcooking or to slow cooling. Extreme discoloration of the yolk was seen in eggs boiled for 20 min and allowed to cool in the cooking water (Maurer, 1975b). These results suggest that eggs should be cooked for the minimum length of time and cooled quickly with cold water.

To hard cook large eggs, the American Egg Board (Anon., 1987) recommends covering a single layer of eggs with tap water, bringing it to a boil, turning off the heat, allowing the eggs to sit for 15–17 min, and cooling in cold water. For larger or smaller sized eggs the time is adjusted by approximately 3 min. Sheldon and Kimsey (1985) compared the effects of the standard method of simmering (20 min), steaming, and boiling (20 min) on the properties of hard-cooked eggs. Boiled eggs were firmer than those cooked by the other two methods, which may reflect the higher end-point temperature of the boiled eggs. Steamed and standard method eggs had weaker egg odor and off-flavors, respectively, than did boiled eggs. The eggs that were steamed had fewer sulfur volatiles than the eggs cooked by the two water methods, resulting in a difference in egg aroma intensity.

Hard-cooked, peeled eggs stored in a preservative solution are readily available commercially. Production of these eggs is based on the same principles as those for preparation of small quantities of hard-cooked eggs. Ease of peeling is important in quantity production because appearance after peeling determines use. Eggs that have less than perfect appearance (salad egg grade) are sold for use in products such as salad, which allows for chopping of the eggs, and No. 1 grade hard-cooked eggs are used for situations in which appearance is important (sliced, wedges, whole, or pickled) (Stadelman, 1986b, p. 394). Stadelman and Rhorer (1984) studied production of hard-cooked eggs and found that to obtain eggs with albumen pH of 8.8 or higher, which facilitates peeling, eggs can be held in circulating air at 15°C for 4 days. Britton and Fletcher (1987) reported that holding eggs in a sealed container that also contained a solution of NaOH resulted in an average albumen pH of 8.49 compared with the pH of 8.11 for eggs stored in an air environment. The eggs with the higher pH were then easier to peel as reported by other investigators. Stadelman and Rhorer (1984) also indicated that rotation of the eggs while heating at 85°C for 25 min minimized shell cracking and yolk discoloration and maximized yolk centering. Hard-cooked eggs usually are preserved in a solution of organic acid with 0.1% mold inhibitor added (sodium benzoate or potassium sorbate). The formula of the preservative solution will influence the flavor and tenderness of the egg. Addition of salt (Fischer and Fletcher, 1983), a tripolyphosphate salt with NaCl (Stadelman and Rhorer, 1984), and lactose (Stadelman and Rhorer, 1984) improves flavor. As organic acid level was reduced, tenderness of the white increased (Stadelman and Rhorer, 1984). Type of organic acid may influence the quality of the egg. In a comparison of 0.5% levels of citric, malic, and fumaric acids, Sheldon (1986) found that albumen of eggs stored in malic acid for only 24 hr had a typical fresh egg flavor. However, with storage the flavor of the egg declined and differences with respect to the kind of acid used for preservation were minimized. Eggs held in organic acids must be stored at refrigerator temperatures, unless thermally processed. Thermal processing of hard-cooked eggs in a retortable container allows shelf storage for more than 5 months (Stadelman, 1986b).

2. Poached and Fried Eggs

Poached eggs are prepared by dropping broken out eggs into boiling water. If the white is thin, it may be difficult to obtain a satisfactory product with any method. The addition of vinegar or salt to the cooking water hastens coagulation, and hence may improve the shape of poached eggs when they have thin whites (Andross, 1940). The shape of a fried egg also depends on the viscosity of the white and on the temperature at which it is cooked. If the temperature of the cooking utensil is too low (115°C), the egg will spread excessively, and if it is too high, (146°C), the egg will be overcooked. Temperatures of 126 to 137°C are satisfactory (Andross, 1940).

3. Scrambled Eggs

The amount of liquid added to the egg determines whether a small, firm mass or a larger, soft mass will be formed. Andross (1940) studied the coagulation and curdling temperatures of scrambled eggs made with 10–70 ml of milk per egg. The greatest difference between the coagulation and curdling temperatures occurred when 20 ml of milk was added per egg. Overcooking results in a firm, rubbery mass if little or no liquid is added, or in a curdlike mass that may separate during cooking if much liquid is used.

For food service use, frozen scrambled egg mixes are available. Formulation of these products is discussed in the section on processed eggs. Reheating of cooked scrambled eggs from one of three mixes may be desirable in ready-prepared institutional food systems. Microwave reheating is of interest in such systems for saving of energy. Cremer and Chipley (1980) showed that color, texture, and flavor of microwave-reheated 100-g portions of scrambled eggs were scored higher than were those of conventionally reheated scrambled eggs. These differences were attributed to the longer heating time required and browning that may occur in the conventional oven during reheating.

4. Stirred and Baked Custards

Coagulation of the protein of egg is responsible for the thickening of a stirred custard and the gel structure of a baked custard. One egg (50 g) is sufficient to thicken 250 ml of milk containing 25 g of sugar for either type of custard. Small amounts of salt and vanilla also are added. The effects of different types of milk on the quality and preparation of custards are discussed in Chapter 8.

Success in making custards depends on avoiding the application of excessive heat. Excessive heat results in overcoagulation and syneresis of the protein, characterized by curdling in stirred custards and porosity in baked custards. Stirred custards are likely to curdle if they are cooked above boiling water in a double boiler. Andross (1940) found that custards cooked over rapidly boiling water thickened at 87°C and curdled at 90°C, leaving a narrow margin for safety. However, when water in the bottom of the double boiler was cold when cooking started and was brought to a boil slowly, the custard thickened at 82°C and curdled at 87°C. Thus, the cold-water start produced custard that thickened at a lower temperature and that could be cooked with a greater margin of safety. A similar result can be achieved by keeping the water at a temperature just below the boiling point for the entire heating period.

Scalding the milk used in custards shortens cooking time and also may improve the flavor and texture of the product. Baked custards made from milk preheated to 90°C were better and more uniform in appearance than custards made from milk heated to 50°C (Jordan et al., 1954). Even when baked at relatively low oven temperatures, custards are improved by being placed in a pan of hot water for protection from oven heat. An oven temperature of 177°C was preferred to 163 or 204°C because the baking period was unduly long at 163°C and so short at 204°C that the custard could be easily overcooked. In addition, custards baked at 204°C were less desirable in appearance than those baked at the other two temperatures (Jordan et al., 1954).

Coagulation temperature depends on a number of factors. The temperature can be studied by suspending a thermometer or thermocouple in the mixture during baking. The temperature inside the custard rises rapidly during the early portion of the baking period. Since the coagulation process is an endothermic reaction, the heating curve will become level as gelation starts (Palmer, 1972, p. 537). As with stirred custards, a slower rate of heating will lower the end-point temperature. When coagulation is complete, the temperature rises again and a knife inserted in the custard comes out "clean" (Longree et al., 1961). Depending on the formula and the rate of heating, the end-point temperature for baked custards is about 86–92°C. As the end-point temperature within the range of 86–92°C increased, the custards became firmer and showed less syneresis on standing (Miller et al., 1959). Increasing the proportion of egg lowers the coagulation temperature and increases the firmness of the custard. Because egg white coagulates at a lower temperature than yolk, baked custards made with whites become firm at a lower temperature than those made with yolks.

Sucrose influences baking time or end-point temperature and quality of baked custards. The addition of 10% sucrose increased heat stability of proteins (Woodward and Cotterill, 1983). This principle is reflected in several studies. Seideman and co-workers (1963) found that sucrose raised the temperature necessary for coagulation of the egg white, especially at a pH above 8.5. Wang and co-workers (1974) studied the effect of sucrose on the quality of custards baked to a temperature in the range of 84–86°C. Increasing sucrose levels increased baking times and crust tenderness. However, these characteristics were shown to depend more on percentage of protein than on the presence of sucrose.

Salts are necessary for gelation of an egg custard mixture. Substitution of deionized water for milk prevents gelation. Adding salts prior to heating will result in normal gel formation. If salts are added after heating, curdled custard will be formed. As the valence of the cations present increases, the amount of salt required is decreased (Lowe, 1955). For example, 1.2, 0.3, and 0.09 g of NaCl, $CaCl_2$, and $FeCl_3$ were required for maximum gelation of a mixture of 48 g egg, 244 g milk, and 25 g sugar. Milk supplies the necessary salts in custards made without NaCl.

B. Emulsification

Egg yolk, which is itself an emulsion, is a good emulsifying agent for fats or oils and water. It performs this function in many foods, including shortened cakes, salad dressing, and mayonnaise. The emulsifying qualities of eggs often are studied in mayonnaise because it is a relatively simple system. The role of emulsifiers in stabilization of emulsions is discussed in Chapter 15. Commercial production of mayonnaise was discussed by Potter (1986, pp. 461–462).

In mayonnaise, particles of yolk interact at the surface of oil droplets to form a layer (Chang *et al.*, 1972) that protects the emulsion by inhibiting coalescence of the oil droplets. Participants in the layer include both low- and high-density lipoproteins (Powrie and Nakai, 1985, p. 848).

C. Foaming

The foaming of egg whites is important in many foods, because it makes the products light in texture and contributes leavening. Egg white foam is a colloidal suspension consisting of bubbles surrounded by albumen that has undergone some denaturation at the liquid–air interfaces. This denaturation, which is caused by the drying and stretching during beating, makes some of the albumen insoluble, thus stiffening and stabilizing the foam.

Johnson and Zabik (1981a) described the functions of various albumen proteins in foam formation for an angel food cake system. Six protein fractions were isolated and whipped individually and in combinations. Cakes then were baked to study the performance in the food system. The globulins exhibited greater foaming ability than did ovomucin, lysozyme, conalbumin, ovomucoid, and ovalbumin. As foams were formed, the number of sulfhydryl groups decreased, reflecting sulfhydryl-disulfide interchanges during foaming. Volumes

of cakes ranging from largest to smallest were produced from globulins, ovalbumin, control (a mixture of all proteins), conalbumin, lysozyme, ovomucoid,
and ovomucin. Lack of aeration or instability of foams was responsible for low
volumes with the last four. When combinations of proteins were studied, it was
shown that the inclusion of ovomucin increased viscosity and contributed to
foam formation but depressed cake volume. The addition of lysozyme with the
ovomucin decreased foaming but increased cake volume, suggesting interaction
between ovomucin and lysozyme. The interactions among the proteins of egg
albumen were shown to be complex. The ultrastructure of foams from these
protein fractions also was studied and reported by Johnson and Zabik (1981b).

During denaturation the protein molecules unfold, and their polypeptide
chains lie with their long axes parallel to the surface. Overbeating incorporates
too much air, resulting in extensive denaturation of ovomucin so that the protein films become thin and less elastic. Elasticity is needed, especially in foams
that are to be baked, so that the incorporated air can expand without breaking
the cell walls before the ovalbumin is coagulated by the heat of the oven.

1. Procedures for Evaluation of Egg White Foams

A decision as to when to stop beating a foam is made on the basis of the appearance of the foam in the bowl when the beater is withdrawn and when the bowl is
tilted. The characteristics of egg whites beaten to various stages are described in
Table II. Whip time may be defined as the time required to reach one of these
stages (Seideman *et al.*, 1963; Baldwin *et al.*, 1967).

Foam volume and foam stability also may be measured. The volume of the

TABLE II
Characteristics and Uses of Egg Whites Beaten to Various Stages[a]

Stage	Description	Uses
I. Slightly beaten (foamy)	Frothy, or slightly foamy; large air bubbles; transparent; flows easily	Clarifying, "coating," emulsi-fying, thickening
II. Stiff foam (wet peak)	Frothy quality disappearing; air cells smaller; whiter; flows if bowl is tipped; very shiny, glossy, and moist in appear-ance; if left to stand, liquid separates out readily; when beater is withdrawn, white follows, forming rounded peaks	Soft meringues; angel cake, if made by meringue method
III. Stiff	No longer foamy; air cells very small and very white; may slip slightly if bowl is tipped; still glossy, smooth, and moist in appearance	Cakes, omelets, souffles, sherbets, marshmallows, hard meringues, and ice cream
IV. Dry	Very white but dull, small flakes or curds beginning to show; rigid and almost brittle; particles will be thrown off espe-cially by a whisk beater; if left to stand, liquid separates at bottom slowly	Shirred eggs

[a] From Niles (1937).

foam usually is not measured directly because of the difficulty of packing it into a measuring device. A rough comparative measure is made by inserting a ruler into a foam that is still in the bowl in which it was beaten. Alternatively the volume of the bowl to the depth of the foam may be calculated (Baldwin *et al.*, 1968). Specific gravity is an indirect measure of foam volume. Directions for this measurement are given in Appendix D. The greater the total volume of the foam, the less will be its weight per unit volume and the lower its specific gravity. Measurement of foam stability also is described in Appendix D. Obviously, the volume of a product made from a foam may also be an indicator of foam quality. Volume measurements are discussed in Chapter 3.

2. Factors Affecting Foaming

The time required for foam formation, the foam volume, and the foam stability are affected by many factors. These factors include method, time, and temperature of beating; characteristics of the egg white; pH; and the presence of other substances such as water, fat, salt, sucrose, egg yolk, and additives.

a. Method, Time, and Temperature of Beating.   Bailey (1935) found that the volume of foams obtained with a wire beater moving with a hypocycloidal action in a stationary bowl was greater than that obtained with stationary twin beaters and a moving bowl. The reason for these differences may be that the stationary twin beaters produce a stiff foam that becomes immobilized around the beaters. As a result, the blades rotate in a central cavity. Hand beaters proved to be less effective than electric mixers in foam formation (Barmore, 1934). As beating time increases, volume and stability of the foam first increases, then decreases. The time required depends on the type and speed of the beater. Time is shorter for higher speeds. Maximum stability is reached before maximum volume, however. With prolonged beating, the bubbles become smaller, but more unstable, and break on standing to form a coarse foam (Barmore, 1934). Thus, increased volume of egg white foam is often accompanied by decreased stability (McKellar and Stadelman, 1955). Egg white that is at room temperature can be beaten more readily than that at refrigerator temperature, possibly because of lowered surface tension at the higher temperature.

b. Characteristics of the Egg White.   When the thick and the thin white are separated, the thin white can be beaten more readily than the thick white. Thin white produced greater volume than thick when beaten with either a hand beater (St. John and Flor, 1930–1931) or an electric mixer having twin beaters (Henry and Barbour, 1933). When a mixer with hypocycloidal action was used, the foam from the thick white had greater volume and greater stability than the foam from the thin white (Bailey, 1935). Foam volume was less for high-lysozyme egg white than for egg white containing less lysozyme. The negative correlation between foam volume and lysozyme content was evident for eggs up to 4 wk old and could have been due to inadequate whip time for the thick egg white of high-lysozyme egg white (Sauter and Montoure, 1972). Barmore (1934) reported that if thick white was beaten long enough, it produced angel food cakes of the same quality as did the thin white. Because thick white becomes thinner when eggs are stored, white from stored eggs may foam more quickly than white from fresh eggs.

c. pH.   The pH of the egg white is important to foam formation. Cream of tartar frequently is added to lower the pH because it is more effective in increasing foam stability than either acetic acid or citric acid. Lowering the pH changes the protein concentration at the liquid–foam interface and makes the foam more stable to heat. As foam stability is increased, shrinkage of angel food cake is decreased. The increased stability makes it possible for heat to penetrate the product and coagulate the egg white protein before the air cells collapse (Baldwin, 1986, pp. 361–362). Excessive lowering of the pH also results in foam instability.

The whipping properties of duck egg whites are improved by the addition of lemon juice (Rhodes *et al.*, 1960). The improvement is attributable to a stabilizing effect on the ovomucin, which reduces the time required for whipping. Acidified duck egg whites produce angel food cakes as acceptable as those produced with chicken egg whites.

d. Water and Fat.   The finding that water added to egg white in amounts up to 40% by volume increases the volume of the foam but decreases its stability was reported by Henry and Barbour (1933). The same investigators reported that the addition of cottonseed oil reduced foam volume but did not affect stability unless the amount of oil was 0.5% or greater.

e. Sodium Chloride.   The addition of salt to egg white or whole eggs before beating reduces the stability of the foam. Amounts of salt up to 1.5 g/66 g egg decreased volume and stability and increased whipping time for reconstituted dried egg whites (Sechler *et al.*, 1959). The addition of salt to fresh egg whites also lessened the stability of the foam unless beating time was increased from 6 to 9 min (Hanning, 1945). With whole eggs, the addition of salt before beating resulted in a foam of small volume that would not form peaks. Sponge cakes made from such a foam were smaller in volume and less tender than when salt was omitted or was sifted with the flour (Briant and Willman, 1956).

f. Sucrose.   Sugar added during beating of egg whites retards the denaturation of the egg white proteins. Therefore, the addition of 50% sugar more than doubled the time required for foam formation and produced a foam that was less stiff, more plastic, and more stable than one containing no sucrose (Hanning, 1945). In a study of methods of adding sugar to egg white, it was found that soft meringues were better if the whites were beaten until they held a soft peak than if they were either not beaten at all or beaten to the stiff-but-not-dry stage before the addition of sugar (Gillis and Fitch, 1956).

g. Egg Yolk.   The presence of egg yolk decreases the volume of egg white foam. Cunningham and Cotterill (1972) postulated that the lipid fraction of the yolk prevents the normal functioning of the globular protein in foam formation. Lipovitellin did not reduce the volume of cakes prepared with lysozyme-free egg white, whereas it did reduce the volume of cakes made with normal egg white. This finding suggests the formation of a complex that reduces the volume. The problem of yolk contamination is significant in the commercial processing of eggs because it is impossible to avoid. Therefore, several techniques for improving functionality of egg whites contaminated with yolk have been proposed. The addition of 2% freeze-dried egg white to whites used in the preparation of angel food cake maintained volume and overall acceptability. The freeze-dried egg white probably supplied enough ovoglobulin, ovomucin, and/or lysozyme–ovomucin complex to replace the proteins that were bound to the yolk lipoprotein (Sauter and Montoure, 1975). The complex formed between egg white and a yolk fraction may dissociate with heat as heating improved the foaming properties of yolk-containing egg whites (Cunningham and Cotterill, 1964; Baldwin *et al.*, 1967). The effect of heat was not evident when heat was applied to spray-dried yolk containing egg whites (Cotterill *et al.*, 1965).

b. Additives.   The effects of surfactants and stabilizers on foaming properties of egg whites have been investigated. Baldwin *et al.* (1968) investigated the use of 0.6% guar gum, 0.6% algin, and 0.3% sodium lauryl sulfate (SLS) plus 0.3% triethyl citrate (TEC) on meringues baked conventionally and by microwaves. Guar gum and SLS plus TEC were more useful than algin in improving meringue quality. All additives reduced whipping time and had varying positive effects on foam volume. However, meringues with alginate had a gummy texture. Further discussion on additives is in the section on dehydrated eggs.

3. Products Depending on the Foaming Properties of Egg White

a. Meringues.   Meringues are made by beating egg whites and then baking the foam. Greater leakage was exhibited by meringues made with 31.25 g of sugar per egg white than by meringues made with 25 g of sugar, but the superior appearance and greater ease of cutting meringues made with the higher amount of sugar more than offset the undesirable increase in leakage (Gillis and Fitch, 1956). Leakage, or the liquid that collects on the filling under the meringue on the pie, was reduced by spreading meringue on hot rather than cold fillings (Hester and Personius, 1949). In addition, spreading meringue on hot filling results in more even heat distribution, reducing the chance of survival of *Salmonella*. A wide range of oven temperatures (163–218°C) can be used if baking times are adjusted so that meringues of similar color are obtained for each temperature. As baking temperature increased from 163 to 191 to 218°C, the internal temperature reached in equally browned meringues decreased. Leakage was less at the lower temperature, but there was less stickiness and greater tenderness at higher temperatures (Hester and Personius, 1949).

Upchurch and Baldwin (1968) reported that the addition of guar gum to fresh egg whites reduced leakage in freshly prepared meringues. Similar results were reported by Morgan and co-workers (1970) for spray-dried egg white meringues with carrageenan. There was no difference in leakage between pies made with and without a stabilizer after holding at refrigerator temperatures for 20–22 hr. Stored meringues containing carrageenan were more tender than those that did not contain the stabilizer.

Meringues heated by microwaves are more compact than meringues heated conventionally. Browning does not occur, which may result in a more delicate flavor than in conventionally heated meringues (Baldwin *et al.*, 1968).

Hard meringues usually contain 50 g of sugar per egg white or twice as much as soft meringues. The sugar contributes to the sweet flavor. A low oven temperature (100–120°C) is used for hard meringues. Baking times are 1–2 hr followed by holding in the oven after it is turned off to complete the drying process.

b. Angel Food Cakes.

b. Angel Food Cakes.   The foaming power of egg white is important in making angel food cakes. Therefore, the factors affecting foaming, discussed previously, are significant. Underbeating or incorporation of too little air to stretch the cell membranes to their potential capacity during baking results in a small cake with thick cell walls. However, overmixing results in low volume due to the rupture of the overextended cell walls during the baking process (Palmer, 1972, p. 540). The egg whites can be beaten to a soft peak and sugar beaten in, or they can be beaten until stiff but not dry and the sugar folded into the foam. A small amount of sugar usually is sifted with the flour to facilitate blending. The flour is folded gently into the foam to avoid loss of air from the beaten egg whites.

The proportion and type of ingredients in angel cakes are important factors in quality. If the amount of flour is increased, the cake becomes tougher and drier, whereas additional amounts of sugar make the cake more tender until, with an excessive amount, it falls. An excessive amount of sugar prevents adequate coagulation of the egg white protein. Palmer (1972, p. 542) suggested that 0.2–0.4 g of flour and 1–1.25 g of sugar can be used per gram of egg white. The larger amount of sugar is used with the larger amount of flour. Sugar must be decreased in cakes baked at higher altitudes (Lorenz, 1974). Cake flour is more satisfactory for angel cakes than is all-purpose flour because all-purpose flour contributes more gluten than cake flour and, therefore, contributes to shrinkage of the structure during baking and cooling (Palmer, 1972, p. 542).

Cream of tartar also is added to egg whites for angel food cake to make the cake whiter, finer in grain, and more tender than it would be without the acid. This effect can be attributed to the effect of the cream of tartar on the pH of the cake because similar results were obtained with the substitution of citric, malic, or tartaric acid if the pH of the cake was the same (Grewe and Child, 1930). The acid probably stabilizes the foam by shifting the pH to the isoelectric point of the globulins, thus coagulating some of the protein that surrounds the air cells of the foam. Stabilization prevents the foam from collapsing to produce thick cell walls before the temperature of coagulation is reached in the oven (Barmore, 1934). Several explanations for increased whiteness with the use of cream of tartar have been proposed. Palmer (1972, p. 542) suggested that it is effective because flavone pigments contributed by the flour are colorless at neutral or acidic pH but yellow at a higher pH. An angel food cake made with cream of tartar may appear to be whiter than one made without because of a difference in the character of light reflected (Francis and Clydesdale, 1975, p. 178) from the finer grained cake made with the cream of tartar than from a coarser grained cake.

Maximum temperature attained in cakes prepared from the same formula does not vary with oven temperature if baking times are adjusted so that the same crust color is obtained in all cakes (Barmore, 1934). When angel food cakes were baked at 177, 191, 204, 218, and 232°C, the volume and tenderness increased up to 218°C. At 232°C, cakes were smaller and very moist. High palatability scores for cakes baked at 204 and 218°C suggested that one of these oven temperatures should be used (Miller and Vail, 1943).

c. Sponge Cakes.   Sponge cakes can be made from the yolks and whites beaten separately, from whole eggs, or from yolks only. A simplified method in which unseparated eggs are used was published (Briant and Willman, 1956). The eggs and lemon juice are beaten in an electric mixer until soft peaks are formed and the sugar is added while beating continues. The flour and salt are added with gentle mixing. Cakes made by this method compared favorably to those made by the usual method of using separated eggs. The success of the method depends on thorough beating of the eggs. Egg yolk serves as a stabilizer of the fragile egg white foam produced from egg whites in sponge cake production (Graham and Kamat, 1977).

IV. PROCESSED EGGS AND THEIR PERFORMANCE IN FOOD SYSTEMS

Processed eggs are marketed as refrigerated, dried, and frozen eggs and egg products for use in the food industry and in quantity food production. Research has led to the development of new processing techniques and the adoption of regulations to ensure the safety of processed egg products. The Food and Drug Administration (FDA) is responsible for regulating the processing of eggs. To ensure destruction of potentially pathogenic microorganisms, particularly *Salmonella*, all liquid, frozen, and dried whole eggs, yolks, and whites must be pasteurized or otherwise treated (Anon., 1975). Pasteurization temperatures range from 56 to 63°C, depending on the part(s) of the egg and additive(s) present. For example, *Salmonella* are reduced in white (pH 9.0) with a less severe heat treatment than is required for whole egg (pH 7.6). Resistance to heat is even greater in yolk with its high solids content (Cunningham, 1986, p. 249).

Pasteurized eggs have a shelf life of up to 14 days at refrigerator temperatures. Limited shelf life is attributable to the fact that although pasteurization is adequate to destroy *Salmonella*, microorganisms that can cause spoilage will survive the treatment. Therefore, if eggs are not to be used within 14 days, further treatment may be necessary. Ball *et al.* (1987) suggested that eggs be ultrapasteurized and aseptically packaged to prolong refrigerator storage to as long as 3–6 months. Processes tested involved heat treatments ranging from 63.7 to 72.2°C for 2.7 to 192.2 sec. Ball and co-workers concluded that the functional performance of the eggs as measured in custards and sponge cakes was not

affected by the processing. This approach may be used in the future for the preservation of eggs. Currently to extend shelf life eggs may be dehydrated, frozen,
incorporated into egg substitute products, or frozen as prepared products.

A. Dehydrated Eggs

Although dried egg products usually are not available to consumers at the retail
level, they are widely used in mixes and in quantity food production. Removal of
water lowers a_w and therefore retards chemical reactions that affect quality and
inhibits the growth of microorganisms. Spray drying is the most commonly used
method for drying whole eggs, egg yolks, or egg whites. The eggs are preheated
to 60°C, then sprayed into a drying chamber through which air between 121 and
149°C is passing. The powder separates from the air and is continuously removed from the drying chamber. Prior to drying, egg white usually is subjected
to a controlled bacterial fermentation process (Hill and Sebring, 1986, pp. 276–
277). Alternative systems for removal of glucose to prevent carbonyl-amine
browning during storage of the dried product are addition of glucose oxidase
and fermentation with yeast. Whole eggs and yolk are treated with a glucose
oxidase–catalase system for glucose removal (Hill and Sebring, 1986, p. 281). A
modification of the spray-dried process, foam-spray drying, involves the incorporation of carbon dioxide, air, or nitrogen into the liquid egg before it is
sprayed into the drying chamber. Freeze drying is used to remove water from
frozen eggs in a vacuum chamber. Only a few freeze-dried egg products, primarily scrambled egg mixes, are available on the retail market because the process is very expensive.

   An alternative technique used for whites is pan drying. A flakelike product
(12–16% moisture) results when whites are dried on pans in an air oven or
water jacket-type dryer (Bergquist, 1986, p. 315). Particles, flakes, granules, or
milled powder forms of pan-dried whites are available. These products are used
primarily in confectioneries and in some pie and meringue formulas (Toney and
Bergquist, 1983). Instant dried egg white, a fine powder containing about 25%
sugar, is available and is more readily dispersed in water than other dried egg
products (Toney and Bergquist, 1983).

   The effects of drying on the functional properties of eggs have been of interest. Bergquist (1986, p. 299) suggested that the heat coagulation properties of
eggs are not affected by appropriate drying and subsequent storage. Scrambled
eggs from spray-dried, foam-spray-dried, and freeze-dried eggs did not differ
with respect to tenderness and syneresis (Janek and Downs, 1969). However,
foam-spray-dried scrambled eggs received lower flavor scores. Baked custards
from eggs dried by the same three methods also have been evaluated. A series of
studies on the use of these eggs in gels from yolk or albumen and in custards
suggested that drying may adversely affect the white. Gel strength and custard
strength were reduced in products containing, respectively, foam-spray-dried
egg white and whole egg, but were not reduced in gels containing the processed
yolks (Zabik, 1968; Wolfe and Zabik, 1968).

   The foaming properties of egg white may be altered by drying. Degree of alteration of whipping properties is related to pasteurization temperature and time, the addition of acids or whipping agents, and drying procedure and temperature (Bergquist, 1986, pp. 295–299). Two types of spray-dried egg whites, whipping and nonwhipping, are available. For angel food cakes and meringues, a product containing a whipping agent such as sodium lauryl sulfate may be appropriate. A nonwhipping product is suitable for layer cakes (Toney and Bergquist, 1983).

The foaming potential of yolks and whole eggs is reduced by drying (Joslin and Proctor, 1954) and it has been noted that the addition of whipping agents that are effective in whites is not effective in yolks and whole eggs. Toney and Bergquist (1983) noted that the addition of carbohydrate will minimize the loss of foaming and emulsification properties. Carbohydrates added to whole eggs include sucrose and corn syrup; corn syrup may be added to yolks. The carbohydrate also improves dispersibility of dried whole egg and yolk products. Improvements in foaming and foam stability were found by Kim and Setser (1982) when sodium lauryl sulfate and gums were added at higher levels.

Good to very good sensory ratings were given to angel food cakes made with foam-spray-dried, spray-dried, and freeze-dried egg whites. Spray-dried eggs produced the smallest volume and foam spray dried the largest (Franks et al., 1969). In the same study, cakes made from frozen egg whites were given the highest scores for flavor, suggesting that drying affects the flavor.

Although dried desugared egg white is quite stable during storage, dried whole egg and yolk deteriorate rapidly in flavor, color, solubility, and cooking quality when stored at room temperature. Refrigerator storage extends the shelf life to at least 1 year (Bergquist, 1986, p. 319).

B. Frozen Eggs

Freezing effectively preserves the quality of eggs. Whites may be frozen without any addition, but 10% sugar or 10% salt usually is added to yolks before freezing so that they will be free from gummy or lumpy particles when they thaw. Triethyl citrate may be added as a whipping agent (Bergquist, 1986, p. 307). The increased viscosity or gelation of frozen yolks after thawing most likely is due to aggregation of the lipoproteins of yolk to form a network capable of entrapping large quantities of water (Hasiak et al., 1972). The low-density lipoprotein (LDL) participates in the gelation process. Specifically, Kurisaki and co-workers (1980) suggested that gelation occurs when very low-density lipoprotein (VLDL) particles become disrupted from yolk particles. Gelation occurs when the disrupted VLDL particles aggregate to form a "mesh-type structure." Sugar or salt may be added to whole eggs before freezing to minimize the gelling effect. Wakamatu and co-workers (1983) pointed out that NaCl (1–10%) inhibits the gelation of a 40% LDL solution during storage at −20 and −25°C. At temperatures below −40°C concentrations of NaCl above 4% enhanced gelation. The mechanism of action for gelation inhibition was formation of a complex consisting of LDL–water–NaCl in which water is not completely frozen.

Whole eggs frozen with or without sugar or yolks frozen with sugar or corn syrup have been found entirely satisfactory for plain cakes and custards (Jordan *et al.*, 1952). Angel food cakes made from frozen whites compared favorably with cakes made from fresh whites except for a slightly smaller volume (Clinger *et al.*, 1951). In a study of the influence of pasteurization and the use of the additives sodium chloride and fructose on the gelation of egg yolk, the quality of mayonnaise and sponge cakes made from the yolks was evaluated (Jaax and Travnicek, 1968). Sodium chloride and fructose reduced gelation of the yolks. Fructose decreased the emulsification capacity of the yolk, whereas sodium chloride increased it. Sponge cakes made with yolks containing the additives were greater in volume and more easily compressed. Sensory evaluation suggested that sodium chloride was less satisfactory than fructose. Yang and Cotterill (1989) reported that the use of salted, frozen egg yolk in mayonnaise does reduce its stability. Similar effects were seen with yolks that were spray or freeze dried. The use of various additives to improve the quality of scrambled eggs made from frozen eggs has been evaluated (Ijichi *et al.*, 1970). Low levels of salt, sucrose, dextrose, or skim milk combined with homogenization prevented an increase in viscosity and eliminated the undesirable appearance associated with frozen eggs used for scrambling.

Frozen scrambled egg mixes may have added nonfat dry milk, vegetable oil, gums, water to reconstitute dry ingredients, organic acids, and egg white (Cotterill, 1986, pp. 226–237). To minimize color changes in scrambled eggs from frozen mixes, the pH of the whole frozen eggs may be adjusted from 7.6 to 6.8 (Toney and Bergquist, 1983). Heating of egg yolk to about 45°C after thawing partially reverses the gelation that occurs with freezing (Palmer *et al.*, 1970).

C. Egg Substitutes

Several commercial egg substitutes have been introduced to the market in response to recommendations for cholesterol-free diets. The chemical composition and nutritional properties of several of these products have been reported (Childs and Ostrander, 1976; Pratt, 1975; Navidi and Kummerow, 1974).

Several investigators have studied the quality of products made from selected egg substitutes. These products usually contain a high proportion of egg albumen but contain no yolk. Fat-soluble nutrients are provided by product fortification. Leutzinger *et al.* (1977) reported that experienced panelists found that scrambled eggs from a frozen egg substitute had more cooked egg flavor and aroma than scrambled whole eggs. Color and texture did not differ. Consumer panelists liked the flavor of the whole egg product better. Gardner *et al.* (1982) reported that flavor, tenderness, moistness, and overall acceptability of scrambled whole eggs were superior to that of scrambled frozen egg substitute.

Chen (1981) studied weeping in scrambled egg products. Experimental formulas for a whole-egg product (whole egg, water, nonfat dry milk solids, corn oil, and lactic acid) and a yolk-free product (egg white, water, nonfat dry milk solids, corn oil, monosodium glutamate, salt, and coloring) were tested. The yolk-free product exhibited more weeping than did the whole egg product. As cooking time was increased weeping decreased. Addition of carboxymethylcellulose (CMC) reduced weeping in both products. Chen's (1981) formula for the yolk-free product could be used as a basis for cholesterol-free egg products for special diets.

Cakes made with egg substitutes also have been studied (Gardner *et al.*, 1982; Leutzinger *et al.*, 1977). "Scratch" cakes containing substitutes were lower in volume and rated lower in flavor than those containing the whole eggs (Leutzinger *et al.*, 1977). In contrast, cakes from mixes had greater volumes when frozen substitutes were used in place of whole eggs (Gardner *et al.*, 1982). However, flavor and overall acceptability were higher for the whole egg cake.

Gardner *et al.* (1982) also reported that stirred custards were thicker and baked custards sagged less when made with frozen substitutes than with whole eggs. Results from the studies with cakes and custards reflect the fact that the thickeners, stabilizers, and emulsifiers added to the substitutes are effective contributors to the functionality of the substitutes in these products. Lowered flavor ratings reflect the importance of the yolk as a contributor to the flavor of products (Gardner *et al.*, 1982).

D. Cooked–Frozen Egg Products

Demand for convenience has resulted in development of cooked–frozen egg products that may or may not need heating after thawing. A problem associated with the freezing of cooked eggs is that the white becomes tough and rubbery. The problem may be solved with the addition of a water-binding additive such as carrageenan, algin, or starch. These ingredients are used in the production of the cooked egg roll or "long egg." A hollow cylinder of cooked egg white is formed and then filled with yolk. The roll then is cooked and frozen (Cotterill, 1986, pp. 223–224).

Syneresis is a major problem for products that are cooked, frozen, thawed, and reheated (CFTR). Rapid freezing may help to minimize the problem by limiting damage to the product by ice crystals. In a study on CFTR scrambled eggs Feiser and Cotterill (1982) showed that lowering pH of the liquid egg from 7.0 to 6.0 with citric acid increased serum that could be pressed from the product. Addition of salt decreased serum volumes (Feiser and Cotterill, 1983). O'Brien *et al.* (1982) demonstrated that cryogenic freezing and the addition of xanthan gum, CMC, and pregelatinized starches improved texture and water-holding capacity of frozen omelets. Thus, changes in formulation make possible the production of frozen-cooked egg products that will be acceptable upon thawing and reheating.

Suggested Exercises

1. Compare the quality of several products made from fresh, frozen, or dried eggs or egg substitutes. For a basic custard formula see Appendix A. Choose commercially available egg substitutes or use the formula given by Chen (1981) for a yolk-free scrambled egg product.

2. Suspend a thermometer in either a stirred or baked custard mixture. Cook beyond the stage of doneness, recording the temperature of the custard frequently during cooking. Graph temperature against time and mark the points at which coagulation and curdling are first observed. Is coagulation an endothermic reaction?

3. Calculate the amount of protein supplied by the whole egg in the basic formula for custards. Determine the amount of white and yolk required to give an equal amount of protein. Prepare custards using these quantities. Adjust the water and fat content of the custards so that they are equal. Use water and vegetable oil for the adjustments.

4. Vary the amount of egg and part of egg used in stirred or baked custards. Use the formula is Appendix A. For variations (1) increase the whole egg to 100 g, (2) substitute 34 g egg yolk, (3) 66 g egg white, and (4) 50 g frozen egg substitute for the whole egg in the basic formula. Custard made from the basic formula should serve as a control.

5. Study the role of salt in a custard. Follow basic procedures for preparation (Appendix A). Increase egg in the basic formula to 100 g. Omit the sugar, salt, and vanilla for the control and other variations. For variations (1) substitute deionized water for milk, (2) substitute 0.1 M NaCl for the milk, and (3) substitute 0.1 M CaCl$_2$ for the milk.

6. Investigate the influence of variation in cooking time and temperature on custard quality. For stirred custard make three times the basic formula (Appendix A). Divide the mixture equally (310 g/pan) among the tops of three identical double boilers. The water in the lower portion should just touch the upper portion of the double boiler. Cook one custard following the basic procedure. For the second, put cold water in the bottom of the double boiler and bring water to a boil slowly after putting the custard in the top of the double boiler. For the third have the water in the bottom of the double boiler boiling rapidly throughout the cooking period. For each variation, record the time required for the mixture to coat a spoon.

7. Compare the volume or specific gravity and stability of foams (Appendix D) made from fresh, frozen, and reconstituted dried egg white, allowing the water and dry albumen to stand for various lengths of time.

8. Study the effects of variation in beating time on foam stability. For the control, beat 50 g of egg white until stiff enough to hold a peak, but still shiny. Record the length of time required for beating. For the first variation reduce the beating time by 25% and for the second, beat until stiff and dry. Measure foam stability (Appendix D).

9. Study the effects of added ingredients on foam stability. For the control, beat 50 g of egg white until stiff enough to hold a peak, but still shiny. Record the length of time required for beating. Added ingredients could be (1) 1.56 g salt, (2) 0.38 g cream of tartar, (3) 0.76 g cream of tartar, (4) 12 drops egg yolk, (5) 30 ml water, (6) 50 g sugar added at the soft-peak stage, and (7) 30 g water plus 50 g sugar added at the soft-peak stage. Follow the procedure in Appendix D for measuring foam stability.

10. Study the effect of cream of tartar on the quality of angel food cake

(Appendix A). Omit the cream of tartar for one variation and double the amount of cream of tartar for a second.

References

Adams, C. F. 1975. Nutritive value of American foods. Agric. Handbook 456, Agricultural Research Service, USDA, Washington, DC.

Alam, A. U., Creger, C. R., and Couch, J. R. 1968. Petals of Aztec marigold, *Tagetes erecta*, as a source of pigment for avian species. *J. Food Sci.* **33**: 635–636.

AMS. 1983. Egg grading manual. Agric. Handbook 75, Agricultural Marketing Service, USDA, Washington, DC.

Andross, M. 1940. Effect of cooking on eggs. *Chem. and Ind.* **59**: 449–454.

Anon. 1975. Regulations governing the inspection of eggs and egg products. 7CFR, Part 59, June 30, USDA, Washington, D.C.

Anon. 1987. Eggcyclopedia. American Egg Board, Park Ridge, Illinois.

Bailey, M. I. 1935. Foaming of egg white. *Ind. Eng. Chem.* **27**: 973–976.

Baker, R. C. and Darfler, J. 1969. Discoloration of egg albumen in hard-cooked eggs. *Food Technol.* **23**: 77–79.

Baker, R. C. Darfler, J., and Lifshitz, A. 1967. Factors affecting the discoloration of hard-cooked egg yolks. *Poultry Sci.* **46**: 664–672.

Baldwin, R. E. 1986. Functional properties in foods. In "Egg Science and Technology," 3rd edn., Stadelman, W. J. and Cotterill, O. J. (Eds.). Avi Publ. Co., Westport, Connecticut.

Baldwin, R. E. Cotterill, O. J., Thompson, M. M., and Myers, M. 1967. High temperature storage of spray-dried egg white. *Poultry Sci.* **46**: 1421–1425.

Baldwin, R. E. Upchurch, R., and Cotterill, O. J. 1968. Ingredient effects on meringue cooked by microwaves and by baking. *Food Technol.* **22**: 1573–1576.

Ball, H. R., Hamid-Samimi, M., Foegeding, P. M., and Swartzel, K. R. 1987. Functionality and microbial stability of ultrapasteurized aseptically packaged refrigerated whole egg. *J. Food Sci.* **52**: 1212–1218.

Barmore, M. A. 1934. The influence of chemical and physical factors on egg-white foams. Tech. Bull. 9, Colorado Agric. Exp. Stn., Fort Collins.

Bergquist, D. H. 1986. Egg dehydration. *In* "Egg Science and Technology," 3rd edn., Stadelman, W. J. and Cotterill, O. J. (Eds.), Avi Publ. Co., Westport, Connecticut.

Beveridge, T. and Ko, S. 1984. Firmness of heat-induced whole egg coagulum. *Poultry Sci.* **63**: 1372–1377.

Beveridge, T., Arntfield, S., Ko, S., and Chung, J. K. L. 1980. Firmness of heat induced albumen coagulum. *Poultry Sci.* **59**: 1229–1236.

Board, R. G. and Tranter, H. S. 1986. The microbiology of eggs. *In* "Egg Science and Technology," 3rd edn., Stadelman, W. J. and Cotterill, O. J. (Eds.) Avi Publ. Co., Westport, Connecticut.

Briant, A. M. and Willman, A. R. 1956. Whole-egg sponge cakes. *J. Home Econ.* **48**: 420–421.

Britton, W. M. and Fletcher, D. L. 1987. Influence of storage environment on ease of shell removal from hard-cooked eggs. *Poultry Sci.* **66**: 453–457.

Cardetti, M. M., Rhorer, A. R., and Stadelman, W. J. 1979. Effect of storage on the interior quality of shell eggs. *Poultry Sci.* **58**: 1403–1405.

Carlson, C. W. and Halverson, A. W. 1964. Some effects of dietary pigmentors on egg yolks and mayonnaise. *Poultry Sci.* **43**: 654–662.

Chang, C. M., Powrie, W. D., and Fennema, O. 1972. Electron microscopy of mayonnaise. *Can. Inst. Food Sci. Technol. J.* **5**: 134–137.

Chen, T. C. 1981. Observations on the weeping of double boiler scrambled egg products and its prevention. *J. Food Sci.* **46**: 310–311.

Childs, M. T. and Ostrander, J. 1976. Egg substitutes: Chemical and biological evaluations. *J. Amer. Dietet. Assoc.* **68**: 229–234.

Clinger, C., Young, A., Prudent, I., and Winter, A. R. 1951. The influence of pasteurization, freezing, and storage on the functional properties of egg white. *Food Technol.* **5**: 166–170.

Cotterill, O. J. 1986. Freezing egg products. *In* "Egg Science and Technology," 3rd edn., Stadelman, W. J. and Cotterill, O. J. (Eds.). Avl Publ. Co., Westport, Connecticut.

Cotterill, O. J., Seideman, W. E., and Funk, E. M. 1965. Improving yolk-contaminated egg white by heat treatments. *Poultry Sci.* **44**: 228–235.

Cremer, M. L. and Chipley, J. R. 1980. Hospital ready-prepared type foodservice system: Time and temperature conditions, sensory and microbiological quality of scrambled eggs. *J. Food Sci.* **45**: 1422–1424, 1429.

Cunningham, F. E. 1986. Egg product pasteurization. *In* "Egg Science and Technology," 3rd edn., Stadelman, W. J. and Cotterill, O. J. (Eds.). Avi Publ. Co., Westport, Connecticut.

Cunningham, F. E. and Cotterill, O. J. 1964. Effect of centrifuging yolk-contaminated liquid egg white on functional performance. *Poultry Sci.* **43**: 283–291.

Cunningham, F. E., and Cotterill, O. J. 1972. Performance of egg white in the presence of yolk fraction. *Poultry Sci.* **51**: 712–714.

Curtis, P. A., Gardner, F. A., and Mellor, D. B. 1986. A comparison of selected quality and compositional characteristics of brown and white shell eggs. III. Compositional and nutritional characteristics. *Poultry Sci.* **65**: 501–507.

Dawson, E. H., Miller, C., and Redstrom, R. A. 1956. Cooking quality and flavor of eggs as related to candled quality, storage conditions, and other factors. Agric. Inform. Bull. 164, USDA, Washington, DC.

Feiser, G. E. and Cotterill, O. J. 1982. Composition of serum from cooked-frozen-thawed-reheated scrambled eggs at various pH levels. *J. Food Sci.* **47**: 1333–1337.

Feiser, G. E. and Cotterill, O. J. 1983. Composition of serum and sensory evaluation of cooked-frozen-thawed-reheated scrambled eggs at various salt levels. *J. Food Sci.* **48**: 794–797.

Fischer, J. R. and Fletcher, D. L. 1983. The effect of adding salt (NaCl) to the preservation solution on the acceptability of hard cooked eggs. *Poultry Sci.* **62**: 1345 (Abstract).

Francis, F. J. and Clydesdale, F. M. 1975. "Food Colorimetry: Theory and Applications." Avi Publ. Co., Westport, Connecticut.

Franks, O. J., Zabik, M. E., and Funk, K. 1969. Angel cakes using frozen, foam-spray-dried, freeze-dried, and spray-dried albumen. *Cereal Chem.* **46**: 349–356.

Gardner, F. A., Beck, M. L., and Denton, J. H. 1982. Functional quality comparison of whole egg and selected egg substitute products. *Poultry Sci.* **61**: 75–78.

Gillis, J. N. and Fitch, N. K. 1956. Leakage of baked soft-meringue topping. *J. Home Econ.* **48**: 703–707.

Graham, G. E. and Kamat, V. B. 1977. The role of egg yolk lipoprotein in fatless sponge-cake making. *J. Sci. Food Agric.* **28**: 34–40.

Grewe, E. and Child, A. M. 1930. The effect of acid potassium tartrate as an ingredient in angel cake. *Cereal Chem.* **7**: 245–250.

Grunden, L. P., Mulnix, E. J., Darfler, J. M., and Baker, R. C. 1975. Yolk position in hard cooked eggs as related to heredity, age, and cooking position. *Poultry Sci.* **54**: 546–552.

Hale, K. K., Jr. and Britton, W. M. 1974. Peeling hard cooked eggs by rapid cooling and heating. *Poultry Sci.* **53**: 1069–1077.

Hale, K. K., Jr., and Holleran, K. A. 1978. Factors affecting shell cracking, peeling ease and tenderness of hard-cooked eggs. *Poultry Sci.* **57**: 1187 (Abstract).

Hanning, F. M. 1945. Effect of sugar or salt upon denaturation produced by beating and upon the ease of formation and the stability of egg white foams. *Iowa State College J. Sci.* **20**: 10–12.

Hasiak, R. J., Vadehra, D. V., Baker, R. C., and Hood, L. 1972. Effect of certain physical and chemical treatments on the microstructure of egg yolk. *J. Food Sci.* **37**: 913–917.

Henry, W. C. and Barbour, A. D. 1933. Beating properties of egg white. *Ind. Eng. Chem.* **25**: 1054–1058.

Hester, E. E. and Personius, C. J. 1949. Factors affecting the beading and the leakage of soft meringues. *Food Technol.* **3**: 236–240.

Hill, W. M. and Sebring, M. 1986. Desugarization of egg products. *In* "Egg Science and Technology," 3rd edn., Stadelman, W. J. and Cotterill, O. J. (Eds.). Avi Publ. Co., Westport, Connecticut.

Ijichi, K., Palmer, H. H., and Lineweaver, H. 1970. Frozen whole eggs for scrambling. *J. Food Sci.* **35**: 695–698.

Irmiter, T. F., Dawson, L. E., and Reagen, J. G. 1970. Methods for preparing hard cooked eggs. *Poultry Sci.* **49**: 1232–1236.

Jaax, S. and Travnicek, D. 1968. The effects of pasteurization, selected additives and freezing rate on the gelation of frozen-defrosted egg yolk. *Poultry Sci.* **47**: 1013–1022.

Janek, W. A. and Downs, D. M. 1969. Scrambled eggs prepared from three types of dried whole eggs. *J. Amer. Dietet. Assoc.* **55**: 578–582.

Janky, D. M. 1986. Variation in the pigmentation and interior quality of commercially available table eggs. *Poultry Sci.* **65**: 607–610.

Johnson, T. M. and Zabik, M. E. 1981a. Egg albumen protein interactions in an angel food cake system. *J. Food Sci.* **46**: 1231–1236.

Johnson, T. M. and Zabik, M. E. 1981b. Ultrastructural examination of egg albumen protein foams. *J. Food Sci.* **46**: 1237–1240.

Jordan, R., Luginbill, R. N., Dawson, L. E., and Echterling, C. J. 1952. The effect of selected pretreatments upon culinary qualities of eggs frozen and stored in a home-type freezer. I. Plain cakes and baked custards. *Food Research* **17**: 1–7.

Jordan, R., Wegner, E. S., and Hollender, H. A. 1954. Nonhomogenized vs. homogenized milk in baked custards. *J. Amer. Dietet. Assoc.* **30**: 1126–1130.

Joslin, R. P. and Proctor, B. E. 1954. Some factors affecting the whipping characteristics of dried whole egg powders. *Food Technol.* **8**: 150–154.

Kato, A., Ogino, K., Kuramoto, Y., and Kobayashi, K. 1979. Degradation of the O-glycosidically linked carbohydrate units of ovomucin during egg white thinning. *J. Food Sci.* **44**: 1341–1344.

Kim, K. and Setser, C. A. 1982. Foaming properties of fresh and commercially dried eggs in the presence of stabilizers and surfactants. *Poultry Sci.* **61**: 2194–2199.

Kurisaki, J.-I., Kaminogawa, S., and Yamauchi, K. 1980. Studies on freeze-thaw gelation of very low density lipoprotein from hen's egg yolk. *J. Food Sci.* **45**: 463–465.

Leutzinger, R. L., Baldwin, R. E., and Cotterill, O. J. 1977. Sensory attributes of commercial egg substitutes. *J. Food Sci.* **42**: 1124–1125.

Longree, K., Jooste, M., and White, J. C. 1961. Time-temperature relationships of custards made with whole egg solids. III. Baked in large batches. *J. Amer. Dietet. Assoc.* **38**: 147–151.

Lorenz, K. 1974. High altitude food preparation and processing. *CRC Crit. Rev. Food Technol.* **5**: 402–441.

Lowe, B. 1955. "Experimental Cookery." John Wiley and Sons, New York.

Ma, C.-Y. and Holme, J. 1982. Effect of chemical modification on some physicochemical properties and heat coagulation of egg albumen. *J. Food Sci.* **47**: 1454–1459.

MacDonnell, L. R., Feeney, R. E., Hanson, H. L., Campbell, A., and Sugihara, T. F. 1955. The functional properties of the egg white proteins. *Food Technol.* **9**: 49–53.

MacLeod, A. J. and Cave, S. J. 1977. Absorption of a taint aroma by eggs. *J. Food Sci.* **42**: 539–540.

Marusich, W. L. and Bauernfeind, J. C 1981. Oxycarotenoids in poultry feeds. *In* "Carotenoids as Colorants and Vitamin A. Precursors," Bauernfeind, J. C. (Ed.). Academic Press, New York.

Mast, M. G. and Clouser, C. S. 1988. Processing options for improving the nutritional value of poultry, meat, and egg products. *In* "Developing Foods. Animal Product Options in the Marketplace." National Research Council, National Academy Press, Washington, DC.

Maurer, A. J. 1975a. Refrigerated egg storage at two air movement rates. *Poultry Sci.* **54**: 409–412.

Maurer, A. J. 1975b. Hard-cooking and pickling eggs as teaching aids. *Poultry Sci.* **54**: 1019–1024.

McKellar, D. M. B. and Stadelman, W. J. 1955. A method for measuring volume and drainage of egg white foams. *Poultry Sci.* **34**: 455–458.

Mellor, D. B., Gardner, F. A., and Campos, E. J. 1975. Effect of type of package and storage temperature on interior quality of shell treated shell eggs. *Poultry Sci.* **54**: 742–746.

Miller, E. L. and Vail, G. E. 1943. Angel food cakes made from fresh and frozen egg whites. *Cereal Chem.* **20**: 528–535.

Miller, G. A., Jones, E. M., and Aldrich, P. J. 1959. A comparison of the gelation properties and palatability of shell eggs, frozen whole eggs, and whole egg solids in standard baked custard. *Food Research* **24**: 584–594.

Morgan, K. J., Funk, K., and Zabik, M. E. 1970. Comparison of frozen, foam-spray-dried, freeze-dried, and spray-dried eggs. 7. Soft meringues prepared with a carrageenan stabilizer. *J. Food Sci.* **35**: 699–701.

Navidi, M. K. and Kummerow, F. A. 1974. Nutritional value of Egg Beaters® compared with "farm fresh eggs." *Pediatrics* **53**: 565–566.

Niles, K. B. 1937. Egg whites on parade. *U.S. Egg and Poultry Mag.* **43**: 337–340, 376–386.

O'Brien, S. W., Baker, R. C., Hood, L. F., and Liboff, M. 1982. Water-holding capacity and textural acceptability of precooked, frozen, whole-egg omelets. *J. Food Sci.* **47**: 412–417.

Palmer, H. H. 1972. Eggs. *In* "Food Theory and Applications," Paul, P. C. and Palmer, H. H. (Eds.). John Wiley and Sons, New York.

Palmer, H. H., Ijichi, K., and Riff, H. 1970. Partial reversal of gelation in thawed egg proteins. *J. Food Sci.* **35**: 403–406.

Pomeranz, Y. 1985. "Functional Properties of Food Components." Academic Press, Orlando, Florida.

Potter, N. N. 1986. "Food Science." Avi Publ. Co., Westport, Connecticut.

Powrie, W. D. and Nakai, S. 1985. Characteristics of edible fluids of animal origin: Eggs. *In* "Food Chemistry," 2nd edn., Fennema, O. (Ed.). Marcel Dekker, Inc., New York.

Powrie, W. D. and Nakai, S. 1986. The chemistry of eggs and egg products. *In* "Egg Science and Technology," 3rd edn., Stadelman, W. J. and Cotterill, O. J. (Eds.). Avi Publ. Co., Westport, Connecticut.

Pratt, D. E. 1975. Lipid analysis of a frozen egg substitute. *J. Amer. Dietet. Assoc.* **66**: 31–33.

Reinke, W. C., Spencer, J. V., and Tryhnew, L. J. 1973. The effect of storage upon the chemical, physical and functional properties of chicken eggs. *Poultry Sci.* **52**: 692–702.

Rhodes, M. B., Adams, J. L., Bennett, N., and Feeney, R. E. 1960. Properties and food uses of duck eggs. *Poultry Sci.* **39**: 1473–1478.

Sauter, E. A. and Montoure, J. E. 1972. The relationship of lysozyme content of egg white to volume and stability of foams. *J. Food Sci.* **37**: 918–920.

Sauter, E. A. and Montoure, J. E. 1975. Effects of adding 2% freeze-dried egg white to batters of angel food cakes from white containing egg yolk. *J. Food Sci.* **40**: 869–871.

Sauter, E. A., Stadelman, W. J., Harns, J. V., and McLaren, B. A. 1951. Methods for measuring yolk index. *Poultry Sci.* **30**: 629–630.

Sechler, C., Maharg, L. G., and Mangel, M. 1959. The effect of household table salt on the whipping quality of egg white solids. *Food Research* **24**: 198–204.

Seideman, W. E., Cotterill, O. J., and Funk, E. M. 1963. Factors affecting heat coagulation of egg white. *Poultry Sci.* **42**: 406–417.

Sheldon, B. W. 1986. Influence of three organic acids on the quality characteristics of hard-cooked eggs. *Poultry Sci.* **65**: 294–301.

Sheldon, B. W. and Kimsey, H. R., Jr. 1985. The effects of cooking methods on the chemical, physical, and sensory properties of hard-cooked eggs. *Poultry Sci.* **64**: 84–92.

Spencer, J. V. and Tryhnew, L. J. 1973. Effect of storage on peeling quality and flavor of hard-cooked shell eggs. *Poultry Sci.* **52**: 654–657.

Stadelman, W. J. 1986a. Preservation of quality in shell eggs. *In* "Egg Science and Technology," 3rd edn., Stadelman, W. J. and Cotterill, O. J. (Eds.). Avi Publ. Co., Westport, Connecticut.

Stadelman, W. J. 1986b. Hard-cooked eggs. *In* "Egg Science and Technology," 3rd edn., Stadelman, *W. J. and Cotterill, O. J. (Eds.). Avi Publ. Co., Westport, Connecticut.

Stadelman, W. J. and Rhorer, A. R. 1984. Quality improvement of hard cooked eggs. *Poultry Sci.* **63**: 949–953.

St. John, J. L. and Flor, I. H. 1930–1931. A study of whipping and coagulation of eggs of varying quality. *Poultry Sci.* **10**: 71–82.

Toney, J. and Bergquist, D. H. 1983. Functional egg products for the cereal industry. *Cereal Foods World* **28**: 445–447.

Upchurch, R. and Baldwin, R. E. 1968. Gar gum and triacetin in meringues and a meringue product cooked by microwaves. *Food Technol.* **22**: 1309–1310.

USDA. 1981. Revision of shell egg standards and grades. *Federal Register* **46**(149): 39566–39573.

Vadehra, D. V. and Nath, K. R. 1973. Eggs as a source of protein. *CRC Crit. Rev. Food Technol.* **4**: 193–309.

Wakamatu, T., Sato, Y., and Saito, Y. 1983. On sodium chloride action in the gelation process of low density lipoprotein (LDL) from hen egg yolk. *J. Food Sci.* **48**: 507–512, 516.

Wang, A. C., Funk, K., and Zabik, M. E. 1974. Effect of sucrose on the quality characteristics of baked custards. *Poultry Sci.* **53**: 807–813.

Wolfe, N. J. and Zabik, M. E. 1968. Comparison of frozen, foam-spray-dried, freeze-dried, and spray-dried eggs. 3. Baked custards prepared from eggs with corn syrup solids. *Food Technol.* **22**: 1470–1476.

Woodward, S. A. and Cotterill, O. J. 1983. Electrophoresis and chromatography of heat-treated plain, sugared, and salted whole egg. *J. Food Sci.* **48**: 501–506.

Woodward, S. A. and Cotterill, O. J. 1986. Texture and microstructure of heat-formed egg white gels. *J. Food Sci.* **51**: 333–339.

Woodward, S. A. and Cotterill, O. J. 1987. Texture and microstructure of cooked whole egg yolks and heat-formed gels of stirred egg yolk. *J. Food Sci.* **52**: 63–67.

Yang, S.-S. and Cotterill, O. J. 1989. Physical and functional properties of 10% salted egg yolk in mayonnaise. *J. Food Sci.* **54**: 210–213.

Zabik, M. E. 1968. Comparison of frozen, foam-spray-dried, freeze-dried, and spray-dried eggs. 2. Gels made with milk and albumen or yolk containing corn syrup. *Food Technol.* **22**: 1465–1469

MILK AND MILK PRODUCTS

In this chapter, physical properties, components, flavor, and functional properties of milk and of products made from cow's milk are considered. Milk is legally defined as *"the lacteal secretion, practically free of colostrum, obtained by the complete milking of one or more healthy cows. Milk that is in final form for beverage use shall have been pasteurized or ultrapasteurized, and shall not contain less than 8.25% of milk-solids-not-fat and not less than 3.25% milk fat." (FDA, 1988)*

Definitions and standards for other milk products also are included in the "Code of Federal Regulations" (FDA, 1988).

I. PHYSICAL PROPERTIES

A. Physical State

Milk is a complex physicochemical system whose various constituents differ widely in molecular size and solubility. The smallest molecules, those of salts, lactose, and water-soluble vitamins, are in true solution. The proteins, including enzymes, are in colloidal state because of the large size of their molecules. The fat in nonhomogenized milk is present as globules of larger than colloidal size. The size of the fat globules varies with the individual cow, its breed, and the stage of lactation (Keenan *et al.*, 1988, pp. 512–515). Reported diameters range from less than 0.1 to 20 μm or greater (Keenan *et al.*, 1988, p. 513); globules number 15×10^9/ml of milk (Jenness, 1988, pp. 2–3). During homogenization, these fat globules are reduced in size to less than 1 μm and are increased greatly in number by being forced under high pressures through a small slit. The resulting globule size depends on the pressure and the type of homogenizer used. The difference in fat globule size is illustrated in Fig. 8.1

Each fat globule has a membrane composed of a complex mixture of lipids (phospholipids, cholesterol, and glycerides) and protein. Homogenization causes changes in the membrane which prevent coalescence of the fat globules. The membrane exhibits a typical bilayer membrane structure (Keenan *et al.*, 1988, pp. 557–560).

B. pH

Milk is slightly acidic, usually having a pH of 6.6 at 25°C. It is well buffered by proteins and salts, especially the phosphates. The pH of milk is temperature dependent. When milk is heated, its pH decreases because hydrogen ions are liberated when calcium phosphate precipitates (Sherbon, 1988, pp. 410–414).

C. Other Physical Properties

Skim milk and whole milk are essentially Newtonian fluids, which are defined in Chapter 3. Therefore, the viscosity of a given sample depends only on temperature, whereas the viscosities of the non-Newtonian creams, concentrated milks, and butter depend also on shear rate. The quantity of dispersed solids influences the viscosity. Thus, whole milk is more viscous than skim milk, which is more viscous than whey (Sherbon, 1988, pp. 425–428).

The freezing point of milk is slightly lower than that of water because of the

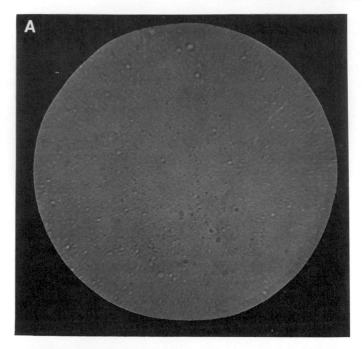

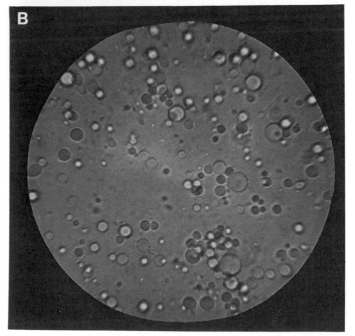

FIGURE 8.1
Microscopic appearance of the fat globules in (A) homogenized and (B) nonhomogenized milk.
(Courtesy of Michigan Agricultural Experiment Station.)

presence of lactose and soluble salts. Reported values range from -0.530 to $-0.570°C$ (Sherbon, 1988, pp. 432–437). Determination of the freezing point is used for detection of milk to which water has been added.

The presence of milkfat, milk proteins, free fatty acids, and phospholipids lowers the surface tension of the water present in milk. At 20°C the surface tension of milk is 50 dyn cm^{-1}, whereas water at that temperature has a surface tension of 72.75 dyn cm^{-1} (Sherbon, 1988, pp. 428–430).

II. COMPONENTS OF MILK

The legal definition of milk includes minimal levels for two components of milk. However, composition varies more than is indicated by the definition. State standards for milk-solids-nonfat vary from 8.0 to 8.5%, and standards for milkfat vary from 3.0 to 3.8% (USDA, 1981). The composition of milk varies with the feed, breed, nutrition, and the physiological condition of the cow (Linn, 1988, pp. 224–241). Therefore, tabular values for the various components of milk represent average values rather than absolute values.

A. Water

Water is the major component of milk, representing 87% of the total. Other components are suspended or dissolved in this medium. A small amount of water is bound to the milk proteins and some is hydrated to the lactose and salts giving milk a water activity (a_w) of 0.993 (Jenness, 1988, p. 2).

B. Protein

Knowledge of milk proteins is essential to understanding changes that take place during the processing of milk, the production of dairy products, and the manufacturing of cheese. Milk proteins can be separated into two major fractions, the caseins and the whey or serum proteins.

1. Caseins
The caseins are a group of phosphoproteins, accounting for 80% of the total proteins in milk. Acidification of raw skim milk to pH 4.6 at 20°C will coagulate this fraction (Eigel *et al.*, 1984). Such coagulation is recognized as curdling in a stirred milk, or gelling in a quiescent milk that is acidified gradually by bacterial action. In the latter case, the coagulated casein first forms a continuous mass that breaks into curds when agitated. The precipitated casein can be rendered soluble by the addition of acid or of alkali, either of which shifts the pH away from the isoelectric point. Casein is prepared for research purposes by precipitating it with acid and resolubilizing it in acid several times. Although casein is spoken of as a single protein, it really is a mixture of proteins, which may be separated by electrophoresis. In this method, the distances that the proteins move in an electrical field are studied. The casein proteins include four groups, α_{s1}-caseins, α_{s2}-caseins, β-caseins, and κ-caseins. The primary structures or

amino acid sequences may be used to identify these components (Eigel *et al.*, 1984). At the pH of milk, about 6.6, casein is present as a colloidal phosphate complex, calcium caseinate, dispersed as particles called micelles. Reflection of light by the micelles is responsible for the white color of milk. Proposed models for micelles or aggregates of the caseins in detail are described by Farrell (1988, pp. 461–493). The percentages of the major caseins in the micelles are α_{s1}, 38%; α_{s2}, 10%; β, 36%; and κ, 13% (Davies and Law, 1980).

κ-Casein has a stabilizing effect on the casein micelle, permitting the existence of the colloidal dispersion. The α- and β-caseins are calcium sensitive or insoluble whereas κ-casein is soluble in the presence of calcium. Therefore, if the protective effect of κ-casein is destroyed by the enzyme rennin, casein reacts with calcium to form a coagulum. Rennin usually is obtained from the fourth stomach of dairy calves. Rennin coagulation is the preferred system for coagulation of casein for cheese production. As previously discussed, acid also is used for coagulation. Acid can be added as such or formed in the milk by natural souring, during which lactose is changed to lactic acid by the acid of bacteria. Although casein usually is not coagulated by boiling, it may be coagulated if the milk is slightly acid or if very high temperatures are used. Thus, fresh milk can be boiled without curdling, but milk that is slightly sour or cooked under pressure is likely to curdle.

2. Whey Proteins

After the caseins are precipitated from milk, the whey proteins or milk serum proteins remain. "Whey" protein is a general term used to refer to milk proteins that are soluble at pH 4.6 at 20°C (Eigel *et al.*, 1984). Whey proteins are denatured with heating above 60°C (Morr, 1975). Heating also causes aggregation of the denatured whey proteins (Morr, 1969). Proteins in the whey fraction include β-lactoglobulins, α-lactalbumins, serum albumin, and the immunoglobulins. In addition the whey fraction includes fragments of β-casein and other heat-stable polypeptides (Eigel *et al.*, 1984). β-Lactoglobulin is the major whey protein, representing 50% of the whey proteins (Farrell, 1988, p. 493). The α-lactalbumins are the next most abundant proteins, constituting 25% of the whey proteins. Unlike the β-lactoglobulins, this group does not contain any free sulfhydryl groups. Further information on the properties of all milk protein fractions is given by Farrell (1988, pp. 461–510) and by Eigel and other members of the American Dairy Association Nomenclature Committee (Eigel *et al.*, 1984).

3. Enzymes

Milk contains many enzymes, most of which are inactivated by pasteurization. If the enzyme alkaline phosphomonoesterase (alkaline phosphatase) is not present in pasteurized milk, the process is considered adequate. This test is called the phosphatase test, and is an effective index of adequacy of pasteurization because the heat necessary to destroy this enzyme will also destroy pathogenic microorganisms. Other enzymes found in cow's milk include lipase, esterases, phosphatases, xanthine oxidase, lactoperoxidase, protease, amylase, catalase, aldolase, ribonuclease, lysozyme, carbonic anhydrase, and others (Whitney, 1988, pp. 105–108). Lipase sometimes causes hydrolytic rancidity in dairy products made from milk that has not been heated enough to inactivate this enzyme.

C. Lipids

Fat is the most variable component of milk. State law regulates the fat content of milk; the required amounts vary from 3.0 to 3.8% (USDA, 1981). The fat content of fluid milk sold may exceed the minimum amount shown in Table I. The milk from Guernsey and Jersey cows contains more fat than the milk from Holstein, Ayrshire, and Brown Swiss cows. Fat content decreases in May, June, and July and with the aging of the cow, but the amount of fat in the feed does not appreciably affect fat content of milk (Jenness, 1988, pp. 27–28).

Milkfat is primarily triglycerides (98%) containing the glycerides of as many as 400 fatty acids, both saturated and unsaturated (Jensen and Clark, 1988); it is distinguished from other food fats by its content of short-chain saturated fatty acids such as butyric, caproic, caprylic, and capric acids. These short-chain fatty acids are important to the flavor of products made from milk and in off-flavors that may develop in milk.

In addition to the triglycerides, milk contains several other lipids in varying amounts, including phospholipids, sterols, carotenoids, and fat-soluble vitamins.

TABLE I

Composition of Selected Dairy Products per 100 g[a]

Food	Water (%)	Protein (g)	Total lipid (g)	Carbo-hydrate (g)	Calcium (mg)	Vitamin A value (IU)
Fluid milk						
Whole	87.99	3.29	3.34	4.66	119	126
Lowfat, 2%	89.21	3.33	1.92	4.80	122	205
Condensed milk						
Sweetened	27.16	7.91	8.70	54.40	284	328
Evaporated milk						
Whole	74.04	6.81	7.56	10.04	261	243
Skim	79.40	7.55	0.20	11.35	290	392
Dried milk products						
Whole	2.47	26.32	26.71	38.42	912	922
Nonfat	3.16	36.16	0.77	51.98	1257	36
Whey, sweet	3.19	12.93	1.07	74.46	796	44
Half and half	80.57	2.96	11.50	4.30	105	434
Cream, heavy						
Whipping	57.71	2.05	37.00	2.79	65	1470
Cheese						
Cheddar	36.75	24.90	33.14	1.28	721	1059
Cottage, creamed	78.96	12.49	4.51	2.68	60	163
Cottage, 2% fat	79.31	13.74	1.93	3.63	68	70

[a] From Posati and Orr (1976).

Factors including type and amount of feed, stage of lactation, and species of animal or breed of cow affect the proportion of these components in milk (Jenness, 1988, pp. 20–30).

D. Carbohydrates

Lactose, the major carbohydrate of milk, is found in cow's milk at levels of approximately 4.8% (Holsinger, 1988, p. 280). This level of sugar does not make the milk unduly sweet because lactose is less sweet than sucrose. In addition to lactose, milk contains small amounts of glucose, galactose, and other saccharides (Jenness, 1988, pp. 3–4).

Upon digestion, lactose yields glucose and galactose. Lactose-intolerant individuals lack the enzyme, β-D-galactoside galactohydrolase, which is responsible for this change in the small intestine. Discomfort results when lactose is fermented by bacteria in the large intestine. Thus a demand exists for products in which the lactose has been removed during processing, as in production of natural cheeses, or has been hydrolyzed during fermentation; and there also is a demand for the enzyme lactase for treatment of milk by consumers. Lactose intolerance and possible uses of dairy products by individuals who have it are described in two recent reviews (Houts, 1988; Savaiano and Kotz, 1988).

Lactose is found as one of two crystalline forms, α-monohydrate and anhydrous-β, or as lactose "glass," a mixture of α- and β-lactose. Although the lactose of milk is in true solution, its low solubility creates some manufacturing and preparation problems, since lactose crystals are objectionably gritty in texture. Lactose crystallizes from sweetened condensed milk because the small amount of water remaining after evaporation is not sufficient to keep the lactose and the added sucrose in solution. Manufacturers attempt to promote the formation of many nuclei for crystal formation so the crystals formed will be very small and imperceptible.

In contrast, evaporated milk contains enough water to dissolve all the lactose. Lactose may crystallize from ice cream since much of the water is frozen and hence is not available to hold the lactose in solution. As with sweetened condensed milk, the formation of very small crystals is promoted by agitation or seeding with a supersaturated solution of lactose. Crystallization of lactose also is responsible for lumping and caking of dried milk during storage if moisture is present. Lactose is the source of lactic acid, formed by bacterial action as milk sours. When milk is coagulated, lactose is present in the whey, from which it can be prepared commercially. For this reason, cheese that is prepared from the curd is low in carbohydrates. Some whey cheeses such as Myost and Primost are high in lactose and therefore sweet in taste.

E. Salts, Trace Elements, and Vitamins

The salts of milk constitute less than 1% of the milk. Anionic components include chlorides, phosphates, sulfate, carbonate, and citrates and the cations are potassium, sodium, calcium, and magnesium (Jenness, 1988, pp. 6–10).

Practically all of the minerals in the soil from which the cow obtains her feed are present in milk, some of them only in trace amounts. Cobalt, copper, and iodine may be low due to deficiencies in soil content. Copper is significant to the sensory quality of milk because it exerts a catalytic effect on the development of oxidized flavor. Other trace elements include iron, magnesium, molybdenum, nickel, and zinc.

Milk contains many vitamins, some of them in abundance, and some in small quantities. Riboflavin is present in significant quantities. This water-soluble compound is responsible for the light yellowish or greenish tint of skim milk. Exposure to light results in degradation of riboflavin. Riboflavin in skim milk is more susceptible to degradation than is the vitamin in whole milk. Palanuk *et al.* (1988) demonstrated that skim milk at the top of a translucent container may, after 5 days storage under light, contain only 58% of the riboflavin initially present.

The fat-soluble vitamin A precursor, carotene, is responsible for the yellowish color of milkfat. Skim milk is fortified with retinyl palmitate to replace the carotene lost with the removal of fat. Exposure of skim milk fortified with retinyl palmitate to fluorescent light for 24 hr resulted in loss of 70% of the all-*trans* retinyl palmitate (Gaylord *et al.*, 1986). Thus, the use of opaque or colored containers or storage in darkened refrigerators is important for preservation of these nutrients in milk.

III. ALTERATION OF MILK AND MILK PRODUCTS BY PROCESSING

Prior to the consumption of milk as fluid milk or as a product prepared from fluid milk, milk is subjected to one or more treatments that may influence the characteristics of the product. Treatment may include one or more heat treatments, coagulation, and/or dehydration and may influence flavor, color, texture, functional properties, and nutritional value.

A. Heat

1. Pasteurization

Processing affects the flavor of milk, usually because of heating. In fluid milk, a change in flavor is caused by the heat necessary to pasteurize the milk. Pasteurization may be accomplished by one of several treatments that meet government requirements (FDA, 1988). These include 62°C for 30 min, 72°C for 15 sec to 100°C for 0.01 sec [high-temperature short time (HTST)], or above 138°C for 2 sec [ultra-high temperature (UHT) or ultrapasteurization]. Pasteurization is used to destroy pathogenic organisms. In addition, enzymes that may cause the development of off-flavors may be inactivated and spoilage organisms may be destroyed, resulting in an increase in shelf life.

Some of the most common off-flavors in milk are rancid and oxidized flavors. Rancidity resulting from activity of lipase is not usually a problem because the enzyme is destroyed by pasteurization. Probably because of the large surface of its fat globules, homogenized milk becomes rancid within a few hours if not adequately pasteurized or if mixed with small amounts of raw milk. A different type of off-flavor, oxidized flavor, is accelerated by traces of copper; this finding has caused a virtual elimination of copper-containing equipment from dairies.

The flavor of fluid milk is changed by the heat of pasteurization. The use of a high temperature for a short time, such as 72°C or higher for 15 sec, changes the flavor less than the holding method of at least 62°C for 30 min. Boiling, of course, changes the flavor of milk more than does pasteurization. Hutton and Patton (1952) reported that sulfhydryl groups of β-lactoglobulin, which give rise to hydrogen sulfide with denaturation, are responsible for the cooked flavor of milk.

2. Evaporation and Canning

To produce evaporated milk, milk is forewarmed and concentrated to slightly more than double the solids content of fluid whole milk (25% total milk solids including ≥ 7.5% milk fat) as required by federal regulations (FDA, 1988). Then it is homogenized, sealed into cans, and sterilized. The characteristic cooked flavor of evaporated milk is caused by the high temperatures required in canning. The milk is sterilized at 115 to 118°C for 15 to 20 min (Morr and Richter, 1988, p. 752). Methyl sulfide, a component that is responsible for a "cowy" flavor in fresh milk (Patton et al., 1956), has been found at elevated levels in evaporated milk, suggesting that it plays a role in the cooked flavor.

Flavor deterioration in concentrated milk in the form of cooked, scorched, and staled notes was greater at 20 and 37°C than at 4°C when concentrated milk was stored for 8 months (Loney et al., 1968). Off-colors may develop in evaporated milk stored at high temperatures for long periods of time. Carbonylamine browning may be responsible.

Sweetened condensed milk also contains a little more than double the milk solids content of fluid whole milk; in addition, sucrose is added after evaporation, resulting in a total carbohydrate concentration of approximately 56%. The flavor of sweetened condensed milk cannot be compared with that of other milks because of its high sugar content. Less change in flavor due to heat can be expected for this milk than for evaporated milk because the added sucrose serves as a preservative, eliminating the need for sterilization.

3. Drying

Several methods are used for the production of whole dry milk powder (WDM), nonfat dry milk (NFDM), and other dried milk products and are described by Knipschildt (1986, pp. 131–234) and Bodyfelt et al. (1988, pp. 436–451). Whole dry milk powder deteriorates more rapidly during storage than NFDM. Oxidation of the milkfat results in a tallowy flavor, and the carbonyl-amine reaction is responsible for the stale flavor that develops. Ingredients in chocolate products, soup mixes, and confections may mask the flavors of WDM (Pomeranz, 1985, pp. 385, 387). In candies, the proteins from WDM provide a chewy matrix, the whey proteins facilitate air incorporation, and the fat provides flavor (Kinsella, 1984). Deterioration of WDM may be delayed by preheating, reducing its moisture content, adding small amounts of an antioxidant, packaging with nitrogen or carbon dioxide in a sealed container, or storing at low temperatures.

Nonfat dry milk of less than 4% moisture can be stored at 21°C for up to 18 months. Nonfat dry milk products are described by a heat treatment classification based on the extent of denaturation of the whey proteins. Low-heat, medium-heat, and high-heat classifications represent increasing degrees of denaturation. Functional performance varies with degree of denaturation, and selection of product depends on the functional properties required. For beverage products, low-heat NFDM is used. High-heat and medium-heat products may be used in candies and in comminuted meat products and ice creams, respectively (Kinsella, 1984). The dispersibility of NFDM is improved by agglomeration, a process that involves rewetting and redrying. Instantized NFDM produced by this process has a light, granular texture and is dispersed easily (Neff and Morris, 1968). The product on the retail market is instantized. Nonfat dry milk is blended with thickeners, sweeteners, flavor components, vitamins, and minerals and then instantized to produce instant beverage products (Kinsella, 1984). Calorie content is varied by selection of sweetener.

Dried buttermilk has a higher fat content than NFDM and thus a shorter shelf life. Dried milk products are useful in the experimental study of foods because they allow minimization of difference in products attributable to milk ingredients.

4. UHT Processing

Ultra-high-temperature processing (UHT) of milk involves heating for 1–8 sec at 135–154°C. Aseptic packaging of UHT milk produces a shelf-stable product. Aseptic packaging involves placing a sterile product in a sterile package. Such processing must take place in a sterile environment. Stability of this product is not limited by bacterial spoilage. However, it may be limited by sedimentation of particulate matter, separation of fat, gelation, and the presence of off-flavors, which may include cooked, bitter, sweet, and stale notes (Hill, 1988). Off-flavors may be attributable to free sulfhydryls, aldehydes, and ketones (Hansen, 1987). Storage at 4°C maintained flavor quality more than storage at 40°C in a study by Hansen (1987). Samples with 3.5% fat received higher scores than 0.5% fat samples on a scale for flavor ranging from unacceptable to excellent (Hansen and Swartzel, 1982). Hill (1988) presented a comprehensive review of UHT milk.

B. Cheese Production

The production of cheese was used more than 4000 years ago as a way of preserving milk for future use. Cheese is made by coagulating milk, cream, skim milk, buttermilk, or a combination of these and then draining part or most of the whey. Processing into cheese serves to concentrate the protein and fat while eliminating the carbohydrate (Hettinga, 1988, pp. 292–296). Cheese is made most commonly from cow's milk but the milk of other mammals may be used. Pasteurized milk is used in most cases. Use of milk concentrated by ultrafiltration can facilitate cheese production by allowing lowering of processing temperatures (Sharma *et al.*, 1989). Ultrafiltration of milk, a high-pressure micro-filtration process, results in removal of water and low-molecular-weight solutes (Kosikowski, 1986; Hettinga, 1988, pp. 292–293). Retentate, the product of ul-trafiltration, will coagulate more rapidly than fresh milk. The resultant curd is firmer than curd from fresh milk (Kosikowski, 1986). The curd that has been separated from the whey may be allowed to ripen by the action of enzymes from microorganisms or animal sources to produce a natural cheese. Structurally, cheese is an oil–water emulsion. Casein proteins serve as emulsifying agents in the cheese (Shimp, 1985).

1. Coagulation

Milk used for cheese production may be clotted with rennin or acid or both. A bacterial culture is added for acid production. Formerly, the clotting of milk was achieved primarily by using rennin (calf gastric enzyme chymosin). Rennin usu-ally is used as a crude extract, rennet. Rennin acts most effectively at pH 6.7 (Bingham, 1975) and a temperature of about 40°C. Enzymatic coagulation of milk involves two phases. Rennin is involved in the first phase, in which a spe-cific bond of κ-casein is cleaved to form insoluble para-κ-casein and a soluble peptide (Bingham, 1975). In the second phase, a clot is formed by the para-κ-casein and calcium (Cheryan *et al.*, 1975). Coagulation time decreases as the pH of the second phase is decreased from 6.7 to 5.6 and as the temperature increases (Cheryan *et al.*, 1975). Heating milk above 65°C and then cooling it prior to treatment with rennin reduces clotting rate and curd firmness. A heat-induced interaction between κ-casein and β-lactoglobulin may delay the action of rennin on κ-casein (Sawyer, 1969).

Because the availability of rennin is limited, proteolytic enzymes from other sources may be used. The term rennet is used for enzymes that may be used for the clotting of milk in cheese production (NRC, 1981, p. 107). Rennet is used from three animal sources (veal calf, adult bovine, and porcine pepsin) and three microbial sources (*Mucor milhei*, *Mucor pusillus*, and *Endothia parasitica*) (Pszczola, 1989). Use of alternative enzymes has necessitated alterations in pro-cessing techniques and in some cases resulted in cheese that is pasty and bitter. More encouraging results may be possible for microbial enzymes because they contain nonspecific proteases which enhance opportunity for formation of a better flavor (Pszczola, 1989). Biotechnology techniques have been used to clone the milk-clotting enzyme in *Escherichia coli* and *Saccharomyces cerevisiae*. However, commercial use of these products must still be approved (Pitcher, 1986).

2. Curd Treatment

Following coagulation, the curd is cut to allow loss of the whey. This process is controlled by cutting the curd into various-sized pieces. The curd is heated at low temperatures to hasten loss of whey and produce a more compact texture. Salts, acids, and bacterial cultures are added at various steps in the process. The curd then is pressed to make the cheese firmer and more compact. Variation in treatments results in the production of various kinds of cheese. These treat-ments are important in the control of moisture. Cheese may be classified on the basis of their moisture content. Hard cheeses, such as Cheddar, contain 30–

40% moisture; semisoft cheeses such as Muenster contain 50–75% moisture.

For the production of unripened cheeses such as cottage cheese and cream cheese, lactic acid starter bacteria are used for coagulation. The curd then is processed to give the characteristic product.

3. Curing

Cheese may be cured for varying lengths of time and at varying temperatures and humidities in order to allow the product to ripen. The chemical and physical changes associated with ripening result in a change from a bland, tough, rubbery mass to a full-flavored, soft, mellow product. Salt level is related to control of ripening. Thakur et al. (1975) reported that omission of salt speeds the ripening process, resulting in a pasty texture and the development of an unnatural, bitter, fruity, or flat flavor in Cheddar cheese. These effects were attributed to increased bacterial activity in the unsalted cheese. During ripening, the carbohydrate, fat, and protein are altered. Lactose is rapidly converted to lactic acid, which helps to inhibit the growth of undesirable microorganisms. Lipolytic enzymes liberate fatty acids, which contribute to the flavor. The typical flavor of Cheddar cheese is not developed when skim milk is used, suggesting an important role for fat in flavor (Ohren and Tuckey, 1969). Proteolytic enzymes such as rennin are responsible for the formation of nitrogenous products of intermediate size, such as peptones and peptides. Enzymes of the microorganisms act on these and other substances to form products like amino acids, amines, fatty acids, esters, aldehydes, alcohols, and ketones. A complex mixture of substances most likely produces the characteristic flavor of cheese. Slight changes in the proportions of the flavor components, caused by variations in the curing process, may alter the flavor of cheese. A strong shift in the components may lead to off-flavors. Proteolysis of the α_{s1}-casein molecules that comprise the microstructure of cheese results in softening of the product during aging (Creamer and Olson, 1982).

4. Processed Cheeses

The uniformity of processed cheese is attributable to its preparation by blending of natural cheeses of varying qualities to achieve the desired flavor and texture. The cheese is ground and then mixed with the aid of heat and emulsifying salts. Salts used for this purpose include sodium citrate and various sodium phosphates (Shimp, 1985). Acid, cream, water, salt, coloring, and flavorings may be added, but the moisture content may not be more than 1% greater than that of the cheese from which it is made. In the total process, enzymes are inactivated and bacteria are destroyed, preventing changes in flavor during storage. Postprocessing contamination is a possibility with processed cheese products. Magrini et al. (1983) demonstrated that Staphylococcus aureus will grow on processed cheese at a temperature of 30°C, indicating the appropriateness of refrigerated storage. Bennett and Amos (1983) also demonstrated enterotoxin production by S. aureus on several commercially produced imitation cheeses, further indicating the need for refrigerated storage.

Processed cheese food resembles processed cheese, but because of the inclusion of dairy ingredients such as cream, skim milk, whey, or lactalbumin it contains more moisture and less fat than processed cheese. Processed cheese spread is even softer than cheese food because it may contain additional moisture. Stabilizing agents such as gums, gelatin, and algin may be added to the spreads.

C. Fermentation

Lactic acid bacteria are added to skim milk, whole milk, or slightly concentrated milk to produce cultured dairy products such as buttermilk, yogurt, sour cream, creamed cottage cheese, and sweet acidophilus milk. Characteristics of fermented milk products are partially dependent on the microorganisms added. These microorganisms are mesophiles and thermophiles with optimal growth temperatures of 22 and 38°C, respectively (Kroger *et al.*, 1989). Cultured buttermilk is skimmed or partially skimmed milk that is fermented with a mixed culture of lactic acid bacteria (*Streptococcus cremoris*, *Streptococcus lactis*, and/ or *Streptococcus* ssp. *diacetylactis* with *Leuconostoc*). Sour cream is produced from cream containing 18% fat. Nonfat dry milk solids may be added to increase viscosity.

The consistency of yogurt, produced by the action of *Streptococcus thermophilus* and *Lactobacillus bulgaricus*, varies from a liquid resembling a stirred custard to a gel. Control of pH in yogurt production is important. If the pH is less than 5.5, the yogurt is judged to be too sour (Kroger, 1976). Yogurt may be made from cream, milk, partially skimmed milk, or skim milk, alone or in combination. Nonfat dry milk, whey, or other lowfat milk products may be added to increase the nonfat solids content of the product. In addition nutritive carbohydrate sweeteners, flavoring, color additives, and stabilizers may be added (FDA, 1988). Plain yogurt is described as having a somewhat "green apple-like flavor" (Bodyfelt *et al.*, 1988, pp. 271–272). Low levels of lactic acids and acetaldehyde are responsible for this flavor. Whalen *et al.* (1988) reported that whey–caseinate blends could be substituted for NFDM at a level of 50%. In addition, the addition of lactase to hydrolyze lactose resulted in yogurt of reduced calorie content because the glucose and fructose added sweetness so that less sucrose was added to the product.

Sweet acidophilus milk contains an active culture of *Lactobacillus acidophilus*. The culture is added to cold whole, lowfat, or skim milk and the mixture is kept cold. Therefore, relatively little fermentation occurs to produce lactic acid as in buttermilk or acidophilus milk, which has a high acid content and tart taste. Sweet acidophilus milk is similar in composition to whole or skim milk except that it contains less lactose and more lactic acid (Bassette and Acosta, 1988, pp. 47–48).

IV. USE OF MILK AND MILK PRODUCTS IN FOOD PRODUCTION

A. Heating Milk

1. Effects on Proteins
As with commerical processing, use of milk in food preparation may be preceded by or otherwise involve heating. The serum proteins are denatured with heating. The implications of this in flavor development, heat stabilization, and

rennin coagulation were discussed previously. In addition, the volume of loaf bread made with heated milk is greater than the volume of bread made with unheated milk. Insufficient heating of milk before spray drying depresses loaf volume. Milk that has been adequately heated would be designated as United States high-heat NFDM. The coagulated material that is deposited on the bottom of the container in which milk is heated is part of the denatured serum protein fraction. The precipitate is likely to scorch when milk is heated directly over a burner.

The casein proteins are not affected by the usual process of heating. However, problems can be encountered when foods containing acids and salts are heated in milk. Thus, the acid from tomatoes may cause cream of tomato soup to curdle, and salt from cured meats will cause coagulation.

2. Comparison of Use of Various Forms in Cooking

Creamed soups and mashed potatoes made with reconstituted evaporated milk were judged superior to those made with fluid milk, whereas few differences were noted when the two milks were used in yeast bread, quick bread, cakes, puddings, or baked custards (Atwood and Ehlers, 1933). Hussemann (1951) mentioned the distinctive color and flavor but the satisfactory texture of custards made with evaporated milk. Chocolate pudding containing evaporated milk ranked well, perhaps because the distinctive flavor of the milk was masked by the chocolate. Curdling occurred in scalloped potatoes more often with evaporated milk than with other forms of milk even though the milks were added as thin white sauces. Such curdling probably is due to the protein of evaporated milk being rendered unstable by the heat of processing and the fact that potatoes are more acidic than milk.

The basic effects of NFDM in food preparation were studied by Morse *et al.* (1950a, b). Nonfat dry milk increased the viscosity of pastes of flour, dry milk, and water in proportion to the amount of milk used. This increased viscosity could be offset by reducing the amount of flour. Nonfat dry milk also increased the thickness of white sauces, eggnog, custards, and chocolate pudding made with fluid milk. Although custards made without added dry milk were the most acceptable, those enriched with a moderate amount of NFDM were quite palatable and had a good flavor. Excessive amounts of dry milk produced pudding-like custards that were good when fresh and somewhat sticky after 24 hr. When NFDM is added to baked products at enrichment levels, changes in the basic recipe may improve the product. Because added milk increases the tendency of the product to brown during baking, it may be necessary to reduce the amount of sugar and to lower the baking temperature. To compensate for the thickening effect of a high solids content, less flour or more liquid can be used. An increase in fat offsets any toughening effects of milk solids.

B. Heating Cheese

Cheese cookery may consist simply of melting cheese, as in melting cheese on a pizza, or of combining it with other ingredients and cooking, as in making cheese sauce. Low temperatures and short times are essential in preventing excessive fat separation, stringing, matting, and toughening. "Aged natural" cheese and processed cheese are better for cooking than natural cheese that has not been aged properly.

A partial explanation of the good cooking qualities of aged cheese was reported by Personius and co-workers (1944). The cooking quality of Cheddar cheese of normal fat content improved as the products of protein hydrolysis, measured by the protein soluble in dilute salt solution, increased with aging. Cheese of high moisture content was superior in cooking quality to samples of normal or low moisture. The cooking quality of cheese that was low in fat did not improve with age, even though the amount of soluble protein increased during storage as much as that of normal cheese. When fat was not present, the tendency for mat formation and stringing was greatly exaggerated. The fat of normal samples melted during cooking, thus contributing to the consistency and tenderness of the heated product and preventing the formation of a continuous mass.

The harmful effects of excessive heat in cheese cookery also have been demonstrated. Greater fat separation, longer strings, and more toughening were observed in samples heated over water at 100°C than at 75°C and in samples exposed to direct heat of 177°C for the longest periods.

C. Whipping of Milk Products

Whipped cream is an air-in-water foam in which air cells are surrounded by a film containing fat droplets stabilized by a film of protein. Partial denaturation of this protein occurs as the cream is whipped. There is some clumping of the fat globules in the cell walls of the foam, and the fat is partially solidified, preventing collapse of the cell walls. When whipped cream is heated, the fat is melted and the foam collapses. If whipped cream is beaten too long, further clumping of the fat globules occurs and butter is formed. The whipping quality of cream has been found to improve with increased butterfat content up to 30%. Further increases in fat content do not improve the quality of the whip, but do improve the standing up quality and decrease the time required to whip the cream. Cream with less than 22% fat is not satisfactory for whipping.

Evaporated milk can be successfully whipped if it is thoroughly chilled prior to whipping. Air bubbles incorporated during whipping are trapped in the viscous liquid to form a foam. The difference in initial viscosity partially accounts for the difference in stability of foams formed from milk and from the more concentrated evaporated milk.

Nonfat dry milk can be whipped if reconstituted to a higher than normal solids content. Homogenization of milk prior to drying improves the quality of foams prepared from reconstituted NFDM (Tamsma et al., 1969).

V. USE OF MILK COMPONENTS IN FORMULATED FOODS

Dairy-based ingredients frequently are used to improve flavor and nutritive value of products. In addition, these ingredients may provide functional properties including foaming, emulsification, browning, stabilization, and water absorption. Processing methods used affect functional performance, and thus selection of ingredients from those available may be complex. Morr (1984) presented overviews of the production and use of milk proteins.

A. Caseins and Caseinates

On an annual basis, approximately 120,000 metric tons of casein products is produced (Swaisgood, 1985, p. 817), reflecting the importance of these products as functional ingredients in food products. Variation in the composition and functional properties of caseins depends on the process used for production. Therefore, careful choice to meet the needs for a specific product is necessary. Caseins may be prepared by acid or rennet precipitation. Rennin casein is higher in ash content and lower in protein content than is acid casein. Lactose is not present in rennet casein but is present in very small amounts in acid casein. Neither type of casein is very soluble and they are best suited for use in products where they contribute dough-forming characteristics and texture such as baked products, pasta products, snack foods, and breakfast cereal (Swaisgood, 1985, p. 817).

In contrast to the caseins, the caseinates are very soluble. Caseinate is produced by neutralizing acid casein with alkali before drying. Protein and ash composition are intermediate between rennet and acid casein and lactose content is equal to that of acid casein. Functional properties of these products include emulsification, water binding, thickening, whipping–foaming, and gelling (Swaisgood, 1985, p. 817). The wide range of products in which caseinates may be found includes margarine, beverages, coffee whiteners, frozen desserts, whipped topping, cream substitutes, and formulated meat products.

B. Whey Proteins

Sweet whey is a by-product of ripened cheese production; from cottage cheese and cream cheese production, acid whey is produced. Whey contains many valuable nutrients, particularly sulfur-containing amino acids. In addition to the "whey" proteins, caseins may also be present in whey products that are available commercially, particularly those that are produced at the pH that is associated with cheese making (Eigel et al., 1984). Although cheese whey often is used for animal feed, more whey is produced than can be used in animal feed and whey cheeses. As a result, cheese whey has become an environmental pollutant, as well as a wasted resource, and much research effort has been devoted to practical means of concentrating and using it. The primary problem associated with use of whey is the loss of functional properties of the proteins in the isolation process. As less harsh methods of isolating the proteins from whey become more economically feasible, use of these proteins as functional ingredients may increase. Ultrafiltration, a technique that involves separation of components by molecular size through a membrane, offers a means of fractionating undenatured whey proteins and lactose with minimal alteration in the proteins. Resultant products include a protein concentrate (retentate) and a lactose-containing fraction (permeate). Evaporation and spray drying may be used in addition to ultrafiltration in the production of whey protein concentrates (WPC) (Mortensen, 1985, pp. 109–119). Whey product concentrates may contain 25–95% protein (Kinsella, 1984). Fat, lactose, and calcium and phosphorus contents also vary with the preparation process (deWit, 1985, pp. 183–195; Kim *et al.*, 1989).

Earlier workers used dried whey successfully in starch puddings (Hanning *et al.*, 1955), cakes (de Goumois and Hanning, 1953), and baked products (Scanlon, 1974). Kulp *et al.* (1988) found that whey with a high-heat treatment performed better in pan bread than whey that had a low-heat treatment. Effect varied with level of inclusion. Its addition to soft drinks has been tested (Holsinger *et al.*, 1973). More recently, deWit (1985, pp. 183–195) suggested that ultrafiltered whey proteins containing milk fat may replace the emulsifying properties of eggs; this would be beneficial in the formulation of reduced cholesterol products. If lactose is not removed in the processing of WPC, color of cakes may be a problem attributable to the Maillard reaction (deWit, 1985, p. 191). Whey protein concentrates in amounts up to 3.5% have been approved by the USDA (1982) for use as a binder in sausage. Ensor and co-workers (1987) demonstrated that WPC could be successfully used for binding in knockwurst by showing that emulsions containing WPC were more stable than all-meat controls. They showed that levels of WPC below the allowable 3.5% produced a product with better sensory properties. Ultrafiltered skim milk retentates may also find use as coffee whiteners (Jimenez-Flores and Kosikowski, 1986).

Whey protein concentrates may be used for their gel-forming properties. Several factors affect these properties. As with other gel-forming ingredients, as protein concentration increases, gel strength increases and time required for gelation is decreased. The presence of lactose and lipids affects gelling properties; therefore, whey products processed in different ways will differ in gelling properties. A review of these factors is presented by Mulvihill and Kinsella (1987). β-Lactoglobulin plays a major role in gel formation. However, for formation of self-supporting gels, sodium chloride or calcium chloride must be present. Mulvihill and Kinsella (1988) indicated that 200 and 10 mM sodium and calcium chloride, respectively, gave gels of maximum compressive strength. The greater effectiveness of calcium chloride obviously is favorable from a dietary standpoint.

Many investigators have compared use of various milk fractions in products. Cooper *et al.* (1984) concluded that lactalbumins, lactic casein, rennet casein, and low-calcium and high-calcium coprecipitates (casein and whey) could be used in the production of cookies that would be acceptable to consumers. Cookie quality was most affected by particle size and method of drying the protein. Phil *et al.* (1982) reported that WPC and sodium caseinate could be substituted for milk-solids-nonfat (MSNF) in frozen desserts sweetened with lactose and fructose.

Selection of a WPC for use in a food product should be based on desired functional properties. For example, Peltonen–Shalaby and Mangino (1986) showed that fat in a WPC was not detrimental to foaming in a high-fat topping but was detrimental in a simple foaming test.

C. Hydrolyzed Milk Products

Hydrolyzed milk products offer a potential for fortification of food products. Bitter peptides produced during enzymatic hydrolysis may contribute undesirable taste to products. However, Ma and co-workers (1983) reported that debittering and desalting of hydrolysates resulted in a product that could be added at levels of 10% solids (3% protein) to soft drinks and fruit juices. Color, clarity, viscosity, and taste were not adversely affected. Hydrolysates of skim milk or NFDM produced by the action of lactase and a bacterial protease may be used for enteral hyperfeeding (Hernandez and Asenjo, 1982).

D. Lactose

Lactose is another product that may be used as a functional ingredient. It is useful when browning is desirable. Because lactose is not as sweet as sucrose it can be used in products such as toppings and icing where viscosity or body is desired but extensive sweetness is not desirable (Swaisgood, 1985, p. 820). Lactose crystals may be used as a dispersing agent because they are not hydroscopic but are free flowing (Pomeranz, 1985, p. 384). It also may serve as a carrier for flavor and color ingredients (Holsinger, 1988, pp. 330–332).

The permeate that results from ultrafiltration of whey is rich in lactose. Ogunrinola *et al.* (1988) noted that use of this fraction is limited and proposed that a hydrolyzed form of it could be used in white pan bread as a replacement for sucrose.

Suggested Exercises

1. Enrich some foods with NFDM and/or whey at different levels. Use as high a level of NFDM or whey as will make a palatable product. Formulas for custard and cake that could be used are included in Appendix A.
2. Prepare baked custards or white sauce using various forms of milk such as fluid whole milk, 2% milk, skim milk, evaporated whole milk or skim milk, UHT milk, and reconstituted nonfat dry milk. The custard formula in Appendix A could be used.
3. Add varying amounts of disodium phosphate or potassium citrate to samples of natural Cheddar cheese. Levels of 1, 3, and 5% of the weight of the cheese might be used for the two salts. Use these samples and one of processed cheese with no additions to make cheese sauces, and judge smoothness, viscosity, and palatability. Determine pH if possible.
4. Use one or more of the milk substitutes that are on the market in a product such as custard or cake (Appendix A).
5. Purchase milk in an opaque container. Place half of the milk in a clear glass container and allow to sit near a fluorescent light bulb for 30 min. Conduct a triangle test to see if class members can tell the difference in the two products.

6. Obtain a copy of Section 131 of the "Code of Federal Regulations" (FDA, 1988). Compare allowable additives to those listed on the labels of dairy products and discuss their functions.

7. Yogurt may be prepared in the laboratory using a commercial starter or a small amount of yogurt purchased on the retail market. For the latter, allow the yogurt to stand at room temperature for 3–4 hr to increase the population of microorganisms. Skim milk or reconstituted NFDM may be used and should be heated to 50°C and cooled to 40–45°C just before production. Combine 1 liter of milk and 75 g yogurt. Pour into individual containers and place in a water bath at 40–45°C for 3–3.5 hr. Variations to be tested could be addition of various levels of sugar or sugar and alternative sweeteners. Levels of sugar from 0 to 8% could be tested.

References

Atwood, F. J. and Ehlers, M. S. 1933. The use of evaporated milk in quantity cookery. *J. Amer. Dietet. Assoc.* **9**: 306–314.

Bassette, R. and Acosta, J. S. 1988. Composition of milk products. *In* "Fundamentals of Dairy Chemistry," 3rd edn., Wong, N. P., Jenness, R., Keeney, M., and Marth, E. H. (Eds.). Van Nostrand Reinhold, New York.

Bennett, R. W. and Amos, W. T. 1983. *Staphylococcus aureus* growth and toxin production in imitation cheeses. *J. Food Sci.* **48**: 1670–1673.

Bingham, E. W. 1975. Action of rennin on κ-casein. *J. Dairy Sci.* **58**: 13–18.

Bodyfelt, F. W., Tobias, J., and Trout, G. M. 1988. "The Sensory Evaluation of Dairy Products." Van Nostrand Reinhold, New York.

Cheryan, M., Van Wyk, P. J., Olson, N. F., and Richardson, T. 1975. Secondary phase and mechanism of enzymatic milk coagulation. *J. Dairy Sci.* **58**: 477–481.

Cooper, H. R., Patten, J. D., and Fletcher, R. H. 1984. Effect of some insoluble milk proteins on sensory characteristics of appearance and texture in cookies. *J. Food Sci.* **49**: 376–379, 392.

Creamer, L. K. and Olson, N. F. 1982. Rheological evaluation of maturing cheddar cheese. *J. Food Sci.* **47**: 631–636, 646.

Davies, D. T. and Law, A. J. R. 1980. Content and composition of protein in creamery milks in South-West Scotland. *J. Dairy Research* **47**: 83–90.

de Goumois, J. and Hanning, F. 1953. Effects of dried whey and various sugars on the quality of yellow cakes containing 100% sucrose. *Cereal Chem.* **30**: 258–267.

deWit, J. N. 1985. New approach to the functional characterization of whey proteins for use in food products. *In* "Milk Proteins '84, Proceedings of the International Congress of Milk Proteins," Galesloot, T. E. and Tinbergen, B. J. (Eds.). Pudoc, Wageningen, The Netherlands.

Eigel, W. N., Butler, J. E., Ernström, C. A., Farrell, H. M., Jr., Harwalkar, V. R., Jenness, R., and Whitney, R. McL. 1984. Nomenclature of the proteins of cow's milk: Fifth revision. *J. Dairy Sci.* **67**: 1599–1631.

Ensor, S. A., Mandigo, R. W., Calkins, C. R., and Quint, L. N. 1987. Comparative evaluation of whey protein concentrate, soy protein isolate and calcium-reduced nonfat dry milk as binders in an emulsion-type sausage. *J. Food Sci.* **52**: 1155–1158.

Farrell, H. M. Jr. 1988. Physical equilibria: proteins. *In* "Fundamentals of Dairy Chemistry," 3rd edn., Wong, N. P., Jenness, R., Keeney, M., and Marth, E. H. (Eds.). Van Nostrand Reinhold, New York.

FDA. 1988. Milk. *In* "Code of Federal Regulations," Title 21, Section 131.110. U.S. Govt. Printing Office, Washington, DC.

Gaylord, A. M., Warthesen, J. J., and Smith, D. E. 1986. Effect of fluorescent light on the isomerization of retinyl palmitate in skim milk. *J. Food Sci.* **51**: 1456–1458.

Hanning, F., Bloch, J. deG., and Siemers, L. L. 1955. The quality of starch puddings containing whey and/or nonfat milk solids. *J. Home Econ.* **47**: 107–109.

Hansen, A. P. 1987. Effect of ultra-high-temperature processing and storage on dairy food flavor. *Food Technol.* **41**(9): 112–114, 116.

Hansen, A. P. and Swartzel, K. R. 1982. Taste panel testing of UHT fluid dairy products. *J. Food Qual.* **4**: 203–206.

Hernandez, R. and Asenjo, J. A. 1982. Production and characterization of an enzymatic hydrolysate of skim milk lactose and proteins. *J. Food Sci.* **47**: 1895–1898, 1911.

Hettinga, D. H. 1988. Processing technologies for improving the nutritional value of dairy products. *In* "Designing Foods. Animal Product Options in the Marketplace." National Research Council, National Academy Press, Washington, DC.

Hill, A. R. 1988. Quality of ultra-high-temperature processed milk. *Food Technol.* **42**(9): 92–97.

Holsinger, V. H. 1988. Lactose. *In* "Fundamentals of Dairy Chemistry," 3rd edn., Wong, N. P., Jenness, R., Keeney, M., and Marth, E. H. (Eds.). Van Nostrand Reinhold, New York.

Holsinger, V. H., Posati, L. P., De Vilbiss, E. D., and Pallansch, M. J. 1973. Fortifying soft drinks with cheese whey proteins. *Food Technol.* **27**(2): 59–60, 64–65.

Houts, A. R. 1988. Lactose intolerance. *Food Technol.* **42**(3): 110–113.

Hussemann, D. L. 1951. Effect of altering milk solids content on the acceptability of certain foods. *J. Amer. Dietet. Assoc.* **27**: 583–586.

Hutton, J. T. and Patton, S. 1952. The origin of sulfhydryl groups in milk proteins and their contributions to "cooked" flavor. *J. Dairy Sci.* **35**: 699–705.

Jenness, R. 1988. The composition of milk. *In* "Fundamentals of Dairy Chemistry," 3rd edn., Wong, N. P., Jenness, R., Keeney, M., and Marth, E. H. (Eds.). Van Nostrand Reinhold, New York.

Jensen, R. G. and Clark, R. M. 1988. The lipids of milk: Composition and properties. *In* "Fundamentals of Dairy Chemistry," 3rd edn., Wong N. P., Jenness, R., Keeney, M., and Marth, E. H. (Eds.). Van Nostrand Reinhold, New York.

Jimenez-Flores, R. and Kosikowski, F. V. 1986. Nonfat dairy coffee whitener made from ultrafiltered skimmilk retentates. *J. Food Sci.* **51**: 235–238.

Keenan, T. W., Mather, I. H., and Dylewski, D. P. 1988. Physical equilibria: Lipid phase. *In* "Fundamentals of Dairy Chemistry," 3rd edn., Wong, N. P., Jenness, R., Keeney, M., and Marth, E. H. (Eds.). Van Nostrand Reinhold, New York.

Kim, S.-H., Morr, C. V., Seo, A., and Surak, J. G. 1989. Effect of whey pretreatment on composition and functional properties of whey protein concentrate. *J. Food Sci.* **54**: 25–29.

Kinsella, J. E. 1984. Milk proteins: Physicochemical and functional properties. *CRC Crit. Rev. Food Sci. Nutr.* **21**: 197–262.

Knipschildt, M. E. 1986. Drying of milk and milk products. *In* "Modern Dairy Technology. Vol. 1. Advances in Milk Processing," Robinson, R. K. (Ed.). Elsevier Applied Science Publishers, New York.

Kosikowski, F. 1986. New cheese-making procedures utilizing ultrafiltration. *Food Technol.* **40**(6): 71–77, 156.

Kroger, M. 1976. Quality of yogurt. *J. Dairy Sci.* **59**: 344–350.

Kroger, M., Kurmann, J. A., and Rasic, J. L. 1989. Fermented milk—past, present, future. *Food Technol.* **43**(1): 92, 93–97, 99.

Kulp, K., Chung, H. Doerry, W., Baker, A., and Olewnik, M. 1988. Utilization of whey as a white pan bread ingredient. *Cereal Foods World* **33**: 440, 442–443, 445, 446–447.

Linn, J. G. 1988. Factors affecting the composition of milk from dairy cows. *In* "Designing Foods. Animal Product Options in the Marketplace." National Research Council, National Academy Press, Washington, DC.

Loney, B. E., Bassette, R., and Claydon, J. J. 1968. Chemical and flavor changes in sterile concentrated milk during storage. *J. Dairy Sci.* **51**: 1770–1775.

Ma, C. Y., Amantea, G. F., and Nakai, S. 1983. Production of nonbitter, desalted milk hydrolysate for fortification of soft drinks and fruit juices. *J. Food Sci.* **48**: 897–899.

Magrini, R. C., Chirife, J., and Parada, J. L. 1983. A study of *Staphylococcus aureus* in model systems and processed cheese. *J. Food Sci.* **48**: 882–885.

Morr, C. V. 1969. Protein aggregation in conventional and ultra-high temperature heated skim milk. *J. Dairy Sci.* **52**: 1174–1180.

Morr, C. V. 1975. Chemistry of milk proteins in food processing. *J. Dairy Sci.* **58**: 977–984.

Morr, C. V. 1984. Production and use of milk proteins in food. *Food Technol.* **38**(7): 39–42, 44, 46–48.

Morr, C. V. and Richter, R. L. 1988. Chemistry of processing. *In* "Fundamentals of Dairy Chemistry," 3rd edn., Wong, N. P., Jenness, R., Keeney, M., and Marth, E. H. (Eds.). Van Nostrand Reinhold, New York.

Morse, L. M., Davis, D. S., and Jack, E. L. 1950a. Use and properties of non-fat dry milk solids in food preparation. I. Effect on viscosity and gel strength. *Food Research* **15**: 200–215.

Morse, L. M., Davis, D. S., and Jack, E. L. 1950b. Use and properties of non-fat dry milk solids in food preparation. II. Use in typical foods. *Food Research* **15**: 216–222.

Mortensen, B. K. 1985. Recent developments in the utilization of milk proteins in dairy products. *In* "Milk Proteins '84, Proceedings of the International Congress of Milk Proteins," Galesloot, T. E. and Tinbergen, B. J. (Eds.). Pudoc, Wageningen, The Netherlands.

Mulvihill, D. M. and Kinsella, J. E. 1987. Gelation characteristics of whey proteins and β-lactoglobulin. *Food Technol.* **41**(9): 102, 104, 108, 110–111.

Mulvihill, D. M. and Kinsella, J. E. 1988. Gelation of β-lactoglobulin: Effects of sodium chloride and calcium chloride on the rheological and structural properties of gels. *J. Food Sci.* **53**: 231–236.

Neff, E. and Morris, H. A. L. 1968. Agglomeration of milk powder and its influence on reconstitution properties. *J. Dairy Sci.* **51**: 330–338.

NRC. 1981. "Food Chemicals Codex." Food and Nutrition Board, National Research Council, National Academy Press, Washington, DC.

Ogunrinola, O. A., Jeon, I. J., and Ponte, J. G., Jr. 1988. Functional properties of hydrolyzed whey permeate syrups in bread formulations. *J. Food Sci.* **53**: 215–217.

Ohren, J. A. and Tuckey, S. L. 1969. Relation of flavor development in Cheddar cheese to chemical changes in the fat of the cheese. *J. Dairy Sci.* **52**: 598–607.

Palanuk, S. L., Warthesen, J. J., and Smith, D. E. 1988. Effect of agitation, sampling location, and protective films on light-induced riboflavin loss in skim milk. *J. Food Sci.* **53**: 436–438.

Patton, S., Forss, D. A., and Day, E. A. 1956. Methyl sulfide and the flavor of milk. *J. Dairy Sci.* **39**: 1469–1470.

Peltonen-Shalaby, R. and Mangino, M. E. 1986. Compositional factors that affect the emulsifying and foaming properties of whey protein concentrates. *J. Food Sci.* **51**: 91–95.

Personius, C., Boardman, E., and Asherman, A. R. 1944. Some factors affecting the behavior of Cheddar cheese in cooking. *Food Research* **9**: 304–311.

Phil, M. A., Stull, J. W., Taylor, R. R., Angus, R. C., and Daniel T. C. 1982. Characteristics of frozen desserts sweetened with fructose and lactose. *J. Food Sci.* **47**: 989–991.

Pitcher, W. H. 1986. Genetic modification of enzymes used in food processing. *Food Technol.* **40**(10): 62–63, 69.

Pomeranz, Y. 1985. "Functional Properties of Food Components." Academic Press, Orlando, FL.

Posati, L. P. and Orr, M. L. 1976. Composition of foods. Dairy and egg products. Agric. Handbook 8-1, Agricultural Research Service, USDA, Washington, DC.

Pszczola, D. E. 1989. Rennet containing 100% chymosin increases cheese quality and yield. *Food Technol.* **43**(6): 84, 88–89.

Savaiano, D. A. and Kotz, C. 1988. Recent advances in the management of lactose intolerance. *Contemporary Nutr.* **13**(9, 10): 1–4.

Sawyer, W. H. 1969. Complex between β-lactoglobulin and κ-casein. *J. Dairy Sci.* **52**: 1347–1355.

Scanlon, J. 1974. A new sweet whey-soy protein blend for the baker. *Bakers Digest* **48**(5): 30–33, 56.

Sharma, S. K., Ferrier, L. K., and Hill, A. R. 1989. Effect of modified manufacturing parameters on the quality of cheddar cheese made from ultrafiltered milk (UF). *J. Food Sci.* **54**: 573–577.

Sherbon, J. W. 1988. Physical properties of milk. *In* "Fundamentals of Dairy Chemistry," 3rd edn., Wong, N. P., Jenness, R., Keeney, M., and Marth, E. H. (Eds.). Van Nostrand Reinhold, New York.

Shimp, L. A. 1985. Process cheese principles. *Food Technol.* **39**(5): 63–64, 66, 69, 70.

Swaisgood, H. E. 1985. Characteristics of edible fluids of animal origin: Milk. *In* "Food Chemistry," 2nd edn., Fennema, O. (Ed.). Marcel Dekker, Inc., New York.

Tamsma, A., Kontson, A., and Pallansch, M. J. 1969. Production of whippable nonfat dried milk by homogenization. *J. Dairy Sci.* **52**: 428–431.

Thakur, M. K., Kirk, J. R., and Hedrick, T. I. 1975. Changes during ripening of unsalted Cheddar cheese. *J. Dairy Sci.* **58**: 175–180.

USDA. 1981. Federal and state standards for the composition of milk products. Handbook 51, AMS, United States Department of Agriculture, Washington, DC.

USDA. 1982. Approval for use of whey and whey products in sausages, bockwurst, chili concarne, and pork and beef with barbecue sauce. *Federal Register* **47**: 26371-26373.

Whalen, C. A., Gilmore, T. M., Spurgeon, K. R., and Parsons, J. G. 1988. Yogurt manufactured from whey-caseinate blends and hydrolyzed lactose. *J. Dairy Sci.* **71**: 299–305.

Whitney, R. McL. 1988. Proteins of milk. *In* "Fundamentals of Dairy Chemistry," 3rd edn., Wong, N. P., Jenness, R., Keeney, M., and Marth, E. H. (Eds.). Van Nostrand Reinhold, New York.

MEAT

Red meat (beef, veal, pork, and lamb) is a major component of the American diet, supplying approximately 15% of the calories, 27.6% of the protein, and 27.7% of the fat. In addition, significant quantities of the B vitamins and iron are supplied (NRC, 1988, pp. 18–44). Per capita consumption of beef, pork, veal, and lamb and mutton was 72.7, 63.1, 1.4, and 1.4 lb/year, respectively (USDA, 1989). Research regarding red meat, beef in particular, has been extensive; therefore, this chapter must be considered as an introduction to the topic. For more information on meat science the reader is referred to books by Price and Schweigert (1987) and Lawrie (1985).

I. EATING QUALITY OF MEAT AS DETERMINED BY MUSCLE TISSUE COMPONENTS AND STRUCTURE

The gross composition of several red meats and products derived from red meats is shown in Table I. Water, protein, fat, carbohydrate, and other components are important to the eating quality of meat. The contributions of muscle tissue will be considered in terms of their contributions to three quality attributes: color, texture, and flavor.

A. Color

1. Myoglobin and Hemoglobin

The heme pigment myoglobin and its various chemical forms are responsible for the color of muscle tissue. Hemoglobin, the blood pigment, also contributes to meat color. Of the pigment present 20–50% may be hemoglobin (Fox, 1987, pp. 197–199). Both pigments combine reversibly with oxygen to supply it for metabolic processes. Hemoglobin transports oxygen in the blood stream,

TABLE I
Composition of Selected Raw Red Meat Cuts[a]

Muscle food	Water (%)	Protein (%)	Fat (%)
Beef			
Chuck	70.3	21.2	7.4
Loin	47.5–55.7	13.0–17.0	27–35
Round	66.6	20.1	12.2
Lamb			
Leg	64.8	17.8	16.2
Loin	57.5	16.3	24.9
Pork			
Loin	57.2	17.1	25.0
Shoulder cuts	59.3	15.4	24.4

[a]Values calculated from Adams (1975); Carbohydrate listed as 0 for all muscle foods in Adams (1975).

whereas myoglobin holds it in the tissues. In muscle such as heart, which has a high oxygen demand, myoglobin is present in relatively high amounts and contributes to intense color. Muscles of older animals contain increased amounts of myoglobin. Thus, veal is lighter in color than beef. There also is a species difference; for example, pork, which is marketed at a later age than veal, contains approximately the same amount of myoglobin as veal. The iron–porphyrin compound heme is combined with a simple protein of the globin class to form the conjugated protein myoglobin. Hemoglobin is also a conjugated protein containing heme and globin. Although the heme portion of these two molecules is identical, the globins are different, and the hemoglobin molecule contains four heme protein units whereas myoglobin has only one. The structure of heme is shown in Fig. 9.1. The iron atom of heme is bound covalently to four nitrogens, each in a pyrrole ring of the porphyrin structure. In myoglobin the iron atom also is bound to a nitrogen of a histidyl residue in the globin moiety.

A sixth position is open for the formation of complexes with several compounds (Bandman, 1987, pp. 73–76). Variations at the sixth position are in part responsible for differences in color of meat. An abbreviated formula for myoglobin is shown in Fig. 9.1. The uppermost vertical line in the abbreviated formula shows where the element sharing the sixth pair of electrons will be positioned.

a. Oxymyoglobin.   In living tissue, myoglobin, which is purplish red, exists in equilibrium with its bright red oxygenated form, oxymyoglobin, as shown in Fig. 9.2. After slaughter, the oxygen in the tissues is used quickly and the pig-

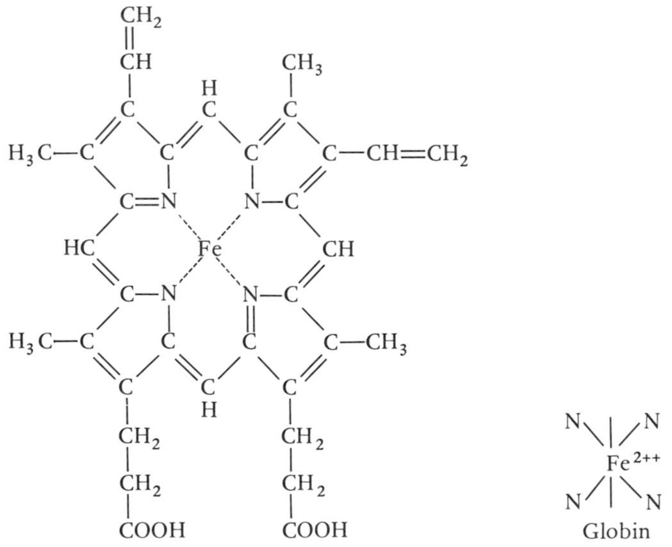

FIGURE 9.1
Complete formula for heme (left) and abbreviated formula for myoglobin (right).

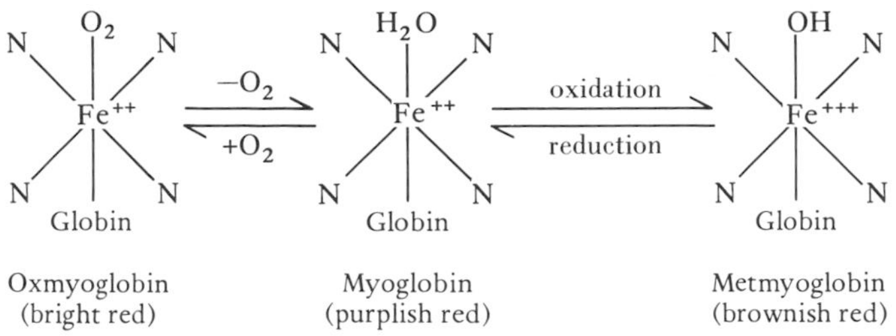

FIGURE 9.2
Pigment changes in red meat.

ment exists almost entirely in the purplish reduced form. In the interior of the meat this pigment is stable at low temperatures for long periods of time because free oxygen is not present. When cut, meat surfaces become bright red as oxygen from the air combines with myoglobin to form oxymyoglobin. (*Note:* the iron is still reduced.) It remains red for a limited time if sufficient oxygen is present, a condition that is met in most prepackaged meats by the use of packaging film that is permeable to oxygen. Exclusion of oxygen from meat, as by vacuum packaging, results in reversion to a purple color as the pigment is deoxygenated to myoglobin. Such packaging extends shelf life of prepackaged beef because it suppresses microbiological growth and retards possible spoilage (Seideman and Durland, 1983). Such packaging is currently used for much of the meat that goes to restaurants, institutions, and manufacturing plants for further processing (Young *et al.*, 1988). Packages of this type may or may not be back flushed with gas to control the initial atmosphere. In either case, the technique is called modified atmosphere packaging because the process or the changes that are effected by the meat itself during storage modify the enclosed atmosphere. Recently, retail cuts of beef in vacuum packages have been introduced to the market. Whether consumers will accept this product remains a question. The meat in these packages will have the purple color of myoglobin, not the bright red color of oxymyoglobin to which consumers are accustomed. It is important to note that although vacuum packaging does extend the shelf life of meat, refrigeration and sanitary handling are still essential.

b. Metmyoglobin.   The major discoloration problem of fresh meat is the formation of the brownish red pigment metmyoglobin as the iron of myoglobin is oxidized from the ferrous to the ferric state (Fig. 9.2). As long as there are reducing substances present in the meat, metmyoglobin that is formed is changed back to myoglobin; but when the reducing power of the muscle is lost, the color of the meat becomes brownish. High temperatures and microorganisms that use oxygen accelerate metmyoglobin formation. It may also be increased in the presence of fluorescent and incandescent light of 250 fc (foot candle) intensity as used in meat departments of some retail stores (Satterlee and Hansmeyer, 1974). After the formation of metmyoglobin, further oxidative changes in meat pigment, caused by enzymes and bacteria, produce a series of brown, green, and faded looking compounds (Watts, 1954).

2. Changes during Cooking

The red pigment of meat cooked to "rare" is the same oxymyoglobin that is present in raw meat. In the range 40–50°C, the globin of oxymyoglobin is denatured and the ferrous iron is oxidized to the ferric state. This denatured, oxidized form of the pigment probably is in all well-done meat. Meat cooked just to the point of pigment denaturation and then cooled may revert from brown to red as denaturation is reversed and reoxygenation occurs (Fox, 1987, pp. 205–207). The progressive change in color that occurs with heating is illustrated in photographs by the National Livestock and Meat Board and published as a center insert in Romans *et al.* (1985). Longissimus steaks broiled and roasted to end-point temperatures were scored with those photographs used as standards in a study by Bowers *et al.* (1987). According to the data presented, visual changes in color were evident between 60 and 65°C and between 75 and 85°C.

Other reactions and physical changes also contribute to the color of cooked meat. Browning and concentration of pigments as the surface is dehydrated contribute to dark colors on the surface (Fox, 1987, pp. 205–207). Surface browning is attributed to three components. Fats decompose and react with the products of carbohydrate and protein decomposition. One reaction that occurs is carbonyl-amine browning, or the Maillard reaction.

3. Changes during Curing

During curing, the pigment of meat changes to a form that remains red during cooking. The traditional curing process involved the preservation of meat by the addition of salt, sodium or potassium nitrate, and sugar. After it was learned that the nitrate functioned in pigment conversion only after being reduced to nitrite, both nitrate and nitrite were added. Public health considerations have resulted in regulations limiting the use of nitrites and nitrates (Townsend and Olson, 1987, pp. 431–439) The use of nitrate is limited to specific products (dry-cured country ham and bacon and dry sausage) and processes because of the difficulty in controlling residual levels when nitrate is the source of nitrite. In addition to fixation of the color of cured meat, nitrite also inhibits the outgrowth of *Clostridium botulinum*, helps to prevent oxidative rancidity, and may contribute to the flavor of some cured products (Townsend and Olson, 1987, pp. 436–437).

Reactions involving myoglobin and occuring during curing are shown in Fig. 9.3. In the older methods of curing, which are still being used to cure solid pieces of meat like hams, salts diffuse into the tissues from brine and dry salt mixtures applied to the outside of the meat; in newer methods, a curing solution is pumped into the tissues through arteries or injected into the meat in several locations. The pigment characteristic of cured meat is formed by the combining of nitric oxide (produced by reduction of nitrite) with myoglobin to form an unstable red pigment, nitrosyl or nitric oxide myoglobin. When the meat is heated, as it often is during the curing process, the more stable pink pigment, nitrosyl hemochrome, is formed. This pigment is responsible for the color of ham, bacon, and corned beef as well as that of many table-ready processed meats. Its color is stable during cooking, but becomes brown when exposed to light and air because the iron of the pigment is oxidized from the ferrous to the ferric state, thus changing the pigment to denatured nitrosyl hemichrome. Be-

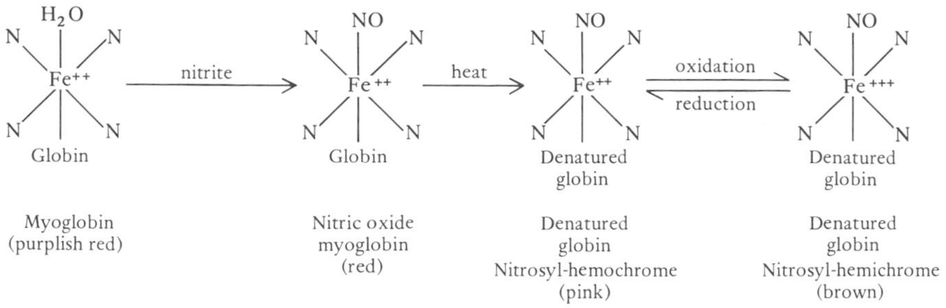

H_2O

N | N   N   NO   N   N   NO   N   N   NO   N

Fe^{++} — nitrite → Fe^{++} — heat → Fe^{++} ⇌ oxidation / reduction Fe^{+++}

N | N   N | N   N | N   N | N

Globin   Globin   Denatured globin   Denatured globin

Myoglobin (purplish red)   Nitric oxide myoglobin (red)   Denatured globin Nitrosyl-hemochrome (pink)   Denatured globin Nitrosyl-hemichrome (brown)

FIGURE 9.3
Pigment changes in cured meat.

cause the pigments of cured meat fade rapidly in transparent packages unless oxygen is excluded, they are often vacuum packaged and/or placed upside down in the meat case to exclude light.

B. Flavor

The odor and taste of cooked meat vary with preslaughter factors as well as time, temperature, and method of cooking. Cooking is important because the desirable attributes of flavor develop during cooking. Carbonyl-amine browning is important to the development of meat flavors (Hornstein and Wasserman, 1987, p. 342). Specific components responsible for meat flavor were reviewed by Chang and Peterson (1977), who suggested that flavor is a complex characteristic arising from the presence of lactones, acyclic sulfur-containing compounds, and aromatic and nonaromatic heterocyclic compounds containing sulfur, nitrogen, and oxygen. Differences in the precursors of these compounds may account for differences in flavor among species (Hornstein and Wasserman, 1987, pp. 329–331). Cramer (1983) postulated that differences in sulfur compounds are responsible for the difference between beef and lamb flavor. Diet of the animal also influences flavor. Grain-finished meat has a more desirable "beefy fat" flavor than meat from animals fed on pasture. The undesirable flavor has been described as "milky–oily" and may be accompanied by sour notes (Melton, 1983a). Beef and lamb flavor were reviewed by Moody (1983) and Cramer (1983), respectively.

C. Texture

Texture of meat is considered by many to be synonymous with tenderness, or the ease of chewing a piece of meat. However, texture is more complex than that. Cover et al. (1962a, b, c) described tenderness as a multicomponent sensation including juiciness, softness, muscle fiber, and connective tissue components of tenderness. Thus, understanding of tenderness, or more correctly texture, requires an understanding of water and water-binding capacity, muscle fiber proteins and ultrastructure, connective tissue proteins and organization, and fat content and distribution.

1. Water and Water-Holding Capacity

Water is quantitatively the most important muscle constituent, as shown in Table I. Since it does represent about 75% of the weight of the tissue it is important to the texture of meat as well as the color and flavor.

A very small portion of the water in muscle tissue is bound very closely to the muscle proteins via hydrogen and hydrophobic bonding (Wismer-Pedersen, 1987, pp. 141–142). Most of the remaining or so-called free water is located within muscle fibers, with smaller amounts located within connective tissue and the sarcoplasm. Mechanical immobilization of water is possible because of spaces in the three-dimensional structure of the muscle fibers. Disrupting this structure reduces water-holding capacity (WHC) or the ability of the meat to hold inherent or added water when force is applied. Grinding, proteolysis, freezing, salting, pH alteration, and heating will reduce WHC.

Water-holding capacity is thought to be related to perception of juiciness of meat. Two phases of juiciness perception can be identified; the initial release of juices from the meat with early chews and the continued impression of juiciness with further release of serum and increased flow of saliva resulting from the stimulation by fat in the meat.

2. Muscle Fibers

The fibers of skeletal muscle tissue are long, slender multinucleated cells ranging in length from a few millimeters up to 34 cm and averaging 60 μm in diameter (Bailey, 1972). Fibers are composed of smaller structures, myofibrils. Approximately 2000 myofibrils, 1.0 μm in diameter, are found in each average-sized fiber. Molecules of the proteins actin, tropomyosin, and troponin are arranged in an orderly manner as illustrated by Fig. 9.4. Microscopically, muscle tissue samples appear to have a pattern of cross-striations as shown in Fig. 9.5.

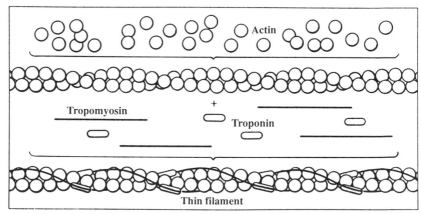

FIGURE 9.4
Schematic diagram showing assembly of the thin filament from actin, tropomyosin, and troponin, and the molecular architecture of the assembled thin filament. (From "The Cooperative Action of Muscle Proteins," by John M. Murray and Annemarie Weber. Copyright © February, 1974 by Scientific American, Inc. All rights reserved.)

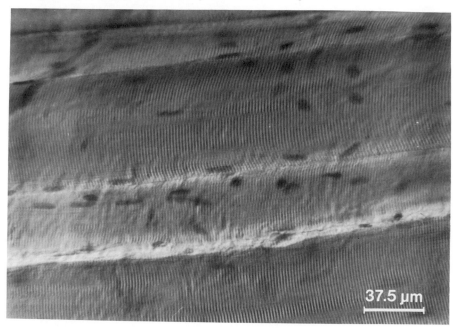

FIGURE 9.5
Light micrograph of raw beef semitendinosus tissue showing cross-striations and nuclei.

The light-and-dark banding pattern is produced by the orderly arrangement of the thin actin filaments and the thicker myosin filaments. Tropomyosin is located along the length of the thin filaments and probably is located, along with α-actinin, in the Z-disk, a structure that apparently holds the thin filaments together at the Z-line. The Z-lines are visible in Fig. 9.6. Actin and myosin form actomyosin as they slide past each other as the muscle contracts. The nucleotide adenosine triphosphate (ATP) supplies the energy for contraction. Other proteins found in muscle tissue are described by Bechtel (1986, pp. 2–35).

Muscle fiber sizes and the effect of heat on them are related to the tenderness of the cooked meat. Slender muscle fibers appear to be associated with tenderness, since as fiber diameter increases, tenderness decreases (Hiner *et al.*, 1953). The diameter of the fiber increases as the animal increases in age and decreases as heat is applied (Bendall and Restall, 1983).

Sarcomere length has been measured as an indicator of muscle fiber characteristics. Shortened sarcomeres are associated with toughness (Marsh and Leet, 1966). Z-Lines, the boundaries between sarcomeres, are clearly visible in the three-dimensional photograph from a scanning electron microscope shown in Fig. 9.6. Electron microscopy in the study of meat texture was described by Jones (1977) and Hearne *et al.* (1978).

Direct evidence of the effect of heating was obtained by heating isolated fibers (Wang *et al.*, 1956). Heating of meat causes denaturation and coagulation of the proteins with an unfolding of the peptide chains and the formation of new salt and hydrogen bonds. At temperatures below 40°C denaturation is mild, but it

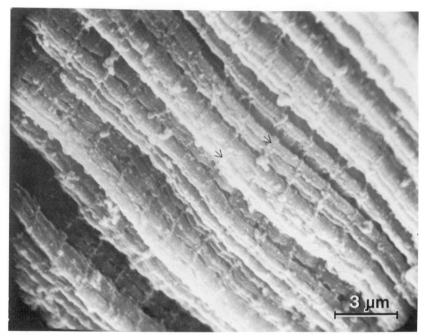

FIGURE 9.6
Scanning electron micrograph of raw beef semitendinosus tissue. Z-Line structures are indicated
by arrows. (Courtesy of Tennessee Agricultural Experiment Station, The University of Tennessee,
Knoxville.)

becomes greater as the temperature increases and is almost complete at 65°C
(Hamm and Deatherage, 1960). Bowers *et al.* (1987) reported that denaturation
is mostly complete by 80°C. Coagulation tightens the network of protein structure of meat, and increases the pH by decreasing the acidic but not the basic
groups of the protein.

3. Connective Tissue

Muscle fibers are supported in the body by connective tissue. Each fiber is enclosed in the cell or plasma membrane, or sarcolemma, and groups of fibers are
organized into bundles by an extremely thin network of connective tissue called
the endomysium. A bundle is surrounded by a sheet of connective tissue called the
perimysium. A group of such bundles, surrounded by an outer layer of connective
tissue called the epimysium, forms the muscle. The relationship of muscle fibers
to these three types of connective tissue is shown in Fig. 9.7. This harnessing of
the muscle by connective tissue and its attachment by means of a tendon to bone
can be seen clearly in a chicken leg or in the heel of a round of beef.

a. Collagen.   Collagen is one of two fibrous connective tissue proteins. It is the
principal component of tendons, which attach muscle to bone, and the connective tissue within and around muscles. Collagen frequently is described as a
"coiled coil," because three helical polypeptide chains are entwined. Approxi-

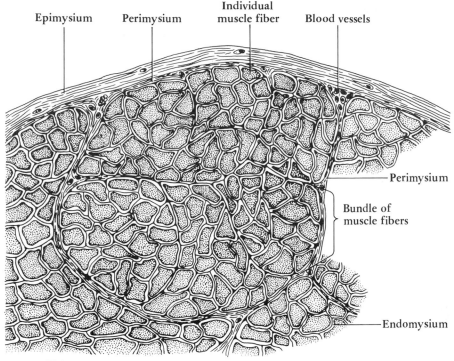

FIGURE 9.7
Diagram of a cross-section of muscle tissue showing bundles of individual fibers.

mately 13% of the amino acid residues in collagen are hydroxyproline. So unique is this hydroxyproline content of collagen that determinations of hydroxyproline are used for estimating collagen levels in meat samples. Collagen fibers contract to about one-third of their original length when heated to the collagen shrink temperature (60–75°C). Although collagen shrink temperature has been reported to vary with species, Gustavson (1956) noted that variation in values within species may be greater than the variation in values between species. Solubilization of collagen occurs around 65°C. When heated to higher temperatures in the presence of water, collagen may be converted to gelatin.

b. Elastin.   Elastin or yellow connective tissue is the second fibrous connective tissue protein and is important in ligaments, which perform various functions, such as connecting two bones or cartilages. Elastin fibers stretch, but return to their original shape when pulling forces are removed. Changes in elastin with heating may occur but are probably not significant because the elastin content of muscles is very low.

c. Reticulin Fibers and Ground Substance.   Muscle tissue also contains thin reticulin fibers, which may be a form of immature collagen and are often associated with the endomysium. Collagen and elastin are embedded in an amorphous material called ground substance, which contains glycoproteins, lipids, carbohy-drates, and water (Cassens, 1987, pp. 16–17).

d. Influence on Texture.   Because connective tissue is an important factor in the tenderness of cooked meat, the content of both collagen and elastin of vari-ous cuts of meat has been studied. McKeith *et al.* (1985) listed collagen content of several muscles from the chuck, round, rib, and loin. Variation in content among the muscles is shown in Table II. A low but significant negative correla-tion between sensory scores and connective tissue content was reported for 12 of the muscles. Johnson *et al.* (1988) also reported a low but significant correla-tion coefficient between shear values and collagen content of 23 muscles. How-ever, the relationship was positive for only 8 of the 23 individual muscles. For the most part, muscles with large amounts of connective tissue are less tender. Muscles that are used constantly may contain more collagen than do muscles used less frequently (Hiner *et al.*, 1955). Although the amount of collagen in muscle tissue does not increase with chronological age, the number and strength

TABLE II
Warner–Bratzler Shear Values, Collagen Content, and Sensory Scores for Tenderness of 13 Beef Muscles [a,b]

			Sensory scores [c]		
Wholesale cut muscle	Shear force (kg)	Collagen (mg/g, wet basis)	Muscle fiber tenderness	Connective tissue amount	Overall tenderness
Chuck					
Infraspinatus	3.28ab	17.81a	6.74b	6.23c	6.50b
Pectoral	6.34d	8.79cd	4.77f	4.02g	4.57f
Supraspinatus	4.45c	8.72cd	5.40e	5.08ef	5.19e
Triceps	3.89bc	10.45c	5.48e	5.55de	5.57de
Loin					
Gluteus medius	3.48abc	8.13cd	5.63e	5.75d	5.63de
Longissimus—loin	3.46abc	5.04ef	6.26cd	6.43bc	6.17bc
Psoas major	2.64a	3.23f	7.30a	7.37a	7.10a
Rib					
Longissimus—rib	3.78bc	4.66f	6.46bc	6.74b	6.42b
Round					
Adductor	4.16bc	5.52ef	5.45e	5.34def	5.42de
Biceps femoris	5.49d	9.60cd	5.83e	4.91f	5.44de
Rectus femoris	3.68bc	13.96b	5.88de	5.75d	5.74cd
Semimembranosus	3.94bc	7.30de	5.40e	5.14ef	5.52de
Semitendinosus	3.57abc	8.32cd	5.48e	5.60de	5.53de

[a] From McKeith *et al.* (1985). (Adapted from *Journal of Food Science*, **50**, No. 5, pp. 869–872. Copyright © 1985 Institute of Food Technologists.)

[b] Means in a column followed by like letters do not differ ($p > 0.05$).

[c] Muscle tissue tenderness and overall tenderness: 1 = extremely tough; 8 = extremely tender. Connective tissue amount: 1 = abundant; 8 = none.

of cross-bonds between peptide chains do increase (Hill, 1966). This increased bonding decreases the amount of collagen that may be solubilized during heating and contributes to decreased tenderness (Herring *et al.*, 1967; Hill, 1966; Goll *et al.*, 1964).

The tenderness of isolated connective tissue increases with heating in water at 65°C for 16 min or at higher temperatures for shorter periods of time (Winegarden *et al.*, 1952). In steak, softening may occur at as low a temperature as 61°C, as shown by loss from the meat of about one-fourth of the collagen during broiling (Irvin and Cover, 1959). Collagen becomes shorter and thicker during heating, a change that may account for the plump appearance of some cooked meat in which the length decreased and the width increased. Clearly, any changes occurring in connective tissue, primarily collagen, during cooking are for the better, since the softening of connective tissue contributes to tenderness of meat.

4. Fat

As an animal is fattened, the fat is deposited first around certain organs and under the skin, and later as marbling within the muscles. Intramuscular fat cells are located in the spaces in the perimysium, most frequently along small blood vessels. Both quantity and distribution of marbling are important factors in the grading system used by the USDA (1975). Photographs of steaks with the varying degrees of marbling are used in grading by the USDA and are reproduced in a center insert in Romans *et al.* (1985). Marbling groups and mean extractable fat levels (%) as determined by Savell *et al.* (1986) are practically devoid (1.77), traces (2.48), slight (3.43), small (4.99), modest (5.97), moderate (7.34), slightly abundant (8.56), and moderately abundant (10.42). Researchers have failed to show a strong relationship between marbling group and sensory scores for flavor desirability, juiciness, tenderness, and overall desirability (Smith *et al.*, 1984). However, Cross *et al.* (1980) reported that as fat increased from 16 to 28% in ground beef patties, panelists perceived greater juiciness and tenderness. Little difference between sensory values for 20 and 28% fat patties suggests that increases above 20% would not be advantageous. Savell *et al.* (1987) conducted a home placement study to determine if consumers could find differences among steaks of different marbling levels. Consumers prepared steaks as they normally prepared steaks and evaluated each steak (one per week for 7 weeks) on a nine-point scale ranging from extremely desirable to extremely undesirable. Mean scores suggested that desirability was decreased as level of marbling was decreased. The magnitude of the effect, however, seemed to depend on city of testing. Although scores decreased as marbling level was decreased, all means were greater than 6.5 on the nine-point scale.

Fat is sometimes yellowish due to the presence of carotenoids. The color is more pronounced in dairy than in beef animals, in older animals, and in those consuming feed high in carotene.

The characteristics of fatty tissue vary with species because fatty acids in the triglycerides (neutral lipids) vary in degree of saturation and in chain length. Most of the fatty acids are monounsaturated and saturated. As the degree of saturation increases, hardness of the fat increases. Thus, lamb fat is harder than beef fat. Details of the fatty acid composition have been reported by many investigators (Duncan and Garton, 1967; Stinson *et al.*, 1967; Terrell *et al.*, 1967).

Animal tissue also contains small quantities of phospholipids. As membrane components, the phospholipids are of structural and functional importance to muscle tissues. The phospholipids are less saturated than the neutral lipids and probably affect flavor and keeping quality of meat and meat products (Dugan, 1987, pp. 109–111). Cholesterol is found associated with phospholipids in the membranes of some animal cells (Dugan, 1987, pp. 111–112).

Microscopic study of the distribution of fat in meat during cooking shows that fat is released from the cells by heating and is dispersed finely in locations where collagen has been hydrolyzed during cooking. The effect of this melted and dispersed fat on tenderness has not been demonstrated although several theories have been proposed and reviewed by Savell and Cross (1988, pp. 345–355). For the "bite theory," fat itself takes less force to chew; therefore, its presence as marbling in meat serves to dilute the muscle fibers and connective tissue, which take more force to chew. Lubrication of the muscle fibers and connective tissue and the resultant increase in perceived tenderness form the basis of the "lubrication theory." The "strain theory" suggests that the presence of fat has a thinning effect on connective tissue walls. The fourth or "insurance theory" indicates that meat that contains marbling will not be overcooked to cause toughening of the muscle fibers. The fat content of meat may influence its perceived juiciness as discussed in the section on water (Section I, C, 1).

5. Carbohydrate

The very small quantity of carbohydrate in meat is important to the texture and other eating qualities. Conversion of glycogen stores to lactic acid during postmortem aging determines the final pH of the meat and therefore influences water-holding capacity, firmness, and color.

II. MUSCLES AND MEAT QUALITY

Muscles that make up a retail cut of meat may differ in tenderness. Muscle locations within beef and pork carcasses are shown diagrammatically by Tucker *et al.* (1952), Kauffman and St. Clair (1965), and Kinsman and Gilbert (1972). Muscles of conventional meat cuts are shown in Fig. 9.8A–F. Those most often used in meat research are included. Some muscles appear in several cuts. For example, longissimus extends the length of the spine from the neck to the pelvis, thus appearing in the chuck, rib, short loin, and sirloin wholesale cuts, whereas psoas major begins in the short loin and extends through the sirloin. The biceps femoris extends from the sirloin into the round and makes up most of the lean meat of the rump.

McKeith and co-workers (1985) presented data on chemical and sensory properties of 13 major beef muscles. Data are based on values obtained for steaks broiled to 70°C. The muscles studied are shown in Fig. 9.8. Data are

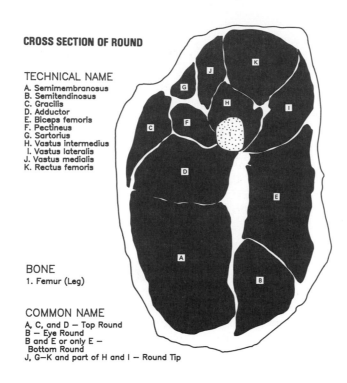

CROSS SECTION OF ROUND

TECHNICAL NAME

A. Semimembranosus
B. Semitendinosus
C. Gracilis
D. Adductor
E. Biceps femoris
F. Pectineus
G. Sartorius
H. Vastus intermedius
I. Vastus lateralis
J. Vastus medialis
K. Rectus femoris

BONE

1. Femur (Leg)

COMMON NAME

A, C, and D — Top Round
B — Eye Round
B and E or only E —
  Bottom Round
J, G—K and part of H and I — Round Tip

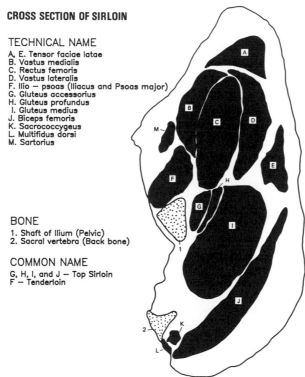

CROSS SECTION OF SIRLOIN

TECHNICAL NAME

A, E. Tensor faciae latae
B. Vastus medialis
C. Rectus femoris
D. Vastus lateralis
F. Ilio — psoas (Iliacus and Psoas major)
G. Gluteus accessorius
H. Gluteus profundus
I. Gluteus medius
J. Biceps femoris
K. Sacrococcygeus
L. Multifidus dorsi
M. Sartorius

BONE

1. Shaft of Ilium (Pelvic)
2. Sacral vertebra (Back bone)

COMMON NAME

G, H, I, and J — Top Sirloin
F — Tenderloin

FIGURE 9.8

The names and locations of the major muscles in several retail beef cuts. (Adapted from Kinsman and Gilbert, 1972)

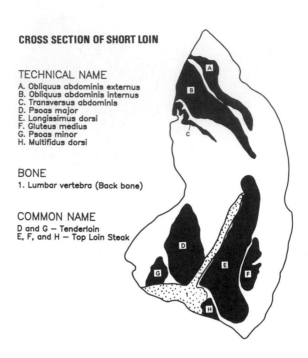

CROSS SECTION OF SHORT LOIN

TECHNICAL NAME
A. Obliquus abdominis externus
B. Obliquus abdominis internus
C. Transversus abdominis
D. Psoas major
E. Longissimus dorsi
F. Gluteus medius
G. Psoas minor
H. Multifidus dorsi

BONE
1. Lumbar vertebra (Back bone)

COMMON NAME
D and G — Tenderloin
E, F, and H — Top Loin Steak

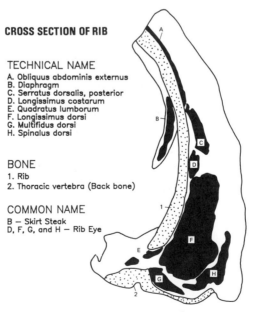

CROSS SECTION OF RIB

TECHNICAL NAME
A. Obliquus abdominis externus
B. Diaphragm
C. Serratus dorsalis, posterior
D. Longissimus costarum
E. Quadratus lumborum
F. Longissimus dorsi
G. Multifidus dorsi
H. Spinalus dorsi

BONE
1. Rib
2. Thoracic vertebra (Back bone)

COMMON NAME
B — Skirt Steak
D, F, G, and H — Rib Eye

FIGURE 9.8 (continued)

CROSS SECTION OF CHUCK BLADE FACE

TECHNICAL NAME

A. Triceps brachii, long head
B. Infraspinatus
C. Teres major
D. Subscapularis
E. Supraspinatus
F. Trapezius
G. Scalenus dorsalis
H. Serratus ventralis
I. Rhomboideus
J. Splenius
K. Longus colli
L. Spinalis dorsi
M. Complexus
N. Multifidus dorsi
O. Longissimus dorsi and
   longissimus costarum

BONE

1. Scapula (Blade)
2. Thoracic vertebra (Back bone)
3. Ligamentum nuchae (Neck strap)
4. Rib

COMMON NAME

A, B, and C — Chuck Top Blade
C, D, G, H, L, M, and O —
Chuck Under Blade
E — Mock Tender

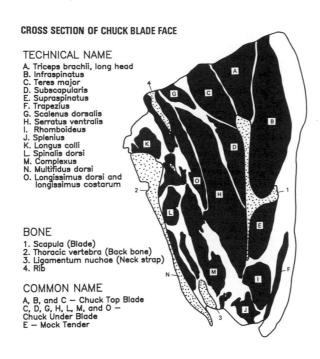

CROSS SECTION OF CHUCK ARM FACE

TECHNICAL NAME

A. Brachiocephalicus
B. Biceps brachii
C. Brachialis
D. Triceps brachii, lateral head
E. Triceps brachii, long head
F. Tensor faciae antibrachial
G. Cutaneus trunci
H. Coracobrachialis
I. Triceps brachii, medial head
J. Deep pectoral
K. Superficial pectoral
L. Rectus thoracis
M. Intercostal muscles
N. Serratus ventralis

BONE

1. Humerus (Arm)
2. Ribs

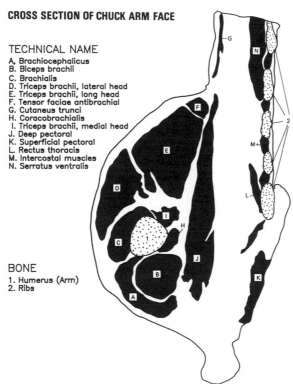

FIGURE 9.8 (continued)

summarized in Table II. Shear values (see Chapter 3) indicated that the psoas major and longissimus from the loin, the infraspinatus from the chuck, and the gluteus medius and semitendinosus from the round were ranked highest and were similar in tenderness. The toughest groups of muscle as determined by shear values included the biceps femoris from the round and the pectoral from the chuck. As indicated in sensory scores for tenderness (Table II), no striking differences were found in the tenderness of the principal muscles of the round. Retail cuts contain muscles of widely varying tenderness. For example, the chuck contains some very tender as well as some very tough muscles (Johnson *et al.*, 1988).

For research in meat cookery, a large uniform muscle is preferred so that adequate samples will be available for sensory evaluation and chemical and/or physical testing. The cut that is used frequently for roasting is the standing rib roast. Because of its relative uniformity, the semitendinosus or eye of round muscle also is used frequently for meat studies. Paired cuts, or identical cuts from the right and left sides of the same animal, are used in comparisons of two treatments if possible. A second choice for research material is adjacent cuts from the same animal. Variation must, however, be expected among muscles of the same cut (Paul *et al.*, 1970) as well as among different parts of the same muscle (Williams *et al.*, 1983). There are, of course, wide variations in the meat of different animals.

III. POST-MORTEM AGING

Processing of meat prior to preparation for consumption involves slaughter and a period of post-mortem aging. Several changes that influence the eating quality of meat occur in that period.

A. Rigor Mortis

A few hours after the slaughter of the animal, the carcass becomes rigid, a condition known as rigor mortis. This development is accompanied by the disappearance of adenosine triphosphate from the tissues and the interaction of actin and myosin filaments as they slide past one another to form actomyosin (Huxley, 1969). Tension is created in the muscle fibers by the formation of actomyosin. Histological examination (Paul *et al.*, 1952) of the fibers reveals dense nodes of concentration alternated with areas of extreme stretch that make some of the fibers look like an accordian, as shown in Fig. 9.9. A day or two after the onset of rigor, the muscle becomes soft again, the fibers become straighter, and breaks appear in the muscle fibers. The breaks occur when the actin filaments separate from their connections to the Z-line. The degradation of the muscle fibers, which continues throughout the aging period, may be caused by the hydrolysis of the muscle fiber proteins by catheptic enzymes, but further investigation of this area is needed.

Zero storage
(Note straight
unbroken fibers)

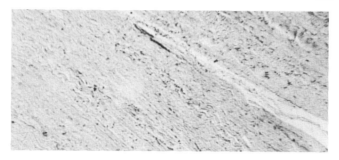

24-hour storage
(Note contracted
appearance of fibers)

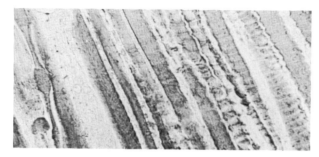

144–149 hours
storage (Note
breaks in fibers)

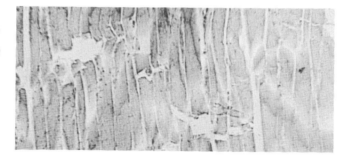

FIGURE 9.9
Histological appearance of raw beef after various periods of cold storage. (From Paul *et al.*, 1952; Reprinted from *Food Research*, **17**, No. 6, pp. 504–510. Copyright © 1952 by Institute of Food Technologists.)

B. Decreased pH

The pH of muscle tissue decreases after slaughter. The pH of living tissue is usually in the range of 7.0–7.2. With depletion of the supply of oxygen to the tissues after slaughter, lactic acid, which is produced from glycogen, accumulates, resulting in a decrease in pH. Normally pH drops to about 5.5.

C. Changes in Quality

At first, quality decreases, as shown by the toughness of meat that is cooked while in rigor. Because of this, all meats are best when aged until rigor has passed before they are frozen or cooked. After rigor has passed, meat gradually becomes more tender and flavor is improved. Currently, the vacuum packaging of primal and subprimal cuts eliminates the problem of shrinkage and microbial spoilage during an aging period of several weeks (Pearson, 1987, pp. 182–183).

IV. PROCESSED MEATS

Forming and restructuring of meat products as well as fish and poultry products provide opportunity for control of fat, connective tissue, and protein contents of products as well as portion control for institutional and restaurant operations. Processed meat products range from comminuted products to flaked and restructured products to whole-muscle formed items. Additional processing may include cooking of the product so that the consumer must only reheat before consumption. For more detailed information on processed meats the reader is referred to Pearson and Tauber (1984).

A. Comminuted Products

The functionality of muscle proteins and the roles of ingredients in comminuted meat products were reviewed by Whiting (1988). Production of such products involves chopping, which disrupts the muscle cellular structure. Finely divided fat is suspended in the water–protein matrix to form a meat batter (previously called meat emulsion). Smoking and cooking the batter results in denaturation of the proteins, leading to gelation or formation of the characteristic texture (Foegeding, 1988) as well as skin formation and flavor development.

Concern for salt in the diet has led to several studies on the lowering of salt, an essential ingredient in comminuted products. If the level of salt is reduced to below 2% from the normal 2.5% in frankfurters, WHC is lost and the product becomes softer (Sofos, 1983a, b). Flavor and texture acceptability are reduced. KCl may be substituted for NaCl but the bitterness of KCl may be a problem (Whiting and Jenkins, 1981). Frye *et al.* (1986) reported that partial replacement of NaCl with KCl gave acceptable properties of tumbled hams. $MgCl_2$ was not as useful. Phosphates are used in these products to improve the functionality of the proteins. Sofos (1986) reviewed their function. No more than 0.05% may be added. Offer and Trinick (1983) demonstrated that in the presence of pyrophosphate in a NaCl solution swelling of the myofibrils is enhanced. Many characteristics of meat products are affected by the use of phosphate. These include increased juiciness, improved binding and WHC, and better texture. These compounds also contribute sodium content to the product. Breidenstein (1982) described calculation of the sodium content of processed meat products. Terrell (1983) reviewed the role of salt and reduction of levels in processed meats.

Fat also plays an important role in the texture of these products. To improve the "healthfulness" of meat products, St. John *et al.* (1986) developed a frankfurter with an increased monounsaturated/saturated (M/S) ratio and 25% less total fat. To alter the M/S ratio, fat from pigs fed varying concentrations of canola oil was used. The resultant products were shown to have sensory properties that were comparable to those of regular frankfurters. Concern for improved products (NRC, 1988, pp. 115–132) will result in additional research of this nature.

B. Restructured Products

The process of restructuring involves reduction of particle size by grinding or flaking, mixing, and forming. Formulations for these products may include sodium chloride and phosphates. Both of these facilitate extraction of the myosin proteins, which allows for binding of the final product. Extraction is accomplished as components of the product are mixed to extract myosin to the surface. Forming may be accomplished by one of three approaches as described by Mandigo (1986). These include the stuff-freeze-temper-slice method, slide plate-forming machines, and cavity fill machines. A review of factors involved in restructuring was presented by Mandigo (1986).

Many studies on formulation and cooking of restructured products may be found in the literature. Level of connective tissue will influence the quality of the finished product. Berry et al. (1986) noted that high levels of connective tissue were associated with more distortion, hardness, and cohesiveness than low levels. Trained panelists were used in the study. Reico et al. (1987) found in a home placement study that consumers preferred restructured steaks made from extensively trimmed (ET) beef shoulder clod over intermediate-trim (IT) and no-trim (NT) products. The degree of trim controlled the amount of connective tissue in the product. Intermediate-trim and ET steaks were scored similarly for juiciness, tenderness, and flavor as well as desirability of juiciness, tenderness, and amount of "gristle" present. Therefore, the authors concluded that intermediate trimming (removal of large pieces of connective tissue) would be adequate for production of products with acceptable levels of connective tissue.

Fat level also may be controlled in restructured products. For example, Costello et al. (1985) found that chewiness and hardness decreased while moisture release and greasiness increased with increased fat level. Steaks containing 15, 20, and 25% fat were equally desirable to consumers. Thus, some lowering of fat in restructured products is feasible.

Flake size is an additional variable that influences the quality of restructured meat products. For example, Bernal et al. (1988) reported that steaks with 13-mm flakes were harder, more chewy, and more cohesive than those with 3- or 6-mm flakes. Salt level also influences the characteristics of these products. Research of this type will enable processors to select conditions that will make the products most acceptable to consumers.

C. Precooked Products

Another important trend in marketing is the sale of precooked meats to be reheated in a food service operation or by the consumer. Reheating may be in boiling water, conventional oven, or microwave. One problem associated with these products, which may or may not be vacuum packaged, is the development of warmed-over flavor (WOF). Warmed-over flavor is caused by the development of oxidative rancidity in cooked meats and also may occur in processed raw meats that have had tissue structures disrupted (Asghar *et al.*, 1988). Compounds related to the development of WOF were studied by St. Angelo *et al.* (1987) and Spanier *et al.* (1988). Major differences in flavor notes between freshly prepared meat and cooked, stored, and reheated meat are decreased cooked beef lean and beef fat and browned notes along with increased intensity of cardboard and oxidized/rancid/painty notes (Johnson and Civille, 1986). The development of WOF frequently is monitored with the 2-thiobarbituric acid test (TBA test), a measure of lipid oxidation. Use of the test was reviewed by Melton (1983b).

White *et al.* (1988) cooked samples of semimembranous muscle and stored them for 0, 1, 4, and 7 days at 1°C. Although TBA numbers increased throughout the storage period, 100 consumer panelists showed no change in their evaluation of the product until it was stored for 7 days. If products are to be accepted by consumers after cooking and storage, methods for preventing the development of WOF must be established.

Nitrites, phosphates, ethylenediaminetetraacetic acid (EDTA), citric acid, ascorbic acid, synthetic phenolic antioxidants, and natural antioxidants are among the compounds that may slow WOF development. Details of their action were discussed by Bailey (1988). Many studies on their effectiveness may be found in the literature.

Paterson and Parrish (1988) studied the effects of four treatments on roasts from the infraspinatus muscle of beef chuck. Test roasts were pumped with 10% of their weight in water, 4.75% sodium tripolyphosphate (TP), or 10% NaCl and 4.75% TP. A roast with no addition served as the control. After thermal processing in a smoke house to 63°C roasts were dipped in a 3% acetic acid solution and vacuum packaged for storage at 2°C or packaged and stored without the dip. Roasts were tested at intervals over a 45-day period. As expected, the roasts with salt and phosphate were shown to be more tender, juicy, flavorful, and palatable than the other roasts; TBA values were equally low for both phosphate-containing products. The acetic acid did not affect sensory properties of the roasts but was effective in lowering bacterial counts.

Jones *et al.* (1987) studied precooked pork roasts prepared as cook-in-the-bag products and found that precooking after packaging in a vacuum bag limited growth of microorganisms. Although the intensity of the pork flavor decreased over 28 days of storage, no off-flavors were noted. This was credited to the antioxidant properties of sodium tripolyphosphate. Huffman *et al.* (1987) also demonstrated the flavor-protecting function of polyphosphates on stability of restructured beef and pork nuggets that were breaded, cooked, and stored in polyethylene bags at −23°C. Less off-flavor was noted by panelists in phosphate-treated nuggets than in control products.

V. MEAT COOKERY

Meat is cooked to produce acceptable color, flavor, and texture. Many studies on the influence of heat on meat quality have been reported. Variations in sizes, shapes, and kinds of cuts and heating equipment are seen in the research literature. Thus, the discussion that follows must be considered an overview. Further reports on the effects of heat on the quality of meat can be found easily in the food science literature.

In general, it is accepted that muscle fibers harden and become less tender with heating, and connective tissue softens with heating. Thus an ideal cooking method will minimize hardening but maximize softening of the connective tissue. Appropriate methods vary with the cut of meat. The hypothesis that very tender cuts of meat containing little or no collagen decrease in tenderness during heating whereas less tender cuts of meat become more tender has been tested many times. Time and temperature of heating also must be considered. Time is more important for collagen softening whereas hardening of muscle fibers is more dependent on temperature.

A. Roasting

Roasting, or more specifically oven roasting, is the cooking of meat uncovered in dry heat as contrasted to pot roasting, which is a form of braising, a moist-heat method. For roasting, meat is placed on a flat open pan, with a rack if the bony structure is not sufficient to keep the meat out of the drip as the meat is heated in the oven. The meat is not basted and no water is added. Such dry roasting results in meat that is quite different from that produced by heating in a covered pan, because covering the pan holds in the moisture that is lost from the meat or from added liquid, resulting in moist-heat cookery. The methods of heat transfer in various meat cookery methods differ, resulting in variations in the characteristics of the product. In roasting, heat energy is transferred through the air by convection and the meat is heated by conduction.

1. Oven Temperature
Early workers (Cline *et al.*, 1930) reported that beef ribs roasted at low temperatures (125 and 163°C) were more tender, juicy, flavorful, and uniformly done throughout, required longer heating times, and showed less cooking loss than meat heated at higher temperatures (218 and 260°C). A temperature of 163°C usually is recommended for roasting of all meats.

2. End-Point Temperature
Control of end-point temperature is important in experimental work. To monitor temperature, thermometers or thermocouples should be placed in the geometric center of the piece of meat. Location should be checked periodically because the probes may be displaced as meat expands and contracts during heating. Suggested end-point temperatures generally are 55°C (very rare), 60°C (rare), 65°C (medium rare), 70°C (medium), 75°C (well), and 80°C (very well done). Internal temperature may continue to rise for a few minutes after meat is removed from the oven. The rise is greater if the oven temperature is relatively high and if the meat is rare, because the difference in temperatures between the outside and the center of the roast is relatively large when the roast is removed from the oven. The internal temperature rises as heat is transferred from the

warmer outside of the roast to its cooler center and the increase is greater for a large roast than for a small one.

As the internal temperature of tender cuts of meat increases during roasting, the meat becomes less tender (Cover and Hostetler, 1960) and less juicy (Bowers *et al.*, 1987), while cooking losses increase (Bayne *et al.*, 1973; Bowers *et al.*, 1987). In contrast, Bowers *et al* (1987) reported no differences in shear values of longissimus steaks roasted or broiled to end points ranging from 50 to 80°C.

3. Cooking Losses

In meat research, cooking losses usually are measured. For roasted meat, these losses consist of drip that remains in the pan and evaporative loss, which is the water that evaporates from the meat and drip during cooking. Losses may be determined by the procedure outlined on the form in Table III. Losses increase as the end-point temperature of the meat or the oven temperature is increased (Bayne *et al.*, 1973). This is of considerable economic importance in commercial food service, because with higher cooking losses, fewer servings will be obtained per pound of raw meat.

4. Oven Roasting of Less Tender Cuts of Meat

Roasting at low temperatures (66–121°C) has been studied extensively as a method of increasing the tenderness of less tender cuts of meat. Bayne *et al.* (1969) reported that large and small top round roasts heated at 93°C to an internal temperature of 67°C were more tender than roasts heated at 149 to 70°C. Laakkonen *et al.* (1970) reported that a major increase in tenderness occurred as semitendinosus samples were heated from 50 to 60°C. The time in that temperature range is almost 3.5 times as long during heating at 93°C as during heat-

TABLE III
Cooking Loss Data for Heating of Meat

Factor measured	Treatment	
	x	y
A. Before cooking		
1. Weight of pan (g)	_____	_____
2. Weight of roast (g)	_____	_____
3. Weight of pan and roast (g)	_____	_____
B. On removal from oven		
1. Weight of pan, roast, and drip (g)	_____	_____
2. Weight of pan and drip (g)	_____	_____
C. Losses		
1. Loss due to evaporation (g) $(A_3 - B_1)$	_____	_____
2. Loss due to drip (g) $(B_2 - A_1)$	_____	_____
3. Total cooking loss (g) $(C_1 + C_2)$	_____	_____
D. Losses as percentages of weight of uncooked roast		
1. Due to evaporation (%) $[100 \times (C_1/A_2)]$	_____	_____
2. Due to drip (%) $[100 \times (C_2/A_2)]$	_____	_____
3. Total loss (%) $(D_1 + D_2)$	_____	_____

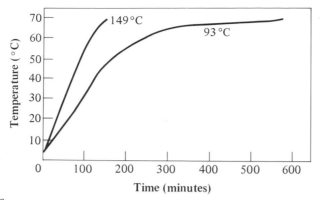

FIGURE 9.10
Time–temperature curves from oven roasting of top round roasts at 93 and 149°C (From Penfield and Meyer, 1975; Reprinted from *Journal of Food Science*, **40**, No. 1, pp. 150–154. Copyright © 1975 by Institute of Food Technologists.)

ing at 149°C, as illustrated in Fig. 9.10. Long, slow heating results in greater solubilization of collagen, and shear values decrease as the percentage of collagen solubilized increases (Penfield and Meyer, 1975). However, it should be noted that extensive collagen solubilization during heating of meat with high collagen content may not be reflected in increased tenderness because hardening of muscle fibers may contribute to decreased tenderness (Paul *et al.*, 1973; Penfield and Meyer, 1975).

Top rounds of choice beef were found desirable for institutional use when roasted at 149°C (Marshall *et al.*, 1959). Reduction of the oven temperature to 93, 107, or 121°C for these roasts was not recommended because cooking times were long and variable, and because the roasts cooked to the well-done stage at 93°C crumbled when sliced (Marshall *et al.*, 1960). The use of lower temperatures for heating sirloin roasts in forced convection ovens was studied by Davenport and Meyer (1970). In a comparison between 93 and 149°C, the lower temperature reduced cooking losses and cost per serving and increased yield and cooking time. Shear values and sensory ratings of tenderness, juiciness, and flavor were not affected. Since the roasts heated at both temperatures were rated "good" to "very good," the benefits of the lower temperature were obvious.

B. Braising

Braising is cooking meat in a small amount of liquid or steam in a covered utensil. The rate of heat penetration is more rapid in braising than in roasting because steam conducts heat more rapidly than does air. Braising methods include braising on the range, in an oven, in a pressure cooker, or in a slow cooker, as well as in aluminum foil or oven film in an oven.

Browning in a small amount of fat before braising is optional but desirable because it improves both the color and the flavor of the meat. Browning in a pan preheated to 246°C proved more satisfactory than browning at a lower temperature (Paul and Bean, 1956) because at a low temperature the meat tended to stew in liquid that escaped from the meat rather than to brown. The palatability and cooking losses of meat that was oven braised at 121, 149, or 177°C were similar but 121°C was selected as best because less watching was required to prevent scorching of the meat. Bowers and Goertz (1966) reported that eating quality, cooking loss, and time were not affected by the addition of water to skillet-braised pork chops. Brownness was decreased, however. Oven temperature did not affect the quality of pork chops that were oven braised. When meat is braised on the range, a temperature below the boiling point usually is recommended. However, on the basis of their research, Cover and Hostetler (1960) reported that a gentle boil is as satisfactory as a simmering temperature, which is often difficult to maintain.

1. End-Point Temperature

There is no general agreement concerning what internal temperature is best for braising meat. Most consumers probably braise meat until it is tender when tested with a fork rather than to a certain end-point temperature. This stage has been reproduced in the laboratory by bringing the meat to an internal temperature of 100°C and holding it there for 25 min. Cover (1959) observed that meat is very tender at this point. The connective tissue is soft, and muscle fibers often seem crumbly or friable when cut with a fork or chewed. The texture of such muscle fibers is very different from that of rare meat, which seems soft to the teeth and tongue. Very friable meat offers less resistance to teeth or fork and therefore seems very tender. It has the disadvantage, however, of seeming dry and crumbly. Bowers and Goertz (1966) reported that the cooking losses and juiciness decreased as the internal temperature of both oven- and skillet-braised pork chops increased.

2. Cooking of Tender and Less Tender Cuts
   of Meat

Tender and less tender cuts of meat do not respond alike to cooking, as shown by Cover (1959). Samples from the loin and bottom round were broiled to two internal temperatures, 61 and 80°C, and braised to two internal temperatures, 85 and 100°C, plus 25 min. Judges gave high scores to the tenderness of connective tissue of the loin cuts, indicating that there was little possibility for improvement with increased doneness or braising. The loin contained only a small amount of soft connective tissue. By contrast, the biceps femoris of the round had a large amount of firm connective tissue that became tender in both methods as doneness increased. It was tough when broiled rare, about equally tender when broiled well done and braised to 85°C, but very tender when braised for the longest period. Muscle fibers of bottom round differed from those of the loin, becoming softer and more friable when braised to the higher temperature. The muscle fibers of the loin, which are soft when rare, may be stringy when braised because little connective tissue is present to hold them together. As the cooking time was increased in the Cover (1959) study, making the meat more well done, the muscle fibers of the loin seemed to toughen, while those of the round became more friable and therefore more tender. Schock and co-workers

(1970) reported that semitendinosus roasts heated to 70°C by deep-fat frying, oven roasting, oven braising, and pressure braising did not differ in flavor, tenderness, or overall acceptability. Juiciness differed, but it did not influence the overall acceptability scores as much as flavor and tenderness did.

3. Pressure Cookery
Cooking time is reduced by pressure cooking, a modification of braising. In a comparison of pressure braising with deep-fat frying, oven broiling, and oven roasting of beef semitendinosus muscle samples, Schock et al. (1970) observed that pressure-braised pieces were scored least juicy and had the greatest cooking losses. As noted earlier, other characteristics did not differ among the methods. Cooking losses and sensory scores for color, tenderness, flavor, juiciness, and overall acceptability did not differ for microwave-heated, pressure-cooked, oven-braised, and slow cooker-braised beef semitendinosus roasts (Brady and Penfield, 1977).

4. Slow Cookers
Few studies on the quality of meat heated in a slow cooker have been published. Rump roasts heated in a slow cooker for 9 hr reached an internal temperature of 81°C as compared to internal temperatures of 85°C at a low oven temperature of 107°C for 9 hr and 77°C at an oven temperature of 177°C. Cooking losses were similar at the low oven temperature and in the slow cooker, but greater than at the high oven temperature (Sundberg and Carlin, 1976). Brady and Penfield (1977) reported a trend toward greater cooking losses from semitendinosus roasts heated in slow cookers than from roasts heated in a conventional oven, a microwave oven, or a pressure cooker. Results of the sensory evaluation were discussed previously. Peters et al. (1983) indicated that quality of meat loaf cooked in a conventional oven was superior to that of meat loaf cooked in a slow cooker.

One area of concern with slow cookers is the microbiological aspect. Several researchers (Brackett and Marth, 1977; Ritter et al., 1979; Peters et al., 1983) concluded that the use of slow cookers as recommended is safe. Overloading of the slow cooker may result in the survival of potentially harmful microorganisms.

5. Foils and Films
Meat can be wrapped tightly in aluminum foil to retain steam as effectively as a covered pan. Bramblett et al. (1959) compared two oven temperatures, 63 and 68°C, for heating of foil-wrapped muscles from beef round. The lower temperature produced the more tender, juicy meat. Beef roasts cooked in foil may have a steamed flavor (Blaker et al., 1959). Baity et al. (1969) reported that at a low oven temperature (93°C), beef loaves cooked more rapidly when foil was tightly sealed than when the loaves were not wrapped or loosely wrapped. At a higher temperature (232°C), cooking times increased with foil wraps. Quality attributes of the loaves were not discussed.

Polyester cooking films and bags are also available for use in the heating of meat. Results of studies of the quality of meat prepared in these films and bags suggest that the quality is comparable to that of meat heated in foil. Beef top round roasts heated in bags to 80°C in ovens at 177 and 205°C took less time but had greater cooking losses than the roasts heated similarly but without bags (Shaffer *et al.*, 1973). Roasts heated in film also have a more well-done appearance (Ferger *et al.*, 1972; Shaffer *et al.*, 1973).

C. Broiling and Frying

Oven broiling, pan broiling, and frying are quick methods of cooking meat by dry heat. In broiling, radiant heat is applied directly to the meat with minimal heating by conduction of heat from the pan, whereas in panbroiling the meat is put into a hot pan and heated by conduction. No fat is added during pan broiling, and fat that accumulates during cooking is poured off. Frying is cooking in fat, either a small amount or a deep layer. For all of these quick methods, characteristic results depend on the use of enough heat to brown the outside of the meat without overcooking the inside. The high temperatures commonly used in these methods do not seem to toughen the meat, perhaps because the cooking time is short, or because tender meat is used.

1. Broiling
Broiling of pork chops has been studied frequently. Oven-broiled chops were more tender, flavorful, and acceptable than deep-fat fried chops (Flynn and Bramblett, 1975). Juiciness was not affected by heating method. Broiling also has been compared with braising. Cooking time was shorter and cooking losses greater with broiling of pork chops than with braising (Weir *et al.*, 1962). Sensory evaluation suggested that the broiled chops were more tender, more juicy, and less flavorful than the braised chops. The thickness of broiled chops and steaks influences their quality. Thick, broiled chops (3.75 cm thick) were less juicy than thin chops (1.88 cm thick) (Weir *et al.*, 1962). Holmes *et al.* (1966) reported similar results.

Because broiled meat had not been compared directly with roasted meat, Batcher and Deary (1975) compared beef round steaks heated by the two methods. They concluded that the steaks prepared by roasting were more tender, more juicy, and less mealy than the broiled steaks. Increased cooking time for roasting may have influenced the increased tenderness.

Broiling of ground beef patties of varying fat levels (9, 21, and 28.5%) to two end-point temperatures (71 and 77°C) was studied by Kregel *et al.* (1986). Total cooking losses were greater at 77°C than at 71°C. Juiciness and tenderness decreased and beef flavor intensity increased as the end-point temperature increased. As fat level increased, tenderness, mouth coating, and juiciness were increased while connective tissue amount and beef flavor were decreased. The investigators reported that on a per patty basis, cholesterol was highest in the lowest fat patty. A comparable study and similar results were reported for ground pork patties (Reitmeier and Prusa, 1987). Thus the effects of cooking method as well as composition of the raw product should be considered in meat selection for a cholesterol-controlled diet. Similar conclusions were drawn by

Hoelscher *et al.* (1987) in a study of broiled and pan-fried beef patties containing 0, 5, 10, 15, 20, 25, or 30% fat.

2. Deep-Fat Frying
Meat heated by deep-fat frying is heated by transfer of heat from the oil to the meat. Because the temperature of the oil is very high, a crust forms rapidly on the outside of the meat. Deep-fat fried roasts lost more moisture and were scored as less juicy and more well done in appearance than oven-roasted semitendinosus roasts (Schock *et al.*, 1970). Differences in flavor, tenderness, and overall acceptability were not found. Carpenter *et al.* (1968) reported that beef rib steaks that had been deep-fat fried were more tender than those heated by microwaves or broiled conventionally.

Quick methods of heating such as broiling and frying are important in institutional and restaurant cooking. As restructured meat products become more important in those operations choice of cooking method will be needed. Quenzer *et al.* (1982) studied four methods of cooking restructured steaks that were or were not breaded. Breaded and unbreaded deep-fat fried steaks were lower in moisture and higher in fat that those cooked by oven roasting, grill frying, and broiling. The texture of broiled and grilled unbreaded steaks was preferred. Breading was needed to give desirable characteristics to deep-fat fried products. Overall, grill frying was found to be preferred.

D. Microwave Heating

The increasing popularity of microwave ovens has encouraged many studies on this cooking method. Principles of the method are described in Chapter 6. Generalizations on the effect of microwaves on meat quality are difficult because so many models of ovens with varying power levels are available. Methods for monitoring end-point temperature in microwave heating are less than satisfactory; therefore, for most studies preliminary work is used to determine how long to cook a product to reach the desired end-point temperature, which is measured when the product is removed from the oven. The application of fiber optics to temperature measurement will help to minimize this problem. However, the systems are expensive and general use is not expected.

Many investigators have reported that cooking losses are greater with microwave heating than with conventional heating of meat (Carpenter *et al.*, 1968; Ream *et al.*, 1974). Greater cooking losses most frequently are attributed to greater evaporative losses. Denaturation of proteins within the muscle may cause juices to be forced out of the meat (Carpenter *et al.*, 1968). However, Korschgen and co-workers (1976) found that cooking losses were greater for beef longissimus but not for pork longissimus or deboned leg of lamb when heated by microwaves. Law *et al.* (1967) reported that losses from beef loin and top round steaks were greatest during conventional broiling, intermediate with microwave heating, and lowest with oven roasting. Berry and Leddy (1984a) evaluated patties of ground round (14% fat), ground chuck (19% fat), and regular ground beef (24% fat) cooked by six methods (electric broiling, charbroiling, conventional and convection roasting, frying, and microwave heating). The effect of cooking method depended on the fat level. As in other studies, cooking loss was greatest with the microwave method. High levels of cooking losses with this method may be attributed to lack of crust formation. The cooking losses from microwave-heated patties contained more fat than patties cooked by other methods. This additional loss of fat produced patties with the lowest caloric value. Original fat level did not affect calorie content.

Juiciness scores are relatively low for microwave-heated beef roasts (Ream *et al.*, 1974). Berry and Leddy (1984b) reported that hamburger patties cooked in a microwave were equal in juiciness to fried patties but less juicy than patties that were broiled, charbroiled, or oven roasted in a conventional or convection oven.

Low surface temperatures and the associated lack of surface browning may influence eating quality of microwave-cooked foods. For example, Berry and Leddy (1984b) found that microwave-heated patties were comparable in flavor intensity to conventionally oven-roasted patties but had less intense flavor than broiled and fried patties. Korschgen and co-workers (1976) found that flavor scores for the outside portions of pork and lamb roasts were higher for conventionally heated roasts than for those heated in a microwave oven, whereas scores for internal portions did not differ. Beef roasts heated by microwaves have been described as less flavorful (Ream *et al.*, 1974).

In addition to juiciness and flavor, tenderness of microwave-heated meat has been compared to that of conventionally heated meat. Differences in tenderness seem to depend on the kind of meat, the size of the piece heated, and the conventional method of heating used in the study. In studies comparing conventional roasting and microwave heating, beef arm and rib roasts (Ream *et al.*, 1974) heated by microwaves were shown to be less tender by sensory evaluation. However, Headley and Jacobson (1960) reported that legs of lamb prepared by the two methods did not differ in tenderness. Fulton and Davis (1983) reported that roasted rib eye and braised chuck and round roasts prepared in a microwave were as tender as their conventionally prepared counterparts. Less than full cooking power was used for roasts, which may account for the differences between the results of this study and some earlier ones. Hines *et al.*, (1980) demonstrated that power level can influence product quality. Pork chops (1.9 cm thick) were cooked at four power levels (low, 205 W; medium, 270 W; roast, 350 W; and high, 505 W). Broiled chops cooked to 77°C served as the control. Rotation of chops in the microwave was necessary to ensure uniform heating. Postcooking rises in temperature occurred for the high and roast setting but not the other two. Total cooking losses were greatest for the broiled chops, which had the most desirable flavor. The low power level produced the most juicy chops. Broiled chops were equal in overall acceptability to low, medium, and roast chops and equal in tenderness to the low microwave chops. From these results one can conclude that the highest power level in a microwave may not be desirable for meat cookery.

Microwave heating produced less tender beef rib steaks and pork chops than oven broiling (Carpenter *et al.*, 1968). Microwave-heated beef patties were similar in tenderness to fried beef patties but were less tender than those broiled, charbroiled, and roasted in a conventional or convection oven (Berry and Leddy, 1984b).

Studies on the use of film bags in microwave cooking and conventional heating have indicated that both wraps and microwave heating increase cooking losses as compared with conventional methods (Ruyack and Paul, 1972). Armbruster and Haefele (1975) reported that plastic film covers improved appearance, color, mouth feel, moistness, flavor, and overall acceptability of meatloaf and turkey, and appearance, color, and flavor of pork loin. Cooking times were reduced and foods were more uniformly done because of the more even heat distribution achieved with the films.

A concern with microwave heating is the potential survival of *Trichinella spirales* larvae in pork (Kotula, 1982; Zimmerman and Beach, 1982). On the basis of several studies on the problem, Zimmerman (1983) recommended that small (2 kg or less) bone-in roasts be cooked at 50% power or less. Allowing the roasts to sit for 10 min in the oven or covered after cooking also is recommended. It is critical to make sure that the roast is adequately done throughout. This may be accomplished by measuring the temperature in several places and reheating if any portion has not reached 76.7°C. If pink or red portions are present when the roast is cut, it should be reheated.

As with conventional methods of heating, it is difficult to draw conclusions about the effects of heating by microwaves on the quality of meat. It can be said that comparable flavor and textural characteristics are possible but differences in external color are to be expected.

VI. PROCESSES TO TENDERIZE MEAT _____

The effectiveness of some of the agents used to tenderize meat are discussed in this section. These include mechanical methods and the addition of substances such as enzyme, salt, and vinegar.

A. Mechanical Methods

Meat can be tenderized by mechanical methods that cut or break the muscle fibers and connective tissues. Methods such as these allow "family" steakhouses to use meat from lower quality carcasses but maintain a low price (Romans *et al.*, 1985, pp. 541–543). Cubing with a commercial device, a more rigorous treatment than scoring or pounding, probably is even more effective. Needling or blade tenderization is a more recently introduced technique for tenderization of meat. Needle-like blades approximately 0.5 cm wide are inserted into the meat automatically as it passes through the tenderizer (Glover *et al.*, 1977). Glover and co-workers (1977) reported that sensory scores for tenderness were improved for tenderized round steaks but not for round roasts, chuck roasts, or loin steaks. Roasts cooked in less time, lost more drip, and were less juicy when tenderized; these effects were not seen with tenderized steaks. Savell *et al.* (1977)

reported that steaks from the semitendinosus, biceps femoris, gluteus media, and longissimus muscles could be tenderized as many as two times in order to improve tenderness without decreasing overall palatability or increasing cooking losses. The tenderizing effect of grinding meat is evident to anyone who has broiled or roasted ground meat that was suitable only for soup or stew prior to grinding.

B. Enzymes

Proteolytic enzymes including papain from the green fruit of the papaya plant, bromelin from pineapple, ficin from figs, trypsin from the pancreas of animals, and fungal proteases may be used to tenderize meat. Because muscle fiber protein accounts for 75% of the edible portion, its hydrolysis is probably the most important mechanism in enzymatic tenderization of beef. Enzymes also attack the connective tissue fibrous proteins, producing granulation of collagen and segmentation of elastin fibers (Wang *et al.*, 1958). Kang and Rice (1970) investigated the effects of several enzymes on meat components and reported that the extent of enzymatic activity varied with the enzyme as well as the substrate. The proteolytic enzymes that may be used as meat tenderizers affect muscle fibers and connective tissue differently.

The way in which enzymatic tenderizers are used influences their effectiveness. Since papain penetrates only 0.5–2.0 mm when applied to the surface, a physical method of incorporation such as forking or injecting the enzyme into the meat may be more effective than mere surface treatment. Directions for the use of some commercial enzyme tenderizers suggest that the meat stand for a time, usually at room temperature, after application of the enzyme and before cooking. This probably is unnecessary, as shown by an experiment in which no differences were found in the tenderness of meat cooked immediatley after treatment and that of meat held for 1–5 hr. The fact that meat cooked immediately after treatment is more tender than untreated meat and that the action of papain is greatest at temperatures of 60–80°C indicate that the enzyme acts during the cooking process (Tappel *et al.*, 1956). Furthermore, Hinrichs and Whitaker (1962) reported that collagen is not affected by the proteolytic enzymes used in meat tenderizers at low temperatures, but that the enzyme will attack collagen that has been denatured by heat.

Commercially, papain may be injected into the veins of an animal about 10 min prior to slaughter (Robinson and Goesner, 1962). The animal's circulatory system distributes the enzyme throughout its body. The tenderizing action starts when the meat is heated. Aging is reduced 1–3 days. Overtenderization or decrease in juiciness and a mushy, crumbly texture may result from the use of excessive amounts of enzymes, from the use of enzymes in cooking tender steaks or meat made tender by mechanical means, or from a cooking method in which the enzyme-treated meat is held for long periods in the temperature range favorable for enzyme activity. Fogle *et al.* (1982) indicated that a lower level of enzyme activity is needed if a long, slow cooking method is used than if a fast method is used. Papain also was shown to be more effective than bromelin for effecting a change in shear values. The enzyme is inactivated when the temperature is high enough to denature it.

C. Other Methods

Salt may, under certain conditions, have an effect on tenderness. Draudt (1972) suggested that sodium chloride and sodium bicarbonate decreased the severity of the hardening of muscle fibers during heating, thus increasing tenderness. These salts increase the water-holding capacity of the muscle fiber proteins, which affects tenderness (Lawrie, 1985, p. 198). Some of the effects of salt on the hydration of meat were described by Hamm (1960) and Deatherage (1963, pp. 45–68). The amounts of sodium chloride normally used with ground meat probably increase its ability to hold water, as shown by an experiment in which the addition of 1.3% sodium chloride to ground beef decreased the amount of juice lost from it during cooking (Wierbicki et al., 1957a). Larger amounts of juice were lost from meat cooked without salt or with concentrations of salt beyond the limit of taste acceptability (Wierbicki et al., 1957b). The possible effects of water retention on the tenderness of meat are best observed on meat that has not been ground. So these effects can be studied, salt has been incorporated in several ways. Tenderness was improved when freeze-dried steaks were rehydrated in 2% sodium chloride instead of water (Wang et al., 1958), when a solution of sodium and magnesium chlorides was injected into beef rounds (Wierbicki et al., 1957b), and when steaks were injected with a sodium tripolyphosphate and sodium chloride solution prior to freeze drying (Hinnergardt et al., 1975).

Another way to increase the hydration, and thus perhaps the tenderness, of meat is to make the meat more acidic or more basic. Decreasing or increasing the pH of meat reduces cooking losses (Hamm and Deatherage, 1960). Raising the pH had the disadvantage of darkening the meat. The addition of acid may be feasible, as in the German dish called "sauerbraten," for which meat is soaked in vinegar before cooking. The strong buffering effect of meat necessitates rather large amounts of acid and base for changing its pH. Therefore, the amount of vinegar required to have an effect may make the meat sensorially unacceptable. Lind and co-workers (1971) reported that soaking commercial-grade top round steaks in a wine vinegar solution for 48 hr prior to braising increased juiciness and tenderness but decreased odor, taste, and overall acceptability scores.

VII. VEGETABLE PROTEINS AS MEAT EXTENDERS

Textured vegetable proteins (TVP) have been used to extend ground meat as a means of reducing the cost without reducing nutritional value (Lachance, 1972). Federal school lunch regulations allow for inclusion of up to 30% TVP in ground meat for that program (USDA, 1971). Addition of TVP influences the quality of ground meat products, as indicated in several studies. The effect on quality depends on the level of TVP added. Textured soy protein (TSP) products have been used as the extender in most studies although the use of powdered forms also has been explored.

The hydrophilic nature of soy proteins results in retention of moisture in beef patties (Anderson and Lind, 1975; Seideman et al., 1977). Reports indicate that fat content may or may not increase with the addition of soy proteins (Miles et al., 1984, Ziprin and Carlin, 1976). Soy extenders also were addressed by Miles et al. (1984), who reported that as refinement of the soy product increased, retention of moisture in meat patties decreased. More moisture was retained in flour-containing patties than in patties containing isolate. No difference was found for fat retention. Mineral content of the soy ingredient was reflected in the mineral content of the cooked patties; more refined soy products have lower mineral concentrations.

In general, as the level of soy increases, flavor desirability and overall palatability decrease. Levels of 10% (Seideman et al., 1977) and 20% (Cross et al., 1975) in beef have been reported to be acceptable. Increased moisture content in extended patties does not increase panel scores for juiciness (Bowers and Engler, 1975). Ali et al. (1982) substituted 30% TSP or cooked ground soybeans for ground beef in meat loaves. As in the other studies, cooking losses were decreased and sensory properties were adversely affected. Loaves containing soybeans were more desirable than those containing TSP. Product quality was better maintained in frozen, raw loaves than in frozen cooked loaves.

Reheating of patties containing TSP and all-meat patties reduced flavor differences (Bowers and Engler, 1975). Stale taste and aroma were almost absent in reheated patties containing soy, suggesting that if components responsible for stale taste and aroma were present, other components masked the stale flavors.

The addition of flavorful ingredients such as vinegar, pineapple, and soy sauces may minimize the effects of soy proteins on the flavor of combination dishes (Baldwin et al., 1975). Cooking method also may influence quality of soy and beef products. Microwave heating of beef–soy loaves resulted in increased loss of moisture and reduced beef flavor (Ziprin and Carlin, 1976).

Suggested Exercises

1. Prepare ground beef patties from ground beef of varying fat contents. Cook one-half of the patties and freeze. Freeze the other half raw. Store for at least 1 wk. Cook the raw frozen patties from the frozen state. Thaw and heat the cooked, frozen patties in a microwave oven on defrost. Collect data to calculate cooking losses as shown in Table III. Compare flavor and texture of the two products. Compare cooking losses. Compare results with those of Berry et al. (1981).
2. As indicated by Parizek et al. (1981), the demand for ground beef suggests the need for a less expensive alternative. Thus it seems appropriate to mix pork with beef in ground meat patties. Mix ground pork and beef in vary-

ing proportions, cook, and evaluate cooking losses and sensory properties. Other tests that could be done if equipment is available would be shear tests and chemical analysis for fat content of raw and cooked patties to determine fat retention.

3. To evaluate the uniformity of heating in a microwave oven, place equal weights (110 g should be adequate) of ground beef from the same lot into nine custard cups. Place the nine cups in three rows and columns in the oven, leaving space between. Heat on high power for 5 min. Remove from the oven and invert dish to remove meat. Cut through the center and compare for differences in degree of doneness.

4. Study the effect of meat tenderizer on the tenderness of broiled round steak. Use adjacent cuts of meat. Sprinkle a weighed amount of tenderizer on each side of one steak (2.0 g/side of a 450-g steak). Fork in 50 strokes per side. Weigh the steak. Treat a second steak in a like manner, except omit the tenderizer. Broil each steak for 12 min/side or until desired degree of doneness is reached. After the steaks cook, cool for 10 min. Weigh each one. Calculate cooking losses. If possible do shear values. For sensory evaluation, cut samples of equal size from the same position in both steaks for each of the judges. Ask the judges to record the number of chews that are required before the meat is ready to swallow. Ask them to describe the texture of each sample.

5. Compare the quality and cooking losses of meat loaves baked to 60, 71, 77, and 80°C. Evaluate the flavor, color, and texture of each of the loaves, noting differences among them.

6. To study the effects of salt on the retention of water in meat during heating, divide a well-mixed sample of ground beef into three equal portions. To one portion, add no salt; to the second, add 5.5 g of salt/450 g of meat; and to the third add 15 g of salt/450 g of meat. Cook as meat patties or loaves. Bake loaves at 163°C to the end-point temperature 77°C. Weigh the patties or loaves before and after cooking. Calculate cooking losses.

7. Compare oven roasting of beef semitendinosus with heating in a microwave oven. One-half of a muscle may be used for each cooking method. Unless a special thermometer is available for use in the microwave oven, the roast must be removed from the oven periodically and a thermometer or thermocouple inserted to see if the roast has reached the desired degree of doneness. Compare cooking losses and sensory properties. If equipment is available, determine Warner–Bratzler shear values.

References

Adams, C. 1975. Nutritive value of American foods. Agric. Handbook 456, Agricultural Research Service, USDA, Washington, DC.

Ali, F. S., Perry, A. K., and Van Duyne, F. O. 1982. Soybeans vs. textured soy protein as meat extenders. *J. Amer. Dietet. Assoc.* **91**: 439–444.

Anderson, R. H. and Lind, K. D. 1975. Retention of water and fat in cooked patties of beef and of beef extended with textured vegetable protein. *Food Technol.* **29**(2): 44–45.

Armbruster, G. and Haefele, C. 1975. Quality of foods after cooking in 915 MHz and 2450 MHz microwave appliances using plastic film covers. *J. Food Sci.* **40**: 721–723.

Asghar, A., Gray, J. I., Buckley, D. J., Pearson, A. M., and Booren, A. M. 1988. Perspectives on warmed-over flavor. *Food Technol.* **42**(6): 102–108.

Bailey A. J. 1972. The basis of meat texture. *J. Sci. Food Agric.* **23**: 995–1007.

Bailey, M. E. 1988. Inhibition of warmed-over flavor, with emphasis on Maillard reaction products. *Food Technol.* **42**(6): 123–132.

Baity, M. R., Ellington, A. E., and Woodburn, M. 1969. Foil wrap in oven cooking. *J. Home Econ.* **61**: 174–176.

Baldwin, R. E., Korschgen, B. M., Vandepopuliere, J. M., and Russell, W. D. 1975. Palatability of ground turkey and beef containing soy. *Poultry Sci.* **54**: 1102–1106.

Bandman, E. 1987. Chemistry of animal tissues. Part 1. Proteins. *In* "The Science of Meat and Meat Products," 3rd edn., Price, J. F. and Schweigert, B. S. (Eds.). Food and Nutrition Press, Inc., Westport, Connecticut.

Batcher, O. M. and Deary, P. A. 1975. Quality characteristics of broiled and roasted beef steaks. *J. Food Sci.* **40**: 745–746.

Bayne, B. H., Meyer, B. H., and Cole, J. W. 1969. Response of beef roasts differing in finish, location, and size to two rates of heat application. *J. Animal Sci.* **29**: 283–287.

Bayne, B. H., Allen, M. B., Large, N. F., Meyer, B. H., and Goertz, G. E. 1973. Sensory and hislogical characteristics of beef rib cuts heated at two rates to three end point temperatures. *Home Econ. Research J.* **2**: 29–34.

Bechtel, P. J. 1986. Muscle development and contractile proteins. *In* "Muscle as Food," Bechtel, P. J. (Ed.). Academic Press, Orlando, Florida.

Bendall, J. R. and Restall, D. J. 1983. The cooking of single myofibres, small myofibre bundles and muscle strips from beef *M. psoas* and *M. sternomandibularis* muscles at varying heating rates and temperature. *Meat Sci.* **8**: 93–117.

Bernal, W. V. W., Bernal, V. M., Gullett, E. A., and Stanley, D. W. 1988. Sensory and objective evaluation of a restructured beef product. *J. Texture Studies* **19**: 231–246.

Berry, B. W. and Leddy, K. 1984a. Beef patty composition: Effects of fat content and cooking method. *J. Amer. Dietet. Assoc.* **84**: 654–658.

Berry, B. W. and Leddy, K. 1984b. Effects of fat level and cooking method on sensory and textural properties of ground beef patties. *J. Food Sci.* **49**: 870–875.

Berry, B. W., Marshall, W. H., and Koch, E. J. 1981. Cooking and chemical properties of raw and precooked flaked and ground beef patties cooked from the frozen state. *J. Food Sci.* **46**: 856–859.

Berry, B. W., Smith, J. J., and Secrist, J. L. 1986. Effects of connective tissue levels on sensory, Instron, cooking and collagen values of restructured beef steaks. *J. Food Protection* **49**: 455–460.

Blaker, G. G., Newcomer, J. L., and Stafford, W. D. 1959. Conventional roasting vs. high-temperature foil cookery. *J. Amer. Dietet. Assoc.* **35**: 1255–1259.

Bowers, J. A. and Engler, P. P. 1975. Freshly cooked and cooked, frozen, reheated beef and beef-soy patties. *J. Food Sci.* **40**: 624–625.

Bowers, J. A. and Goertz, G. E. 1966. Effect of internal temperature on eating quality of pork chops. I. Skillet- and oven-braising. *J. Amer. Dietet. Assoc.* **48**: 116–120.

Bowers, J. A., Craig, J. A., Kropf, D. H., and Tucker, T. J. 1987. Flavor, color, and other characteristics of beef longissimus muscle heated to seven internal temperatures between 55° and 85°C. *J. Food Sci.* **52**: 533–536.

Brackett, R. E. and Marth, E. H. 1977. Heating patterns of products in crockery cookers. *J. Food Protection* **40**: 664–667.

Brady, P. L. and Penfield, M. P. 1977. A comparison of four methods of heating beef roasts: Conventional oven, slow cooker, microwave oven, and pressure cooker. *Tennessee Farm and Home Science* Issue 101: 15–19.

Bramblett, V. D., Hostetler, R. L., Vail, G. E., and Draudt, H. N. 1959. Qualities of beef as affected by cooking at very low temperatures for long periods of time. *Food Technol.* **13**: 707–711.

Breidenstein, B. C. 1982. Understanding and calculating the sodium content of your products. *Meat Processing* **21**(5): 62, 64, 67.

Carpenter, Z. L., Abraham, H. C., and King, G. T. 1968. Tenderness and cooking loss of beef and pork. I. Relative effects of microwave cooking, deep-fat frying, and oven broiling. *J. Amer. Dietet. Assoc.* **53**: 353–356.

Cassens, R. J. 1987. Structure of muscle. *In* "The Science of Meat and Meat Products," 3rd edn., Price, J. F. and Schweigert, B. S. (Eds.). Food and Nutrition Press, Inc., Westport, Connecticut.

Chang, S. S. and Peterson, R. J. 1977. Symposium: The basis of quality in muscle foods. Recent developments in the flavor of meat. *J. Food Sci.* **42**: 298–305.

Cline, J. A., Trowbridge, E. A., Foster, M. T., and Fry, H. E. 1930. How certain methods of cooking affect the quality and palatability of beef. Bull. 293, Missouri Agric. Exp. Stn., Columbia.

Costello, C. A., Penfield, M. P., and Riemann, M. J. 1985. Quality of restructured steaks: Effects of days on feed, fat level, and cooking method. *J. Food Sci.* **50**: 685–688.

Cover, S. 1959. Scoring for three components of tenderness to characterize differences among beef steaks. *Food Research* **24**: 564–573.

Cover, S., and Hostetler, R. L. 1960. An examination of some theories about beef tenderness by using new methods. Bull. 947, Texas Agric. Exp. Stn., College Station.

Cover, S., Ritchey, S. J., and Hostetler, R. L. 1962a. Tenderness of beef. I. The connective tissue component of tenderness. *J. Food Sci.* **27**: 469–475.

Cover, S., Ritchey, S. J., and Hostetler, R. L. 1962b. Tenderness of beef. II. Juiciness and the softness components of tenderness. *J. Food Sci.* **27**: 476–482.

Cover, S., Ritchey, S. J., and Hostetler, R. L. 1962c. Tenderness of beef. III. The muscle fiber components of tenderness. *J. Food Sci.* **27**: 483–488.

Cramer, D. A. 1983. Chemical compounds implicated in lamb flavor. *Food Technol.* **37**(5): 249–257.

Cross, H. R., Stanfield, M. S., Green, E. C., Heinemeyer, J. M., and Hollick, A. B. 1975. Effect of fat and textured soy protein content on consumer acceptance of ground beef. *J. Food Sci.* **40**: 1331–1332.

Cross, H. R., Berry, B. W., and Wells, L. H. 1980. Effects of fat level and source on the chemical, sensory and cooking properties of ground beef patties. *J. Food Sci.* **45**: 791–793.

Davenport, M. M. and Meyer, B. H. 1970. Forced convection roasting at 200° and 300°F: yield, cost, and acceptability of beef sirloin. *J. Amer. Dietet. Assoc.* **56**: 31–33.

Deatherage, F. E. 1963. The effect of water and inorganic salts on tenderness. *In* "Proceedings Meat Tenderness Symposium." Campbell Soup Co., Camden, New Jersey.

Draudt, H. N. 1972. Changes in meat during cooking. *Recip. Meat Conf. Proc.* **25**: 243–259.

Dugan, L. R., Jr. 1987. Chemistry of animal tissues. Part 2. Fats. *In* "The Science of Meat and Meat Products," 3rd edn., Price, J. F. and Schweigert, B. S. (Eds.). Food and Nutrition Press, Inc., Westport, Connecticut.

Duncan, W. R. H. and Garton, G. A. 1967. The fatty acid composition and intramuscular structure of triglycerides derived from different sites in the body of the sheep. *J. Sci. Food Agric.* **18**: 99–102.

Ferger, D. C., Harrison, D. L., and Anderson, L. L. 1972. Lamb and beef roast cooked from the frozen state by dry and moist heat. *J. Food Sci.* **37**: 226–229.

Flynn, A. W. and Bramblett, V. D. 1975. Effects of frozen storage, cooking method and muscle quality on attributes of pork loins. *J. Food Sci.* **40**: 631–633.

Foegeding, E. A. 1988. Thermally induced changes in muscle proteins. *Food Technol.* **42**(6): 58, 60–62, 64.

Fogle, D. R., Plimpton, R. F., Ockerman, H. W., Jarenback, L., and Persson, T. 1982. Tenderization of beef: Effect of enzyme, enzyme level, and cooking method. *J. Food Sci.* **47**: 1113–1118.

Fox, J. B., Jr. 1987. The pigments of meat. *In* "The Science of Meat and Meat Products," 3rd edn., Price, J. F. and Schweigert, B. S. (Eds.). Food and Nutrition Press, Inc., Westport, Connecticut.

Frye, C. B., Hand, L. W., Calkins, C. R., and Mandigo, R. W. 1986. Reduction or replacement of sodium chloride in a tumbled ham product. *J. Food Sci.* **51**: 836–837.

Fulton, L. and Davis, C. 1983. Roasting and braising beef roasts in microwave ovens. *J. Amer. Dietet. Assoc.* **83**: 560–563.

Glover, E. E., Forrest, J. C., Johnson, H. R., Bramblett, V. D., and Judge, M. D. 1977. Palatability and cooking characteristics of mechanically tenderized beef. *J. Food Sci.* **42**: 871–874.

Goll, D. E., Hoekstra, W. G., and Bray, R. W. 1964. Age-associated changes in bovine muscle connective tissue. II. Exposure to increasing temperature. *J. Food Sci.* **29**: 615–621.

Gustavson, K. H. 1956. "The Chemistry and Reactivity of Collagen." Academic Press, New York.

Hamm, R. 1960. Biochemistry of meat hydration. *Adv. Food Research* **10**: 365–463.

Hamm, R. and Deatherage, F. E. 1960. Changes in hydration, solubility and charges of muscle proteins during heating of meat. *Food Research* **25**: 587–610.

Headley, M. E. and Jacobson, M. 1960. Electronic and conventional cookery of lamb roasts. *J. Amer. Dietet. Assoc.* **36**: 337–340.

Hearne, L. E., Penfield, M. P., and Goertz, G. E. 1978. Heating effects on bovine semitendinosus: Phase contrast microscopy and scanning electron microscopy. *J. Food Sci.* **43**: 13–16.

Herring, H. K., Cassens, R. G., and Briskey, E. J. 1967. Factors affecting collagen solubility in bovine muscles. *J. Food Sci.* **32**: 534–538.

Hill, F. 1966. The solubility of intramuscular collagen in meat animals of various ages. *J. Food Sci.* **31**: 161–166.

Hiner, R. L., Hankins, O. G., Sloane, H. S., Fellers, C. R., and Anderson, E. E. 1953. Fiber diameter in relation to tenderness of beef muscle. *Food Research* **18**: 364–376.

Hiner, R. L., Anderson, E. E., and Fellers, C. R. 1955. Amount and character of connective tissue as it relates to tenderness in beef muscle. *Food Technol.* **9**: 80–86.

Hines, R. C., Ramsey, C. B., and Hoes, T. L. 1980. Effect of microwave cooking rate on palatability of pork loin chops. *J. Animal Sci.* **50**: 446–451.

Hinnergardt, L. C., Drake, S. R., and Kluter, R. A. 1975. Grilled freeze-dried steaks: Effects of mechanical tenderization plus phosphate and salt. *J. Food Sci.* **40**: 621–623.

Hinrichs, J. R. and Whitaker, J. R. 1962. Enzymatic degradation of collagen. *J. Food Sci.* **27**: 250–254.

Hoelscher, L. M., and Savell, J. W., Harris, J. M., Cross, H. R., and Rhee, K. S. 1987. Effect of initial fat level and cooking method on cholesterol content and caloric value of ground beef patties. *J. Food Sci.* **52**: 883–885.

Holmes, Z. A., Bowers, J. A., and Goertz, G. E. 1966. Effect of internal temperature on eating quality of pork chops. *J. Amer. Dietet. Assoc.* **48**: 121–123.

Hornstein, I. and Wasserman, A. 1987. Sensory characteristics of meat. Part 1—Chemistry of meat flavor. *In* "The Science of Meat and Meat Products," 3rd edn., Price, J. F. and Schweigert, B. S. (Eds.). Food and Nutrition Press, Inc., Westport, Connecticut.

Huffman, D. L., Ande, C. F., Cordray, J. C., Stanley, M. H., and Egbert, W. R. 1987. Influence of polyphosphate on storage stability of restructured beef and pork nuggets. *J. Food Sci.* **52**: 275–278.

Huxley, H. E. 1969. The mechanism of muscular contraction. *Science* **164**: 1356–1366.

Irvin, L. and Cover, S. 1959. Effect of a dry heat method of cooking on the collagen content of two beef muscles. *Food Technol.* **13**: 655–658.

Johnson, P. B. and Civille, G. V. 1986. A standardized lexicon of meat WOF descriptors. *J. Sensory Studies* **1**: 99–104.

Johnson, R. C., Chen, C. M., Mueller, T. S., Costello, W. J., Romans, J. R., and Jones, K. W. 1988. Characterization of the muscles within the beef forequarter. *J. Food Sci.* **53**: 1247–1249, 1257.

Jones, S. B. 1977. Ultrastructural characteristics of beef muscle. *Food Technol.* **31**(4): 82–85.

Jones, S. L., Carr, T. R., and McKeith, F. K. 1987. Palatability and storage characteristics of precooked pork roasts. *J. Food Sci.* **52**: 279–281, 285.

Kang, C. K. and Rice, E. E. 1970. Degradation of various meat fractions by tenderizing enzymes. *J. Food Sci.* **35**: 563–565.

Kauffman, R. G. and St. Clair, L. E. 1965. Porcine myology. Bull. 715, University of Illinois College of Agriculture, Agric. Exp. Stn., Urbana.

Kinsman, D. M. and Gilbert, R. A. 1972. "Some Major Beef Muscles and Their Characteristics." Cooperative Extension Service, University of Connecticut, Storrs.

Korschgen, B. M., Baldwin, R. E., and Snider, S. 1976. Quality factors in beef, pork, and lamb cooked by microwaves. *J. Amer. Dietet. Assoc.* **69**: 635–639.

Kotula, A. W. 1982. Destruction of *Trichinella spiralis* by microwave heating. *Recip. Meat Conf. Proc.* **35**: 77–80.

Kregel, K. K., Prusa, K. J., and Hughes, K. V. 1986. Cholesterol content and sensory analysis of ground beef as influenced by fat level, heating and storage. *J. Food Sci.* **51**: 1162–1165, 1190.

Laakkonen, E., Wellington, G. H., and Sherbon, J. W. 1970. Low-temperature, long-time heating of bovine muscle. I. Changes in tenderness, water-binding capacity, pH and amount of water soluble components. *J. Food Sci.* **35**: 175–177.

Lachance, P. A. 1972. Update: Meat extenders and analogues in child feeding programs. *Proc. Meat Ind. Research Conf.* **24**: 97–103.

Law, H. M., Yang, S. P., Mullins, A. M., and Fielder, M. M. 1967. Effect of storage and cooking on qualities of loin and top round steaks. *J. Food Sci.* **32**: 637–641.

Lawrie, R. A. 1985. "Meat Science." Pergamon Press, New York.

Lind, J. M., Griswold, R. M., and Bramblett, V. D. 1971. Tenderizing effect of wine vinegar marinade on beef round. *J. Amer. Dietet. Assoc.* **58**: 133–136.

Mandigo, R. W. 1986. Restructuring of muscle foods. *Food Technol.* **40**(3): 85–89.

Marsh, B. B. and Leet, N. G. 1966. Studies in meat tenderness. III. The effects of cold shortening on tenderness. *J. Food Sci.* **31**: 450–459.

Marshall, N., Wood, L., and Patton, M. B. 1959. Cooking choice grade, top round beef roasts: Effects of size and internal temperature. *J. Amer. Dietet. Assoc.* **35**: 569–573.

Marshall, N., Wood, L., and Patton, M. B. 1960. Cooking choice grade, top round beef roasts: Effects of internal temperature on yield and cooking time. *J. Amer. Dietet. Assoc.* **36**: 341–345.

McKeith, F. K., DeVol, D. L., Miles, R. S., Bechtel, P. J., and Carr, T. R. 1985. Chemical and sensory properties of thirteen major beef muscles. *J. Food Sci.* **50**: 869–872.

Melton, S. L. 1983a. Effect of forage feeding on beef flavor. *Food Technol.* **37**(5): 239–248.

Melton, S. L. 1983b. Methodology for following lipid oxidation in muscle foods. *Food Technol.* **37**(7): 105–111, 116.

Miles, C. W., Ziyad, J., Bodwell, C. E., and Steele, P. D. 1984. True and apparent retention of nutrients in hamburger patties made from beef or beef extended with three different soy proteins. *J. Food Sci.* **49**: 1167–1170.

Moody, W. G. 1983. Beef flavor—a review. *Food Technol.* **37**(5): 227–232.

Murray, J. M. and Weber, A. 1974. The cooperative action of muscle proteins. *Scientific American* **230**(2): 58–71.

NRC. 1988. "Designing Foods. Animal Product Options in the Market Place." National Research Council, National Academy Press, Washington, DC.

Offer, G. and Trinick, J. 1983. On the mechanism of water holding in meat: The swelling and shrinking of myofibrils. *Meat Sci.* **8**: 245–281.

Parizek, E. A., Ramsey, C. B., Galyean, R. D., and Tatum, J. D. 1981. Sensory properties and cooking losses of beef/pork hamburger patties. *J. Food Sci.* **46**: 860–862, 867.

Paterson, B. C. and Parrish, F. C., Jr. 1988. Factors affecting the palatability and shelf life of precooked, microwave-reheated beef roasts. *J. Food Sci.* **53**: 31–33.

Paul, P. C. and Bean, M. 1956. Method for braising beef round steaks. *Food Research* **21**: 75–86.

Paul, P. C., Bratzler, L. J., Farwell, E. D., and Knight, K. 1952. Studies on the tenderness of beef. I. Rate of heat penetration. *Food Research* **17**: 504–510.

Paul, P. C., Mandigo, R. W., and Arthaud, V. H. 1970. Textural and histological differences among 3 muscles in the same cut of beef. *J. Food Sci.* **35**: 505–510.

Paul, P. C., McCrae, S., and Hofferber, L. M. 1973. Heat-induced changes in extractability of beef muscle collagen. *J. Food Sci.* **38**: 66–68.

Pearson, A. M. 1987. Muscle function and postmortem changes. *In* "The Science of Meat and Meat Products," 3rd edn., Price, J. F. and Schweigert, B. S. (Eds.). Food and Nutrition Press Inc., Westport, Connecticut.

Pearson, A. M. and Tauber, F. W. 1984. "Processed Meats." Avi Publ. Co., Westport, Connecticut.

Penfield, M. P. and Meyer, B. H. 1975. Changes in tenderness and collagen of beef semitendinosus muscle heated at two rates. *J. Food Sci.* **40**: 150–154.

Peters, C. R., Sinwell, D. D., and Van Duyne F. O. 1983. Slow cookers vs. oven preparation of meat loaves. *J. Amer. Dietet. Assoc.* **83**: 430–435.

Price, J. F. and Schweigert, B. S. 1987. "The Science of Meat and Meat Products," 3rd edn., Price, J. F. and Schweigert, B. S. (Eds.). Food and Nutrition Press, Inc., Westport, Connecticut.

Quenzer, N. M., Donnelly, L. S., and Seideman, S. C. 1982. Institutional cookery of restructured beef steaks. *J. Food Qual.* **5**: 301–309.

Ream, E. E., Wilcox, E. B., Taylor, F. G., and Bennett, J. A. 1974. Tenderness of beef roasts. *J. Amer. Dietet. Assoc.* **65**: 155–160.

Reico, H. A., Savell, J. W., Branson, R. E., Cross, H. R., and Smith, G. C. 1987. Consumer ratings of restructured beef steaks manufactured to contain different residual contents of connective tissue. *J. Food Sci.* **52**: 1461–1463, 1470.

Reitmeier, C. A. and Prusa, K. J. 1987. Cholesterol content and sensory analysis of ground pork as influenced by fat level and heating. *J. Food Sci.* **52**: 916–918.

Ritter, J., O'Leary, J., and Langlois, B. E. 1979. Fate of selected pathogens inoculated into foods prepared in slow cookers. *J. Food Protection* **42**: 872–876.

Robinson, H. E. and Goesner, P. A. 1962. Enzymatic tenderization of meat. *J. Home Econ.* **54**: 195–200.

Romans, J. R., Jones, K. W., Costello, W. J., Carlson, C. W., and Ziegler, P. T. 1985. "The Meat We Eat." Interstate Printers and Publishers, Inc., Danville, Illinois.

Ruyack, D. F. and Paul, P. C. 1972. Conventional and microwave heating of beef: Use of plastic wrap. *Home Econ. Research J.* **1**: 98–103.

Satterlee, L. D. and Hansmeyer, W. 1974. The role of light and surface bacteria in the color stability of prepackaged beef. *J. Food Sci.* **39**: 305–308.

Savell, J. W. and Cross, H. R. 1988. The role of fat in the palatability of beef, pork, and lamb. *In* "Designing Foods. Animal Product Options in the Market Place." National Research Council, National Academy Press, Washington, DC.

Savell, J. W., Smith, G. C., and Carpenter, Z. L. 1977. Blade tenderization of four muscles from three weight-grade groups of beef. *J. Food Sci.* **42**: 866–870, 874.

Savell, J. W., Cross, H. R., and Smith, G. C. 1986. Percentage ether extractable fat and moisture content of beef longissimus muscle as related to USDA marbling score. *J. Food Sci.* **51**: 838, 840.

Savell, J. W., Branson, R. E., Cross, H. R., Stiffler, D. M., Wise, J. W., Griffin, D. B., and Smith, G. C. 1987. National consumer retail beef study: Palatability evaluations of beef loin steaks that differed in marbling. *J. Food Sci.* **52**: 517–519, 532.

Schock, D. R., Harrison, D. L., and Anderson, L. L. 1970. Effect of dry and moist heat treatments on selected beef quality factors. *J. Food Sci.* **35**: 195–198.

Seideman, S. C. and Durland, P. R. 1983. Vacuum packaging of fresh beef: A review. *J. Food Qual.* **6**: 29–47.

Seideman, S. C., Smith, G. C., and Carpenter, Z. L. 1977. Addition of textured soy protein and mechanically deboned beef to ground beef formulations. *J. Food Sci.* **42**: 197–201.

Shaffer, T. A., Harrison, D. L., and Anderson, L. L. 1973. Effects of end point and oven temperatures on beef roasts cooked in oven film bags and open pans. *J. Food Sci.* **38**: 1205–1210.

Smith, G. C., Carpenter, Z. L., Cross, H. R., Murphey, C. E., Abraham, H. C., Savell, J. W., David, G. W., Berry, B. W., and Parrish, F. C., Jr. 1984. Relationship of USDA marbling groups to palatability of cooked beef. *J. Food Qual.* **7**: 289–308.

Sofos, J. N. 1983a. Effects of reduced salt (NaCl) levels on the stability of frankfurters. *J. Food Sci.* **48**: 1684–1691.

Sofos, J. N. 1983b. Effects of reduced salt (NaCl) levels on the sensory and instrumental evaluation of frankfurters. *J. Food Sci.* **48**: 1692–1695, 1699.

Sofos, J. N. 1986. Use of phosphates in low-sodium meat products. *Food Technol.* **40**(9): 52, 58, 60, 62, 64, 66, 68.

Spanier, A. M., Edwards, J. V., and Dupuy, H. P. 1988. The warmed-over flavor process in beef: A study of meat proteins and peptides. *Food Technol.* **42**(6): 110, 112–118.

St. Angelo, A. G., Vercellotti, J. R., Legendre, M. G., Vinnett, C. H., Kuan, J. W., James, C., Jr., and Dupuy, H. P. 1987. Chemical and instrumental analyses of warmed-over flavor of beef. *J. Food Sci.* **52**: 1163–1168.

Stinson, C. G., deMan, J. M., and Bowland, J. P. 1967. Fatty acid composition and glyceride structure of piglet body fat from different sampling sites. *J. Amer. Oil Chemists' Soc.* **44**: 253–255.

St. John, L. C., Buyck, M. J., Keeton, J. T., Leu, R., and Smith, S. B. 1986. Sensory and physical attributes of frankfurters with reduced fat and elevated monounsaturated fats. *J. Food Sci.* **51**: 1144–1146, 1179.

Sundberg, A. D. and Carlin, A. F. 1976. Survival of *Clostridium perfringens* in rump roasts cooked in an oven at 107° or 170°C or in an electric crockery pot. *J. Food Sci.* **41**: 451–452.

Tappel, A. L., Miyada, D. S., Sterling, C., and Maier, V. P. 1956. Meat tenderization. II. Factors affecting the tenderization of beef by papain. *Food Research* **21**: 375–383.

Terrell, R. N. 1983. Reducing the sodium content of processed meats. *Food Technol.* **37**(7): 66–71.

Terrell, R. N., Lewis, R. W., Cassens, R. G., and Bray, R. W. 1967. Fatty acid compositions of bovine subcutaneous fat depots determined by gas-liquid chromatography. *J. Food Sci.* **32**: 516–520.

Townsend, W. E. and Olson, D. G. 1987. Cured meat and cured meat products processing. *In* "The Science of Meat and Meat Products," 3rd edn., Price, J. F. and Schweigert, B. S. (Eds.). Food and Nutrition Press, Inc., Westport, Connecticut.

Tucker, H. Q., Voegeli, M. M., and Wellington, G. H. 1952. "A Cross Sectional Muscle Nomenclature of the Beef Carcass." Michigan State College Press, East Lansing, Michigan.

USDA. 1971. Textured vegetable protein products (B-1). Notice 219, Food and Nutrition Service, United States Department of Agriculture, Washington, DC.

USDA. 1975. Grades of carcass beef: Slaughter cattle. *Federal Register* **40**(49): 11535–11547.

USDA. 1989. Livestock and Poultry. Situation and Outlook. Report. USDA. **LPS-35**: 42–44.

Wang, H., Doty, D. M., Beard, F. J., Pierce, J. C., and Hankins, O. D. 1956. Extensibility of single beef muscle fibers. *J. Animal Sci.* **15**: 97–108.

Wang, H., Weir, C. E., Birkner, M. L., and Ginger, B. 1958. Studies on enzymatic tenderization of meat. III. Histological and panel analysis of enzyme preparations from three distinct sources. *Food Research* **23**: 423–438.

Watts, B. M. 1954. Oxidative rancidity and discoloration in meat. *Adv. Food Research* **5**: 1–42.

Weir, C. E., Slover, A., Pohl, C., and Wilson, G. D. 1962. Effect of cooking procedures on the composition and organoleptic properties of pork chops. *Food Technol.* **16**(5): 133–136.

White, F. D., Resurreccion, A. V. A., and Lillard, D. A. 1988. Effect of warmed-over flavor on consumer acceptance and purchase of precooked top round steaks. *J. Food Sci.* **53**: 1251–1252, 1257.

Whiting, R. C. 1988. Ingredients and processing factors that control muscle protein functionality. *Food Technol.* **42**(4): 104, 110–114, 210.

Whiting, R. C., and Jenkins, R. K. 1981. Partial substitution of sodium chloride by potassium chloride in frankfurter formulations. *J. Food Qual.* **4**: 259–269.

Wierbicki, E., Kunkle, L. E., and Deatherage, F. E. 1957a. Changes in the water-holding capacity and cationic shifts during heating and freezing and thawing of meat as revealed by a simple centrifugal method for measuring shrinkage. *Food Technol.* **11**: 69–73.

Wierbicki, E., Cahill, V. R., and Deatherage, F. E. 1957b. Effects of added sodium chloride, potassium chloride, calcium chloride, magnesium chloride, and citric acid on meat shrinkage at 70°C and of added sodium chloride on drip losses after freezing and thawing. *Food Technol.* **11**: 74–76.

Williams, J. C., Field, R. A., and Riley, M. L. 1983. Influence of storage times after cooking on Warner-Bratzler shear values of beef roasts. *J. Food Sci.* **48**: 309–310, 312.

Winegarden, M. W., Lowe, B., Kastelic, J., Kline, E. A., Plagge, A. R., and Shearer, P. S. 1952. Physical changes of connective tissues of beef during heating. *Food Research* **17**: 172–184.

Wismer-Pedersen, J. 1987. Chemistry of animal tissues. Part 5—water. *In* "The Science of Meat and Meat Products," 3rd edn., Price, J. F. and Schweigert, B. S. (Eds.). Food and Nutrition Press, Inc., Westport, Connecticut.

Young, L. L., Reviere, R. D., and Cole, A. B. 1988. Fresh red meats: A place to apply modified atmospheres. *Food Technol.* **42**(9): 65–69.

Zimmerman, W. J. 1983. An approach to safe microwave cooking of pork roasts containing *Trichinella spiralis. J. Food Sci.* **48**: 1715–1718, 1722.

Zimmerman, W. J. and Beach, P. J. 1982. Efficacy of microwave cooking for devitalizing trichinae in pork roasts and chops. *J. Food Protection* **45**: 405–409.

Ziprin, Y. A. and Carlin, A. F. 1976. Microwave and conventional cooking in relation to quality and nutritive value of beef and beef-soy loaves. *J. Food Sci.* **41**: 4–8.

POULTRY

Poultry consumption in the United States has increased over the past few years. From 1986 to 1988, consumption of chicken and turkey increased 5.9 and 2.7 lb/capita to 67.1 and 16.4 lb/capita, respectively. The total of 83.5 lb/capita exceeded that of beef (72.7 lb). Further increases are predicted (USDA, 1988, 1989). Values for poultry consumption for 1988 represent 32% of total meat consumption. Increased consumption of poultry may be related to price in comparison to other meats and recommendations to reduce dietary intake of fat. Increased consumption may also be related to the availability of a reliable supply of fresh and frozen poultry products. The introduction of chicken nuggets and breast portions in fast food establishments also may have influenced the increase in consumption (Huffman *et al.*, 1986).

I. COMPOSITION AND STRUCTURE

Proximate composition of poultry is included in Table I. The values presented in the table represent general values for the components. More detailed information on composition of poultry cooked by different methods may be found in the USDA Handbook 8-5 (Posati, 1979).

TABLE I

Composition of Selected Poultry and Poultry Products[a]

Product	Water (%)	Protein (%)	Fat (%)
Chicken			
Light meat[b]	63.8	31.6	3.4
Dark meat[b]	64.4	28.0	6.3
Turkey			
Light meat[b]	62.1	32.9	3.9
Dark meat[b]	60.5	30.0	8.2
Duck[b]	64.2	23.5	11.2
Goose[b]	57.2	29.0	12.7
Frankfurters			
Chicken	57.5	12.9	19.5
Turkey	63.0	14.3	17.7

[a] From Adams (1975) and Posati (1979).
[b] Cooked.

Much of the general information on muscle structure included in Chapter 9 also is applicable to poultry because its structure does not differ from that of other muscle foods, as illustrated by Johnson and Bowers (1976). Like mammalian muscles, changes that effect the conversion of muscle to meat occur after slaughter. In poultry the changes are basically the same with one major difference. The time required for conversion is much shorter. Breasts and legs go into rigor 30–60 and 15–30 min post-mortem, respectively. Commercial practices are designed to ensure tenderness by freezing poultry at the appropriate time. However, Addis (1986, pp. 379–383) points out that a few very tough turkeys do exist but that researchers have not been able to completely explain the cause. He postulates that a combination of three factors contributes to extreme toughness when it occurs. These are freezing turkeys before onset of rigor, storing for a very short period during marketing, and cooking from the frozen state. The last factor would eliminate the opportunity for postthawing tenderization. Holding inadequately aged poultry at freezer temperatures does not improve its tenderness, but holding it at 1.7°C after frozen storage has as much tenderizing effect as an equal period before freezing (Klose *et al.*, 1959). To aid in understanding the literature on poultry, a diagram of the musculature of chicken is shown in Fig. 10.1.

II. QUALITY ATTRIBUTES

A. Color

As with red meats, the color of poultry is attributable to the hemoglobin and myoglobin contents. Values for these components for dark and light meat of chicken and turkey and the meat of duckling were reported by Saffle (1973). Obviously, dark meat contains significantly more myoglobin and hemoglobin than

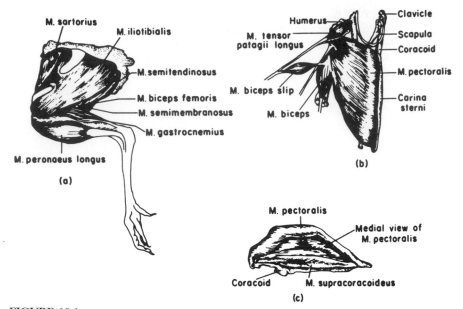

FIGURE 10.1

Musculature of chicken (*Gallus gallus*): (a) muscles in the hind limb, (b) lateral view of the muscle pectoralis, and (c) muscles in the breast of the chicken. (From Hsu and Jacobson, 1974; *Home Economics Research Journal*, American Home Economics Association, Washington, DC. Reprinted with permission.)

does light meat. Saffle (1973) indicated that the amount of pigment in both light and dark meat increases dramatically with age of the bird.

During the cooking of broiler chickens, a pink color may develop in the leg and thigh muscles. Nash and co-workers (1985) suggested that consumers may find this undesirable and associate it with undercooking or improper bleeding during processing. Nash *et al.* (1985) further indicated that the color may be attributed to the formation of nitrosomyoglobin, which changes to the nitrosomyochromagen during cooking. Nitrates or nitrites in the water used in processing or ice packs may serve as reactants for this process.

B. Flavor

Raw chicken has little flavor. Flavor develops during cooking as the concentration of carbonyl compounds increases (Dimick *et al.*, 1972). These compounds are thought to be major contributors to "chickeny" flavor (Minor *et al.*, 1965). Sulfides also are contributors to the flavor (Minor *et al.*, 1965; Wu and Sheldon, 1988a). The small amount of phospholipids present is important in flavor deterioration that may occur during refrigerator or frozen storage of cooked chicken (Lee and Dawson, 1976; Wu and Sheldon, 1988b). Factors affecting poultry flavor include pH, feed, age, and sex of the bird and were reviewed by Ramaswamy and Richards (1982).

TABLE II
Twelve Sensory Descriptive Terms with Definitions
Developed for Evaluation of Chicken Flavor[a]

Term	Definition
	Aromatic/taste sensation associated with:
Chickeny	Cooked white chicken muscle
Meaty	Cooked dark chicken muscle
Brothy	Chicken stock
Liver/organy	Liver, serum, or blood vessels
Browned	Roasted, grilled, or broiled chicken patties (not seared, blackened, or burned)
Burned	Excessive heating or browning (scorched, seared, charred)
Cardboard/musty	Cardboard, paper, mold, or mildew; described as nutty stale
Warmed over	Reheated meat; not newly cooked nor rancid/painty
Rancid/painty	Oxidized fat and linseed oil
	Primary taste associated with:
Sweet	Sucrose, sugar
Bitter	Quinine or caffeine
	Feeling factor or tongue associated with:
Metallic	Iron/copper ions

[a] From Lyon (1987).

An important consideration in the study of the flavor of any food is the selection of terms that can be used to describe the quality attribute. Lyon (1987) used sensory techniques to collect and multivariate statistical techniques to analyze 45 terms associated with poultry flavor. When chicken patties of 50% light and 50% dark meat were evaluated with these terms it was possible to narrow the list to 12 terms that could be used to evaluate chicken flavor. These terms could be used as a basis for class study of the flavor of chicken and chicken products. The terms and definitions are shown in Table II. Note that terms for flavor changes that may occur as chicken is stored are included. "Warmed over" would obviously be applied to cooked poultry held in the refrigerator after cooking. Rancid usually would be applied to changes that occur in frozen raw poultry held for a long time.

Rancidity in poultry is of interest because highly unsaturated fats are present. Warmed over flavor (WOF) is attributed to rancidity resulting from oxidation of muscle phospholipids.

C. Texture

The dark meat of poultry usually is more juicy but less tender than the light meat. The greater perceived juiciness of the dark meat may be attributable to the higher fat content (Table I) and its role in perception of juiciness.

Because poultry has a bland flavor, tenderness is the important factor in consumer satisfaction. Aged birds are less tender than younger birds. As for beef, this difference has been attributed to the formation of cross-links in collagen during the aging process (Nakamura *et al.*, 1975).

III. PROCESSING AND PRESERVATION _____

Most chicken is sold fresh whereas most turkeys are frozen for future use. The steps involved in inspecting, grading, and processing of poultry are described in detail by Moreng and Avens (1985, Chapter 10). All poultry that is to be distributed in interstate commerce must be inspected for wholesomeness. In addition poultry may be assigned a United States Grade of A, B, or C on the basis of several factors including shape, quantities of meat and fat under the skin, and possible defects such as broken bones, missing parts, exposed flesh, and discoloration.

To meet consumer demands for convenient food products, the poultry industry further processes poultry, primarily turkey, into a variety of products including boneless roasts, ground meat, and cured products including frankfurters, sausages, and luncheon meats. Use of poultry in place of beef and/or pork in frankfurters changes the proportion and nature of the fat content of the product, as shown in Fig. 10.2. The level of sodium used in production of these products is of concern. Normally, 2.5% salt is used in frankfurters. Maurer (1983) reviewed possible reduction in salt content of these products and indicated that although it is possible to reduce sodium chloride or replace it partially with potassium chloride in products such as frankfurters and poultry loaves, salt plays important functional (binding) roles and the amount of reduction is limited by sensory property changes and safety. Barbut *et al.* (1988) indicated that sodium acid pyrophosphate (SAPP) may provide functional properties lost with a reduction in sodium level in frankfurters. Sodium acid pyrophosphate performed better than tripolyphosphate and hexametaphosphate.

Additionally intact turkey breast muscles may be roasted, smoked, or bar-

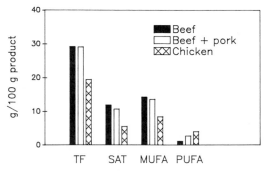

FIGURE 10.2
Total fat (TF) content and distribution of saturated (SAT), monounsaturated (MUFA), and polyunsaturated (PUFA) fatty acids in frankfurters from meat of varying species. (From Richardson *et al.*, 1980.)

becued and then vacuum packaged for retail and institutional marketing. The smaller muscle (supracoracoideus) frequently is used for this purpose. The larger breast muscle (pectoralis) may be made into boneless roasts. Location of these muscles is similar in chicken and turkey (Fig. 10.1). Stability of vacuum-packaged uncooked rolls during 87 days of storage at 4°C was studied by Smith and Alvarez (1988). No changes in weight parameters, pH, and tensile strength were noted, and decreased shear values and a slight increase in thiobarbaturic acid (TBA) values were observed. Mesophilic anaerobic counts indicated that temperature control during storage is critical to the life of these products. Although a shelf-life limit of 30 days normally is assigned to vacuum-packaged turkey rolls by the producer, Smith and Alvarez indicated that because there were so few changes in their study, longer shelf life is feasible. Sensory properties were not studied.

To make more complete use of the meat from poultry carcasses, mechanical deboning processes have been developed. Neck, backs, and frames are crushed or ground and pushed through a sieve. Meat and bone are separated as the meat is pushed through the sieve. The product, which has a pastelike consistency, is used in processed products such as frankfurters, luncheon meats, and sausages. Lipid oxidation is a problem in mechanically deboned poultry (MDP) because of the presence of unsaturated fatty acids. Processing at low temperatures, limiting exposure to oxygen during processing, vacuum packaging of products, use of antioxidants, and low-temperature storage are effective means of controlling lipid oxidation in products containing MDP (Dawson and Gartner, 1983).

IV. INFLUENCE OF COOKING METHODS ON QUALITY

A review of poultry cooking methods should include discussion of methods used by the consumer at the point of consumption. In addition, restaurant and institutional use of poultry and poultry products is of considerable interest because of the quantities of poultry used in such establishments.

Satisfactory determination of end-point temperatures is difficult with poultry because the muscles are smaller than those of beef and pork and tend to separate during cooking, sometimes leaving the bulb of the thermometer exposed to air or liquid. Because of their large size, however, turkey breast and thigh muscles are fairly satisfactory for this purpose. An end-point temperature of 95°C in the thigh or 90°C in the breast produced turkeys of optimum doneness (Goertz *et al.*, 1960). Martinez *et al.* (1984) published heat penetration curves for roasting of turkeys and noted that the recommendation for 82–85°C in the thigh would result in a temperature of only 77–79°C in the breast, which may not be adequate. Curves from Martinez *et al.* (1984) for the effect of initial temperature and stuffed and unstuffed turkey of different sizes clearly show the effect of these factors on heating rates (Fig. 10.3). Recommendations for end-point temperatures for chicken and turkey are 82–85°C (USDA, 1984a). These temperatures ensure the destruction of *Salmonella* microorganisms, which may be found in raw poultry. Because poultry, particularly turkey, may be cooked one day, cooled, and reheated the next, care must be taken to ensure that the carcass

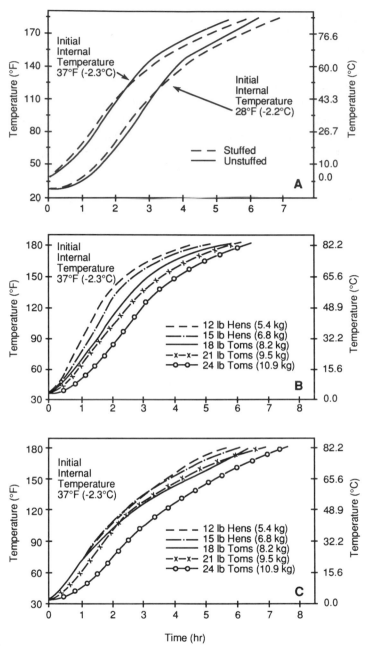

FIGURE 10.3
Rate of heating for (A) 6.8-kg turkeys as influenced by initial temperature, (B) unstuffed turkeys of different weights, and (C) stuffed turkeys of different weights. (From Martinez *et al.*, 1984.)

cools rapidly to prevent growth of any microorganisms that may have been introduced after cooking. Care also should be taken with poultry to avoid postcooking contamination.

A. Chicken

An oven temperature of 163°C is now recommended for roasting poultry as well as other meats. However, Hoke (1968) suggested that if time is a factor, 191°C is suitable although yields will be lower than at 163°C. Shear values and cooking losses did not vary when chickens were broiled at temperatures varying from 175 to 230°C (Goertz *et al.*, 1964).

Several recent studies on chicken cookery have focused on the effects of microwave heating on quality. The first studies on microwave heating of poultry were done with ovens capable of operation only at a fixed power level.

The introduction of variable power level ovens provided new questions to be answered about the effect of microwave heating on quality of poultry meat. Oltrogge and Prusa (1987) studied the effects of microwave heating of hen breasts (506–800 g) at four power levels, 100, 80, 60, and 40% (660, 442, 332, and 229 W, respectively). All samples were heated to an end-point temperature of 82°C. Sensory juiciness, mealiness, and chicken flavor scores did not vary with power level. The samples heated at 60% power were shown to be more tender than the others by both sensory evaluation and Instron measurements.

Broilers roasted in conventional ovens lost more thiamine than did broilers roasted in microwave ovens. Thiamine retention was greatest in light meat. The difference was attributed to the contribution of a lower pH of light meat to thiamine stability (Hall and Lin, 1981).

A popular use of the microwave is for thawing of frozen foods. Younathan *et al.* (1984) investigated the effect of microwave thawing on quality of broilers. Breasts and thighs thawed in the microwave were more and less tender, respectively, than parts thawed in the refrigerator.

Peters *et al.* (1983) compared cooking in a "slow" cooker with oven roasting of chicken. The slow cooker was set on "high" for 1 hr and on "low" for the remainder of the 9-hr cooking period. Total cooking losses were greater in the slow cooker although evaporative losses were greater in the conventional oven. Thiamine content was reduced to a greater extent in the slow cooker than in the conventional oven probably because of the greater drip loss in the slow cooker. The authors concluded that the chickens cooked by both methods had an internal temperature above 65°C for a long enough time period to ensure destruction of pathogenic microorganisms. Reduction of microorganisms present in the chicken was one log cycle greater in the slow cooker than in the conventional oven. Chicken cooked in the conventional oven had more desirable aroma and flavor, was more juicy, was scored closer to the golden brown exterior color and desirable ends of the scales for interior color, and was more desirable overall than that cooked in the slow cooker. Tenderness did not differ between the two methods. One would expect the reported differences on the basis of the inherent difference in the two cooking methods: one was a moist method and the other a dry method of heating.

Bowers *et al.* (1987) indicated that a large portion of chicken consumed per person is consumed as fried chicken prepared in fast-food operations, and they studied the effect of the cooking processes used in such establishments on proximate composition and vitamin content. Statistical differences in levels of some vitamins were found among samples from several chains. The practical importance of the numerically small differences also was noted. A difference of importance was the fact that pressure-fried chicken contained 4–5% less fat than open vat-fried chicken from one chain. This suggests that judicious selection of cooking method can be effective in providing foods with lower fat contents.

Because fried chicken is a popular product, several other researchers have studied factors affecting its eating quality. Lane *et al.* (1982) investigated the effect of variation in frying fat temperature on the composition and yield of chicken thighs. Thighs were deep-fat fried at temperatures of 163, 177, or 191°C to an internal temperature of 93°C. Samples heated at 163°C retained more moisture than those fried at the other temperatures. Yields did not, however, differ. This finding was attributed to possible differences in amount and loss of breading. Fat content increased with frying but did not differ with frying temperature. The coating on fried chicken is an important contributor to the sensory properties of the product. Nakai and Chen (1986) compared effects of methods of coating on composition and yield of fried thighs, breasts, and drumsticks. Their methods included flour predust-batter-flour, batter breading, breading only, and no coating. Product but not meat yields were lower for breading only and no coating than for the others. Fat content of the meat did not differ with coating method. In an earlier study Hanson and Fletcher (1963) reported that adhesion of the batter and fat content were influenced by the ingredients in the batter. For example, batters containing cornstarch and waxy cornstarch adhered better than did those containing wheat starch or flour. Egg yolk content of the batters also was varied. As yolk content increased fat content of the coating increased beyond the level contributed by the yolk.

The effect of cooking method on cholesterol content of chicken also has been investigated (Prusa and Lonergan, 1987). Breaded breast fillets, an item of growing popularity in fast-food restaurants, were broiled (216 ± 4°C) in a convection oven or heated in a conventional oven at 163°C. Raw meat with skin contained more fat and moisture than samples without it. Inclusion of the skin in samples for analysis did not increase cholesterol. However, samples cooked and analyzed with the skin had more cholesterol and fat and less moisture than those heated without skin. Cooking losses were greater in the convection oven. Fat and cholesterol contents were not affected by cooking methods.

B. Turkey

Early studies on turkey were focused on questions related to the roasting of turkeys in conventional ovens. For example, the question of roasting turkey covered or uncovered was studied. Wrapping turkey in aluminum foil is recommended frequently because the oven temperature can be raised to shorten the cooking time without excessive browning. Baity and co-workers (1969) reported that turkeys tightly sealed in foil cooked in less time than unwrapped or loosely wrapped turkeys. Juiciness of turkeys may be influenced more than are flavor and tenderness by cooking in foil. Deethardt *et al.* (1971) reported that juiciness decreased. The use of film cooking bags in oven roasting of turkeys was compared with foil wraps, paper bags, and open pans (Heine *et al.*, 1973). Cooking time was longest with the paper bag and shortest with the film bag. Cooking losses were smaller with foil than with other methods. Juiciness was greatest in the foil-wrapped and conventionally roasted turkey. Panelists found that flavor did not differ but that samples from the dark muscles of carcasses cooked in paper bags and open pans were more tender than samples cooked by the other methods.

Recommendations for roasting vary with respect to position of the turkey during cooking. Salmon *et al.* (1988) reported that white meat quality did not differ with position of the bird. Dark meat, on the other hand, was more tender and flavorful in birds that were roasted breast down. However, cooking losses were greater in that position.

Research on frozen stuffed turkeys indicated that the practice of covering birds with cheesecloth, basting with fat, and roasting at 163°C results in a cooked product of high quality and attractive appearance (Esselen *et al.*, 1956). Reducing the temperature to 121°C lengthened the roasting time and markedly increased the period during which the meat and stuffing were in the incubation zone for food spoilage microorganisms. In order to ensure the destruction of organisms that may cause food-borne illness, primarily *Staphylococcus aureus*, the stuffing should reach a temperature of not less than 74°C (USDA, 1984a, b). Procedures for ensuring safety of stuffed turkeys were studied by Woodburn and Ellington (1967) and Bramblett and Fugate (1967).

As indicated earlier, toughness in turkey may be attributed to cooking from the frozen state a carcass that did not reach the onset of rigor prior to freezing (Addis, 1986, pp. 379–383). Cornforth and co-workers (1982) compared methods of cooking turkeys from the frozen state, suggesting that elimination of the need to thaw before roasting might increase home consumption of turkey. Sensory scores for turkeys cooked by six methods from the frozen state are illustrated in Fig. 10.4. Aluminum foil was used for tents and wraps, and cooking temperature was varied. The sixth method involved use of a microwave oven. In general, overall ratings suggested that no one best method exists although the turkeys roasted at the low temperature (93.3°C) and tented did differ from the bag, high-temperature tent, and microwave methods. Some differences in individual characteristics were reported as shown in the figure. It is interesting to note that the microwave samples were similar to others for flavor and textural attributes but differed from all for color and appearance factors. It also is interesting to note that scores for tenderness are all in the upper half of the seven-point scale representing tenderness. Thus cooking from the frozen state does not necessarily result in a tough turkey. Thigh meat was pink for all methods, reflecting a lack of adequate time at temperatures needed to denature the protein portion of the pigments. End-point temperature in this study was 71°C.

Two additional studies on roasting turkeys in microwave and conventional ovens have been reported (McNeil and Penfield, 1983; Prusa and Hughes,

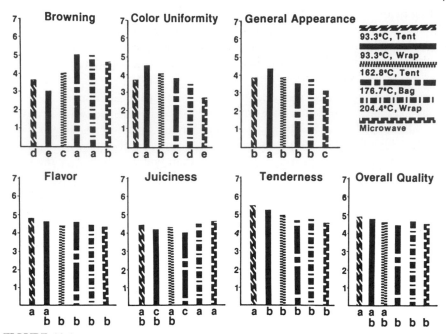

FIGURE 10.4

Mean sensory scores for turkey roasted by six methods. Within each attribute, columns with the same letter are not significantly different ($p < 0.05$, Duncan's multiple range test). (From Cornforth *et al.*, 1982; Reprinted from *Journal of Food Science*, **47**, No. 4, pp. 1108–1112. Copyright © 1982 by Institute of Food Technologists.)

1986). Convection ovens also were used in both studies. In the earlier study (McNeil and Penfield, 1983) few differences were found among boneless turkey breast roasts heated in the three ovens to 77°C. Sensory scores and Warner–Bratzler shear values indicated that the microwave roasts were less tender than roasts heated in the convection and conventional ovens. However, a consumer panel did not indicate a preference for any of the roasts. Prusa and Hughes (1986) used the same three types of ovens but cooked the roasts to two endpoint temperatures, 77 and 82°C. In contrast to the earlier study, drip losses were greater from the microwave-heated roasts than from roasts heated in the other two ovens. Shear values were not affected by oven type. Heating to 82°C increased the amount of drip. Prusa and Hughes (1986) also studied the sodium and cholesterol content of the turkey roasts. No differences in sodium or cholesterol contents were found among heating treatments. Sodium was decreased 36% by cooking. Roasts cooked to 82°C had higher cholesterol contents than those cooked to 77°C.

Results for the McNeil and Penfield (1983) and Prusa and Hughes (1986) studies are primarily applicable to household situations. Cremer and Richman (1987) studied the effects of roasting turkey breasts at two temperatures (105 and 163°C) in an institutional gas convection oven on cooking losses and shear values. Energy consumption also was monitored. The lower oven temperature reduced energy consumption. Total cooking loss was greater in breasts roasted at 163°C. Shear values did not differ.

Engler and Bowers (1975) studied the use of slow cookers for turkey preparation and found that tenderness was greater and cooking losses lower for "slow-cooked" turkey breast halves than for halves roasted conventionally. Thus these methods may offer added convenience with quality as acceptable as is obtained by conventional methods.

C. Processed Poultry Products

One processed poultry product that has been available for some time is the turkey roll. Because many institutions use commercially prepared rolls and roasts, Brown and Chyuan (1987) studied the effect of convection heating on the sensory quality of commercially prepared turkey rolls. Three cooking temperatures (105, 135, and 165°C), two holding conditions (not chilled and chilled for 24 hr), and three hot-holding times (0, 60, and 120 min) were used. As expected, the highest oven temperature took less time and the lowest temperature took less energy than the others. Sensory properties varied but no clearly defined trends were obvious for most characteristics. The researchers indicated that when the roast was cooked at 105°C, not chilled, and held hot for 60 min or less, panelists indicated that the product was juicier than for other treatment combinations. They suggested that this difference might also be obvious to consumers and therefore should be considered along with the findings on energy and time in making a decision about cooking method.

Cremer and Hartley (1988) also studied the effects of roasting conditions on quality of turkey rolls as the first step in a study on reheating. Temperatures for roasting the rolls in a gas forced air convection oven were 105 and 165°C. Cooking losses were greater for the higher roasting temperature.

D. Reheating Poultry Products

Warmed-over flavor, which was previously discussed, is of concern in precooked convenience products. Lyon (1988) explored the sensory changes that occurred when precooked chicken patties were stored and reheated. The sensory terms for WOF that are shown in Table II were used by a trained panel to describe the flavor of products tested. Results are presented in Fig. 10.5. The results are presented as sunray or spider web plots. The center of the figure represents weak intensity and the points along a ray from the center to the end represent increasing intensity of an attribute. The plot for the fresh cooked control is on each of the figures to facilitate comparison with each treatment. With prompt cooling, freezing, and reheating, chickeny (CHY), meaty (MTY), brothy (BRO), liver/organy (LIV), and sweet (SWT) decreased in intensity. Other attributes increased. The profile for the sample that was stored for 1 day before freezing is similar to the promptly frozen sample, except that cardboardy (CBD), warmed over (WMO), and rancid (RAN) increased, indicating that the onset of development of warmed over flavor occurs soon after cooking. After 3 days of storage, dramatic changes are seen in the profile. Decreases in desirable characteristics (CHY, MTY, BRO) occurred while obviously less desirable (BUR, CBD, WMO, and RAN) notes increased in intensity.

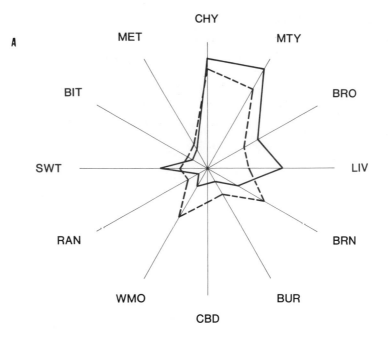

A

CHY · MET · MTY · BIT · BRO · SWT · LIV · RAN · BRN · WMO · BUR · CBD

—— FRESH COOKED
- - - COOKED, COOLED, FROZEN, REHEATED

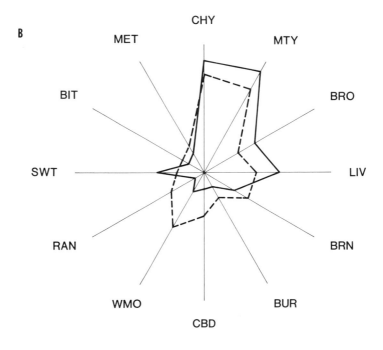

B

CHY · MET · MTY · BIT · BRO · SWT · LIV · RAN · BRN · WMO · BUR · CBD

—— FRESH COOKED
- - - COOKED; STORED 1 DAY; FROZEN; REHEATED

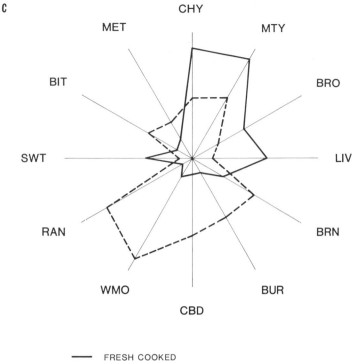

FRESH COOKED
COOKED; STORED 3 DAYS; FROZEN; REHEATED

FIGURE 10.5

Plots of means scores on 12 descriptive attributes comparing freshly cooked product with (A) cooked, 0-day stored product, (B) cooked, 1-day stored product, and (C) cooked, 3-day stored product. Abbreviated terms: CHY, chickeny; MTY, meaty; BRO, brothy; LIV, liver/organy; BRN, browned; BUR, burned; CBD, cardboard; WMO, warmed over; RAN, rancid; SWT, sweet; BIT, bitter; MET, metallic. (From Lyon, 1988; Reprinted from *Journal of Food Science*, **53**, No. 4, pp. 1086–1090. Copyright © 1988 by Institute of Food Technologists.)

Warmed-over flavor also is a problem in institutional settings where poultry may be prepared ahead of time when equipment is available and held refrigerated and reheated before serving. Method of reheating has been the subject of several studies. Similarly to results with red meat, turkey reheated by microwaves had greater turkey flavor and less stale flavor than conventionally reheated turkey (Cipra and Bowers, 1971). Turkey scored immediately after conventional heating and turkey reheated in a microwave had higher meaty brothy flavor and lower rancidity scores than conventionally reheated turkey (Bowers, 1972).

Cremer and Richman (1987) reheated 75-g slices of turkey breast in three types of ovens: microwave, convection, and infrared. Freshly heated turkey was scored higher than the reheated samples. Fresh aroma was equal in fresh and convection-reheated samples and roasted aroma equal in fresh and microwave-reheated samples. The investigators recommended that turkey should be freshly prepared before serving. If reheating is necessary, the microwave should be used because quality did not differ among ovens and energy can be saved by using the microwave. Cremer also studied reheating of turkey rolls in the three ovens (Cremer and Hartley, 1988). Temperature of roasting in a convection oven also was an independent variable in this study. Samples were reheated with no sauce, a starch-thickened sauce, or broth. Use of a sauce or broth increased sensory scores for most characteristics. Scores for stale off-notes (stale, rancid, or other off-flavor) suggest that all samples did have some degree of off-flavor. Samples roasted at 105 or 165°C, reheated in any of the three ovens with broth or sauce were equally acceptable, suggesting that the use of sauce was a major factor in determination of quality of reheated turkey roast.

The chemical changes that occur in the development of WOF were studied recently by Wu and Sheldon (1988a, b). Products associated with fatty acid oxidation (primarily carbonyls) were identified and related to TBA values and sensory scores for off-odor and flavor. Wu and Sheldon (1988a) noted that flavor changes may begin within hours of cooking.

Suggested Exercises

1. The principles of meat binding and factors affecting binding can be studied using poultry meat in the production of chicken loaves in the laboratory. (*Note:* This experiment illustrates principles discussed in Chapter 9 as well as this chapter.) The idea for this experiment is adapted from one designed by Maurer (1977). Place 454 g of cubed (1.25–2.5 cm) chicken (turkey or beef could be used) in 600-ml beakers, cover with foil, and bake in a 180°C oven to 85°C. After cooling in a refrigerator the loaves can be sliced and evaluated for binding properties. Variations that could be used are mixing (0 or 3 min in an electric mixer with paddle attachment), salt (0 or 2% salt mixed before baking), phosphate (0 or 0.5% before mixing), and pressure (place a smaller beaker on top of the 600-ml beaker to weight the cubes and add no beaker to a second product). Obviously combinations of these treatments could be used.
2. Prepare chicken breasts with and without skin in several types of ovens. Evaluate the effects of the treatments on tenderness, juiciness, color, and flavor.
3. Prepare patties of varying proportions of ground poultry and ground beef. Broil to 75°C. Measure cooking losses. Evaluate the effect of proportion of poultry meat on flavor, color, tenderness, and juiciness of the patties. Relate the differences to the differences in fat contents of the patties.

References

Adams, C. 1975. Nutritive value of American foods. Agric. Handbook 456, Agricultural Research Service, USDA, Washington, DC.

Addis, P. B. 1986. Poultry muscle as food. *In* "Muscle as Food," Bechtel, P. J. (Ed.). Academic Press, Orlando, Florida.

Baity, M. R., Ellington, A. E., and Woodburn, M. 1969. Foil wrap in oven cooking. *J. Home Econ.* **61**: 174–176.

Barbut, S., Maurer, A. J., and Lindsay, R. C. 1988. Effects of sodium chloride and added phosphates on physical and sensory properties of turkey frankfurters. *J. Food Sci.* **53**: 62–66.

Bowers, J. A. 1972. Eating quality, sulfhydryl content, and TBA values of turkey breast muscles. *J. Agric. Food Chem.* **20**: 706–708.

Bowers, J. A., Craig, J. A., Tucker, T. J., Holden, J. M., and Posati, L. P. 1987. Vitamin and proximate composition of fast-food fried chicken. *J. Amer. Dietet. Assoc.* **87**: 736–739.

Bramblett, V. D. and Fugate, K. W. 1967. Choice of cooking temperature for stuffed turkeys. I. Palatability factors. *J. Home Econ.* **59**: 180–185.

Brown, N. E. and Chyuan, J.-Y. A. 1987. Convective heat processing of turkey roll: Effects on sensory quality and energy usage. *J. Amer. Dietet. Assoc.* **87**: 1521–1525.

Cipra, J. S. and Bowers, J. A. 1971. Flavor of microwave and conventionally reheated turkey. *Poultry Sci.* **50**: 703–706.

Cornforth, D. P., Brennand, C. P., Brown, R. J., and Godfrey, D. 1982. Evaluation of various methods for roasting frozen turkeys. *J. Food Sci.* **47**: 1108–1112.

Cremer, M. L. and Hartley, S. K. 1988. Sensory quality of turkey rolls roasted at two temperatures and reheated with varying sauces in three types of institutional ovens. *J. Food Sci.* **53**: 1605–1609.

Cremer, M. L. and Richman, D. K. 1987. Sensory quality of turkey breasts and energy consumption for roasting in a convection oven and reheating in infrared, microwave, and convection ovens. *J. Food Sci.* **52**: 846–850.

Dawson, L. E. and Gartner, R. 1983. Lipid oxidation in mechanically deboned poultry. *Food Technol.* **37**(7): 112–116.

Deethardt, D., Burrill, L. M., Schneider, K., and Carlson, C. W. 1971. Foil-covered versus open-pan procedures for roasting turkey. *J. Food Sci.* **36**: 624–625.

Dimick, P. S., MacNeil, J. H., and Grunden, L. P. 1972. Poultry product quality. Carbonyl composition, and organoleptic evaluation of mechanically deboned poultry meat. *J. Food Sci.* **37**: 544–546.

Engler, P. P. and Bowers, J. A. 1975. Eating quality and thiamin retention of turkey breast muscle roasted and "slow-cooked" from frozen and thawed states. *Home Econ. Research J.* **4**: 27–31.

Esselen, W. B., Levine, A. S., and Brushway, M. J. 1956. Adequate roasting procedures for frozen stuffed poultry. *J. Amer. Dietet. Assoc.* **32**: 1162–1166.

Goertz, G. E., Hopper, A. S., and Harrison, D. L. 1960. Comparison of rate of cooking and doneness of fresh-unfrozen, defrosted turkey hens. *Food Technol.* **14**: 458–462.

Goertz, G. E., Meyer, B. H., Weathers, B., and Hooper, A. S. 1964. Effect of cooking temperatures on broiler acceptability. *J. Amer. Dietet. Assoc.* **45**: 526–529.

Hall, K. N. and Lin, C. S. 1981. Effect of cooking rates in electric or microwave oven on cooking losses and retention of thiamin in broilers. *J. Food Sci.* **46**: 1292–1293.

Hanson, H. L. and Fletcher, L. R. 1963. Adhesion of coating on frozen chicken. *Food Technol.* **17**: 793–796.

Heine, N., Bowers, J. A., and Johnson, P. G. 1973. Eating quality of half turkey hens cooked by four methods. *Home Econ. Research J.* **1**: 210–214.

Hoke, I. M. 1968. Evaluation of procedures and temperatures for roasting chickens. *J. Home Econ.* **60**: 661–665.

Hsu, D. L. and Jacobson, M. 1974. Macrostructure and nomenclature of plant and animal sources. *Home Econ. Research J.* **3**: 24–32.

Huffman, D. L., Marchello, J. A., and Ringkob, T. P. 1986. Development of processed meat items for the fast-food industry. *Recip. Meat Conf. Proc.* **39**: 25–26.

Johnson, P. G. and Bowers, J. A. 1976. Influence of aging on the electrophoretic and structural characteristics of turkey breast muscle. *J. Food Sci.* **41**: 255–261.

Klose, A. A., Pool, M. F., Wiele, M. B., Hanson, H. L., and Lineweaver, H. 1959. Poultry tenderness. I. Influence of processing on tenderness of turkeys. *Food Technol.* **13**: 20–24.

Lane, R. H., Nguyen, H., Jones, S. W., and Midkiff, V. C. 1982. The effect of fryer temperature on yield and composition of deep-fat fried chicken thighs. *Poultry Sci.* **61**: 294–299.

Lee, W. T. and Dawson, L. E. 1976. Changes in phospholipids in chicken tissues during cooking in fresh and reused cooking oil, and during frozen storage. *J. Food Sci.* **41**: 598–600.

Lyon, B. G. 1987. Development of chicken flavor descriptive attribute terms aided by multivariate statistical procedures. *J. Sensory Studies* **2**: 55–67.

Lyon, B. G. 1988. Descriptive profile analysis of cooked, stored, and reheated chicken patties. *J. Food Sci.* **53**: 1086–1090.

Martinez, J. B., Maurer, A. J., and Arrington, L. C. 1984. Heating curves during roasting of turkeys. *Poultry Sci.* **63**: 260–264.

Maurer, A. J. 1977. Teaching meat binding using poultry loaves. *Poultry Sci.* **56**: 323–327.

Maurer, A. J. 1983. Reduced sodium usage in poultry muscle foods. *Food Technol.* **37**(7): 60–65.

McNeil, M. and Penfield, M. P. 1983. Turkey quality as affected by ovens of varying energy costs. *J. Food Sci.* **48**: 853–855.

Minor, L. J., Pearson, A. M., Dawson, L. E., and Schweigert, B. S. 1965. Chicken flavor: The identification of some chemical components and the importance of sulfur compounds in the cooked volatile fraction. *J. Food Sci.* **30**: 686–696.

Moreng, R. E. and Avens, J. S. 1985. "Poultry Science and Production." Reston Publ. Co., Inc., Reston, Virginia.

Nakai, Y. and Chen, T. C. 1986. Effects of coating preparation methods on yields and compositions of deep-fat fried chicken parts. *Poultry Sci.* **65**: 307–313.

Nakamura, R., Sekoguchi, S., and Sato, Y. 1975. The contribution of intramuscular collagen to the tenderness of meat from chickens of different ages. *Poultry Sci.* **54**: 1604–1612.

Nash, D. M., Proudfoot, F. G., and Hulan, H. W. 1985. Pink discoloration in cooked broiler chicken. *Poultry Sci.* **64**: 917–919.

Oltrogge, M. H. and Prusa, K. J. 1987. Microwave-heated baking hen breasts. *Poultry Sci.* **66**: 1548–1551.

Peters, C. R., Sinwell, D. D., and Van Duyne, F. O. 1983. Slow cooker vs. oven preparation of meat loaves and chicken. *J. Amer. Dietet. Assoc.* **83**: 430–435.

Posati, L. P. 1979. Composition of foods. Poultry products. Raw. Processed. Prepared. Agric. Handbook 8-5, Consumer and Food Economics Institute, USDA, Washington, DC.

Prusa, K. J., and Hughes, K. V. 1986. Quality characteristics, cholesterol, and sodium content of turkey as affected by conventional, convection, and microwave heating. *Poultry Sci.* **65**: 940–948.

Prusa, K. J. and Lonergan, M. M. 1987. Cholesterol content of broiler breast fillets heated with and without the skin in convection and conventional ovens. *Poultry Sci.* **66**: 990–994.

Ramaswamy, H. S. and Richards, J. F. 1982. Flavor of poultry meat—A review. *Can. Inst. Food Sci. Technol. J.* **15**(1): 7–18.

Richardson, M., Posati, L. P., and Anderson, B. A. 1980. Composition of foods. Sausages and luncheon meats. Agric. Handbook 8-7, Consumer Nutrition Center, Science and Education Administration, USDA, Washington, DC.

Saffle, R. L. 1973. Quantititave determination of combined hemoglobin and myoglobin in various poultry meats. *J. Food Sci.* **38**: 968–970.

Salmon, R. E., Stevens, V. I., Poste, L. M., Agar, V., and Butler, G. 1988. Effect of roasting breast up or breast down and dietary canola meal on the sensory qualities of turkeys. *Poultry Sci.* **67**: 680–683.

Smith, D. M. and Alvarez, V. B. 1988. Stability of vacuum cook-in-bag turkey breast rolls during refrigerated storage. *J. Food Sci.* **53**: 46–48, 61.

USDA. 1984a. The safe food book. Home and Garden Bull. No. 241, Food Safety and Inspection Service, United States Department of Agriculture, Washington, DC.

USDA. 1984b. Talking about turkey. Home and Garden Bull. No. 243, United States Department of Agriculture, Washington, DC.

USDA. 1988. Livestock and Poultry. Situation and Outlook. Report. USDA. **LPS-29**: 40–43.

USDA. 1989. Livestock and Poultry. Situation and Outlook. Report. USDA. **LPS-35**: 42–44.

Woodburn, M. and Ellington, A. E. 1967. Choice of cooking temperature for stuffed turkeys. II. Microbiological safety of stuffing. *J. Home Econ.* **59**: 186–190.

Wu, T. C. and Sheldon, B. W. 1988a. Flavor components and factors associated with development of off-flavors in cooked turkey rolls. *J. Food Sci.* **53**: 49–54.

Wu, T. C. and Sheldon, B. W. 1988b. Influence of phospholipids on the development of oxidized off-flavors in cooked turkey rolls. *J. Food Sci.* **53**: 55–61.

Younathan, M. T., Farr, A. J., and Laird, D. L. 1984. Microwave energy as a rapid-thaw method for frozen poultry. *Poultry Sci.* **63**: 265–268.

FISH

Consumption of fish increased 34% from 10.8 lb/capita in 1965 to 14.5 lb/capita in 1985 (NRC, 1988, pp. 18–44). Consumption can be expected to continue increasing because of recommendations in the dietary guidelines that less fat be consumed (USDA and USDHHS, 1985). Health professionals recommend inclusion of fish rich in $\omega-3$ fatty acids (eicosapentaenoic, 20:5, and docosahexaenoic, 22:6) because of their association with a decrease in coronary heart disease. Common sources of these fatty acids include mackerel (Atlantic and king), herring (Atlantic and Pacific), lake trout, salmon (chinook, Atlantic, sockeye, pink, chum, and coho), bluefish, and Atlantic halibut. Inclusion of fish also helps to decrease the saturated fatty acid content of the diet (Kris-Etherton *et al.*, 1988). In addition, nutrients supplied in significant amounts include protein, niacin, and vitamin B_{12}. Interest in fish also continues to be high because of increased concern for feeding the world's population. With increased interest in fish and the diversity of species available for consumption, research activity also has increased; therefore, the following discussion represents only a few of the many studies that have been reported in the last 10 years in addition to basic studies on quality and on the effects of cooking on composition and eating quality.

I. CLASSIFICATION

Officially, fish may be divided into two groups, finfish and shellfish. Finfish include those fish with a backbone and fins and may be from fresh water (catfish, carp, trout, lake perch, and whitefish) or salt water (salmon, haddock, bluefish, flounder, halibut, tuna, and cod). Shellfish include those without fins and a backbone. Subgroups of shellfish include mollusks with two enclosing shells (clams, oysters, mussels, and scallops) and crustaceans with a soft undershell and a hard upper shell (crabs, shrimp, and lobsters). Price (1985) provides a comprehensive listing of these fish and their characteristics. Table I includes information on the protein, carbohydrate, and fat content of representative fish.

Fish also may be divided into two groups on the basis of their fat content. Classification into these groups is valuable for determining appropriate preparation and storage methods. All shellfish and many fish with bones contain less than 5% fat in the edible portions. Examples of such lean fish with bones are butterfish, bream, sea bass, catfish, cod, flounder, haddock, Pacific halibut, perch, pike, pink salmon, snapper, swordfish, rainbow trout, and whiting (Hepburn *et al.*, 1986; Hearn *et al.*, 1987). Examples of fish containing more than 5% fat are bluefish, coho salmon, Greenland halibut, herring, mackerel, and sar-

TABLE I

Composition of Selected Raw and Cooked Fish[a]

	Protein (%)	Carbohydrate (%)	Total fat (%)
Catfish, channel			
Breaded, fried	18.1	8.0	13.3
Crab, blue			
Crab cake	20.2	0.5	7.5
Cooked, moist heat	20.2	0.0	1.8
Haddock			
Cooked, dry heat	24.2	0.0	0.9
Halibut			
Cooked, dry heat	26.7	0.0	2.9
Oyster, eastern			
Breaded, fried	8.8	11.6	12.6
Canned	7.1	3.9	2.5
Scallop			
Breaded, fried	18.1	10.1	10.9
Shrimp			
Breaded, fried	21.4	11.5	12.3
Cooked, moist heat	20.9	0.0	1.1
Surimi (pollack)	15.2	6.9	0.9
Tuna			
Light, oil pack	26.5	0.0	8.1
White, water pack	29.6	0.0	0.5

[a] From Exler (1987).

dines (Hepburn *et al.*, 1986; Hearn *et al.*, 1987). Differences in values for specific fish may be found in the two references cited. For example, pompano would be classified as greater than 5% from one list but less than 5% from the other.

II. QUALITY ATTRIBUTES

A. Sensory Characteristics

A deterrent to consumer acceptance of fish has been the lack of knowledge of or misconception about fish quality in relation to the names of the various species. On the basis of their image of the species, consumers, in a study by Wesson *et al.* (1979), rated the desirability of cod higher than that of catfish. However, when asked to evaluate the actual products that were coded or identified, equal preference for the two samples was shown. Thus mental images associated with the name of a species may be changed when actual products are tried.

Martin *et al.* (1983) described the problem associated with the nomenclature. The complexity of the situation is clearly illustrated by the facts that consumers choose from four commercial meat species, beef, veal, lamb, and pork, and from four commercial poultry species, chicken, turkey, goose, and duck. In contrast, there are more than 500 species of seafood that may be available in the United States and more than 1000 that are available worldwide. As part of a comprehensive plan to develop a meaningful nomenclature system, work at the United States Army Natick Research and Development Laboratories has been directed toward development of sensory and objective methodology for determining the "edibility" characteristics of fish. The work also was directed toward the selection of statistical techniques to analyze the data on characteristics of fish in order to group the fish according to similar qualities. Results of this work are reported in several papers (Cardello *et al.*, 1982, 1983; Prell and Sawyer, 1988; Sawyer *et al.*, 1984, 1988).

Review of this work is appropriate for a book such as this because it illustrates the approach taken to identify the characteristics of a class of food and then use the results to provide information about the relative quality of the members of that group. The first step in the process was to have trained sensory panelists describe the characteristics of various species of fish. In work with trained panelists and a consumer panel 13 flavor and texture attributes were shown to be important for discrimination among various species of fish (Cardello *et al.*, 1983). The 13 characteristics of various species of fish then were evaluated by the sensory procedures developed (Cardello *et al.*, 1982; Sawyer *et al.*, 1988). Definitions for the texture terms are included in Table II. Characteristics were evaluated on seven-point intensity scales ranging from 1 for weak to 7 for strong, and reference standards were established (Sawyer *et al.*, 1988). Individual texture and flavor profiles for 6 of the 18 fish are shown in Fig. 11.1. Inspection of the profiles shows some obvious differences among the species. For example, mackerel, bass, and swordfish appear to be darker than the others but similar to each other in darkness. Grouper is definitely harder while bass is more flaky than the others. Mackerel appears to have a higher flavor intensity than the

TABLE II

Sensory Characteristics and Definitions Used in Natick Laboratories on
Textural Characteristics of Fish[a]

Characteristic	Definition
Hardness	The perceived force required to compress the sample with the molar teeth
Flakiness	The perceived degree of separation of the sample into individual flakes when manipulated with the tongue against the palate
Chewiness	The total perceived effort required to prepare a sample to a state ready for swallowing
Fibrousness	The perceived degree (number × size) of fibers evident during mastication
Moistness	The perceived degree of oil and/or water in the sample during chewing
Oily mouth coating	The perceived degree of oil left on the teeth, tongue, and palate after swallowing

[a]From Cardello *et al.* (1982). (Reprinted from *Journal of Food Science*, **47**, No. 6, pp. 1818–1823. Copyright © 1982 Institute of Food Technologists.)

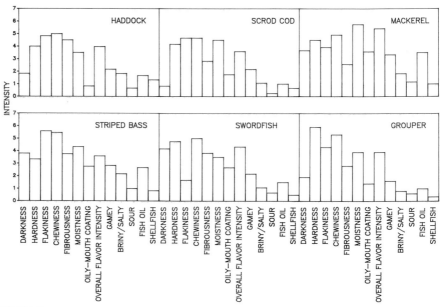

FIGURE 11.1
Consumer sensory profiles for six common species of Atlantic fish. Seven on scale represents strong intensity. (From Sawyer *et al.*, 1988; Reprinted from *Journal of Food Science*, **53**, No. 1, pp. 12–18, 24. Copyright © 1988 by Institute of Food Technologists.)

others. Cluster analysis, a statistical technique that is used to identify groupings of items on the basis of similarities in the intensity of several characteristics, was used to show groupings among 18 different species of fish. Cluster diagrams representing data from consumer and sensory panelists for the 18 species are shown in Fig. 11.2. From this figure it can be concluded that consumers and trained panelists differ somewhat in their assessment of fish so that expert opinion may not reflect that of consumers. The figure shows that swordfish and halibut are alike but different from haddock. Other comparisons can be made and related to individual profiles such as those shown in Fig. 11.1.

Application of these techniques to recommendations for marketing of fish was shown in a study by the same group on snapper and rock fish species (Sawyer *et al.*, 1984). It was shown that fish of the family *Lutjanus* (snappers) are similar in flavor to those from the family *Scorpaenidae* (rockfish). However, those from the rockfish family are firmer, chewier, more fibrous, and more flaky than those from the snapper family. Cluster diagrams like those shown previously

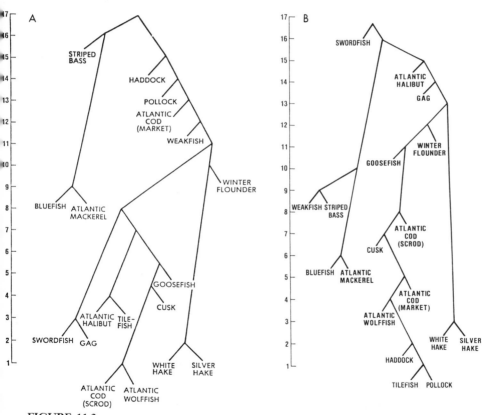

FIGURE 11.2
Grouping of 18 species of fish by cluster analysis applied to consumer panel data (A) and trained panel data (B). (From Sawyer *et al.*, 1988; Reprinted from *Journal of Food Science*, **53**, No. 1, pp. 12–18, 24. Copyright ©1988 by Institute of Food Technologists.)

were used to illustrate these differences. The study showed that the market practice of selling Pacific rockfish as red snapper is misleading to the consumer. Further application of these techniques to nomenclature and marketing of fish is expected to enable consumers to better understand fish quality and be able to select fish that is similar in sensory characteristics to the type to which they are accustomed.

Sensory characteristics of shrimp were studied in depth by Solberg *et al.* (1986, pp. 109–121). As with the terms defined by the Natick group for finfish, the terms for shrimp could be used as a basis for sensory studies on the effect of treatment on quality as well as for quality control.

B. Safety

Although foodborne illnesses are discussed in Chapter 12, additional mention in this chapter is important because of some unique potential problems associated with fish. In addition to *Vibrioaceae* and *Clostridium botulinum*, which are of particular concern in seafood (Hackney and Dicharry, 1988), viral disease transmission (Gerba, 1988) and toxins are of concern (Taylor, 1988). Further information on these may be found in the cited references.

In addition to microbiological concerns, parasites also may be of concern (Olson, 1986, pp. 339–355). Larval anisakid nematodes and cestodes of the genus *Diphyllobothrium* are potentially infective. However, because cooking at 60°C for 1 min and freezing to −20°C for 24 hr will kill the parasites, problems should be minimal. The consumption of raw fish, on the other hand, presents problems. The problem is greater in Asia than in the United States, but the number of cases in the United States has increased as raw fish consumption has increased.

III. STRUCTURE AND COMPOSITION _____

The proteins and muscle structure of fish are described in detail by Suzuki (1981, pp. 1–61). Myofibrils appear to be composed of the same proteins as the myofibrils of muscle tissue from warm-blooded animals (Connell, 1964, pp. 255–293). Fibers lie parallel to each other within myotomes, giving a striated appearance. Fiber diameter and length vary with species (Suzuki, 1981, p. 2). The myotomes or flakes are divided by sheets of connective tissue called myocommata. The arrangement of these structures is shown in Fig. 11.3. When heated, the connective tissue softens and the cells separate, giving a flaky appearance to the fish. Connective tissue is present in very small quantities in fish and contains less proline and hydroxyproline than does red meat collagen (Dunajski, 1979). Dunajski also suggested in his review of fish texture that fish disintegrates easily when heated because of its low connective tissue content and because the collagen is less heat stable than that of red meat. Species variation in composition is found. For example, Feinstein and Buck (1984) reported that flounder has more soluble collagen and a higher pH than cusk. Neither measurement was related to instrumental texture measurements on cooked fish.

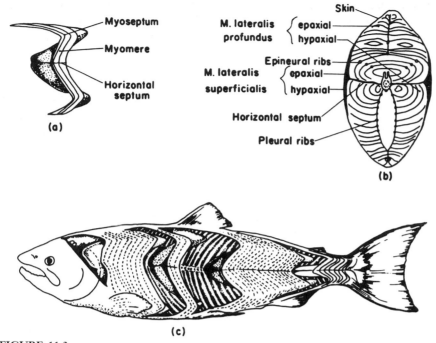

FIGURE 11.3

Musculature of salmon (*Oncorhynchus tschawytscha*). (a) Lateral view of three myomeres from the anterior portion of the trunk. (b) Transverse section through the trunk of the salmon. (c) Lateral view of the salmon. (From Hsu and Jacobson, 1974; *Home Economics Research Journal*, American Home Economics Association, Washington, DC. Reprinted with permission.)

Fish have both red and white muscle, as do land animals. Small areas of red muscle are found even in light-pigmented fish such as halibut and sole (Brown, 1986, pp. 414–416). The quantity of red muscle, which contains more myoglobin than white muscle, increases as the activity level of the fish increases. Red muscle constitutes a large portion of the total muscle tissue in active fish like tuna (Brown, 1986, pp. 414–416). Distribution of the dark muscle tissue is species dependent. In most fish, the dark muscle lies just under the skin at the sides, as seen in Fig. 11.3 (b). In others, dark muscle may be located near the spine. Dark muscle is associated with more connective tissue than is light muscle (Suzuki, 1981, pp. 1–5). Fiber sizes of dark muscle were shown to be smaller than those of light muscle in yellowfin tuna (*Neothunnos albacora*) and shear force for dark muscle was greater than for white muscle after heating. Kanoh *et al.* (1988) suggested that the small fibers of the dark muscle are associated with more collagen than are the larger fibers of ordinary muscle.

Studies of the myofibrillar proteins of fish have shown that myosin from both finfish and shellfish resembles myosin from warm-blooded animals. However, aggregation of the fish myosin occurs readily due to its instability (Brown, 1986, p. 408). The functional properties of fish muscle have not been studied to any great extent as little interest has been shown in comminuted fish products. However, as surimi increases in importance more interest will be shown.

Composition of fish varies with season and influences the quality of the flesh. For example, post-mortem variation in pH in cod occurs with season. Love (1979) indicated that as pH of cod decreases, connective tissue becomes weakened and the muscle cells fall apart, a condition called gaping. Cod muscle with low pH also tends to be very firm. Salmon are not fit for consumption at spawning because of biochemical changes. Lipid deposits are formed to serve as an energy source for migration to the spawning grounds (Brown, 1986, p. 413).

With increasing consumer interest in fish, aquaculture is of interest. Catfish farming meets almost all of the demand for catfish, and nearly all of the rainbow trout and crawfish and about 40% of oysters are supplied in the United States by aquaculture (Martin, 1988). Martin indicated that freshwater prawns, striped and largemouth bass, mussels, penaeid shrimp, Pacific salmon, and abalone also have been successfully cultured. These species may be supplied by aquaculture in the future. However, farming of other species is of interest. As cultivation of various species is initiated, it is important to determine the differences in composition and quality of the pond-raised fish and wild counterparts. For example, Jahncke et al. (1988) compared pond-raised and wild red drum (channel bass) and found that fatty acid profiles differed. Similar eicosapentaenoic acid concentrations were found in both while lower levels of linoleic and docosahexaenoic acids were found in the pond-raised fish. The fatty acid profile of the pond-raised fish was related to that of the feed, indicating that the composition of cultured fish may be controlled by selection of feeding regime. Chanmugam et al. (1986) also reported alterations in the lipid profiles of pond-reared fish. Pond-reared prawn, catfish, and crayfish had lower ω−3 and higher ω−6 fatty acid levels than the wild seafood. Profiles were attributed to the diet given the pond-reared shellfish.

Differences in the sensory properties of pond-raised fish also were reported (Jahncke et al., 1988). The pond-raised fish were milder, sweeter, slightly more earthy, and less sour than the wild bass. Both fish were soft in texture and easy to chew and swallow. Wild fish were less moist, flaky, and fibrous than the pond-raised fish.

Color of the flesh is an important attribute of salmon. Carotenoids are responsible for the red color with astaxanthin the predominant pigment. In farmed Atlantic salmon (Salmo salar) canthaxanthin was found to be the predominant carotenoid (Skrede and Storebakken, 1986). Sensory panelists could differentiate visually between wild and pen-reared fish. In general, raw, baked, and smoked pen-reared salmon were more yellow in hue than wild salmon. Alteration in diet can be used to control the color of salmon flesh (Simpson et al., 1981, pp. 463–538).

Important components of fish tissue are the cathepsins or proteolytic enzymes. These enzymes may be important in flavor and textural changes that occur during storage of fresh fish (Brown, 1986). However, Geist and Crawford (1974) reported that sensory panelists could not differentiate between samples of three species of Pacific sole that exhibited high and low catheptic activity. These results suggested that these enzymes did not play a significant role in degradation while the fish were held on ice for 4 days after being caught. Deng

(1981) studied the effect of protease activity on texture changes during heating and concluded that during heating from 55 to 85°C enzyme hyrolysis contributed to reduction in shear force values. At higher temperatures protein–protein interactions were the major contributors to textural attributes.

IV. PROCESSING

Because the majority of the fish supply is harvested from the ocean, preservation of the product offers unique challenges. As fish farming increases, some of these problems may be reduced. Obviously preservation must start on the vessel immediately after the fish are caught. The major concern is reduction of microbial activity until the product is consumed in the fresh state or is processed by canning, drying, or freezing. Most fish is processed in some manner before sale to the consumer although sale of fresh fish and shellfish is increasing.

A. Fresh Fish Handling and Preservation

Fish goes through rigor mortis just as other animals do. If fish are not chilled after catch and go through rigor mortis with the tissue attached to the skeleton, extensive contraction may cause disruption of the myocommata junctions, which results in gaping and reduces the market value of the fish (Dunajski, 1979).

Filleting is the source of microbial contamination because living fish are sterile. Therefore, it seems logical that storing as gutted fish and filleting as needed would increase shelf life. Shaw et al. (1984) showed that fillets recently prepared were superior to those that had been stored as fillets for 7 days. Delaying filleting of chilled fish also ensures completion of rigor and minimizes gaping (Huss, 1988, pp. 27–29).

Shelf life of mullet (Mugil sp.) was extended from 7 to 10 days by storing in ice (−2°C) containing 2% NaCl or 7% propylene glycol instead of in ice at 0°C. Sensory evaluation was used to determine length of acceptable storage time. The extension was attributed to a reduction in the rate of biochemical change (reduced hypoxanthine and trimethylamine production) and delayed microbial growth. Damage to tissues by ice crystals at −2°C was minimal as compared with that at −20°C. Ampola and Keller (1985) reported that dipping cod fillets in a potassium sulfate dip and wrapping in an oxygen-impermeable film extended shelf life, doubling it in some cases.

B. Freezing

Freezing is probably the best method for preserving fish; however, deterioration will occur during freezer storage at −20°C or above. "Spongy and rubbery" are terms used to describe the resultant texture (Samson et al., 1985). Owusu–Ansah and Hultin (1986) reported that as storage time increased for red hake (Urophy-

cis chuss) the force to deform also increased. Cross-linking of proteins and/or peptides is thought to be the cause for this change in texture. The cross-linking may be caused by formaldehyde (H_2CO) which is a product of the enzymatic breakdown of trimethylamine oxide (TMAO) (Gill *et al.*, 1979), and methods for improving stability have been explored. The type of deterioration is dependent on the fat level of the fish. Loss of water-holding capacity is the major problem in lean fish whereas pigment degradation and lipid oxidation, which result in off-flavor, are of major concern in fatty fish. Several investigators have studied the effect of cryoprotectants on fish quality. For example, Krivchenia and Fennema (1988a) studied the effects of sodium tripolyphosphate (STP), monosodium glutamate, and a high-pH buffer on the quality of whitefish (3–6% fat) fillets stored at −12 and −60°C. Injection or dipping was used to apply the solutions to the fillets. Results indicated that surface application of the solutions was as effective as injection. Samples treated with STP and the high-pH buffer had better texture than did untreated controls. Firmness (Kramer shear), centrifugal drip, and cooking losses were indicators of texture. Although textural differences were noted by sensory panelists, a preference was indicated for the STP-dipped samples stored at −12°C over the −12°C controls but not over the −60°C controls, suggesting that the use of the cryoprotectant STP is useful in maintaining quality of fish stored at higher freezer temperatures. Similar results were obtained in a study of burbot (*Lota lota*), a lean fish with about 1% fat (Krivchenia and Fennema, 1988b).

Josephson *et al.* (1985) indicated that the use of polyethylene pouches for loose packaging and the use of vacuum packaging in barrier films both resulted in reduction of oxidative changes in whitefish as compared with the usual practice of holding without protection. Scaling and filleting prior to freezing increased the development of oxidized flavors.

Solberg *et al.* (1986, pp. 109–121) studied the effects of freezing treatments on the quality of shrimp. They reported that quick freezing in nitrogen, glazing with water, vacuum packaging, and low storage temperatures were best for preserving the quality. Sensory characteristics that were of importance in studying the effects of freezing were red color, flavor intensity, fresh odor, fresh flavor, chewing resistance, and juiciness, which were most intense after 3 months of storage. After 8 months of storage, yellowish discoloration, dehydration, ammonia odor, fishy/TMA (trimethyl amine) flavor, stale flavor, cardboardy flavor, and graininess were most intense. The faster freezing methods preserved good quality.

Improper handling of the product after freezing may be detrimental to the quality of frozen fish and seafood products. Gates and co-workers (1985) indicated that although nutritional value did not change, sensory quality changes occurred in 3–4 months after production of breaded shrimp stored in retail freezers. Over a 13-month storage period freezer burn and old seafood flavors increased in the shrimp stored in retail freezers. Temperatures in wholesale freezers were equal to or less than −20°C on 70% of the days of monitoring. The retail freezers monitored fluctuated more and did not maintain the −20°C level most of the time. Moisture content also was lost in all samples. Thus temperature of storage should be comparable to that in a wholesale freezer ($\leq -20°C$) to maintain quality of frozen breaded seafood products.

Breaded fish portions are a major menu item in restaurants. Loss of the breading is an undesirable characteristic. Tempura and batter mixes were used to coat cod (*Gadus* species) fillets in a study by Corey *et al.* (1987). The influence of several variables on breading adhesion was evaluated. Phosphate predips and breading composition did not affect adhesion. Breading loss increased with storage time and with freeze–thaw cycling. The investigators attributed the loss of breading to dehydration.

C. Surimi

Discussion of fish and seafood would be incomplete without the mention of surimi, a "new" food item that is becoming important in the United States. Surimi is washed, minced, frozen, raw fish that serves as a basis for the production of seafood analogs such as simulated crab flakes and crab legs. Buck and Fafard (1985) described the production of a frankfurter analog from red hake surimi that was comparable to a chicken frankfurter.

Fish used for surimi usually is of low fat content. The processing of raw fish into surimi originated with the Japanese. Lee (1984) described processing of surimi and indicated that the difference between frozen minced fish and surimi is the addition to surimi of cryoprotectants to improve shelf life. Compounds used for this purpose include sugar, sorbitol, and polyphosphates. Most surimi is produced from Alaska pollack (Lee, 1984). However, as supplies of that species are depleted, the need to find other species for surimi production will become important (Lee, 1986). Douglas–Schwartz and Lee (1988) reported that red hake and Alaska pollock exhibit similar surimi-making properties although red hake is more thermally stable than the pollock; they demonstrated that optimal temperatures for washing and chopping of red hake surimi were higher than those for pollack surimi. Heat setting of both types at 40°C followed by heating to 90°C resulted in high gel strength. Limitations to use of other species for surimi may include the presence of proteolytic enzymes that may degrade the proteins during thermal processing. This possibility was illustrated in Atlantic menhaden by Boye and Lanier (1988).

Washing of the deboned fish removes the bulk of the sarcoplasmic proteins and fish contains a very low level of connective tissue; therefore, surimi is basically a myofibrillar protein concentrate that exhibits the functional properties of cohesion and fat and water binding (Lanier, 1986). Lanier (1986) discusses the functional properties of surimi in detail and indicates that the formation of a gel structure is paramount to each of the functional properties exhibited by surimi. Because the myofibrillar proteins of fish are relatively unstable to heat, gel formation occurs at a relatively low temperature of 40°C (Lanier, 1986). The literature contains many reports of studies on factors affecting the properties of surimi gels (Babbitt and Reppond, 1988; Hennigar *et al.*, 1988). Gel-strengthening ingredients such as wheat or potato starch and egg white may be added to improve the texture of the surimi gel (Lee, 1984, 1986). Fat may be added to molded products such as shrimp analogs (Lee, 1986). Extrusion techniques are used to produce surimi-based products of various shapes.

V. EFFECTS OF COOKING ON QUALITY _____
A. Sensory Properties

Few studies have been reported on the influence of cooking method on the sensory properties of fish. Because of the limited research and the diversity of species studied generalizations are difficult. However, general recommendations indicate that fatty fish are desirable for baking and broiling because their fat keeps them from becoming dry. Lean fish may, however, be cooked with these methods if brushed or basted with melted fat, and are less likely to fall apart on boiling or steaming than fatty fish. Both fatty and lean fish can be fried successfully.

Because fish muscle tissue is similar to meat tissue, the same types of changes that occur in meat with heating occur in fish. Thus one would expect, as described by Dunajski (1979), that myofibrils will become tougher and connective tissue softer with heating. A plateau in shear values occurs at an internal temperature of about 70°C. Because fish collagen is very prone to thermal degradation, method of cooking is not as critical for fish as it is for red meats. Thus dry or moist methods may be used and only a few minutes at 70–80°C are needed to cook the fish to well done (Dunajski, 1979).

It is recommended that cooking should be stopped as soon as the fish flakes easily when tested with a fork because overcooking results in dryness, crumbliness, and firmness. A more scientific method for testing doneness was suggested by Charley (1952), who used a meat thermometer in salmon steaks 2.5 cm thick. The steaks were baked to an end-point temperature of 75°C in ovens at 177, 204, 232, and 260°C. Although these temperatures had little effect on palatability scores, 204°C was the most practical of the temperatures tested because there were more spattering and drip at the higher temperatures, while the lowest temperature increased the cooking time and caused a white material to ooze from the fish. Steaks baked at an oven temperature of 204°C to internal temperatures of 70, 75, 80, and 85°C became drier as the end-point temperature increased but did not differ in tenderness. Salmon steaks cooked to 80 and 85°C were more palatable than those cooked to the two lower temperatures. There was no significant difference between the two higher temperatures, except that there were significantly greater cooking losses at 85°C than at 80°C. It will be recalled that salmon may be relatively high in fat. In a later study (Charley and Goertz, 1958), larger pieces (900 g) of salmon were baked at the same four oven temperatures as in the earlier study. Total cooking losses and evaporative losses increased as oven temperatures increased. Anterior cuts were more flaky and tender than posterior cuts.

Greenland turbot (*Reinhardtius hippoglossides*) fillet pieces were baked in microwave and conventional ovens in a study by Madeira and Penfield (1985).

TABLE III

Microwave Cooking Times Determined
for Turbot Fillet Sections of Various
Thicknesses[a,b]

Thickness (mm)	Cooking time (min)
15–16	2.25
17–18	2.75
19–21	3.00
22	3.25
23–24	3.50
25–26	3.75

[a]Heating resulted in end-point temperatures of 77–88°C.

[b]From Madeira and Penfield (1985). (Reprinted from *Journal of Food Science*, **50**, No. 1, pp. 172–177. Copyright Institute of Food Technologists.)

Cooking time required to reach 82°C averaged 28.5 min in the conventional oven, and the time to reach 77–82°C in the microwave oven averaged 2.77 min. Cooking time in the microwave was dependent on thickness of the fillet, as shown in Table III. Microwave samples were softer and less chewy than conventionally baked samples. Consumer panelists found no differences between fish samples cooked by the two methods. Covering during cooking with waxed paper or plastic wrap is necessitated by the fact that fish may spatter while being cooked in the microwave oven. Baldwin and Snider (1971) found that spatter increased as moisture content of carp increased and decreased as the amount of fat increased. Because there is some disintegration of fish muscle during heating, cooking loss measurements are appropriately expressed as solid-drip loss and evaporative loss.

The effects of infrared broiling in a conveyorized tube broiler and baking in a convection oven on sensory quality and thiamine and riboflavin retention in salmon were studied by Takahashi and Khan (1987). Differences in fat, thiamine, and riboflavin content were not found. Appearance and color of the broiled steaks were less desirable and they were more juicy and tender than those cooked in the convection oven. Odor, flakiness, flavor, and overall acceptability did not differ. Because overall acceptability did not differ it appeared that either method would be suitable for food service operations.

B. Composition

Because of the interest in the nutritional value of fish, several studies have been directed toward the effect of cooking methods on the fat components of the fish. Gall and co-workers (1983) evaluated the effects of baking, broiling, deep fat frying, and cooking in a microwave oven on the composition of grouper (*Epi-*

nephelus morio), red snaper (*Lutjanus campechanus*), Florida pompano (*Trachinotus carolinus*), and Spanish mackerel (*Scomberomorus maculatus*). Fillets were cooked to an internal temperature of $71 \pm 2.7°C$. Extensive information on the changes in fatty acids are presented. For grouper, snapper, and pompano deep-fat frying resulted in the lowest and microwave heating in the highest yield. No differences between baking and broiling were found. No differences in yield were found among the four methods for mackerel. Lipid absorption during frying is of concern from a nutritional standpoint as well as from a sensory standpoint. Gall *et al.* (1983) reported that as fat content of the fillet increased the percentage of lipid absorbed was decreased. The raw and deep-fried lipid contents (wet basis) were, respectively, 0.88 and 3.73 g/100-g fillet for grouper, 1.50 and 5.49 g/100-g fillet for red snapper, 5.17 and 8.7 g/100-g fillet for pompano, and 13.75 and 12.42 g/100-g fillet for mackerel. Mai *et al.* (1978) reported similar results for lake trout (*Salvelinus namacush*), white sucker (*Castastomus commersonni*) and bluegill (*Lepomis macrochirus*). Lake trout has a higher initial lipid content and resisted absorption of fat whereas the lower fat white sucker and bluegill absorbed fat during cooking.

Suggested Exercises

1. Purchase several species of fish using Sawyer *et al.* (1988) as a guide for selection of fish of varying characteristics. Prepare by a selected cooking method. If possible test the textural characteristics with an instrument such as a Kramer shear press or cell on a UTM. Have class members evaluate the products with use of the terminology published by Prell and Sawyer (1988) and Cardello *et al.* (1982).
2. Purchase enough of one species of a low-fat fish and enough of a high-fat fish to cook by two or more methods (baking, microwave heating, steaming, and frying). Compare the textural characteristics of the fish. How does the effect of cooking method vary with fat content of the fish?

References

Ampola, V. G. and Keller, C. L. 1985. Shelf life extension of drawn whole Atlantic cod, *Gadus morhua*, and cod fillets by treatment with potassium sorbate. *Marine Fisheries Rev.* **47**(3): 26–29.

Babbitt, J. K. and Reppond, K. D. 1988. Factors affecting the gel properties of surimi. *J. Food Sci.* **53**: 965–966.

Baldwin, R. E. and Snider, S. 1971. Factors affecting spatter of carp during microwave cooking. *Microwave Energy Appl. Newsletter* **4**(4): 9–11.

Boye, S. W. and Lanier, T. C. 1988. Effects of heat-stable alkaline protease activity of Atlantic menhaden (*Brevoorti tyrannus*) on surimi gels. *J. Food Sci.* **53**: 1340–1341, 1398.

Brown, W. D. 1986. Fish muscle as food. *In* "Muscle as Food," Bechtel, P. J. (Ed.). Academic Press. Orlando, Florida.

Buck, E. M. and Fafard, R. D. 1985. Development of a frankfurter analog from red hake surimi. *J. Food Sci.* **50**: 321–324, 329.

Cardello, A. V., Sawyer, F. M., Maller, O., and Digman, L. 1982. Sensory evaluation of the texture and appearance of 17 species of North Atlantic fish. *J. Food Sci.* **47**: 1818–1823.

Cardello, A. V., Sawyer, F. M., Prell, P., Maller, O., and Kapsalis, J. 1983. Sensory methodology for the classification of fish according to "edibility characteristics." *Lebensm.-Wiss. u.-Technol.* **16**: 190–194.

Chanmugam, P., Boudreau, M., and Hwang, D. H. 1986. Differences in the ω3 fatty acid contents in pond-reared and wild fish and shellfish. *J. Food Sci.* **51:** 1556–1557.

Charley, H. 1952. Effects of internal temperature and of oven temperature on the cooking losses and palatability of baked salmon steaks. *Food Research* **17:** 136–143.

Charley, H. and Goertz, G. E. 1958. The effects of oven temperature on certain characteristics of baked salmon. *Food Research* **23:** 17–24.

Connell, J. J. 1964. Fish muscle proteins and some effects on them of processing. *In* "Symposium on Foods: Proteins and Their Reactions," Schultz, H. W., and Anglemeir, A. F. (Eds.). Avi. Publ. Co., Westport, Connecticut.

Corey, M. L., Gerdes, D. L., and Grodner, R. M. 1987. Influence of frozen storage and phosphate predips on coating adhesion in breaded fish portions. *J. Food Sci.* **52:** 297–299.

Deng, J. 1981. Effect of temperatures on fish alkaline protease, protein interaction and texture quality. *J. Food Sci.* **46:** 62–65.

Douglas-Schwartz, M. and Lee, C. M. 1988. Comparison of thermostability of red hake and Alaska pollack surimi during processing. *J. Food Sci.* **53:** 1347–1351.

Dunajski, E. 1979. Texture of fish muscle. *J. Texture Studies* **10:** 301–318.

Exler, J. 1987. Composition of foods: Finfish and shellfish products. Agric. Handbook 8-15, Human Nutrition Information Service, USDA, Washington, DC.

Feinstein, G. R. and Buck, E. M. 1984. Relationship of texture to pH and collagen of yellowtail flounder and cusk. *J. Food Sci.* **49:** 298–299.

Gall, K. L., Otwell, W. S., Koburger, J. A., and Appledorf, H. 1983. Effects of four cooking methods on the proximate, mineral and fatty acid composition of fish fillets. *J. Food Sci.* **48:** 1068–1074.

Gates, K. W., Eudaly, J. G., Parker, A. H., and Pittman, L. A. 1985. Quality and nutritional changes in frozen breaded shrimp stored in wholesale and retail freezers. *J. Food Sci.* **50:** 853–857, 868.

Geist, G. M. and Crawford, D. L. 1974. Muscle cathepsins in three species of Pacific sole. *J. Food Sci.* **39:** 548–551.

Gerba, C. P. 1988. Viral disease transmission by seafoods. *Food Technol.* **42**(3): 99–103.

Gill, T. A., Keith, R. A., and Smith-Lall, B. 1979. Texture deterioration of red hake and haddock muscle in frozen storage as related to chemical parameters and changes in the myofibrillar proteins. *J. Food Sci.* **44:** 661–667.

Hackney, C. R., and Dicharry, A. 1988. Seafood-borne bacterial pathogens of marine origin. *Food Technol.* **42**(3): 104–109.

Hearn, T. L., Sgoutas, S. A., Hearn, J. A., and Sgoutas, D. S. 1987. Polyunsaturated fatty acids and fat in fish flesh for selecting species for health benefits. *J. Food Sci.* **52:** 1209–1211.

Hennigar, C. J., Buck, E. M., Hultin, H. O., Peleg, M., and Vareltzis, K. 1988. Effect of washing and sodium chloride on mechanical properties of fish muscle gels. *J. Food Sci.* **53:** 963–964.

Hepburn, F. N., Exler, J., and Weihrauch, J. L. 1986. Provisional tables on the content of omega-3 fatty acids and other fat components of selected foods. *J. Amer. Dietet. Assoc.* **86:** 788–793.

Hsu, D. L. and Jacobson, M. 1974. Macrostructure and nomenclature of plant and animal food sources. *Home Econ. Research J.* **3:** 24–32.

Huss, H. H. 1988. "Fresh Fish—Quality and Quality Changes." Food and Agriculture Organization of the United Nations, Danish International Development Agency, Rome.

Jahncke, M., Hale, M. B., Gooch, J. A., and Hopkins, J. S. 1988. Comparison of pond-raised and wild red drum (*Sciaenops ocellatus*) with respect to proximate composition, fatty acid profiles, and sensory evaluation. *J. Food Sci.* **53:** 286–287.

Josephson, D. B., Lindsay, R. C., and Stuiber, D. A. 1985. Effect of handling and packaging on the quality of frozen whitefish. *J. Food Sci.* **50:** 1–4.

Kanoh, S., Polo, J. M. A., Kariya, Y., Kaneko, T., Watabe, S., and Hashimoto, K. 1988. Heat-induced textural and histological changes of ordinary and dark muscles of yellowfin tuna. *J. Food Sci.* **53:** 673–678.

Kris-Etherton, P. M., Krummel, D., Russell, M. E., Dreon, D., Mackey, A., Borchers, J., and Wood, P. D. 1988. The effect of diet on plasma lipids, lipoproteins, and coronary heart disease. *J. Amer. Dietet. Assoc.* **88:** 1373–1400.

Krivchenia, M. and Fennema, O. 1988a. Effect of cryoprotectants on frozen whitefish fillets. *J. Food Sci.* **53:** 999–1003.

Krivchenia, M. and Fennema, O. 1988b. Effect of cryoprotectants on frozen burbot fillets and a comparison with whitefish fillets. *J. Food Sci.* **53**: 1004–1008, 1050.

Lanier, T. C. 1986. Functional properties of surimi. *Food Technol.* **40**(3): 107–114, 124.

Lee, C. M. 1984. Surimi process technology. *Food Technol.* **40**(11): 69–80.

Lee, C. M. 1986. Surimi manufacturing and fabrication of surimi-based products. *Food Technol.* **40**(3): 115–124.

Love, R. M. 1979. The post-mortem pH of cod and haddock muscle and its seasonal variation. *J. Sci. Food Agric.* **30**: 433–438.

Madeira, K. and Penfield, M. P. 1985. Turbot fish sections cooked by microwave and conventional heating methods—objective and sensory evaluation. *J. Food Sci.* **50**: 172–177.

Mai, J., Shimp, J., Weihrauch, J., and Kinsella, J. E. 1978. Lipids of fish fillets: changes following cooking by different methods. *J. Food Sci.* **43**: 1669–1674.

Martin, R. E. 1988. Seafood products, technology, and research in the U.S. *Food Technol.* **42**(3): 58, 60, 62.

Martin, R. E., Doyle, W. H., and Brooker, J. R. 1983. Toward an improved seafood nomenclature system. *Marine Fisheries Rev.* **45**(7-8-9): 1–20.

NRC. 1988. "Designing Foods. Animal Products Options in the Market Place." National Research Council, National Academy Press, Washington, DC.

Olson, R. E. 1986. Marine fish parasites of public health importance. *In* "Seafood Quality Determination," Kramer, D. E. and Liston, J. (Eds.). Elsevier Science Publishers B.V., Amsterdam.

Owusu-Ansah, Y. J. and Hultin, H. O. 1986. Chemical and physical changes in red hake fillets during frozen storage. *J. Food Sci.* **51**: 1402–1406.

Prell, P. A., and Sawyer, F. M. 1988. Flavor profiles of 17 species of North Atlantic Fish. *J. Food Sci.* **53**: 1036–1042.

Price, R. J. 1985. "Seafood Retailing Manual." Sea Grant Marine Advisory Program, University of California, Davis.

Samson, A., Regenstein, J. M., and Laird, W. M. 1985. Measuring textural changes in frozen minced cod flesh. *J. Food Biochem.* **9**: 147–159.

Sawyer, F. M., Cardello, A. V., Prell, P. A., Johnson, E. A., Segars, R. A., Maller, O., and Kapsalis, J. 1984. Sensory and instrumental evaluation of snapper and rockfish species. *J. Food Sci.* **49**: 727–733.

Sawyer, F. M., Cardello, A. V., and Prell, P. A. 1988. Consumer evaluation of the sensory properties of fish. *J. Food Sci.* **53**: 12–18, 24.

Shaw, S. J., Bligh, E. G., and Woyewoda, A. D. 1984. Effect of delayed filleting on quality of cod flesh. *J. Food Sci.* **49**: 979–980.

Simpson, K. L., Katayama, T., and Chichester, C. O. 1981. Carotenoids in fish feeds. *In* "Carotenoids as Colorants and Vitamin A Precursors," Bauernfeind, J. C. (Ed.). Academic Press, New York.

Skrede, G. and Storebakken, T. 1986. Characteristics of color in raw, baked and smoked wild and pen-reared Atlantic salmon. *J. Food Sci.* **51**: 804–808.

Solberg, T., Tidemann, E., and Martens, M. 1986. Sensory profiling of cooked, peeled and individually frozen shrimps (*Pandalus borealis*), and investigation of sensory changes during frozen storage. *In* "Seafood Quality Determination," Kramer, D. E. and Liston, J. (Eds.). Elsevier Science Publishers B.V., Amsterdam.

Suzuki, T. 1981. "Fish and Krill Protein: Processing Technology." Applied Science Publishers, Ltd., London.

Takahashi, Y. and Khan, M. A. 1987. Impact of infrared broiling on the thiamin and riboflavin retention and sensory quality of salmon steaks for foodservice use. *J. Food Sci.* **52**: 4–6.

Taylor, S. L. 1988. Marine toxins of microbial origin. *Food Technol.* **42**(3): 94–98.

USDA and USDHHS. 1985. Nutrition and your health. Dietary guidelines for Americans. Home and Garden Bull. 232, 2nd edn., United States Department of Agriculture and United States Department of Health and Human Services, Washington, DC.

Wesson, J. B., Lindsay, R. C., and Stuiber, D. A. 1979. Discrimination of fish and seafood quality by consumer populations. *J. Food Sci.* **44**: 878–882.

CHAPTER 12

FOOD MICROBIOLOGY

Discussion of current food science knowledge would be incomplete without mention of microorganisms and their relationship to food and food quality and safety. Obviously, it is not the purpose of this book to present a comprehensive coverage of food microbiology. Textbooks on the topic include Jay (1986) and Frazier and Westhoff (1988). The purpose of this chapter is to provide insight into the aspects of food microbiology that are related to food quality and safety and therefore influence how it should be handled. Microorganisms responsible for foodborne illness are discussed to give insight into how handling influences the occurrence of serious disease which can be unpleasant, fatal, and economically devastating.

I. FACTORS AFFECTING SURVIVAL AND GROWTH OF MICROORGANISMS

Major factors that influence the survival and growth of microorganisms include pH, temperature, gases in the environment, moisture content of the food and the atmosphere, water activity, oxidation–reduction (E_h) potential, available nutrients, and the presence of antimicrobial agents (Jay, 1986; Frazier and Westhoff, 1988). Many of these factors also are related to the sensory and nutritional quality of food and are considered in determining how food will be handled during storage, processing, and preparation. If microorganisms are detrimental to the quality of the product, methods to control their growth or to destroy them are based on making the environment of the microorganisms undesirable with respect to these factors. If a microorganism is beneficial, these factors are controlled to favor growth of the organism. For example, as discussed in Chapter 18, when yeast is used for the leavening of bread, temperature, humidity, and nutrients are factors that are considered. Similar examples could be cited regarding the production of fermented dairy products and cheeses, meats, and pickles. Microorganisms in relation to the quality of food are discussed throughout the book. Methods of controlling their activity are discussed in Chapter 13.

II. MICROORGANISMS AND FOODBORNE ILLNESSES

Archer (1988) indicated that in the past few years illnesses associated with foodborne pathogens have increased. Estimates of millions of cases of foodborne illnesses are reported annually. Although many cases can be attributed to mishandling at the consumer level, individuals at all levels of the food distribution and production systems must take the responsibility of ensuring a safe food supply. In most cases, basic rules of handling are broken by a person with disease resulting (Archer, 1988). Basic knowledge and understanding of where these organisms may exist and factors that affect how they grow will help in the prevention of foodborne illnesses. Bryan (1988a,b) reviewed foods and practices most often involved in outbreaks of foodborne illness.

Microorganisms that cause foodborne illness are of interest because methods used to preserve foods should ensure the destruction of these undesirable microorganisms as well as microorganisms that cause spoilage. Microorganisms of the genera *Staphylococcus* and *Salmonella* and the *Clostridium* species *C. perfringens* and *C. botulinum* have been of interest for many years. In addition, several foodborne pathogens have been recognized more recently as significant causative agents in major outbreaks. These are *Listeria monocytogenes*, *Yersinia enterocolitica*, *Campylobacter jejuni*, *Vibrio cholerae*, and *Escherichia coli* O157:H7 Generally, foodborne illnesses may be divided into two groups, intoxications and infections. Intoxications are caused by toxins that are formed by bacteria in food prior to ingestion. *Clostridium botulinum* and *Staphylococcus aureus*, which are discussed below, cause intoxications. Other organisms discussed cause food infections because the illness results when food containing the pathogenic organisms is consumed. Symptoms of the illness result from the body's reaction to the bacteria or their metabolites (Frazier and Westhoff, 1988, pp. 401–404).

A. *Staphylococcus aureus*

Outbreaks of staphylococcal intoxication caused by certain strains of *S. aureus* are caused more often by inadequate refrigeration of vulnerable foods than by improper use of other methods of food preservation. *Staphylococcus* organisms usually are involved when several people become ill 1–6 hr after eating a meal together. When vulnerable foods such as cooked meats, cream-filled pastries, potato salad, macaroni salad, and salad dressing that have been contaminated with these microorganisms, often by handling, are allowed to stand without proper refrigeration, the organisms produce an extracellular substance, or enterotoxin. When the toxin is ingested it causes nausea, vomiting, and diarrhea. Recovery is rapid (24–48 hr) and complete (Newsome, 1988). The main reservoir of the organism is the nose. Secondary major sources are infected wounds. Thus, care must be taken not to cough or sneeze into food and to wash the hands when starting to prepare food and after using a handkerchief or touching the hair or face. To prevent the growth of and toxin production by *S. aureus* organisms that may have found their way into food, food should be cooled promptly after cooking. Toxin production occurs in the temperature range of 16–46°C. Therefore, hot food should be kept hotter than 46°C and cold food colder than 16°C. Optimal pH for growth is 6–7 although growth will occur at 4.0–9.8. Growth also will occur at salt concentrations of 7–10%.

It is possible to destroy the microorganisms in food. Woodburn and co-workers (1962) reported that 2×10^6 cells/g were destroyed when chicken products in plastic pouches were boiled for 10 min or heated by microwaves for 2 min. However, it is not possible to destroy toxin by heating; therefore, prevention depends on destruction of the microorganisms or inhibition of their growth.

B. *Salmonella* Species

The foodborne illness caused by several members of the genus *Salmonella* is a food infection caused by ingestion of large numbers of the microorganism. Approximately 40% of foodborne illness cases are attributable to *Salmonella* (Flowers, 1988). Symptoms, which frequently are diagnosed as flu, may include nausea, vomiting, diarrhea, headache, and chills and usually develop 12–36 hr after ingestion of the contaminated food and last for 1–7 days. The main sources of the organism are contaminated water and the intestinal tracts of animals such as birds, reptiles, farm animals, and humans. Thus personal sanitation is extremely important in the control of the disease. Eggs, poultry, meat, and mixtures containing these foods are the most common vehicles. Raw milk also may be a vehicle (Flowers, 1988). *Salmonella* will grow between 5 and 45°C (Jay, 1986, pp. 496–498). Therefore, important aspects of control are to keep foods above or below this range and to properly cook vulnerable foods to assure destruction of the organisms. Avoiding postcooking contamination also is important in preventing outbreaks of salmonellosis. Optimal pH is 7.0 with possible growth from 4.0 to 9.0.

C. *Clostridium perfringens*

Clostridium perfringens is the causative agent in another food infection, which is sometimes classed as an intoxication. The microorganism is widely distributed in nature, including the intestinal tract of man and animals, making meat and poultry the food of primary concern. It occurs most frequently in food service establishments, involving meat dishes such as pot pies, stews, sauces, and gravies that are cooked one day, then cooled and reheated prior to serving. Slow cooling rates for large quantities of foods of these types are responsible for their involvement (Labbe, 1988). The mesophilic organism grows best at temperatures between 37 and 45°C. Growth occurs in the pH range of 5.0–8.5. Survival during cooking of roasts at low temperatures is possible (Sundberg and Carlin, 1976), so adequate cooling during holding is imperative. Because some of the spores are heat resistant, control depends primarily on rapid cooling and reheating of vulnerable foods. Microwave reheating of precooked chicken may not be adequate to destroy all cells of *C. perfringens* (Craven and Lillard, 1974). Problems with survival may be encountered in long-time, low-temperature waterbath heating. Smith *et al.* (1981) reported that holding beef at 60°C for 2.3 hr or longer adequately reduced the population of *C. perfringens*.

The symptoms usually appear 8–24 hr after food containing the organisms is ingested and include abdominal pain, diarrhea, and nausea without vomiting. Large numbers of the organism must be ingested. When they reach the intestine, an enterotoxin is released as sporulation occurs (Frazier and Westhoff, 1988, pp. 401–404). The toxin can be inactivated by heating. Bradshaw *et al.* (1982) found that toxin in chicken gravy was inactivated in 23.8 min at 61°C. At 65°C only 1.5 min was required. Controls include proper cooking and refrigeration (Labbe, 1988).

D. *Clostridium botulinum*

Improper methods of food preservation may result in a less common but more serious form of food intoxication called botulism. The organism, *Clostridium botulinum*, is the most heat resistant of the organisms responsible for foodborne illnesses. It is a sporeforming anaerobic bacterium that produces an extremely potent exotoxin. Although the toxin is destroyed by heat, the spores are more heat resistant than most bacteria. No harm results from consuming the spores of the organism, but if the toxin is ingested it damages the central nervous system, resulting in double vision and difficulty in swallowing and speaking, possibly followed by death due to respiratory failure. There may or may not be digestive disturbances. The mortality rate from this illness in the United States was over 65% at one time, but advances in detection have helped to lower the rate (CDC,

1976). *Clostridium botulinum* occurs in the soil and water of many areas and food grown in these areas. Several types of *C. botulinum* exist and vary in their characteristics. Optimal temperatures for growth are around 45°C. Type E, most frequently associated with fish, will grow at lower temperatures (as low as 3.3°C) than other types. Several outbreaks of type E botulism were associated with fish smoked to a light color and vacuum packaged. *Clostridium botulinum* is of great importance in the study of canning methods because if it is not destroyed by the canning process, it produces its deadly toxin under the anaerobic conditions of the can in foods with a pH of 4.5 or greater. Heat treatment, nitrites, salt, refrigeration, and acidification are all agents used for control (Pierson and Reddy, 1988). Further discussion of this organism is included in Chapter 13.

E. *Listeria monocytogenes*

Three listerosis outbreaks with fatalities of 100 or more led to the recognition of *L. monocytogenes* as a significant foodborne pathogen. Foods implicated were coleslaw, cheese, and milk. Food animals as well as humans may be healthy carriers of the organism. Thus meats, poultry, and milk and milk products are potential vehicles. Contamination of cabbage with fertilizer from manure from a flock of sheep with ovine listerosis led to the cole slaw-related outbreak. In addition, the organism has been isolated from unwashed fruits and vegetables. Water also may be a source (Brackett, 1988). Incubation ranges from 2 days to 3 wk. Symptoms are manifested as meningitis, perinatal septicemia, and abortion. Such clinical manifestations would indicate that three groups, newborns, pregnant women, and immunosuppressed individuals, are particularly susceptible to infection. Mortality is greatest in newborns and individuals over 70 years of age. In general the rate is 30%.

Lovett *et al.* (1987) reported that about 4% of the nation's milk supply contains *L. monocytogenes.* However, proper pasteurization of milk will destroy the microorganism (Donnelly and Briggs, 1986; Doyle, 1988a). Recalls for *L. monocytogenes*-contaminated products followed increased surveillance by the FDA. Products contaminated included ice cream mix, ice cream novelties, chocolate milk, and soft-ripened cheeses.

Listeria monocytogenes is a psychrotroph that will grow from 3 to 45°C and at a pH of 5.0–9.8 (Lovett and Twedt, 1988). Tolerance to NaCl has also been reported (Doyle, 1988a). The diversity of sources of this microorganism suggests that adequate pasteurization, as well as avoidance of contamination of raw materials and cross-contamination of processed foods, are critical points in control.

F. *Yersinia enterocolitica*

Although few cases of this foodborne illness have been reported, its occurrence is serious. Onset occurs in 24–36 hr after ingestion. Primary symptoms of the enterocolitis include severe abdominal pain, diarrhea, vomiting, and fever and the illness frequently is misdiagnosed as appendicitis (Doyle, 1988c). The psychrotrophic organism will grow between 1 and 45°C but grows best between 22 and 29°C. It is relatively easily destroyed by heating at 50°C and is sensitive to acidity (pH 4.6) and 5% NaCl (Doyle, 1988c). Outbreaks have been attributed to chocolate syrup added to milk after pasteurization, pasteurized milk contaminated after processing, and tofu processed with unchlorinated spring water. It may be found in meat, particularly pork. However, thorough cooking of pork to eliminate trichinae may minimize occurrences of yersiniosis. Control measures involve destruction by pasteurization or cooking and avoidance of postprocessing contamination.

G. *Campylobacter jejuni*

In contrast to the other foodborne pathogens, *C. jejuni* is not widely distributed. Campylobacteriosis is characterized by abdominal cramps, nausea, and diarrhea. The organism is found in the gastrointestinal tracts of animals. Incidence in poultry is high. Primary vehicles include raw milk, raw hamburger, and undercooked poultry. Growth in foods is limited as the organism is sensitive to air and will grow only at low levels of O_2 (5–10%) and will not grow below 30°C (Doyle, 1988b). Primary controls are proper cooking of food, pasteurization of milk, and avoidance of cross-contamination.

H. *Vibrio*

Madden (1988) lists four species of *Vibrio* that are of significance in foodborne illness. These include *V. cholerae*, *V. parahaemolyticus*, *V. vulnificus*, and *V. mimicus*. Some are more prevalent as causative agents of foodborne illness than others. Some of *Vibrio* species, particularly *V. parahaemolyticus*, are found in seafood so the consumption of raw seafood becomes a problem with respect to these organisms. Shellfish and oysters from waters contaminated with sewage are likely to be contaminated. Although *Vibrio* species will grow slowly below 10°C, they will grow very rapidly in fish held at 30–37°C. They also are tolerant to salt (Madden, 1988). Control involves refrigeration of susceptible foods and adequate cooking.

I. Enteropathogenic *Escherichia coli*

Several strains of *E. coli* are significant diarrhea-causing pathogens (Frank, 1988). *Escherichia coli* O157:H7 is the causative agent of hemorrhagic colitis characterized by severe abdominal cramps and frank blood in the stools. Large numbers of the microorganisms must be ingested, so highly contaminated food or food that has been held under conditions that favor growth must be consumed. The organism grows from 10 to 40°C and from pH 4 to 8.5 (Frazier and Westhoff, 1988, pp. 430–431).

Sources of food contamination vary with the strain. Human feces and contaminated water are most common. Animal gastrointestinal tracts are likely sources for at least one strain. Control of coliform bacteria by proper pasteurization and cooking and by avoidance of cross-contamination is important for avoidance of these illnesses.

J. Summary

Several other foodborne illnesses (Jay, 1986; Frazier and Westhoff, 1988) have been described. In general, principles of control are similar for all. Principles of control become obvious when one reviews the major contributors to outbreaks of foodborne illnesses. Bryan (1988b) summarized factors that contributed to outbreaks of foodborne illness in food service operations, homes, and food processing plants from 1973 to 1982. A summary of the information presented in Bryan's article is in Table I. In all cases, human error, as in failing to apply sound

TABLE I

Factors that Contributed to the Occurrence of Outbreaks of Foodborne Disease that Resulted Because of Mishandling and/or Mistreatment of Food in Homes, Food Service Establishments, and Food Processing Plants, 1973–1982[a,b]

Contributory factor	Homes (n = 345)	Food processing plants (n = 75)	Food service establishments (n = 660)
Contaminated raw food/ingredient	145	14	58
Inadequate cooking/canning/heat processing	108	20	29
Obtain food from unsafe source	99	5	42
Improper cooling	77	11	366
Lapse of 12 or more hours between preparing and eating	44	1	203
Colonized person handled implicated food	34	8	160
Mistaken for foods	24	—	1
Improper fermentations	16	8	—
Inadequate/improper thawing	—	6	—
Inadequate reheating	12	—	130
Toxic containers	12	1	23
Improper hot holding	11	2	107
Cross-contamination	11	1	31
Use of leftovers[c]	9	1	31
Intentional additives	8	5	13
Incidental additives	3	6	9
Contaminated water	2	1	2
Inadequate acidification	2	—	—
Improper cleaning of equipment/utensils	1	8	38
Improper dishwashing/contamination afterward	—	—	1
Poor dry storage practices	1	—	—
Inadequate preservation	1	—	—
Inadequate drying	1	1	—
Faulty sealing	1	—	—
Flies contaminated food	1	—	—
Soil/fertilizer contamination	—	1	—
Microbial growth during germination of seeds	—	1	—

[a] Numbers add to more than *n* because multiple factors contribute to single outbreaks.
[b] From Bryan (1988b).
[c] Also lapse of 12 hr or more.

principles of food preservation, is a contributing factor. For example, a 1987 outbreak of botulism was traced to potato salad made from potatoes that were baked in foil and kept at room temperature for several days prior to preparation of the salad (Foster, 1986). Awareness and application of proper control techniques by all involved in handling of food is essential for the minimizing of foodborne illnesses. Guidelines for consumer practices and food service operations are provided in government publications (USDA, 1984; USDHEW, 1978). Food processing plants are regulated by the FDA and USDA, both of which also provide guidelines for the safe handling of foods.

Suggested Exercises

1. Obtain a recent issue of *Morbidity and Mortality Report* or *Journal of Food Protection*. Select a description of a foodborne illness outbreak. Describe controls that should have been in effect to prevent the outbreak.
2. Observe preparation and handling of food in a food service operation. Discuss practices that have the potential for contributing to outbreaks of foodborne illnesses.
3. Obtain a menu from a restaurant. For several items that are potential vehicles discuss points at which careful control must be exercised in order to ensure that the foods will not be contributors to outbreaks of foodborne illness.

References

Archer, D. L. 1988. The true impact of foodborne infections. *Food Technol.* **42**(7): 53–58.

Brackett, R. E. 1988. Presence and persistence of *Listeria monocytogenes* in food and water. *Food Technol.* **42**(4): 162–164, 178.

Bradshaw, J. G., Stelma, G. N., Jones, V. I., Peeler, J. T., Wimsatt, J. C., Corwin, J. J., and Twedt, R. M. 1982. Thermal inactivation of *Clostridium perfringens* enterotoxin in buffers and in chicken gravy. *J. Food Sci.* **47**: 914–916.

Bryan, F. L. 1988a. Risks associated with vehicles of foodborne pathogens and toxins. *J. Food Protection* **51**: 489–508.

Bryan, F. L. 1988b. Risks of practices, procedures and processes that lead to outbreaks of foodborne diseases. *J. Food Protection* **51**: 663–673.

CDC. 1976. Botulism in 1975—United States. *Morbidity and Mortality Weekly Rpt.* **25**: 76–77.

Craven, S. E. and Lillard, H. S. 1974. Effect of microwave heating of precooked chicken on *Clostridium perfringens. J. Food Sci.* **38**: 211–212.

Donnelly, C. W. and Briggs, E. H. 1986. Psychrotrophic growth and thermal inactivation of *Listeria monocytogenes* as a function of milk composition. *J. Food Protection* **49**: 994–998.

Doyle, M. P. 1988a. Effect of environmental and processing conditions on *Listeria monocytogenes. Food Technol.* **42**(4): 169–171.

Doyle, M. P. 1988b. *Campylobacter jejuni. Food Technol.* **42**(4): 187.

Doyle, M. P. 1988c. *Yersinia enterocolitica. Food Technol.* **42**(4): 188.

Flowers, R. S. 1988. *Salmonella. Food Technol.* **42**(4): 182–185.

Foster, L. M. 1986. Botulism. *Food Technol.* **40**(8): 18–19.

Frank, J. F. 1988. Enteropathic *Escherichia coli. Food Technol.* **42**(4): 192–193.

Frazier, W. C. and Westhoff, D. C. 1988. "Food Microbiology," 4th edn. McGraw-Hill Book Co., New York.

Jay, J. M. 1986. "Modern Food Microbiology," 3rd edn. Van Nostrand Reinhold New York.

Labbe, R. G. 1988. *Clostridium perfringens. Food Technol.* **42**(4): 195–196.

Lovett, J. and Twedt, R. M. 1988. *Listeria. Food Technol.* **42**(4): 188–191.

Lovett, J., Francis, D. W., and Hunt, J. M. 1987. *Listeria monocytogenes* in raw milk: Detection, incidence, and pathogenicity. *J. Food Protection* **50**: 188–191.

Madden, J. M. 1988. *Vibrio. Food Technol.* **42**(4): 191–192.

Newsome, R. L. 1988. *Staphylococcus aureus. Food Technol.* **42**(4): 194–195.

Pierson, M. D. and Reddy, N. R. 1988. *Clostridium botulinum. Food Technol.* **42**(4): 196–198.

Smith, A. M., Evans, D. A., and Buck, E. M. 1981. Growth and survival of *Clostridium perfringens* in rare beef prepared in a water bath. *J. Food Protection* **44**: 9–14.

Sundberg, A. D. and Carlin, A. F. 1976. Survival of *Clostridium perfringens* in rump roasts cooked in an oven at 107° and 177°C or in an electric crockery pot. *J. Food Sci.* **41**: 451–452.

USDA. 1984. The safe food book. Home and Garden Bull. 241, United States Department of Agriculture, Washington, DC.

USDHEW. 1978. Food service sanitation manual. Public Health Serv., HEW Pub. No. (FDA) 78-2081, Food and Drug Admin., Div. of Retail Food Protection, Washington, DC.

Woodburn, M., Bennion, M., and Vail, G. E. 1962. Destruction of salmonellae and staphylococci in precooked poultry products by heat treatment before freezing. *Food Technol.* **16**(6): 98–100.

FOOD PRESERVATION

The purpose of food preservation is to prevent or to at least delay enzymatic and/or microbial changes in food so that the food is available at some future time. That time may be later today, tomorrow, next week, or several months from now. A discussion of food preservation should involve enzymes and microorganisms and the basic principles of preventing their activity in food. The principles are employed in several basic techniques of food preservation, including applying heat; lowering temperatures, water activity, and pH; adding inhibitory substances; and irradiating the food. Controlled or modified atmosphere packaging may be used with any of these basic methods of preservation. Wagner and Moberg (1989) pointed out that any one of these techniques by itself is not very practical because extremes may be needed to ensure safety. Therefore, they frequently are used in combinations.

266

Enzymes, which are discussed in Chapter 6, are readily destroyed by heat, as in canning or in blanching vegetables before freezing or dehydration. If enzymes are not inactivated before freezing, they are likely to cause deterioration because freezing slows but does not stop their action.

Prevention of the growth or destruction of pathogenic microorganisms, which are discussed in Chapter 12, or microorganisms responsible for spoilage is basic to most methods of food preservation. Yeasts and molds are killed easily by heat, but the ease of destruction of bacteria varies with the type of bacteria and with their condition. Actively growing bacteria are destroyed more easily than are bacteria in the spore form. For all methods of preservation it is important that the raw food contain as few microorganisms as possible.

Methods of preservation of food for future use are many. Some are practical at the consumer level; others are not. An understanding of consumer methods available is important for professionals because interest in home production and preservation remained steady in recent years. Understanding of commercially used methods is important to the professional who will answer consumer questions about foods in the marketplace and to the professional in the food industry. In the discussion that follows, scientific principles are emphasized and reasons for some of the procedures commonly used are discussed. Specific directions for home preservation of many foods are given in bulletins published by the United States Department of Agriculture (USDA, 1975, 1976a,b, 1982, 1988).

Methods of food preservation for some specific foods are discussed in chapters on those foods. Space does not permit extensive discussion of the effects of these methods on the nutritive quality of food. For more information on that topic the reader is referred to Karmas and Harris (1988). Reference to those discussions will be made as the principles of the methods are presented.

I. APPLICATION OF HEAT

Heat may be applied to food to make it more palatable as in cooking or it may be applied to preserve it for future use. Three heat processes, blanching, pasteurization, and sterilization, may be applied to preserve food (Lund, 1988). Microorganisms are destroyed by heat, probably because of denaturation of protein and inactivation of enzymes needed for metabolism.

A. Blanching

Blanching reduces the number of microorganisms, removes some air from the tissues, makes them more compact, and enhances their color. However, its most important function is to inactivate enzymes that would otherwise cause deterioration in flavor, texture, color, and nutrients during storage. Reduction of microorganisms is not the major purpose of blanching, and blanching alone is not adequate for food preservation. Canning, freezing, and dehydration are used in conjunction with blanching. Adequacy of blanching may be monitored by measuring activity of indicator enzymes. Peroxidase and catalase are used most frequently because they are heat resistant and their destruction ensures that other enzymes are destroyed. Blanching is accomplished by treating in boiling water or steam. The length of time required depends on the vegetable and the amount of enzymes present. Microwave blanching also has been studied. Lane *et al.* (1985) reported that blanching green beans, yellow squash, and mustard greens by microwave, water, or steam did not affect ascorbic acid content and that steam and microwave blanching did not ensure complete destruction of peroxidase. Further blanching would have resulted in overcooked vegetables. Glasscock *et al.* (1983) found that acceptable reduction of peroxidase was not achieved with blanching times recommended by a microwave oven manufacturer for carrots, broccoli, and zucchini. Blanching was adequate for cauliflower and green beans. Microwave-blanched cauliflower and carrots were not different in quality from the water-blanched counterparts. Drake *et al.* (1981) reported that water- and steam-blanched products were similar in most respects, although, in general, microwave-blanched vegetables were less desirable with respect to ascorbic acid, color, and drip loss after thawing. Vegetables studied included asparagus, green beans, green peas, and sweet corn. Differences in results of these studies may be attributed to differences in blanching times as well as differences in the microwave ovens used. Variation in power output from oven to oven contributes to differences in results of studies on use of microwaves.

Lee *et al.* (1988) evaluated water blanching of green beans with varying time–temperature combinations. They concluded that beans could be blanched at temperatures lower than 99–100°C but that temperatures higher than 82°C were required to ensure adequate destruction of enzymes and to maintain good color, flavor, and texture.

B. Pasteurization

The purpose of pasteurization is to destroy pathogenic and spoilage organisms. For example, as discussed in Chapter 8, milk is pasteurized to destroy the pathogens *Mycobacterium tuberculosis* and *Coxiella burnetti*. In addition, molds, yeasts, and some bacteria are destroyed. However, thermoduric and thermophilic organisms may survive. *Streptococcus* and *Lactobacillus* species may survive. Therefore, refrigeration usually is necessary to prevent the growth of these surviving organisms, which will cause spoilage.

The development of aseptic packaging systems has made it possible to maintain better quality of some pasteurized foods. For example, juice products may be pasteurized and then placed in chemically sterilized, aseptic plastic cartons under aseptic conditions. The packages are sterilized with hydrogen peroxide and heat. Ultraviolet light may be used in place of the heat. Nitrogen may be introduced into the headspace of the package to prevent oxidative deterioration during unrefrigerated storage. Hermetic sealing of the package prevents post-processing contamination. Advantages for such products are lack of metallic flavor from a can and better flavor because of reduced exposure to heat. Individual serving-sized packages of some juices have been widely accepted. However,

Sizer *et al.* (1988) indicated that consumer acceptance of orange juice has not been as high as expected. Flavor deterioration does occur and methods are being investigated to prevent that change. Storage at low temperatures is helpful but eliminates the benefit of unrefrigerated storage. Marcy *et al.* (1989) found that aseptically packaged orange juice concentrate could be stored at refrigerator temperatures for 6 months. Low-acid foods with particulate matter present a challenge for food processors. Factors to be considered in determining processing times to ensure microbiological safety and a quality product are discussed by Heldman (1989).

C. Sterilization

Sterilization destroys all viable microorganisms. Complete sterility is not always attained when heat is applied as in canning, but microorganisms are not detected when appropriate methods of enumeration are applied. Microorganisms likely to cause spoilage under ordinary conditions of storage are destroyed. Dormant nonpathogenic microorganisms may remain but will not multiply because other preservation techniques used will establish environmental conditions that are not favorable (Lund, 1988, pp. 320–321). For example, in canning, the reentry of microorganisms is prevented by a hermetically sealed container. The term "commercially sterile" or "effectively sterile" is more accurately applied to heat processing of this nature than the term "sterile."

For success in commercial canning, the heat process must be sufficient to destroy pathogenic microorganisms. Recommendations for home canning (USDA, 1988) as well as procedures established for commercial canning vary with the particular food being canned because the thermal conditions needed to produce "commercial sterility" vary. Five factors affect the conditions needed: the nature of the food, including its pH; the heat resistance of the microorganism; the heat-transfer characteristics of the food and the container; the number of microorganisms present; and the projected conditions of storage for the processed food. The resistance of microorganisms to destruction by heating is indicated by the thermal death time, or the time required to kill the microorganisms at a given temperature. Commercially, procedures for canning each food take into account each of these factors. Likewise, home canning methods must be based on these factors.

The canning process must, of course, heat food in all parts of the can to the temperature necessary to destroy microorganisms. Studies of heat penetration are made by inserting a thermocouple or thermometer into the can and following the change in temperature as a function of time. The temperature of the slowest heating point should be monitored. This point may not be in the center of the can, depending on the nature of the food, which in turn determines whether the food will be heated by conduction or convection (Potter, 1986, pp. 179–180).

Heat penetrates foods such as peas or peaches that are packed in brine or syrup more rapidly than a viscous food like pumpkin. Foods in a watery medium are heated by convection currents that rapidly equalize the temperatures on the outside and the center of the jar. Viscous and solid foods are heated rather slowly by conduction, in which transfer of heat takes place between adjacent molecules rather than by circulation of liquid. For similar reasons, heat penetrates better if brine or syrup is added than if the food is packed too closely into the container. The processing times shown in canning directions and established in commercial operations depend on packing the food in a certain way, since the times would be longer for food packed more closely than specified in the directions and longer for food packed cold rather than hot. Because heat penetrates the center of large containers more slowly than small containers, and glass more slowly than metal, separate processing times are given for jars and for cans of various sizes. Lists of processing times for home canning involving the use of a small number of jar sizes are short (USDA, 1988) as compared with processing data for commercial operations (National Canners Association, 1976), where many different can sizes may be used as described by Stumbo *et al.* (1975). Increasing the temperature of steam by applying pressure accelerates the rate of heat penetration.

Methods for commercial canning of foods cannot be adapted for home use because the equipment used in a home is quite different from that used in a commercial canning operation. A major difference is the use of glass jars, which heat and cool more slowly than the metal cans used commercially. Additional processing takes place as glass jars are removed from the pressure cooker and cooled at room temperature; metal cans, on the other hand, are cooled rapidly in water. Recommendations for home canning of foods were recently updated on the basis of work directed by Kuhn (USDA, 1988).

Different methods are used for canning of acid and low-acid foods because the pH of the food affects the temperature required to destroy any spores of *Clostridium botulinum* that might be present. A pH of 4.5 is the dividing line between low-acid and acid foods. The division is based on the fact that *C. botulinum* will not grow and produce toxin below a pH of slightly above 4.5 (Stumbo, 1973, pp. 9–17). Early workers (Esty and Meyer, 1922) studied the pH–temperature relationship by suspending spores of *C. botulinum* in buffers of varying pH, heating at 100°C, and determining the length of time necessary to destroy the spores. Only a short time is required at a low pH or at a pH higher than that normally associated with food, but a much longer time is necessary around neutrality. The pH values of a number of foods are shown in Table I. Fruits are acid foods, whereas vegetables and foods of animal origin are low-acid foods. Tomatoes vary in pH and may be low-acid or acid foods.

To determine if a process is adequate for low-acid food, a food is inoculated with spores of *C. botulinum* or *Clostridium sporangenes* (putrefactive anaerobe PA 3679). PA 3679 is significantly more heat resistant than *C. botulinum*, nontoxic, and easier to assay. Therefore, it is used to determine whether or not a process is adequate (Potter, 1986, p. 184). The destruction of the organism is taken as a measure of the adequacy of the process because more heat is required for its destruction than for most, if not all, of the other organisms that may be present. In the canning industry, the minimum heat process for low-acid foods must ensure reduction of *C. botulinum* spores from a level equal to 1 spore in each of 12 cans to one spore in a total of 12 cans. This is referred to as the 12D concept (Frazier and Westhoff, 1988, pp. 104–105).

TABLE I

Average pH Values for Some Common Foods[a]

pH	Food	pH	Food
0.0		—	
—		5.0	Pumpkins, carrots
2.9	Vinegar	5.1	Cucumbers
—		5.2	Turnips, cabbage, squash
3.2	Rhubarb, dill pickles	5.3	Parsnips, beets, snap beans
3.3	Apricots, blackberries	5.4	Sweet potatoes
3.4	Strawberries	5.5	Spinach, aged meat
3.5	Peaches	5.6	Asparagus, cauliflower
3.6	Raspberries, sauerkraut	5.8	Mushrooms; meat, ripened
3.7	Blueberries	—	
3.8	Sweet cherries	6.0	Tuna
3.9	Pears	6.1	Potatoes
—		6.2	Peas
4.2	Tomatoes	6.3	Corn
4.4	**Lowest acidity for**	—	
	processing at 100°C	7.0	Meat, before aging
4.6	Figs, pimientos		

[a] From Pritchard (1974).

Adequacy of a processing method for acid foods is determined, as for low-acid food processes, with the use of those microorganisms responsible for spoilage. *Bacillus coagulans*, the microorganism responsible for flat sour in tomato products (Frazier and Westhoff, 1988, p. 303), is used as a test organism. Flat sour is a condition characterized by a slight change in pH and the development of off-odors and flavors. Commercially, the development of flat sour can be of economic significance.

D. Home Canning

1. Low-Acid Foods

Because of the comparative ease of destruction of *C. botulinum* in acid foods, they can be processed at the temperature of boiling water, as in the water-bath method. Low-acid foods, however, require a long time for processing at 100°C, and there is no certainty that all the spores will be destroyed at this temperature even after a long period because of varying conditions affecting heat penetration. As the temperature of heating a buffer of pH 7 containing *C. botulinum* spores increases from 100 to 200°C, the time required to destroy the spores decreases from 300 to 4 min. In order to increase the temperature to above 100°C, pressure must be increased beyond that of the atmosphere. Therefore, to assure destruction of *C. botulinum* with heating for a reasonable length of time, the pressure canner is recommended by the USDA (1988).

In canning low-acid vegetables and meat, care must be taken to observe every precaution given in the directions and to have the pressure canner in good working order (USDA, 1988). Either the hot pack or the raw pack can be used for most vegetables and for meat and poultry. However, for some vegetables that are difficult to pack in the jar, hot packing is recommended. Prepared jars are placed in the pressure canner and processed at the designated pressure for the recommended times. The new recommendations (USDA, 1988) are based on size of jar, method of pack, and altitude.

An argument that sometimes is used against the recommendation that all low-acid foods be processed in a pressure canner is that *C. botulinum* rarely if ever is reported in the soil of some areas. However, considering the possible danger when contaminated food has not been adequately processed, the wisdom of recommending a canning process inadequate to destroy any pathogens that may be present is debatable. It is possible that vegetables shipped from another area may be canned. This reasoning is the basis for the recommendation that the pressure canner always be used for low-acid foods.

Canned foods that show evidence of spoilage or of pressure within the container should be rejected. Unfortunately, however, development of the toxin by *C. botulinum* in canned foods cannot always be detected by changes in appearance and odor. It is therefore important not to taste suspected food for signs of spoilage. A minute amount of the toxin can be fatal. If the food or the canning process is under suspicion, the food can be made safe by boiling it actively for at least 10 min to inactivate the toxin, which is much more sensitive to heat than the spores. With regard to low-acid or tomato products (even those that do not show signs of spoilage) the following statement of caution is given on the first page of the USDA canning guide (USDA, 1988):

> "... to prevent the risk of botulism, low-acid and tomato foods not canned according to the recommendations in this publication or according to other USDA-endorsed recommendations should be boiled even if you detect no signs of spoilage. At altitudes below 1,000 ft, boil foods for 10 min. Add an additional minute of boiling time for each additional 1,000-ft elevation."

2. Acid Foods

Acid foods, those having a pH of 4.5 or less (see Table I) are processed more easily than low-acid foods because acid aids in the destruction of certain microorganisms by heat, as explained previously, and because the microorganisms that usually cause spoilage of such foods are destroyed easily by heat. Acid foods can be processed safely if their internal temperature, as determined by a thermometer or thermocouple inserted into the contents of the can during processing, reaches 82.2°C (Ball, 1938).

Fruits can be packed into jars or cans either raw or hot (USDA, 1988). The hot pack has the advantage that more food can be packed into a jar, but it was reported that palatability (Gilpin *et al.*, 1951; Dawson *et al.*, 1953) of certain fruits and vegetables is better, and their vitamin content (Dawson *et al.*, 1953) at least as good, when they are packed raw as when they are packed hot. The raw pack, sometimes called the cold pack, is especially good for fruits with a delicate texture, such as berries. Fruits packed without added sugar are processed for the same length of time as those packed with sugar or syrup. Sugar helps retain the color and flavor of fruits. However, it can be omitted when the intake of sugar must be restricted.

After fruit is in the jar and the jar is partially sealed, it usually is processed in a container of boiling water that is deep enough to come 2.5–5 cm above the tops of the jars. The USDA (1988) provides recommendations for processing fruits in a pressure canner.

Several methods, sometimes promoted in the popular press, are not recommended by the USDA (1988). Open kettle canning and processing filled jars in an oven, a microwave, or dishwasher may not produce safe, palatable food, and the process may be hazardous to the person doing the canning.

3. Tomatoes

Tomatoes are singled out because the new USDA recommendations include a separate guide for this food. In the 1970s a controversy developed over the pH of tomatoes. Some investigators (Fields *et al.*, 1977; Gould and Gray, 1974; Leonard *et al.*, 1960; Pray and Powers, 1966; Schoenemann *et al.*, 1974) indicated that some tomatoes may have pH levels of greater than 4.5 whereas the USDA (1976c) reported that this is not the case. Some states recommended at that time that citric acid be added to lower the pH (Gould and Gray, 1974). The Food and Drug Administration (FDA) permits the use of organic acids to lower the pH of tomatoes for commercial processing (FDA, 1988) and the USDA accordingly has changed its recommendations for home canning of tomatoes. Changes incude the addition of lemon juice, citric acid, or vinegar (5% acidity) to acidify the tomatoes. It also recommended that a pressure canner be used because higher quality and more nutritious products will be produced.

The growth of mold on the surface of improperly canned tomatoes can result in a potentially dangerous situation. Metabolites produced by the mold can lower the acidity of the tomatoes. The pH may be raised to a point where *C. botulinum* spores, if present, can germinate, multiply, and produce toxin. Therefore, any moldy tomato product should be discarded in its entirety.

II. LOW TEMPERATURE _____

The use of low temperatures will retard enzymatic changes in food, slow chemical reactions, and slow or stop microorganisms.

A. Refrigeration

Refrigeration is a method for short-term preservation of the quality of food as it may delay spoilage and growth of pathogenic microorganisms. Recent concern for several pathogenic or suspected organisms that will grow at refrigerator temperatures ($< 6°C$) has surfaced. These include *C. botulinum* type E, *Vibrio parahaemolyticus*, *Yersinia enterocolitica*, and *Listeria monocytogenes*. These microorganisms are discussed in Chapter 12. Lechowich (1988) suggested the appropriateness of using refrigeration in combination with other preservation techniques.

The refrigerated food industry is increasing the availability of fresh, refrigerated foods to consumers. The presence of delicatessens in supermarkets has increased this market. Products range from fresh salads of all types to entrees that have shelf lives of 21 to 29 days (El-Hag, 1989). Lechowich (1988) indicated that the most vulnerable of these foods have a shelf life of 2 wk, are packaged in oxygen-impermeable materials, are low acid, and must be heated before consumption. Two processes may be used for some of these new refrigerated products (Lechowich, 1988). The "sous-vide" process involves packaging in a vacuum pouch, cooking in the package, cooling rapidly, and refrigerating. Cooking may be in a pressure cooker, moist steam oven, or water bath. The "nouvelle carte" process differs with respect to the type of package used. A plate is included in the pouch and there is a sleeve with an overwrap. The rapid cooking methods used for these products are designed to minimize quality changes. As this industry continues to grow, more problems associated with pathogenic organisms could occur. Problems can be prevented if handlers of these products in the food distribution chain, as well as in institutional settings and at the consumer level, do not allow temperature abuse to take place and if additional methods for ensuring inhibition of the growth of these microorganisms are employed in the production of the products.

As precooked vacuum-packaged meat products become popular with consumers, concern for potential outbreaks of foodborne illnesses also will increase. Anderson *et al.* (1989) examined packages of precooked beef and pork obtained from the retail market and a meat processor. Pathogenic and indicator organisms were not found in the samples. In a storage study, it was shown that storage at 1°C resulted in a longer shelf life than at 5 or 10°C, emphasizing the need for careful control.

The cook/chill system used in food service operations is a system that involves the use of refrigerator storage of precooked foods prior to reheating for consumption. Shelf life is not expected to be as long in this system as for the systems previously described because vacuum packaging and cooking in the packaging are not employed. Many studies have been reported on the microbiological and sensory quality of foods in a cook/chill system. Ollinger–Snyder and Matthews (1988) reported that turkey slices taken from rolls that were cooked conventionally, held in the refrigerator for not more than 2.5 hr, and then reheated for 30 or 40 sec in a microwave did not have higher coliform or aerobic counts than freshly prepared slices. The investigators stressed the need to follow guidelines (USDHEW, 1978) for reheating to an internal temperature of at least 74°C. A critical point in the cook/chill system is the cooling of the product. Bobeng and David (1978) found that 5 hr was required for cooling of 900-g loaves to 7°C. The use of slower cooling methods in a food service operation than in commercial production of refrigerated entrees (Lechowich, 1988) increases the need for careful handling to avoid postcooking, prechilling contamination of products.

B. Freezing

The process of freezing results in only minor changes in the sensory quality of food. The process of freezing has made possible the proliferation of convenience foods for restaurant, institutional, and home use. The success of freezing as a method of food preservation depends on the fact that low temperatures destroy some microorganisms and prevent the growth of others. Thus frozen food often contains fewer bacteria than the corresponding fresh products, but the food is not rendered sterile. The effect of freezing on microorganisms depends on the type of food, the rate of freezing, the kind of microorganism, and freezing temperature (Frazier and Westhoff, 1988, pp. 130–131). Freezing is lethal to some microorganisms because as water freezes it is no longer available to cells, resulting in dehydration (Jay, 1986, pp. 325–327). Whether the bacterial count of the thawed food is low also depends on the conditions of thawing. Since some microorganisms may survive the freezing process, thawing and holding at elevated temperatures may provide opportunity for growth. Although *C. botulinum* endures freezing well, it does not form toxin because it is unable to grow at freezing temperatures or under aerobic conditions. The action of enzymes is slowed but not stopped by freezing temperatures, making prefreezing treatments of some foods appropriate. Several factors affect the quality of frozen foods.

1. Treatments before Freezing

For some foods, no special treatments are necessary prior to freezing to ensure quality. Fruits and vegetables must be treated to prevent undesirable changes.

a. Vegetables. To maintain good flavor, color, and texture of vegetables, blanching (see Section I,A) before freezing is needed to destroy enzymes. Enzymes that are isolated from the substrate in intact tissue may come in contact with the substrate as cellular structures are disrupted in the freezing of unblanched vegetables (Brown, 1976). In vegetables that have not been blanched off-flavors develop rapidly. For example, flavor profiles for green beans change with freezing to include earthy, sour, and bitter notes (Moriarty, 1966). In addition to the development of rancidity and other off-flavors, a loss of chlorophyll and carotene from green vegetables may occur during storage (Lee and Wagenknecht, 1951).

The use of various additives in blanching and cooking vegetables for home freezing was evaluated by Hudson *et al.* (1974a,b). The addition of salt (1.2%) to blanching water improved retention of chlorophyll in spinach and Brussels sprouts (Hudson *et al.*, 1974a). On the basis of sensory evaluation scores it was recommended that 1.2% NaCl be added to the blanching water for runner beans and Brussels sprouts, and 2% for spinach (Hudson *et al.*, 1974b). Studies on methods of blanching are discussed in Section I,A.

b. Fruits. Enzymes cause frozen fruits to turn brown during frozen storage. The enzyme responsible for this color change, polyphenol oxidase, is discussed in Chapter 14. As with vegetables, the enzymes of fruits can be inactivated by blanching, but this usually is not done because it gives the fruit a cooked flavor and soft texture. When such changes are not important, as with apple slices for pies, fruit is sometimes blanched. Several methods for prevention of enzymatic browning are possible, including the use of ascorbic acid, citric acid, or malic acid. As alternatives, sugar or syrup added to fruit before freezing helps to prevent browning by slowing the action of enzymes and by protecting the fruit from air. Syrup usually is preferred to sugar for fruit that is to be eaten without cooking because it retains the texture of the fruit better and excludes air more effectively than sugar. Sugar draws liquid from the fruit by osmosis, forming a syrup and causing the tissues to shrink. Fruit is best if cut directly into syrup or sugar to prevent oxidation and if excess air is excluded from the package.

In a study on improving quality of home-frozen strawberries (Hudson *et al.*, 1975a,b), calcium lactate, ascorbic acid, citric acid, and tartaric acid were added alone or in combination to a 60% (w/v) sugar syrup used with berries held in storage for up to 3 months. Citric acid gave the best color and overall rating whereas calcium lactate and ascorbic acid improved firmness, suggesting that combinations might be best. Ponting and Jackson (1972) reported that Golden Delicious apples were of highest quality when presoaked in a 20–30% sugar solution containing 0.2–0.4% $CaCl_2$ and 0.2–1.0% ascorbic acid.

c. Eggs.   Pretreatment of eggs prior to freezing is needed to minimize gelation of the yolk. Techniques are discussed in Chapter 7.

2. Packaging

Before freezing, food must be packaged in such a way that it cannot dry out during storage. The package, therefore, must be moisture and vapor proof to prevent passage of water or water vapor from the food to the dry air of the freezer. Waxed paper and the type of aluminum foil intended for home use are unsatisfactory for freezing because they may have very small holes that will allow moisture and vapors to pass and microorganisms to enter. In addition, wrapping must ensure that as much air as possible is excluded from the package. A commercially frozen food may have a longer shelf life than one frozen at home because of differences in packaging technique.

3. Rate of Freezing

Because rate of freezing also influences quality, a distinction should be made between rapid and "sharp," or slow, freezing. Rapid freezing is attained commercially by direct immersion in a refrigerating medium, by indirect contact with a refrigerant, as in a contact plate freezer, or by freezing in the moving cold air of a blast freezer. Freezer burn may be a problem in blast freezers if products are not adequately protected by wrap or a glaze. Sharp freezing is accomplished by placing foods in a room or cabinet having a temperature varying from −15 to −29°C. Usually the air is still, although a fan may be used. Freezing may require from several hours to several days under such conditions (Potter, 1986, p. 231). Obviously home freezing is more likely to be sharp than rapid. Freezing rate is dependent on the composition of the product. Solutes in the water portion of the food lower its freezing point, as discussed in Chapter 6, thus altering the freezing curve.

Because rapid freezing minimizes size of ice crystals formed, it minimizes cellular changes (Fennema, 1966). Large crystals may rupture cell walls of plant tissues, resulting in inferior texture (Brown, 1967). When green beans are frozen, there is a critical rate for quality retention, but that rate may not be as rapid as that of liquid nitrogen freezing. Liquid nitrogen freezing of carrots resulted in cracking (Wolford and Nelson, 1971), supporting the idea that there is a critical rate for freezing. Hung and Thompson (1989) demonstrated that slow freezing of cooked peas resulted in more cellular damage and reduced firmness and chewiness than did rapid (immersion) freezing. Storage time and temperature also influenced chewiness. Hung and Thompson postulated that chewiness increased with storage time because the peas lost moisture and therefore required more chewing for rehydration of the peas before swallowing. Although rapid freezing has advantages, the widespread use of home freezers suggests that sharp-frozen foods are satisfactory. There is no doubt, however, of the importance of freezing the food as soon and quickly as possible in order to avoid undesirable changes in palatability, nutritive value, and bacterial count. Specific guidelines for home freezing may be found in USDA publications (USDA, 1975, 1976a,b, 1982).

4. Storage Temperature

A temperature of $-18°C$ usually is recommended for frozen foods. Because food is not completely inert at this temperature, deteriorative changes during storage may be greater than changes occurring during blanching, freezing, and thawing. Variation in storage temperature influences the degree of deterioration. For example, it may take two to five times as long for a given amount of deterioration to occur if the storage temperature is lowered $3°C$. Temperature fluctuations are inevitable as the freezer door is opened and as the freezer operates in the normal mode of temperature rise followed by operation of the compressor to lower the temperature a few degrees before the compressor stops. This process is known as cycling. As cycling occurs, moisture sublimes from the food and accumulates as frost in the package. The effects of temperature fluctuations are not great if they occur below $-18°C$ and if the packages are well wrapped.

Singh and Wells (1985) followed changes in quality of ground beef–soy patties stored at three freezer temperatures for 168 days. In addition, one set of samples stored at $-18°C$ was subjected to higher temperatures four times in the study to simulate fluctuation of temperature. Results of the sensory testing are shown in Fig. 13.1. The horizontal line at a score of 50 represents the reference sample which was stored at $-35°C$. Deviation of the scores for the $-35°C$ sample reflects variability in product and judges. At days 61 and 103 the $-12°C$ samples were more rancid than the reference and $-18°C$ samples, respectively. Results illustrate the effects of both time and temperature of freezer storage on product flavor.

Specific changes that occur during freezer storage are of a chemical or physical nature since microorganisms are not viable at such low temperatures. Migration of water and recrystallization occur during storage. Recrystallization is particularly damaging to frozen desserts, breads, and foods thickened with starch (Fennema, 1966). Another physical change that may occur is freezer burn. Freezer burn occurs on the surface of improperly packaged foods and results from the sublimation of ice during temperature recycling. Prechilling food in a

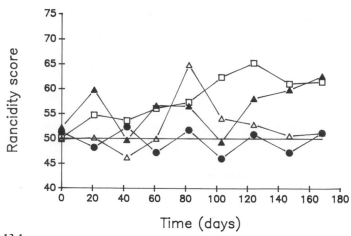

FIGURE 13.1

Mean sensory scores for hamburger–soy patties over time. ●, −35°C, △, −18°C; ▲, variable; □, −12°C. (From Singh and Wells, 1985; Reprinted from *Food Technology*, 39, No. 12, pp. 42–50. Copyright © 1985 by Institute of Food Technologists.)

high-humidity atmosphere at about −4°C before freezing rapidly at low temperatures helps to minimize freezer burn (Potter, 1986, p. 234). Use of suitable packaging material is a must.

5. Storage Time

The shelf life of a frozen food product varies with the food, how it was processed, and storage temperatures. Table II shows approximate maximum storage times at −18°C for home-frozen and commercially frozen foods. At higher temperatures the shelf life is dramatically reduced. For example, at −6.7°C the shelf life of vegetables is reduced to 1 month (Potter, 1986, p. 222).

6. Thawing

Thawing may or may not be a greater cause of damage than freezing (Fennema, 1975). Since the time required for thawing is short in relation to the normal length of frozen storage, changes during thawing may not be as great as those occurring during storage. Fennema (1975) based this conclusion on the observation that food can be frozen and immediately thawed with little loss of quality. To avoid slow thawing, cooking from the frozen state may be advantageous. For example, frozen vegetables are cooked without prethawing, and cooking times are shorter than for fresh vegetables because the tissue of the vegetable has been softened by both blanching and freezing. Cooking poultry from the frozen state is discussed in Chapter 10. Not all microorganisms are destroyed by freezing and are, therefore, ready to multiply when the food is thawed. Therefore, thawing in the refrigerator is preferred to thawing at room temperature.

7. Refreezing

Refreezing of thawed foods should be avoided if possible because it results in reduced quality and increased bacterial count. For the times when food thaws

TABLE II

Storage Life Recommendation to Maintain High-Quality
Frozen Food Stored at $-18°C$ or Lower[a]

Food	Months
Fresh meats	
Beef and lamb roasts and steaks	8–12
Veal and pork roasts	4–8
Chops, cutlets	3–6
Variety meats	3–4
Ground meats (except pork)	3–4
Ground pork	1–3
Sausage	1–2
Cured, smoked, and ready-to-serve meats	
Ham—whole, half, or sliced	1–2
Bacon, corned beef, frankfurters, and wieners	<1
Ready-to-eat luncheon meats	Freezing not recommended
Cooked meat and meat dishes	2–3
Fresh poultry	
Chicken and turkey	12
Duck and goose	6
Giblets	3
Cooked poultry	
Dishes and slices or pieces covered with gravy or broth	6
Fried chicken	4
Sandwiches and pieces or slices not covered with gravy or broth	1
Fresh fish	6–9
Commercially frozen fish	
Shrimp and fillets of lean fish	3–4
Clams, shucked, and cooked fish	3
Fillets of fatty fish and crab meat	2–3
Oysters, shucked	1
Fruits and vegetables, most	8–12
Milk products	
Cheddar cheese (450 g or less, no more than 2.5 cm thick)	≤6
Butter or margarine	2
Frozen milk desserts, commercial	1
Prepared foods	
Cookies	6
Cakes, prebaked	4–9
Combination main dishes and fruit pies	3–6
Breads, prebaked	3
Pastry	1–2

[a] From Allen and Crandall (1974).

because of uncontrolled situations, USDA (1984) guidelines are available. If foods contain ice crystals and no warm portions, it is safe to refreeze. Completely rethawed fruit may be refrozen. Vegetables, fish, and cooked foods should be discarded if any part of the food is warmer than 4°C because it is difficult to tell by odor if these foods have spoiled. Meat and poultry that show no signs of spoilage as indicated by poor odor and color can be refrozen. Increased cooking losses will occur if chicken or beef is repeatedly frozen and thawed (Baker *et al.*, 1976; Locker and Daines, 1973). However, Baker and co-workers (1976) concluded that chicken can be repeatedly refrozen if refrozen as soon as it has thawed. Sensory panel scores for tenderness, juiciness, and flavor were not affected by thawing and refreezing five times. Moisture content and shear value were not affected. Chickens were kept in freezer storage for 4 to 5 days between thawings.

III. LOWERING WATER ACTIVITY

Growth of microorganisms is dependent on the presence of water. Water activity (a_w), as discussed in Chapter 6, is an indicator of the potential for growth of microorganisms in a food product. An a_w of 0.75 or greater is required for molds to grow and yeasts will grow at about 0.88, but most bacteria must have an a_w of 0.91 or above for significant growth to occur. The relationship of a_w to growth of microorganisms is shown in Fig. 6.1. Exceptions to these values include halophilic bacteria, which will grow at an a_w of 0.75, and osmophilic yeast, which will grow at an a_w of 0.60. Thus reduction in water activity to levels below these will improve the stability of the product. Reduction of a_w is achieved by several food preservation techniques, including the addition of solute to the system. For example, salting or curing reduces water activity. Obviously drying lowers a_w. Freezing, which was previously discussed, lowers water activity as ice forms.

A. Dehydration and Concentration

Water is removed from food to preserve it or for economic reasons. For example, preservation of orange juice concentrate is less expensive than preservation of orange juice. One of the oldest methods of food preservation is dehydration, or removal of water by evaporation or sublimation as in freeze drying. Many techniques are available for dehydration but all involve application of heat so that moisture is forced from the food. Surface area, air temperature, velocity of the air, humidity, and atmospheric pressure affect the rate of evaporation. In addition, characteristics such as organization of the constituents and composition of the products affect the drying rate. For example, sugars lower the vapor pressure of water in a product; thus products containing sugars will dry slower than those that do not. Methods of drying include drum drying, tunnel drying, freeze drying, kiln drying, foam-mat drying, spray drying, oven drying, and microwave drying and are outlined by Potter (1986, pp. 261–285).

Apricots have traditionally been dried in the sun. However, mechanical means are necessary to ensure a consistent quality. Quality is evaluated in terms of degree of browning and hardness. Abdelhaq and Labuza (1987) reported that air drying at 50–80°C of sulfited (800–1000 ppm SO_2) apricots was optimal but noted that the sulfite was the critical ingredient. Thus replacement of sulfite is difficult. Dehydrated potatoes also are treated with sulfite to retard browning. Treatment of some products with glucose oxidase to reduce glucose content and thus inhibit enzymatic browning is a possible solution (Low et al., 1989). Dehydration of bananas of varying maturities results in different degrees of browning. As the banana matures, amylose is hydrolyzed to glucose, which may participate in carbonyl-amine browning during drying (Mao and Kinsella, 1981).

One problem associated with drying is shriveling and case hardening, resulting in the need for a long cooking period for rehydration. Explosion puffing is a method that has been developed to reduce the hardening problem. For this process, as described by Sullivan and Craig (1984), a product is partially dried (15–35% moisture), removed from the drying chamber, and placed in a pressure gun. Under pressure the remaining moisture is heated above 100°C. When the pressure is released the product puffs, resulting in a porous piece. The product is returned to the oven for completion of drying. Products dried in this manner rehydrate faster and retain a more desirable texture. Kozempel et al. (1989) reported that mushrooms, celery, onions, peppers, rutabagas, beets, yams, pears, pineapples, strawberries, and cranberries can be processed in this manner.

Freeze drying is a drying method that retains the size and shape of food pieces. Color of vegetables may also be better retained. Foods that are freeze dried will reconstitute more rapidly than those that are dried by conventional means. The frozen product is placed in a vacuum chamber at 0°C or below. Water from the produce sublimes under these conditions. In sublimation, the ice changes to vapor without thawing to water. Because freeze-dried products are porous, they will readily take up moisture from a moist atmosphere. Therefore, they must be packaged to exclude moistness. Some companies may use conventional air drying followed by freeze drying to reduce the cost because freeze drying is more expensive. Freeze drying is used for food that may be used in small quantities such as chives and vegetables for soup and pasta mixes or for special-purpose items such as camping foods.

B. Salt

Salt is used in a number of food products to inhibit the growth of undesirable microorganisms. Inhibition is attributed to the lowering of the water activity as salt draws moisture from tissues. Several other effects of salt also have been described (Frazier and Westhoff, 1988, p. 151). Plasmolysis of cells may occur due to the high osmotic pressure caused by the salt. In addition, upon ionization the chloride of NaCl has harmful effects on microorganisms. The chloride of KCl, which has been studied as a substitute for NaCl, would have the same effect.

Sauerkraut production requires salt to serve in the role of inhibitor and add flavor. Salt draws water from the cabbage leaves. The brine that is formed provides a medium for the growth of lactic acid bacteria, which are responsible for the fermentation process. Undesirable microorganisms that are found on the cabbage leaves cannot tolerate the salt content of the brine (Marsh, 1983). Salt also inhibits the growth of unwanted microorganisms in cheese production and cured meat production. Salt levels in processed meats are discussed in Chapter 9.

C. Addition of Sugar: Jellies and Jams

Jellies and jams are preserved by their sugar content, which is high enough to prevent the growth of microorganisms, except for mold growth on the surface. Commercially, products are protected by a vacuum cap and may be pasteurized after the containers are filled. The USDA (1988) recommends that home-prepared jellies and jams be processed in a boiling water bath.

1. Jellies

Jellies are clear substances since they are made of fruit juice or a water extract of fruit. Jams, however, contain all or most of the insoluble solids of the fruit because whole, crushed, macerated, or pureed fruit is used in their manufacture. Technically, jams and preserves are identical, except the term preserves is used for products containing whole fruit.

Gelation of pectin is brought about by the addition of sugar in the presence of acid. Hydrogen bonding between hydroxyl groups and between hydroxyl and carboxyl groups (Whistler and Daniel, 1985, p. 124) is responsible, at least in part, for the rigidity of fruit jellies. The relationships among the three essential ingredients, pectin, sugar, and acid, are important to the quality of the product. For example, insufficient pectin or acid may prevent gel formation and too little sugar results in a tough jelly (Woodroof, 1986).

Purified pectin is made from apple pomace and the white inner skin, or albedo, of citrus fruit. Pectins with high molecular weight and a relatively high proportion of methyl ester groups have the best jelly-forming ability. The quality of pectin is indicated by its ability to carry sugar when made into jelly and is expressed commercially by grade. Jelly grade is the proportion of sugar that one part of pectin is capable of turning, under prescribed conditions, into a jelly with suitable characteristics. If, for example, 1 lb of pectin will carry 150 lb of sugar to make a standard jelly, it is a 150-grade pectin. The chemistry of pectin is discussed in Chapter 14.

The amount of sugar in a jelly depends on the amount and quality of the pectin used. When the sugar content of a mixture is increased or the pectin content decreased, a weaker jelly will result. Tougher gels result with reduction in sugar or an excess of pectin. In a normal finished jelly, the concentration of sugar is about that of a saturated sugar solution.

The third essential ingredient for jelly is acid. The amount of acid required depends not on total but active acidity or pH. Apparently the acid acts by neutralizing the charge on the carboxyl groups of pectin, thus increasing the tendency of the molecules to associate and hence to form a gel. The pH must be below 3.5 for gel formation. As it is decreased below 3.5, the firmness of the jelly increases. Acids such as vinegar, lemon juice, lime juice, citric acid, lactic acid, malic acid, and tartaric acid often are added in making jellies and jams.

Water, a fourth ingredient in jelly, is taken for granted. In a natural fruit jelly, there are traces of other components, such as salts, proteins, and starches. Such components are not essential, as shown by a formula for test jelly: 450 ml distilled water, 5.2 g of 150-grade pectin, and 775 g of sucrose (Cox and Higby, 1944). These ingredients are cooked until the batch weighs 1200 g and poured into four glasses, each of which contains 2.0 ml of tartaric acid solution (22.4 g of acid/50 ml of solution). This product is sometimes called a "synthetic" jelly because only purified ingredients are used in making it. The formula could be used in an experiment for studying the effects of ingredient variation on gel strength of jellies.

To make jelly without added pectin, fruits with adequate pectin levels (crabapples, grapes, apples, currants, sour blackberries, cherries, quinces, lemons, sour oranges, and grapefruit) are best used (Woodroof, 1986, p. 428). Adding one of these fruits to lower pectin fruits (strawberries and apricots) (Woodroof, 1986), using partly underripe fruit, which is higher in pectin (USDA, 1988), or adding a commercial pectin will contribute needed pectin. Pectin content of juice is indicated by its viscosity because large molecules increase viscosity.

To make jelly, fruit is heated, and the juice is extracted and filtered. The mixture of sugar and juice is cooked in a kettle until the desired concentration of sugar (usually 65–69%) is achieved. In commercial operations, a vacuum system may be used for heating; the resultant lower cooking temperature minimizes hydrolysis of the pectin and maintains color and flavor (Woodroof, 1986, pp. 428–430). Methods of determining when jelly has cooked long enough can be divided into two groups: measurement of gelation and measurement of soluble solids content. Gelation can be measured by chilling a small amount of boiling jelly in a refrigerator or by the sheeting off test, in which the mixture is allowed to drip from a large cool spoon. If the syrup separates into two streams of drops that sheet off together, the jelly is done. The success of these crude tests depends on the experience of the operator.

The other tests are measures of soluble solids content. Measurements of solids are useful because if all the ingredients and other conditions are right, the mixture will form a jelly when it reaches a certain solids content. Juices poor in pectin or acid may not form jelly until the solids content is higher than normal, and juice rich in these constituents may gel at an unusually low solids content. The measures of soluble solids content used in making jelly are boiling point, refractive index, and specific gravity. Boiling point is the only one of these methods available for home cooking. It is a useful guide, especially if applied to a standardized mixture that is cooked at a certain rate so that the extent of hydrolysis is controlled. Of course, the hydrolysis of sucrose caused by the fruit acid will increase the boiling point of the mixture without having a corresponding effect on soluble solids concentration. The boiling point ordinarily used for jelly is 104–105°C, or more accurately, 4–5°C above the boiling point of water (in order to correct for variation in atmospheric pressure and for inaccuracies in the thermometer). A boiling point 4.8°C above that of water corresponds to the

65% soluble solids content required for commercial jellies. The refractometer is used widely commercially because refractive index is the best measure of solids content. Another method that might be used is measurement of specific gravity with a hydrometer (Woodroof, 1986, pp. 429–430).

Acid that is present during boiling hydrolyzes some of the sugar to invert sugar, which helps prevent crystallization of the sugar as the jelly is stored. The presence of acid in the boiling jelly may hydrolyze the pectin. For this reason, commercial manufacturers often add acid after jelly has cooked. Small lots of jelly cook so quickly that pectin hydrolysis seldom presents a problem.

2. Low-Ester Pectins

As explained in Chapter 14, some of the carboxyl groups of pectin are esterified with methyl groups, while others are uncombined or combined with metals to form salts. Pectin in which all of the carboxyl groups are methylated (100% esterified) would contain 14% methoxyl, a degree of methylation that is theoretically possible but has not been attained. High-methoxyl pectins are 50–58% esterified. If pectin is demethylated by enzymes, acid, or alkali until it is 20–40% esterified, low-methoxyl pectin is obtained (Glicksman, 1982, pp. 283–284). The advantage of this pectin is that it will gel with little or no sugar if calcium or another divalent ion is present. Such ions cause ionic bonding by reaction of a calcium ion with two carboxyl groups as in the formula shown in Fig. 13.2. Gel strength depends on pectin and calcium concentrations (Lopez and Li, 1968).

Low-methoxyl pectins may be used commercially for the production of low-calorie jams and jellies and dessert products. For home production of these products low-methoxyl pectins have been introduced to the retail market. These pectins may be used in salads and desserts that contain nearly any desired amount of sugar, including none.

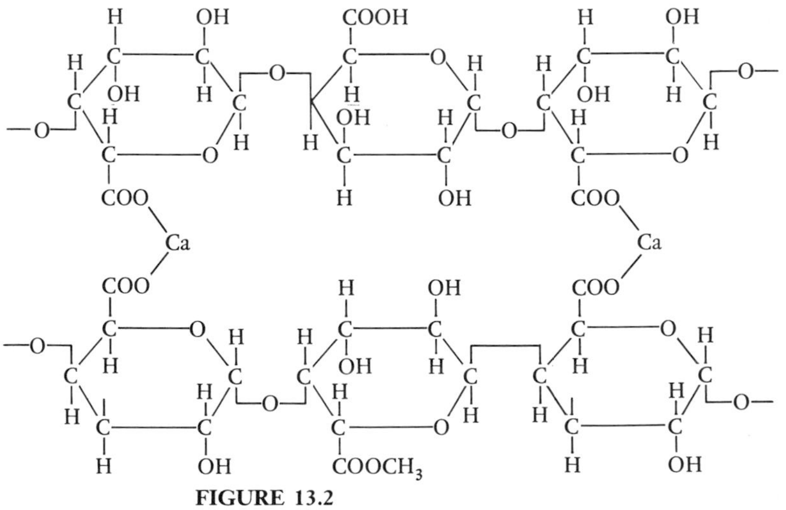

FIGURE 13.2
Low-ester pectin precipitated by calcium.

3. Modified Jam and Jellies

Alternatives for the production of low-calorie and/or sugar-controlled products are outlined by the USDA (1988) and include use of noncaloric sweeteners in products with concentrated fruit pulp. Gelatin and sugar substitutes also may be used in fruit spreads that have a relatively short, refrigerated shelf life of 1 month. Commercial fruit spreads are available that are sweetened with concentrated fruit juices rather than sugar or syrups or contain noncaloric sweeteners (Gross, 1989, pp. 107–109).

IV. CHEMICAL PRESERVATIVES AS ANTIMICROBIAL AGENTS

A wide variety of compounds may be used to inhibit the growth of micro-organisms that may be present in a food. Salt and sugar may be classified in this group. Because their mode of action involves lowering the water activity, they were discussed previously (see Section III,B and C). The following discussion is on selected compounds. Further information on antimicrobial agents may be found in several sources (Dziezak, 1986; Igoe, 1989, pp. 173, 178; Jay, 1986, Chapter 11; Frazier and Westhoff, 1988, Chapter 9).

Some of these compounds are defined as chemical preservatives by law; others are not. For example, salt, sugar, spices and oils extracted from them, vinegar, and wood smoke are excluded from the definition in the Food Additives Amendment of the Food, Drug, and Cosmetic Act. Others are categorized as GRAS (generally recognized as safe for addition to food). Limits may be placed on some of these by the FDA and USDA. These include some organic acids (propionic, sorbic, and benzoic) and their derivatives, parabens, sulfites, and SO_2, sodium diacetate, dehydroacetic acid, sodium nitrite, caprylic acid, and ethyl formate (Igoe, 1989, pp. 173, 178).

Action of many of these additives is related to the pH of the food. For example, sodium benzoate is most active at a low pH because to be effective it must be in its undissociated form. At a pH of 4.0, 60% of the compound is undissociated whereas more is dissociated at higher pH values. Benzoate and benzoic acid are used, therefore, in low-pH foods for mold and yeast inhibition.

It is interesting to note that some of these compounds are found naturally in foods and may serve as antimicrobials. For example, propionic acid is found in Swiss cheese as a result of natural processing. Benzoic acid is found in cranberries. Naturally occurring antimicrobials are reviewed by Beuchat and Golden (1989).

Some of these preservatives are discussed in other chapters. For example, the sodium and potassium salts of nitrate and nitrites are used as curing agents in meats as described in Chapter 9. Nitrite also acts as an antimicrobial agent in such products as it retards growth of *C. botulinum* during storage in vacuum packages. Levels of usage are limited by law to minimize potential formation of nitrosamines, carcinogenic compounds.

Classes of some antimicrobials are shown in Table III. Primary applications also are listed. The effectiveness of the preservative in the foods and against the

TABLE III
Selected Antimicrobial Agents and Their Application[a]

Compounds	Primarily effective against	Food in which used
Acetic acid	Bacteria	Breads
Sodium acetate	Molds	As vinegar in mayonnaise, pickles, and pickled meats
Sodium diacetate		
Calcium acetate		
Benzoic acid	Yeasts	Apple cider, catsup, fish dips, fruit juices, fruit salads, jams
Sodium benzoate	Molds	and jellies, margarine, olives, orange juice products, pickles, preserves, salad dressings, soft drinks, syrups
Sodium propionate	Molds	Bread (to prevent ropiness), cakes, cheese foods and
Calcium propionate		spreads, syrups
Sorbic acid	Yeasts	Baked goods (not yeast-leavened dough), beverages, cakes,
Sorbate	Molds	cheese, cheese foods, cheese spreads, creamed cottage
	Bacteria	cheese, dried fruits, figs, fruit cocktails, fruit juices, fruit salads, icings, jams and jellies, margarine, pickles, salad dressing, salads, syrups
Methylparaben	Molds	Bakery products, pickles, salad dressing, soft drinks
Propylparaben	Yeasts	
Caprylic acid	Mold	Cheese wraps
Sodium nitrite	Clostridia	Cured meat products
Sulfite (SO_2)	Yeast	Dried fruit, lemon juice, molasses, wine
	Mold	
	Bacteria	

[a] From Dziezak (1986); Jay (1986); Frazier and Westhoff (1988).

organisms listed is in some cases dependent on pH of the products. It is important to remember when reviewing the information in the table that it is a general presentation and that specific limits and applications may be regulated by law.

V. CONTROLLED OR MODIFIED ATMOSPHERE STORAGE AND PACKAGING

The shelf life of products may be extended by controlling the gases in the environment of the product. Several approaches and applications of this may be seen. The removal of O_2 from a package and addition of N_2 to the head space provides what is termed an inert atmosphere. This approach is used with instant coffee and dry milk to inhibit oxidation.

Deterioration via senescence of fresh fruits and vegetables may be delayed during shipping or storage by depleting the atmosphere of O_2 and enriching it with CO_2. This may be achieved by placing the product in a sealed carton and allowing the product to create a modified atmosphere (MA) by using up O_2 and adding CO_2. Controlled atmosphere (CA) storage involves controlling the amounts of gases in the atmosphere. An early application of CA storage was in the apple industry. Deák *et al.* (1987) found that shrink wrapping of corn resulted in increased CO_2 and reduced O_2 within the package and reduced moisture loss. Irradiation reduced the microbial population so that they would not grow in the moist environment. The combination of packaging, irradiation, and refrigeration extended the shelf life of the corn. Application to the preservation of minimally processed fruits and vegetables also may be found. Minimally processed fruits and vegetables include shredded lettuce, peeled potatoes, and other partially prepared products (Myers, 1989).

Meats also may be packaged in MA systems. Vacuum packaging of retail cuts of meat has been used for some time to maintain the quality of meat during the shipping and storage that precede use at the retail level. More recently, vacuum-packaged fresh meats have been introduced to the retail market. These MA packages prevent changes in the myoglobin, thus maintaining a fresh appearance in the product and also preventing the growth of spoilage microorganisms. Flushing the package with CO_2, N_2, or O_2 also may be used as a means of maintaining product quality (Young *et al.*, 1988).

Many questions must be answered before this packaging methodology will be used extensively. Questions regarding the potential growth of *C. botulinum* and other foodborne pathogens that will grow at refrigerator temperatures remain to be answered (Hotchkiss, 1988). Potential abuse at any point in the food distribution also is of concern particularly with respect to refrigerated prepared foods. Most of these packaging techniques are used with refrigeration. Failure to keep products refrigerated may result in problems. Application of known principles of food safety should be applied at all levels to those products that now are available.

VI. IRRADIATION _____

In the 1950s study of the effects of radiation preservation on food began. Interest waned but was renewed in the 1980s when the Food and Drug Administration issued regulations allowing irradiation to control insects, to kill *Trichinella spiralis* in pork, and to extend the shelf life of some fruits and vegetables (FDA, 1986). Earlier approval had been given for control of insects in wheat and wheat flour; for control of insects and microorganisms in spices, herbs, and other dehydrated foods; and for prevention of sprouting in potatoes (Newsome, 1987). Levels for all uses are limited. Recent reviews include details of irradiation of food (IFT, 1983; Newsome, 1987; Rogan and Glaros, 1988).

Because the temperature rise in food preserved by irradiation is very small, the process is considered to be a "cold" process. Therefore, changes in the flavor, color, texture, and nutritive value are minimal. Effects of nutritional value were summarized by the IFT (1986), Skala *et al.* (1987), and Thomas (1988, Chapter 18). Terms used in discussion of irradiation are defined in the glossary in Table IV. Two methods can be used for irradiation: electron and gamma radiation. The latter is used most frequently. The Council for Agricultural Science

TABLE IV
Terms for Irradiation[a]

Term	Definition
Electron radiation	Corpuscular radiation, consisting of streams or beams of electrons accelerated to energies of up to 10 MeV
Gamma radiation	Electromagnetic radiation of a very short wavelength, of the same nature as "short" X rays. Gamma rays are emitted by isotopes of such elements as cobalt and cesium as they disintegrate spontaneously
Gray (Gy)	The unit (or level) of energy absorbed by a food from ionizing radiation as it passes through in processing. One Gray (Gy) equals 100 rads; 1000 Gy equal 1 kilogray (kGy)
Rad	Another name or unit for "radiation energy absorbed" by food being processed with radiation. One thousand rads = 1 krad = 10 Gy (see above); 1,000,000 rads = 1 Mrad = 10 kGy. (The rad is being superseded by the Gray)
Radappertization	Sterilization by radiation processing. The resulting processed food can be stored at room temperature in the same way as thermally sterilized foods (canned foods). Precooked food in hermetically sealed packaging is exposed to radiation at levels high enough to kill all organisms of food spoilage or public health significance. Doses used are typically greater than 1 Mrad
Radicidation	Radiation pasteurization intended to kill or render harmless all disease-causing organisms (except viruses and spore-forming bacteria) in food. Processing takes place at dose levels generally below 1 Mrad, and the processed foods usually must be stored under refrigeration (as in heat pasteurization)
Radurization	Radiation pasteurization designed to kill or inactivate food-spoilage organisms, thus extending the shelf life of a given food product. Again, processing takes place at dose levels generally below 1 Mrad, and the product usually must be stored under refrigeration, as in the case of radicidized food (see above)
X Rays	Electromagnetic radiation of a wide variety of short wavelengths. They are usually produced by a machine in which a beam of fast electrons in a high vacuum bombards a metallic target. X Rays are sometimes called "Roentgen rays," after their discoverer, Lord von Roentgen

[a]From IFT (1983). (Reprinted with permission from *Food Technology*, 37, No. 2, pp. 55–60. Copyright 1983 by Institute of Food Technologists.)

and Technology concluded that there has been no evidence shown that irradiated foods are causative agents of toxicity, carcinogenicity, mutagenicity, or teratogenicity (CAST, 1986). Elias (1989) also addressed the issue of wholesomeness.

The mechanism for action of irradiation involves interaction of the gamma or X rays and negatively charged electron rays with the molecules in the food, causing ionization and production of free radicals. Interaction of the free radicals with material of the nucleus of the cells of microorganisms, parasites, and insects prevents reproduction. Killing depends on the dosage level as described in the definitions for radappertization, radicidation, and radurization. Further details of the process are outlined by Potter (1986, pp. 303–320).

It has been indicated that irradiated products are closer in quality to fresh products than products processed by other means. Many studies of the effects of irradiation on food quality can be found in the literature. Doses of 5 and 10 kGy were found to be adequate to destroy molds and bacterial spores, respectively, in pepper, tumeric, chili, and coriander (Munasiri *et al.*, 1987). Irradiated spices retained their quality over a 6-month period better than unirradiated spices. Irradiation of chicken delayed deterioration of the odor over a 14-day storage period (Hansen *et al.*, 1987). Deák *et al.* (1987) reported that irradiation of shrink-wrapped fresh corn decreased initial microbial counts and, therefore, contributed to extension of shelf life.

Studies have demonstrated that irradiation is a beneficial method of food preservation. However, consumer understanding of the process and its safety with regard to the food supply must be improved to ensure acceptance of irradiated foods (Bruhn and Schutz, 1989).

Suggested Exercises

1. Determine the effect of variation in amount of sugar on the quality of jelly and the relationship of sugar and the volume of the jelly. Follow instructions in the USDA guide (USDA, 1988) for the preparation of juice for apple jelly without pectin. Add acid to the juice as specified.

   Prepare four lots of jelly using 240 ml of juice in each and the following levels of sugar: (1) 100 g, (2) 150 g, (3) 200 g, and (4) 250 g. Prepare according to the basic directions, heating the jelly to 104°C.

   To evaluate the quality of the jelly, measure the volume of each, determine percentage sag after 24 hr, and evaluate the sensory properties on a scorecard designed by the class.

2. Determine the effect of boiling temperature on the volume and quality of jelly. Prepare jelly using 150 g sugar and 7 ml lemon juice/240 ml of juice. Prepare three batches, cooking them to 1, 4, and 7°C above the boiling point of water. Evaluate as in exercise 1. Note any differences in the sheeting off tests at the three temperatures.

3. Study the effects of various factors on the speed of heat penetration during canning; insert a thermometer into the center of the food. A maximum registering thermometer or thermocouple is necessary if a pressure cooker is used. The thermometer is inserted through a cork placed in a hole bored in the center of the jar lid or the top of the can.

   If a metal screwband and a flat metal plate are used, a large rubber stopper with a hole in the center to hold the thermometer can be inserted in the screwband in place of the flat metal plate.

   Compare the heat penetration rates of apple sauce and sliced apples in syrup. Compare the rates of heat penetration for whole-kernel corn and cream-style corn.

4. Wrap samples of food, such as ground beef patties, in various ways or various materials, freeze them, and weigh them periodically for several weeks. Observe the amount of freezer burn and calculate the loss of weight as percentage of the raw sample weight.

5. Freeze a vegetable without blanching and after the recommended blanching treatment (USDA, 1982). After storage of at least a month, compare the palatability of the two samples. Samples can be prepared by one class and judged by the following class in order to extend the storage period.
6. Compare the quality of apples or peaches frozen with and without ascorbic acid, stored as long as possible, and thawed under various conditions.
7. Compare drying of carrots in a conventional oven and in a microwave oven.
8. Using fruit from the same lot, prepare uncooked jam and cooked jam, made with and without pectin. Store as long as possible and compare the quality of the three products.
9. Obtain low-methoxyl pectin jelly mix and prepare products according to the directions supplied by the manufacturer. Prepare jelly with regular pectin and compare the two products.
10. Collect labels of foods. Read carefully and identify antimicrobial agents that are listed on the labels. Discuss their functions and modes of action.

References

Abdelhaq, E. H. and Labuza, T. P. 1987. Air drying characteristics of apricots. *J. Food Sci.* **52**: 342–345, 360.

Allen, M. and Crandall, M. L. 1974. The cold facts about freezing. In "Shopper's Guide—1974 Yearbook of Agriculture." USDA, Washington, DC.

Anderson, M. L., Keeton, J. T., Acuff, G. R., Lucia, L. M., and Vanderzant, C. 1989. Microbiological characteristics of precooked, vacuum-packaged uncured beef and pork. *Meat Sci.* **25**: 69–79.

Baker, R. C., Darfler, J. M., Mulnix, E. J., and Nath, K. R. 1976. Palatability and other characteristics of repeatedly refrozen chicken broilers. *J. Food Sci.* **41**: 443–445.

Ball, C. O. 1938. Advancement in sterilization methods for canned foods. *Food Research* **3**: 13–55.

Beuchat, L. R. and Golden, D. A. 1989. Antimicrobials occurring naturally in foods. *Food Technol.* **43**(1): 134–142.

Bobeng, B. J. and David, B. D. 1978. HACCP models for quality control of entree prodution in hospital foodservice systems. *J. Amer. Dietet. Assoc.* **73**: 530–535.

Brown, M. S. 1967. Texture of frozen vegetables: Effect of freezing rate on green beans. *J. Sci. Food Agric.* **18**: 77–88.

Brown, M. S. 1976. Effects of freezing on fruit and vegetable structure. *Food Technol.* **30**(5): 106–109, 114.

Bruhn, C. M. and Schutz, H. G. 1989. Consumer acceptance and outlook for acceptance of food irradiation. *Food Technol.* **43**(7): 93–94, 97.

CAST. 1986. Ionizing energy in food processing and pest control. I. Wholesomeness of food treated with ionizing energy. Rept. No. 109, Council for Agricultural Science and Technology, Ames, Iowa.

Cox, R. E. and Higby, R. H. 1944. A better way to determine the jellying power of pectins. *Food Ind.* **16**: 441–442, 505–507.

Dawson, E. H., Gilpin, G. L., Warren, H. W., and Toepfer, E. W. 1953. Canning snap beans; precooked and raw pack. *J. Home Econ.* **45**: 165–168.

Deák, T., Heaton, E. K., Hung, Y. C., and Beuchat, L. R. 1987. Extending the shelf life of fresh sweet corn by shrink wrapping, refrigeration, and irradiation. *J. Food Sci.* **52**: 1625–1631.

Drake, S. R., Spayd, S. E., and Thompson, J. B. 1981. The influence of blanch and freezing methods on the quality of selected vegetables. *J. Food Qual.* **4**: 271–278.

Dziezak, J. D. 1986. Preservative: Antimicrobial agents. *Food Technol.* **40**(9): 104–111.

El-Hag, N. 1989. The refrigerated food industry: Current status and developing trends. *Food Technol.* **43**(3): 96–98.

Elias, P. S. 1989. New concepts for assessing the wholesomeness of irradiated foods. *Food Technol.* **43**(7): 81–83.

Esty, J. R. and Meyer, K. F. 1922. The heat resistance of the spores of *Bacillus botulinus* and allied anaerobes. XI. *J. Infectious Diseases* **31**: 650–659.

FDA. 1986. Irradiation in the production and handling of food: Final rule. *Federal Register* **51**: 13375–13799.

FDA. 1988. Canned tomatoes. *In* "Code of Federal Regulations," Title 21, Section 155.190. U.S. Govt. Printing Office, Washington, DC.

Fennema, O. 1966. An over-all view of low temperature food preservation. *Cryobiology* **3**: 197–213.

Fennema, O. 1975. Freezing preservation. *In* "Principles of Food Science, Part II. Physical Principles of Food Preservation," Karel, M., Fennema, Fennema, O. R., and Lund, D. B. (Eds.). Marcel Dekker, Inc., New York.

Fields, M. L., Zamora, A. F., and Bradsher, M. 1977. Microbiological analysis of home-canned tomatoes and green beans. *J. Food Sci.* **42**: 931–934.

Frazier, W. C. and Westhoff, D.C. 1988. "Food Microbiology," 4th edn. McGraw-Hill Book Co., New York.

Gilpin, G. L., Hammerle, O. A., and Harkin, A. M. 1951. Palatability of home-canned tomatoes. *J. Home Econ.* **43**: 282–284.

Glasscock, S. J., Axelson, J. M., Palmer, J. K., Phillips, J. A., and Taper, L. J. 1983. Microwave blanching of vegetables for frozen storage. *Home Econ. Research J.* **11**: 149–158.

Glicksman, J. 1982. Food applications of gums. *In* "Food Carbohydrates," Lineback, D. R. and Inglett, G. E. (Eds.). Avi Publ. Co., Westport, Connecticut.

Gould, W. A. and Gray, E. 1974. Canning tomatoes in the home. Publ. L-170, Cooperative Ext. Serv., Ohio State University Columbus, Ohio.

Gross, D. R. 1989. Fruit spreads: Developments in non-standardized preserves. *In* "1989 Food and Beverage Technology International," Twigg, B. A., and Turner, A. (Eds.). Sterling Publ. Ltd., London.

Hansen, T. J., Chen, G.-C., and Shieh, J. J. 1987. Volatiles in skin of low dose irradiated fresh chicken. *J. Food Sci.* **52**: 1180–1182.

Heldman, D. R. 1989. Establishing aseptic thermal processes for low-acid foods containing particulates. *Food Technol.* **43**(3): 122–123, 131.

Hotchkiss, J. H. 1988. Experimental approaches to determining the safety of food packaged in modified atmospheres. *Food Technol.* **42**(9): 55, 56–62, 64.

Hudson, M. A., Sharples, V. J., Pickford, E., and Leach, M. 1974a. Quality of home frozen vegetables. I. Effects of blanching and/or cooling in various solutions on organoleptic assessments and vitamin C content. *J. Food Technol.* **9**: 95–103.

Hudson, M. A., Sharples, V. J., and Gregory, M. E. 1974b. Quality of home frozen vegetables. II. Effects of blanching and/or cooling in various solutions on conversion of chlorophyll. *J. Food Technol.* **9**: 105–114.

Hudson, M. A., Leach, M., Sharples, V. J., and Pickford, E. 1975a. Home frozen strawberries. I. Influences of freezing medium, fanning, syrup temperature, soaking time, storage time and temperature, and rates of freezing and thawing on sensory assessments. *J. Food Technol.* **10**: 681–688.

Hudson, M. A., Holgate, M. E., Gregory, M. E., and Pickford, E. 1975b. Home frozen strawberries. II. Influences of additives in syrup on sensory assessments and texture measurements. *J. Food Technol.* **10**: 689–698.

Hung, Y.-C. and Thompson, D. R. 1989. Changes in texture of green peas during freezing and frozen storage. *J. Food Sci.* **54**: 96–101.

IFT. 1983. Radiation preservation of foods. A scientific status summary by the Institute of Food Technologists' Expert Panel on Food Safety and Nutrition and the Committee on Public Information. *Food Technol.* **37**(2): 55–60.

IFT. 1986. Effects of food processing on nutritive values. A scientific status summary by the Institute of Food Technologists' Expert Panel on Food Safety and Nutrition and the Committee on Public Information. *Food Technol.* **40**(12): 109–116.

Igoe, R. S. 1989. "Dictionary of Food Ingredients," 2nd edn. Van Nostrand Reinhold, New York.

Jay, J. M. 1986. "Modern Food Microbiology," 3rd edn. Van Nostrand Reinhold, New York.

Karmas, E. and Harris, R. S. 1988. "Nutritional Evaluation of Food Processing." Van Nostrand Reinhold, New York.

Kozempel, M. F., Sullivan, J. F., Craig, J. C., Jr., and Konstance, R. P. 1989. Explosion puffing of fruits and vegetables. *J. Food Sci.* **54**: 772–773.

Lane, R. H., Boschung, M. D., and Abdel-Ghany, M. 1985. Ascorbic acid retention of selected vegetables blanched by microwave and conventional methods. *J. Food Qual.* **8**: 139–144.

Lechowich, R. V. 1988. Microbiological challenges of refrigerated foods. *Food Technol.* **42**(12): 84–85, 89.

Lee, C. Y., Smith, N. L., and Hawbecker, D. E. 1988. Enzyme activity and quality of frozen beans as affected by blanching and storage. *J. Food Qual.* **11**: 279–287.

Lee, F. A. and Wagenknecht, A. C. 1951. On the development of off-flavor during storage of frozen raw peas. *Food Research* **16**: 239–244.

Leonard, S., Luh, B. S., and Pangborn, R. M. 1960. Effect of sodium chloride, citric acid, and sucrose on pH and palatability of canned tomatoes. *Food Technol.* **14**: 433–436.

Locker, R. H. and Daines, G. J. 1973. The effect of repeated freeze-thaw cycles on tenderness and cooking loss in beef. *J. Sci. Food Agric.* **24**: 1273–1275.

Lopez, A. and Li, L.-H. 1968, Low-methoxyl pectin apple gels. *Food Technol.* **22**: 1023–1028.

Low, N., Jiang, Z., Ooraikul, B., Dokhani, S., and Palcic, M. M. 1989. Reduction of glucose content in potatoes with glucose oxidase. *J. Food Sci.* **54**: 118–121.

Lund, D. 1988. Effect of heat processing on nutrients. *In* "Nutritional Evaluation of Food Processing," 3rd edn. Van Nostrand Reinhold, New York.

Mao, W. W. and Kinsella, J. E. 1981. Amylase activity in banana fruit: Properties and changes in activity with ripening. *J. Food Sci.* **46**: 1400–1403, 1409.

Marcy, J. E., Hansen, A. P., and Graumlich, T. R. 1989. Effect of storage temperature on the stability of aseptically packaged concentrated orange juice and concentrated orange drink. *J. Food Sci.* **54**: 227–228, 230.

Marsh, A. C. 1983. Processes and formulations that affect the sodium content of foods. *Food Technol.* **37**(7): 45–49.

Moriarty, J. H. 1966. Flavor changes in frozen foods. *Cryobiology* **3**: 230–235.

Munasiri, M. A., Parte, M. N., Ghanekar, A. S., Sharma, A., Padwal-Desai, S. R., and Nadkarni, G. B. 1987. Sterilization of ground prepacked Indian spices by gamma irradiation. *J. Food Sci.* **52**: 823–824, 826.

Myers, R. A. 1989. Packaging considerations for minimally processed fruits and vegetables. *Food Technol.* **43**(2): 129–131.

National Canners Association. 1976. Process for low-acid canned foods in metal containers. Bull. 26-L, 11th edn., NCA, Washington, DC.

Newsome, R. L. 1987. Perspective on food irradiation. *Food Technol.* **41**(2): 100–101.

Ollinger-Snyder, P. A. and Matthews, M. E. 1988. Cook/chill foodservice systems with a microwave oven: Coliforms and aerobic counts from turkey rolls and slices. *J. Food Protection* **51**: 84–86.

Ponting, J. D. and Jackson, R. 1972. Pre-freezing processing of Golden Delicious apple slices. *J. Food Sci.* **37**: 812–814.

Potter, N. N. 1986. "Food Science," 4th edn. Avi Publ. Co., Westport, Connecticut.

Pray, L. W. and Powers, J. J. 1966. Acidification of canned tomatoes. *Food Technol.* **20**: 87–91.

Pritchard, I. 1974. Can's and can'ts for canners. *In* "Shoppers Guide—1974 Yearbook of Agriculture." USDA, Washington, DC.

Rogan, A. and Glaros, G. 1988. Food irradiation: The process and implications for dietitians. *J. Amer. Dietet. Assoc.* **88**: 833–838.

Schoenemann, D. R., Lopez, A., and Cooler, F. W. 1974. pH and acidic stability during storage of acidified and nonacidified canned tomatoes. *J. Food Sci.* **39**: 257–259.

Singh, R. P. and Wells, J. H. 1985. Use of time-temperature indicators to monitor quality of frozen hamburgers. *Food Technol.* **39**(12): 42–50.

Sizer, C. E., Waugh, P. L., Edstam, S., and Ackermann, P. 1988. Maintaining flavor and nutrient quality of aseptic orange juice. *Food Technol.* **42**(6): 152, 154–159.

Skala, J. H., McGown, E. L., and Waring, P. P. 1987. Wholesomeness of irradiated foods. *J. Food Protection* **50**: 150–160.

Stumbo, C. R. 1973. "Thermobacteriology in Food Processing," 2nd edn. Academic Press, New York.

Stumbo, C. R., Purohit, K. S., and Ramamrishnan, T. V. 1975. Thermal processing lethality guides for low-acid foods in metal containers. *J. Food Sci.* **40**: 1316–1323.

Sullivan, J. F. and Craig, J. C. 1984. The development of explosion puffing. *Food Technol.* **38**(2): 52–55, 131.

Thomas, M. H. 1988. Use of ionizing radiation to preserve food. *In* "Nutritional Evaluation of Food Processing," 3rd edn., Karmas, E. and Harris, R. S. (Eds.). Van Nostrand Reinhold, New York.

USDA. 1975. Home freezing of poultry. Home and Garden Bull. 70, United States Department of Agriculture, Washington, DC.

USDA. 1976a. Freezing combination main dishes. Home and Garden Bull. 40, United States Department of Agriculture, Washington, DC.

USDA. 1976b. Freezing meat and fish in the home. Home and Garden Bull. 93, United States Department of Agriculture, Washington, DC.

USDA. 1976c. Modern tomato varieties as safe to can as older varieties. News USDA 1397–76, United States Department of Agriculture, Washington, DC.

USDA. 1982. Home freezing of fruits and vegetables. Home and Garden Bull. 10, United States Department of Agriculture, Washington, DC.

USDA. 1984. The safe food book. Home and Garden Bull. 241, United States Department of Agriculture, Washington, DC.

USDA. 1988. Complete guide to home canning. Principles of home canning. Agriculture Information Bull. No. 539, Extension Service, United States Department of Agriculture, Washington, DC.

USDHEW. 1978. Food service sanitation manual. Public Health Serv., HEW Pub. No. (FDA) 78-2081, Food and Drug Admin., Div. of Retail Food Protection, Washington, DC.

Wagner, M. K. and Moberg, L. J. 1989. Present and future use of traditional antimicrobials. *Food Technol.* **43**(1): 143–147, 155.

Whistler, R. L. and Daniel, J. R. 1985. Carbohydrates. *In* "Food Chemistry," 2nd edn., Fennema, O. R. (Ed.). Marcel Dekker, Inc., New York.

Wolford, E. R. and Nelson, J. W. 1971. Comparison of texture of carrots frozen by airblast, food freezant-12 and nitrogen vapor. *J. Food Sci.* **36**: 959–961.

Woodroof, J. G. 1986. Other products and processes. *In* "Commercial Fruit Processes," 2nd edn., Woodroof, J. G. and Luh, B. S. (Eds.). Avi Publ. Co., Westport, Connecticut.

Young, L. L., Reviere, R. D., and Cole, A. B. 1988. Fresh red meats: A place to apply modified atmospheres. *Food Technol.* **42**(9): 65–66, 68–69.

FRUITS AND VEGETABLES

Many parts of different plants are used as fruits or vegetables. They may be roots, tubers, bulbs, stems and shoots, leaves, flowers and fruits, or pods and seeds. An understanding of the plant cell and compounds contributing to the

texture, color, and flavor is important to a thorough understanding of the influence of various treatments on these quality attributes.

I. TEXTURE

The characteristic texture of a fruit or vegetable depends on the presence and
the relative proportions and arrangement of the various types of cells. For example, raw celery is fibrous because of the arrangement of collenchyma, schler-
enchyma, and xylem (Reeve, 1970). The structure of fruits and vegetables can be
studied at several levels. Some understanding of the cellular structure and the
microstructure is basic to the study of the influence of various treatments on the
texture as well as color and flavor of fruits and vegetables.

A. Structure

1. Cells
Plant tissues are composed of small units, the cells. The components of a plant
cell are represented schematically in Fig. 14.1. Cell constituents are divided into
two groups, protoplasmic and nonprotoplasmic. The protoplasm is the living,
active part of the cell and contains several organelles, including mitochondria,
microsomes, lysosomes, plastids, vacuoles, and nuclei. The nucleus is the reg-
ulator of metabolic activities of the cell. The mitochondria are metabolically important because of their role in respiration. The cytoplasm is an undifferentiated
part of the protoplasm, surrounding the nucleus and forming a rather thin layer
inside the cell wall. Cytoplasm, a transparent substance, contains a high propor-

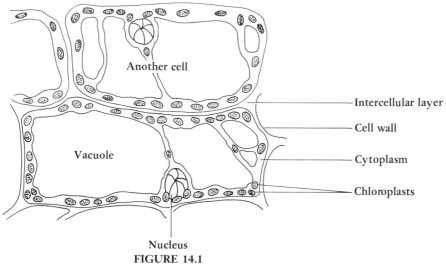

FIGURE 14.1
Food-synthesizing cells of a green plant.

tion of water and organic and inorganic substances in colloidal dispersion, or in some cases, true solution.

Within certain types of cells are organized bodies called plastids. Three types of plastids are recognized: leucoplasts, chloroplasts, and chromoplasts. Leucoplasts lack pigmentation. Specialized leucoplasts, amyloplasts, produce and store starch in potatoes, peas, beans, and other starch-forming tissues. Starch is laid down within the plastid in concentric layers (Chapter 16). The pattern of the layers is characteristic of the plant and may be an aid in the microscopic identification of starch from various sources. Chloroplasts in green plants contain chlorophyll, the green pigment essential for carbohydrate synthesis. Chlorophyll is located between layers of lipids and proteins in the chloroplasts with hydrophobic portions oriented toward lipid layers (Clydesdale *et al.*, 1970). Chromoplasts contain xanthophylls or carotenes and usually are orange or yellow in color. They occur in such vegetables as carrots and sweet potatoes.

Nonprotoplasmic components of the cell are referred to as ergastic components and include cavities called vacuoles, which contain cell sap. In the mature plant, a vacuole is large in relation to the size of the nucleus and the cytoplasm. Cell sap is a watery substance containing sugars, salts, organic acids, polysaccharides, phenolic derivatives, flavones, and the red or blue pigments called anthocyanins. These substances are in true solution or are colloidally dispersed. Substances in the cell sap include nutrients and products of metabolism.

The water in the vacuole of the cell is responsible, in part, for the texture of fruits and vegetables. In crisp raw fruits and vegetables, the pressure exerted on the inside of the cell wall, turgor, is equal to the pressure exerted on the outside of the cell wall. Reduction in the relative humidity of the storage atmosphere may result in loss of crispness due to loss of moisture and thus of turgor. Turgor also is lost as water diffuses through cell membranes during heating. Bourne (1986) recently demonstrated that texture profiles (Chapter 3) of apples change as water activity is lowered.

2. Cell Walls

Cellular components are enclosed by a wall that is responsible for the texture of the tissue. The walls of adjoining cells are held together by the intercellular layer or middle lamella. This layer of cementing substance is composed of pectin in one or more of its forms.

When the cell is immature, the outer, or primary, wall is produced first. In soft tissues, such as those of some fruits, this is the only wall. The primary wall is composed of cellulose, hemicelluloses, and some pectin. In some tissues, a secondary layered wall of cellulose and hemicellulose is laid down inside the primary wall. In vegetables that seem woody, lignin may be present. It stiffens the plant, making it less flexible. An understanding of the properties of the cell wall constituents is essential to the understanding of changes that occur in fruits and vegetables during ripening, storage, processing, and preservation.

a. Cellulose.   Cellulose, which is present in large quantity, is responsible for the firmness of the cell wall. Its purpose is structural, since it does not furnish a reserve food supply for the plant. Like starch, which is discussed in Chapter 16, cellulose is a polysaccharide composed of glucose units. It differs from starch in that the glucose units are combined with β-glucosidic linkages rather than α-glucosidic linkages. As many as 12,000 glucose units may be found in a molecule (Haard, 1985, p. 860). The linkage of β-D-glucose units in cellulose is as follows:

Cellulose

Cellulose molecules are combined in some regions in an orderly crystalline arrangement that is stabilized by hydrogen bonding. In other areas close association of molecules is not possible. These amorphous regions may bind water whereas the crystalline regions cannot (Whistler and Daniel, 1985, p. 109). Derivatives of cellulose are important food additives because they serve as bulking agents and increase viscosity.

b. Hemicelluloses. Hemicelluloses, which are not very similar chemically to cellulose, also may be found in the cell wall. They are insoluble in water but soluble in alkali and are hydrolyzed more readily than cellulose by both alkali and dilute acid. The ease of hydrolysis of the hemicelluloses by alkali is responsible for the mushiness of vegetables heated in water containing baking soda. Hemicellulose content of vegetables is reduced by cooking.

Hemicelluloses vary from plant to plant and are not completely defined in all cases. They are polysaccharides, among which are found xylans, galactans, mannans, glucomannans, and arabinogalactans.

c. Lignin. Another important constituent of some cell walls is lignin, a group of related noncarbohydrate compounds. Lignin molecules are three-dimensional polymers of phenyl propane derivatives (Haard, 1985, p. 859). The amount of lignin in high-quality fruits and vegetables is small. "Stringless beans" represent development of varieties of beans with low lignin content (Bourne, 1976, p. 283). All three layers of a cell wall may contain lignin. Once deposited in the cellulose framework of the cell wall, it is not reutilized by the plant. It resists the action of chemicals, enzymes, and bacteria and is not softened during cooking.

d. Gums. Another group of cell wall carbohydrates that may be present is composed of gums. Formation of gums seems to be stimulated by the presence of microorganisms or by the occurrence of disease or mechanical damage to the cell. Generalizations regarding the composition of these polysaccharides are difficult. They may be composed of a mixture of several sugars or sugar derivatives, including D-galactose, L-arabinose, L-rhamnose, D-mannose, D-xylose, D-glucuronic acid, and others. Gums may swell to many times their original volume in water. Therefore, gums such as gum arabic, gum karaya, gum tragacanth, guar gum, and locust bean gum are used frequently in formulated foods as thickening agents and stabilizers. Gums are discussed in Chapter 6.

e. Pectic Substances.

Pectic substances are probably present in all higher plant tissues, although sometimes the quantity is so small that they are detected only with difficulty. Apples and the albedo of citrus fruits, which contain an abundance of pectic material, furnish the pectin that is purified and sold commercially. The pectic substances are pectic acid, pectinic acid, pectin, and protopectin.

Pectic acid is the simplest of the pectic substances. The molecule is a polyuronide, composed of galacturonic acid units combined by α 1,4 glycosidic linkages. The formulas for the sugar α-D-galactose and galacturonic acid, the uronic acid derived from it, are as follow:

α-D-galactose α-D-galacturonic acid

Pectic acid is soluble in water and contains an abundance of carboxyl (—COOH) groups, making it acidic and capable of forming water-soluble salts. The pectinic acids are similar to pectic acid except that some of the carboxyl groups are esterified with methyl groups (Kritchevsky, 1985, p. 297). Like pectic acid, pectinic acids are capable of reacting with metallic ions to form salts. Indeed, many of the pectic substances exist in plants as calcium or magnesium salts. The term pectin designates those pectinic acids that are capable of forming jelly with sugar and acid. Pectins are esterified to varying degrees, with the remainder of the carboxyl groups present uncombined or combined to form salts.

Pectin

The nature of protopectin is not fully understood. It is a large molecule from which pectin is formed by restricted hydrolysis. It is present in the cell wall but is probably not chemically associated with cellulose, as some early workers believed. In contrast to the other pectic substances, it is insoluble in water.

3. Changes during Ripening

Changes in the textural characteristics of fruits occur during ripening and over-ripening. These changes are related to changes in the proportions of the pectic substances. For example, during normal ripening that occurs in the early storage period as apples are stored at 4°C, soluble pectin increases at the expense of pro-topectin, resulting in decreased hardness. The importance of the proportions of pectic substances to the texture of fruit has been shown in studies on peaches. As tissue firmness decreased with ripening of several varieties of freestone peaches, pectinic acid content increased from an initial concentration of 24–50% to a final concentration of over 70% of the total pectic substances, pectic acid fluctu-ated at levels of less than 10%, and protopectin decreased significantly (Shewfelt et al., 1971). Ripe melting fleshed peaches contain relatively low-molecular-weight pectins (Chang and Smit, 1973). The enzyme polygalacturonase was shown to be active during the ripening period, when firmness of Elberta peaches decreased rapidly. The decrease in firmness was associated with a decrease in the molecular weight of the pectin, resulting from depolymerization of the poly-galacturonide chains (Pressey et al., 1971). Hinton and Pressy (1974) reported that cellulase also may be important in ripening because it also increases in ac-tivity during ripening.

4. Changes during Cooking and Processing

Changes in the proportions of pectic substances during steaming of carrots and parsnips were studied by Simpson and Halliday (1941). The changes resembled those occurring during ripening of fruit in that pectin increased at the expense of protopectin, and total pectic substances decreased, suggesting a degradation of pectin. Hughes and co-workers (1975) reported that potatoes reach a cooked stage when a specific amount of cell wall pectic material has been solubilized. This solubilization decreases adhesion between cells but does not result in breakage of cell walls.

Heat processing of fruits and vegetables is an important method of preserva-tion. Studies have been conducted to determine techniques for minimizing the softening that occurs with heating. As the pectic substances in cell walls break down, softening of the wall and subsequent cell separation occur. The presence of divalent ions increases firmness of canned fruit (Deshpande et al., 1965), canned tomatoes (Hsu et al., 1965), and cooked carrots (Sterling, 1968). Van Buren et al. (1988) showed that soaking green canned beans in calcium chloride increased firmness whereas sodium chloride reduced firmness. Increasing the pH decreased firmness of the beans unless calcium was present. Main et al. (1986) demonstrated that treatment with calcium lactate did not increase firm-ness of frozen strawberry slices. However, heating fruit did increase firmness, suggesting that heating probably reduced bonds on the pectin molecules, mak-ing more sites available for cross-bonding. The divalent calcium ions form cross-links between carboxyl groups of pectinic acid molecules, resulting in in-creased rigidity of the middle lamella and primary cell wall.

The enzyme pectin methyl esterase (PME) may serve an important role in the firming of tissues with divalent ions. Pectin methyl esterase in potatoes is active in the range of 50–70°C and catalyzes the removal of methyl groups from pectin molecules, increasing the number of free carboxyl groups. As plant tissue is heated, intracellular ions such as calcium and magnesium may come in contact and react with cell wall components such as the free carboxyl groups to form bridges that strengthen the tissue so that it may resist degradation during heating (Bartolome and Hoff, 1972). A like role of PME in firming of canned tomatoes (Hsu *et al.*, 1965) and in preheating firming of potatoes (Reeve, 1972) has been described. Another important factor in loss of firmness during heating is the pH of the heating medium. As pH increases from 3 to 8, the firmness of carrots decreases because of increased cell separation (Sterling, 1968). Tomato juice consistency was higher if the pH was reduced to below 3 before heating of the tomatoes. The effectiveness of the added acid was attributed to its inhibitory action on pectic enzymes (Wagner *et al.*, 1969). Davis *et al.* (1976) reported that the effect of heating varies with the type of tissue, such as phloem and xylem. This may be attributable to differences in the amounts and kinds of chemical components of tissues.

B. Types of Cells and Tissues

The preceding discussion was concerned with cells and plant tissue in general. A plant contains several types of cells combined into tissues with specific functions. Extensive discussion of all cells and tissues is beyond the scope of this book. However, some understanding of this topic is essential to comprehension of research literature on plant foods. Therefore, cells and tissues of importance to this field of study are briefly described below.

Cells form tissues that serve several roles in plants. Haard (1985, pp. 872–877) described four types of tissues: protective, parenchyma, supporting, and vascular. Mechanical protection is provided by several types of cells constituting the epidermal layers, peel, or skin. The epidermal layers vary in thickness from thick (watermelon and citrus fruits) to moderately thin (cherries) to very thin (lettuce and strawberries) (Bourne, 1976, p. 283). A cuticle or layer of fatty material sometimes is deposited on the epidermis. The bloom of some fruits such as grapes or blueberries is caused by deposits of wax on the surface of the cuticle. In some plant structures the outer protective layer is the periderm. For example, the outer skin of a potato is peridermic tissue (Hsu and Jacobson, 1974). Parenchyma cells are components of the tissues of several plant structures, including the cortex or tissues located between vascular and dermal tissues, leaves, vascular systems, and secretory or excretory structures. Functions of parenchyma include structure, photosynthesis, storage, and wound healing. Parenchyma cells vary in size from small in compact tissue to larger cells such as those in apples with a spongy texture. In citrus fruits, juice sacs are individual parenchyma cells (Bourne, 1976, pp. 276–279). Collenchyma cells have thicker walls than parenchyma cells and function as support tissue for growing organs as well as in photosynthesis. Sclerenchyma tissue is hard tissue containing sclereids, cells with heavily lignified walls. Stone cells are sclereids and are responsible for the gritty texture of some pears. Xylem and phloem tissues are components of the vascular systems found throughout a plant. Functions of these tissues include conduction of water (xylem), transport of food (phloem), storage of food, and support. Parenchyma and sclereid cells are located within these tissues. Tracheid and vessel members are water-conducting cells, whereas sieve cells and sieve tube members are food-conducting cells.

C. Nutritional Implications of Textural Components

The plant cell wall materials described in this section are important from a nutritional standpoint. It has been hypothesized that a diet rich in foods containing plant cell wall materials is desirable (Slavin, 1987). The term dietary fiber is not consistently defined but is used in reference to cell wall materials, materials that hold cells together, and substances produced by plants in response to injury (Schneeman, 1986). These materials are resistant to the action of digestive enzymes of humans (Hurt and Crocco, 1986). Recent reviews on dietary fiber include information on structure of these materials (Schneeman, 1986) and food sources (Slavin, 1987; Schneeman, 1986). Because of this interest, incorporation of fiber-rich components of fruits and vegetables into food products is of interest to the food scientist (Hurt and Crocco, 1986). For example, Wang and Thomas (1989) studied the inclusion of apple pomace, which is a by-product of juice production, in bakery products. A comprehensive review of all aspects of dietary fiber including food product development is presented by Dreher (1987).

D. Juiciness

An important textural characteristic of plant foods is juiciness, which pertains to the quality of having extractable liquid in cells and tissues. The term succulent may be more frequently applied to vegetables (Szczesniak and Ilker, 1988). Juiciness appears to be a complex sensory characteristic but has not been studied to any extent. Szczesniak and Ilker (1988) recently published results of a comprehensive study of this characteristic. They contend that various types of juiciness can be illustrated by food ranging from melons, which are characterized by release of juice with every bite, to mushrooms, which release moisture as their tight structure is altered by chewing or heating. In this study, sensory panelists ranked raw fruits and vegetables from banana (not at all juicy) to carrot, mushrooms, snap beans, tomato wedge, cucumber, apple, strawberry, honeydew melon, orange, and watermelon (very juicy). Cell size increased in the same order. This tentative scale (Szczesniak and Ilker, 1988) of plant foods of increasing juiciness could be used to train panelists about juiciness in these materials. The degree of juiciness obviously is influenced by the cultivar, growing conditions, storage conditions, and postharvest handling of the product. Further work could be done in this area of study.

II. COLOR _____

Much of the appeal of fresh fruits and vegetables is in their varied colors. The attractive color of the raw fruit or vegetable is, however, subject to change under various conditions associated with use, often resulting in unattractiveness. In this section the nature of the pigments and factors affecting changes in the pigments are considered.

A. Chlorophyll

The green pigment of plants, chlorophyll, is contained in the chloroplasts, which are present in many parts of a plant. Chlorophyll is present in the leaves because their large surface area is ideal for absorption of the sun's rays and the exchange of gases necessary for photosynthesis. Photosynthesis is a process by which CO_2 and water are converted to sugar. In the process, light energy is converted to chemical energy, which makes possible both plant and animal life.

1. Structure

The formula for chlorophyll is shown in Fig. 14.2. The structure is similar to that of heme shown in Fig. 9.1. Both have four pyrrole rings, connected to form a porphyrin nucleus. Resonance of the conjugated double bonds is responsible for the color of chlorophyll. Chlorophyll contains a nonionic magnesium atom in the center of the porphyrin ring. Two ester groups, one phytyl and one methyl, are part of the molecule of chlorophyll *a*. The presence of the phytyl group is

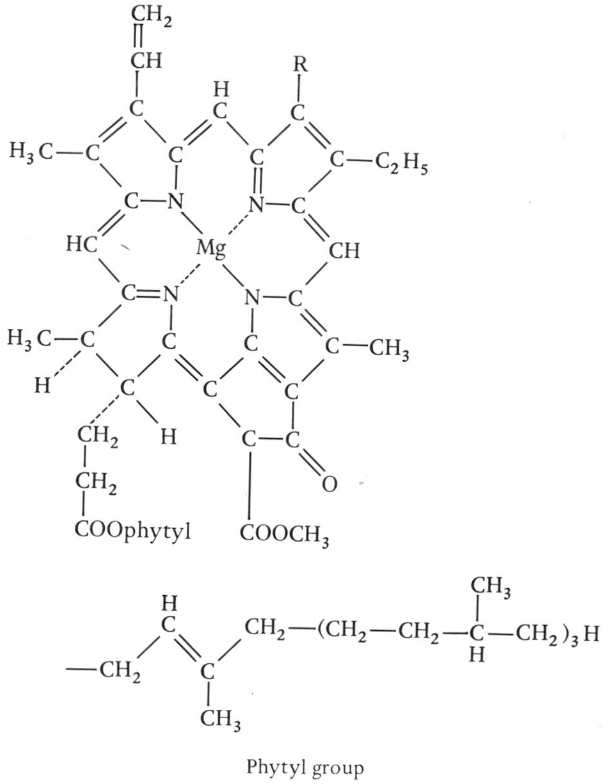

Phytyl group

FIGURE 14.2
Chlorophyll. In chlorophyll *a*, R is —CH₃; in chlorophyll *b*, R is —CHO.

responsible for the insolubility of chlorophyll in water; it cannot be dissolved in cell sap of the plant or in cooking water. Chlorophyll is soluble in fat and solvents such as ethyl ether, ethanol, acetone, chloroform, carbon disulfide, and benzene.

All higher plants and most lower plants contain two types of chlorophyll, *a* and *b*, in the ratio of about three parts of chlorophyll *a* to one part chlorophyll *b*. As indicated in Fig. 14.2, a formyl group (—CHO) is substituted in chlorophyll *b* for a methyl group (—CH₃) of chlorophyll *a*. Chlorophyll *a* is blue–green in color, whereas chlorophyll *b* is yellow–green.

2. Chemical Reactions

Chlorophyll is a complex molecule and can participate in several reactions. The magnesium of chlorophyll is easily replaced by two hydrogen atoms in the presence of mild acids such as oxalic or acetic to produce a compound called pheophytin. Pheophytin *a* is a grayish green and pheophytin *b* is a dull yellowish green. The mixture results in an olive–green to olive–brown color. Loss of the carbomethoxy group from pheophytin results in the formation of pyropheophytin (Schwartz *et al.*, 1981). Schwartz and von Elbe (1983) showed that these degradation products of chlorophyll are present in commercially canned spinach, beans, asparagus, and peas and concluded that they are contributors to the olive–green color of these products. Thus the effects of these changes in the chlorophyll molecule are seen in green vegetables that have been canned or overcooked. If copper is present when acid is released during heating, the metal replaces magnesium, forming a bright green derivative of chlorophyll. Pickles boiled in a copper kettle have a bright green color for this reason but cannot be used because the amount of copper they contain may be toxic.

The derivatives formed by the removal of the phytyl group by the enzyme chlorophyllase or by alkali are water soluble, which accounts for the greenish color sometimes seen in water that has been used for cooking green vegetables. Chlorophyllase activity results in formation of phytyl alcohol and chlorophyllide, a green, water-soluble derivative of chlorophyll. The enzyme occurs in many vegetables at varying levels. Spinach is rich in chlorophyllase, but the amount changes both with the season and with the variety (Weast and Mackinney, 1940). Some vegetables such as snap beans may not contain chlorophyllase (Jones *et al.*, 1963). Chlorophyllase is more resistant to heat than many enzymes. It is quite active in water as hot as 66–75°C, but it is destroyed by boiling (Clydesdale and Francis, 1968). Chlorophyllase also removes the phytyl group from pheophytin to form pheophorbide, which is similar in color to pheophytin. Pheophorbide may be formed during the brining of cucumbers. The enzyme is rapidly inactivated in those vegetables at 100°C and in spinach at 84.5°C (Resende *et al.*, 1969). Treatment of chlorophyll with alkali removes both the phytyl and methyl groups by saponification, thus forming chlorophyllin and two alcohols, phytol and methanol. Chlorophyllin is green and water soluble.

3. Changes during Heating

The changes that occur in green pigments of vegetables can be related to the properties of chlorophyll. In the raw vegetable, chlorophyll is protected from acid in the cell sap by its location in the chloroplasts. When a green vegetable is dropped into boiling water the green color brightens as expulsion of air and collapse of the intercellular spaces occur. During cooking, the chloroplasts become shrunken and clumped in the center of a mass of coagulated protoplasm. The chlorophyll remains in the chloroplasts but is no longer protected by the plastid membranes from the acid-containing cell sap. Consequently, the dull olive–green pheophytin may be formed.

The extent of color change depends on the acidity of the cooking medium, the pH of the vegetable, the chlorophyll content, and the time and temperature of cooking. Organic acids are produced in the plant primarily as intermediates of the Kreb's tricarboxylic acid cycle. Large quantities of acids, predominantly citric and malic, may accumulate in the vacuoles of the cells of fruits or vegetables. Their proportions differ from plant to plant. The pH of all common vegetables is less than 7 because these acids are present in the cell sap.

Both volatile and nonvolatile acids are released during the cooking of vegetables. Therefore, measurements of the pH of a distillate collected as vegetables are cooked and that of the cooking water before and after cooking may be made in vegetable cookery studies (Halliday and Noble, 1943, pp. 11–73). The color of green vegetables is sometimes better if the vegetable is cooked without a cover so that the volatile acids can escape from the pan.

The acids released from the cell vacuole during heating cannot affect the color of vegetables if they are neutralized by the cooking water. The pH of water may be increased by boiling if carbon dioxide (which has dissolved in the water to form carbonic acid) is released, if bicarbonates are changed to carbonates with the release of carbon dioxide, or if hydrogen sulfide is lost. Water that is heavily chlorinated may become more acidic on boiling because of hydrochloric acid. Several of these reactions may take place simultaneously. In food experimentation, the source and treatment of water should be controlled carefully to avoid variation in results due to impurities or variation in pH.

The amount of acid that can be neutralized by an alkaline cooking water depends on the alkalinity and volume of the water. Large amounts of water may result in a desirable green color by neutralization or at least by dilution of plant acids. In a study of the effects of pH, frozen green beans, peas, Brussels sprouts, broccoli, lima beans, and spinach were cooked in buffers of varying pH (Sweeney and Martin, 1961). As the pH increased from 6.2 to 7.0, the green color of the cooked vegetables improved, but buffers of pH greater than 7 caused marked deterioration in flavor and little further improvement in color. Baking soda sometimes is added to increase the alkalinity of cooking water, but its use usually is not recommended because it is difficult to avoid adding an excess. Sodium bicarbonate not neutralized by the acids in the cooking water adversely affects the flavor and texture of the vegetable. The use of other additives has been studied in attempts to retain the color of cooked green vegetables. The addition of a small quantity of a mixture of calcium carbonate and magnesium carbonate to the cooking liquid resulted in an increase in the pH of green beans held on a steam table after heating. The increased pH reduced conversion of chlorophyll to pheophytin and increased preference scores for the beans (Sweeney, 1970). Chlorophyll retention also was improved by the use of ammonium bicarbonate in small volumes of water in the cooking of fresh and frozen green beans, Brussels sprouts, broccoli, kale, and in fresh cabbage. Some of the undesirable textural changes associated with the addition of ammonium bicarbonate were prevented by the addition of calcium acetate (Odland and Eheart, 1974). Color retention is not as difficult in frozen as in fresh vegetables. The blanching process facilitates loss of the acids so that the pH of the vegetable is increased. However, retention of color during blanching is a problem. The use of ammonium bicarbonate improved chlorophyll retention when green vegetables were blanched in acid (Eheart and Odland, 1973b), in steam (Odland and Eheart, 1975), and by microwaves (Eheart and Odland, 1973a).

The susceptibility of green vegetables to color changes during cooking is affected by their chlorophyll content and their pH, as indicated previously. Frozen vegetables containing more chlorophyll initially retained a higher percentage when cooked, and those of comparatively high pH, spinach and peas, retained more pigment than green beans and Brussels sprouts, which had lower pH levels (Sweeney and Martin, 1961).

Color is best if green vegetables are heated through as rapidly as possible and cooked for only a short time. The conversion from chlorophyll to pheophytin is so rapid that from 50 to 75% of the chlorophyll is likely to be lost during the usual cooking period. No chlorophyll remained in snap beans after 1 hr, whereas 62% remained after 10 min and 28% after 20 min (Mackinney and Weast, 1940). Percentages of chlorophyll retained in broccoli cooked for 5, 10, and 20 min were 82.5, 58.9, and 31.3%, respectively (Sweeney and Martin, 1958). Much of this loss is of chlorophyll *a*. Chlorophyll *b* has been found to be relatively stable during cooking (Sweeney and Martin, 1961). The dramatic change of color of green vegetables during the canning process also is a result of conversion of chlorophyll to pheophytin.

B. Carotenoids

Most of the yellow and orange colors and some of the red colors of fruits and vegetables are due to the carotenoids that are located in the chromoplasts of the cells. In addition to their occurrence as the sole or predominating pigments in some fruits and vegetables, they always accompany chlorophyll in the ratio of about three or four parts chlorophyll to one part carotenoid. As fruits ripen, the ratio of chlorophyll to carotene decreases. The carotenoids are insoluble in water and soluble in fats and organic solvents. Members of this large group of related compounds differ in solubility characteristics, offering a convenient method of separating the carotenoids into two groups. The carotenes contain only hydrogen and carbon and are soluble in petroleum ether. Xanthophylls, oxygen-containing carotenoids, are soluble in alcohol.

1. Structure

Most carotenoids contain 40 carbon atoms in configurations such as those for β-carotene, α-carotene, and lycopene shown in Fig. 14.3. The conjugated double bonds are responsible for the intense color of foods containing carotenoids. Yellow color becomes apparent when there are seven conjugated double bonds in the molecule. Although many different cis and trans configurations of the structure are possible, most naturally occurring carotenoids appear to have an

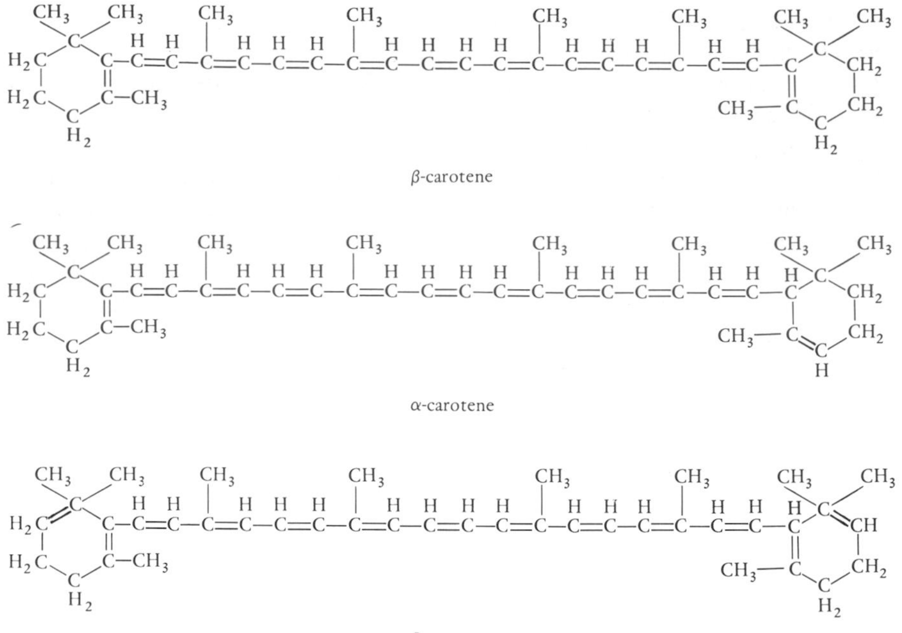

β-carotene

α-carotene

Lycopene

FIGURE 14.3
Structures of β-carotene, α-carotene, and lycopene.

all-trans configuration. As the number of double bonds increases, the hue becomes redder. Lycopene, with two additional double bonds, is redder than β-carotene and is responsible for the color of red tomatoes. A decrease in the number of conjugated double bonds increases yellowness. Consequently, α-carotene is less orange than β-carotene.

Lutein and zeaxanthin are xanthophylls that are similar in structure to α- and β-carotene except for the addition of two hydroxyl groups. Cryptoxanthin contains only one hydroxyl group and is an important pigment in yellow corn, mandarin oranges, and paprika.

In addition to their contribution to the appealing colors of fruits and vegetables, most carotenoids are precursors of vitamin A. The symmetrical β-carotene molecule forms two molecules of vitamin A. Carotenoids such as α- and γ-carotene, in which half of the molecule is exactly like half of the β-carotene molecule, form one molecule of vitamin A. Lycopene has an open ring rather than the closed ring structure that characterizes β-carotene and is therefore not a precursor of vitamin A. Most xanthophylls also are not precursors of vitamin A; the exception is cryptoxanthin, which has only one hydroxyl group.

2. Changes during Preparation, Processing, and Storage

Ordinary cooking methods have little effect on the color or nutritive value of carotenoids. The pigments are little affected by acid, alkali, the volume of water, or the cooking time. Eheart and Gott (1964) reported complete retention of the carotene in peas heated in water in a conventional oven and with or without water in a microwave oven. Although total carotene content does not change as vegetables are heated in water, there is a shift in the visualized color. For example, the orange of carrots may become yellow and the red of tomatoes may become orange-red. Borchgrevink and Charley (1966) attributed this reduction in color intensity in carrots cooked in several ways to an increase in cis isomers of β-carotene during cooking. Carrots that appeared to be orange–red had a higher concentration of all-trans β-carotene than those carrots that appeared yellow. However, DellaMonica and McDowell (1965) reported that color changes cannot be explained in terms of isomerization. Shifts in the color of carrots are caused also by the solution of carotene in cellular lipids after release from disintegrated chromoplasts (Purcell *et al.*, 1969). Crystals of lycopene are formed in tomatoes during heating so that shift in color is not as pronounced in tomatoes as in carrots and sweet potatoes.

Noble (1975) reported that a conversion of *trans*-lycopene to *cis*-lycopene was not responsible for the loss of redness that occurs as tomato paste is concentrated. This loss was attributed to an actual degradation of lycopene during the extended heating process. Dietz and Gould (1986) reported that 40% of β-carotene in tomato juice was lost in processing and during storage. The loss was not significant with respect to nutritional value. Lycopene and α- and β-carotene were lost when exposed to light in a study by Pesek and Warthesen (1987). Photodegradation of the carotenes was more extensive than that of lycopene, resulting in a decrease of yellowness.

The high degree of unsaturation of the carotenoids makes them susceptible to oxidation, with resulting loss of color, after the food containing them has been dried. Loss of or reduction in color is the result of the reaction of peroxides and free radicals with the carotenoids. Goldman *et al.* (1983) demonstrated that oxygen and free radicals promote decolorization of β-carotene in model systems, whereas antioxidants and increasing water activity slowed the process. Thus it is understandable that precooked, dehydrated carrot flakes canned in a nitrogen atmosphere lost little β-carotene, whereas flakes canned in an oxygen atmosphere lost approximately 74% of their β-carotene content (Stephens and McLemore, 1969). Because carotenoids are widely used as food colorants (Francis, 1985, pp. 580–582), the effect of other processing techniques on it are of interest. Extrusion cooking (Chapter 20) that involved mixing β-carotene with starch in a model system and then mixing and heating the paste at 180 $\pm 2°C$ in the extruder and extruding the mixture was studied by Marty and Berset (1988). More degradation occurred during processing in this manner than during heating in sealed tubes in a water bath for 2 hr. The changes during extrusion were attributed to the presence of air during mixing.

3. Uses as Colorants

Plant Extracts containing carotenoids may be used as colorants in food products. Extracts of paprika, alfalfa, *Tagetes*, tomatoes, and carrots and annatto are examples of such materials (Gross, 1987, p. 121). Use of natural food coloring materials was recently reviewed (IFT, 1986b). β-Carotene also is synthesized for use as a food colorant.

C. Anthocyanins

Most of the red, purple, and blue colors of fruits and vegetables are caused by anthocyanin pigments. In some fruits, such as some varieties of cherries, apples, and plums, the anthocyanins occur in the cells of the skin but not in those of the flesh. Other fruits and vegetables containing anthocyanins include raspberries, blueberries, grapes, strawberries, blackberries, peach skins, red potato skins, radishes, eggplant, and red cabbage. Some fruits have only one or two anthocyanins while others may have a complex mixture of 20 or more. Over 200 naturally occurring anthocyanins have been identified (Gross, 1987, p. 76).

1. Structure

The basic structure of these phenolic compounds is composed of the flavan nucleus, C_6—C_3—C_6. Two aromatic rings are joined by an aliphatic three-carbon chain. Hydroxyl (—OH), methoxyl (—OCH_3), or mono- or disaccharides are attached at various points on the skeleton. Most naturally occurring forms are glycosides, meaning that a sugar moiety is present. The combinations of and location of added groups influence the color of the pigment.

Anthocyanins contain hydroxyl groups at positions 3, 5, and 7 and most are glycosides. The sugar moiety usually is attached to the hydroxyl group at position 3. Sugars found in the glycosides are glucose, rhamnose, galactose, arabinose, fructose, and xylose. The sugar portion is responsible for the solubility (Shirkhande, 1976). Substitutions on ring B result in formation of various anthocyanins. The formula for one common anthocyanin, cyanidin, is shown below:

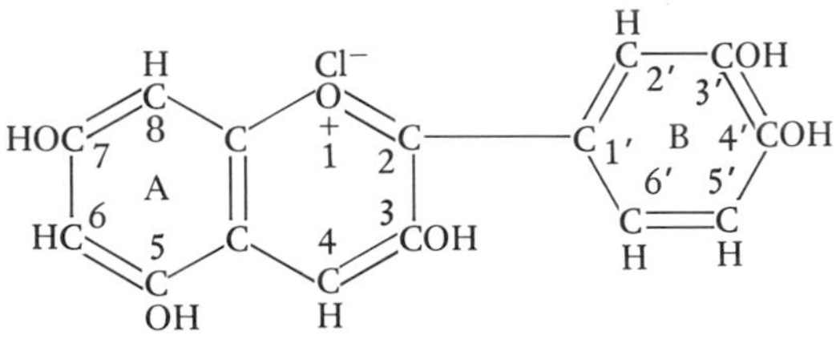

Cyanidin chloride

Derivatives of cyanidin, such as the corresponding glycoside, cyanin, occur in apples, cherries, cranberries, currants, elderberries, purple figs, peaches, plums, raspberries, rhubarb, and purple turnips. Cyanidin has two hydroxyl groups attached to the phenyl ring in positions 3′ and 4′. Pelargonidin, a strawberry pigment, has a hydroxyl group at position 3′ and delphinidin, a pigment found in pomegranate, blueberries, and eggplant, has three hydroxyl groups that are located in the 3′, 4′, and 5′ positions. Other related anthocyanins have methoxyl groups in place of one or more of the hydroxyl groups on the molecule. The color of the anthocyanin varies with variations in molecular structure. As hydroxylation increases, blueness increases. Thus delphinidin is bluer than pelargonidin. As methylation of hydroxyl groups increases, redness increases (Shirkhande, 1976).

Color of the anthocyanins also is influenced by the presence of other phenolic compounds. Flavonoids form complexes with the glycosides of commonly occurring anthocyanins at pH levels ranging from 2 to 5. This interaction increases the blueness of the anthocyanins (Asen *et al.*, 1972). Carotenoids or chlorophyll may be found in tissue containing anthocyanins.

2. Changes during Processing and Storage

Anthocyanins are electron deficient. Therefore, they are very reactive and may undergo detrimental changes during processing and subsequent storage. Color varies with the pH. The molecule assumes the configuration of a cation in acid and is red in color. The color is most intense at very low pH values. Lowering the pH of anthocyanin solutions also decreases the lightness as measured by a color difference meter (Van Buren *et al.*, 1974). The molecule is uncharged at a neutral pH and the pigment fades to colorless; above a pH of 7 the pigment is violet in color. In a mild alkaline medium, the anionic form exists and results in a blue color. Yang and Yang (1987) found that lowbush blueberry puree became darker and more bluish purple as the pH shifted from low to high. The juices of some fruits and vegetables become greenish as alkali is added. This color probably is caused by the presence of the flavones or flavonols with the anthocyanins. With the addition of alkali, the flavones or flavonols turn yellow while the anthocyanins turn blue, and the mixture of the two colors appears green. Such a change can be seen in red cabbage.

Anthocyanins may be degraded by oxidation, hydrolysis, or polymerization (Francis, 1975). Factors influencing degradation include temperature, pH, other cell constituents, enzymes, and the presence of metals. Loss of pigment may be a serious problem in strawberry products such as preserves and grape juice. Furthermore, loss may be rapid. For example, in strawberry preserves, the half-life of red pigment is about 8 wk at 20°C (Francis, 1975). At higher temperatures, the stability is reduced. Therefore, consideration of storage temperature is important. Destruction of the pigment increases as the pH increases, as shown in frozen strawberries (Wrolstad *et al.*, 1970) and Concord grape juice (Skalski and Sistrunk, 1973).

Hydrolysis of the 3-glycosidic group may be caused by an enzyme. The anthocyanidin is unstable to oxidative degradation. Phenolases may be indirectly involved through enzymatic oxidation of catechol to a quinone that oxidizes the anthocyanin. The products of ascorbic acid oxidation, as well as sulfur dioxide, and furfural and hydroxymethylfurfural, products of sugar degradation, exert detrimental effects on the anthocyanins (Shirkhande, 1976).

Metal salts may have a stabilizing effect on color of products containing anthocyanins. The adjacent hydroxyl groups attached to the B ring of cyanidin and delphinidin may form complexes with metals that are very stable at a high pH. The complexes formed may be green, slate, or blue in color. Aluminum chloride complexes with cyanidin-3-glucoside at a pH of 3.0−3.5 to shift color from red to blue−violet (Asen *et al.*, 1969).

Enamel-lined cans are used for foods containing anthocyanins. If an unlined can or a can with an imperfection in the enamel lining is used, the acid of the fruit reacts with the metal of the can to form a salt. The anthocyanin then combines with a metal ion, thus releasing the acid to continue its attack on the can. After the tin has been removed in this way, underlying iron enters into a similar reaction. A pinhole is formed that may make microbial invasion of the contents possible.

Because of their acidity, fruits containing anthocyanins usually do not suffer undesirable color changes in cooking. Red cabbage is the anthocyanin-containing vegetable most often cooked. When steamed or boiled in tap water, red cabbage turns a bluish color that generally is considered undesirable. The color can be changed to an attractive reddish color by the addition during cooking of acid in some form, such as vinegar or apples.

D. Flavonoids

The yellow flavonoids are similar in structure to the anthocyanins and include flavones, flavonols, and flavonones. Skeletal structures for the three are shown in Fig. 14.4. The sugar moiety of the glycosidic form usually is attached on the A ring at position 7, as shown in the structure diagram.

Like the anthocyanins, flavonoids are water soluble and occur in the vacuoles of plant cells. They may occur alone in light-colored vegetables such as potatoes

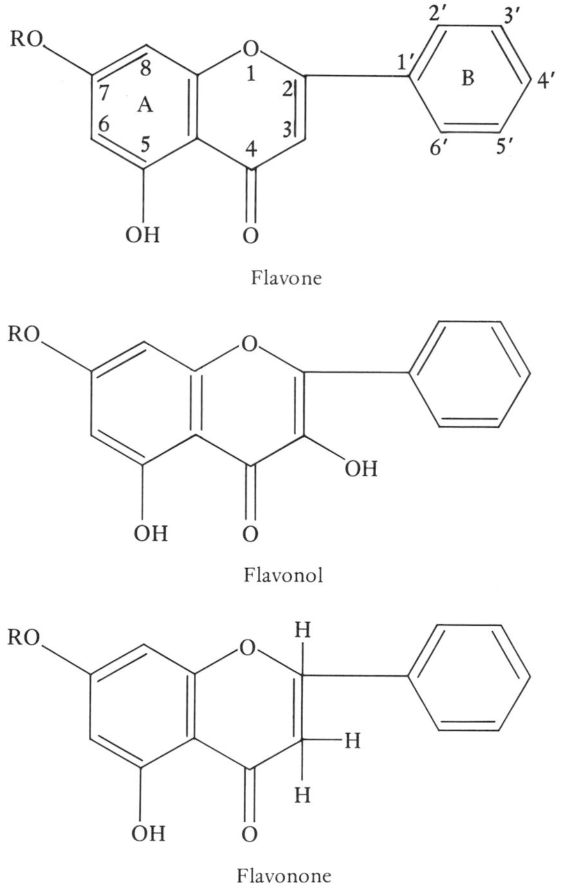

Flavone

Flavonol

Flavonone

FIGURE 14.4
Structure of flavone, flavonol, and flavonone.

and yellow-skinned onions or with other pigments such as anthocyanins. They are so widely distributed that it is exceptional to find a plant in which these pigments are not present. They frequently occur in complex mixtures. The formula for the flavonoid quercetin, which is probably the most widely distributed pigment of this group, is shown below:

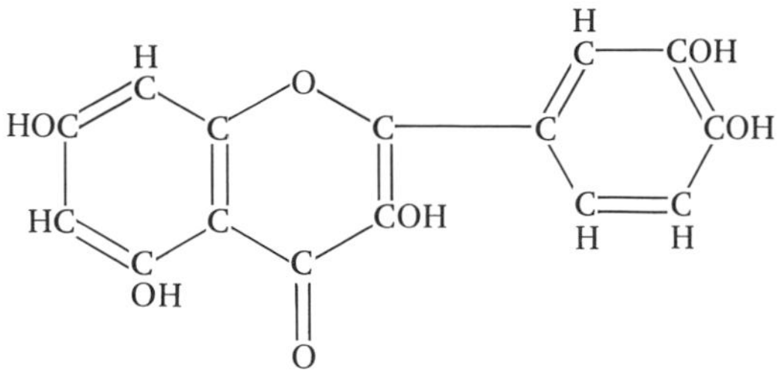

Quercetin

Comparison of this formula with that of cyanidin shows that the hydroxyl groups are similarly located but that quercetin contains more oxygen than the related anthocyanidin. With its derivatives, quercetin occurs in asparagus, yellow-skinned onions, grapes, apples, the rind of citrus fruits, and tea. The compounds turn yellow in the presence of alkali. They also have the ability to chelate metals, which may result in discoloration. Iron salts cause a brownish discoloration. Traces of iron salts in alkaline tap water may react with flavonoids and other related phenolic compounds to produce some of the yellow-to-brown discoloration often seen in cooked white vegetables. Yellow-skinned onions cooked in alkaline water are especially likely to show such a change. Similar results are seen if an aluminum pan is used. The color of white vegetables often can be improved by adding an acidic compound such as cream of tartar to the cooking water, but at the expense of some firming of the tissues.

Blackening of potatoes after cooking is attributed to formation of a dark-colored complex between iron and chlorogenic acid in potatoes with low organic acid content. Organic acids such as citric acid chelate metals so that they are not available to react with chlorogenic acid (Heisler *et al.*, 1964). Cauliflower may become discolored because flavonol glycosides complex with ferrous or stannous ions. Asparagus contains a flavonol, rutin, which precipitates as yellow crystals from the liquid of asparagus canned in glass containers. This precipitate is not found in asparagus canned in tin containers because rutin forms a light yellow complex with stannous ions. The complex is more soluble than rutin so precipitation does not occur (Dame *et al.*, 1959). Sweet potatoes form yellow complexes with tin and dark-greenish complexes with iron. Use of tin cans rather than enamel-lined cans with sweet potatoes provides tin to compete with the iron for complex formation and thus prevents darkening (Scott *et al.*, 1974; Twigg *et al.*, 1974).

E. Proanthocyanidins

Another group of flavonoid compounds important to the color of some canned foods is composed of proanthocyanidins, colorless phenolic compounds previously referred to as leucoanthocyanins. These compounds are converted to anthocyanins when subjected to boiling hydrochloric acid. Some products may become pink or red as a result of this conversion. These compounds also contribute to the astringent taste of some foods and may serve as a substrate for enzymatic browning (Francis, 1985, pp. 565–566).

F. Betalains

The betalains are responsible for the color of the red beet (Francis, 1985, pp. 568–569). This group includes the betacyanin or red pigments and the betaxanthin or yellow pigments. The major red pigment is betanin. Its structure is shown below:

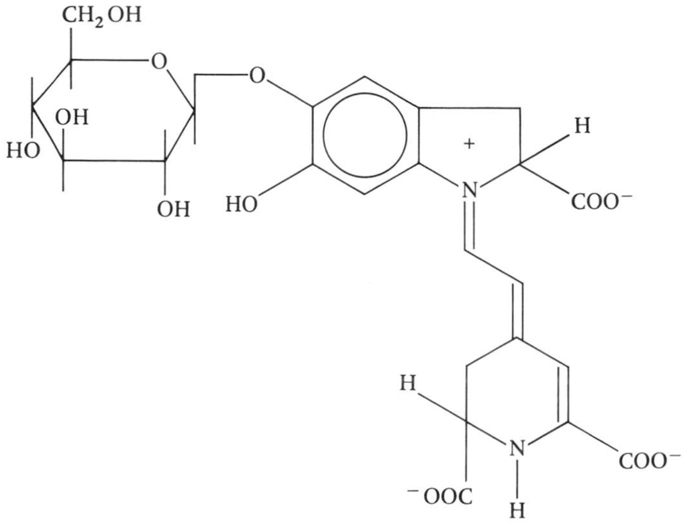

Betanin

The color of betanin depends on pH. The red color is associated with a pH of 4 to 7. Below a pH of 4, the color shifts to violet and above pH 10, the color shifts to yellow (von Elbe, 1975). Loss of betacyanin pigments occurs during heat processing of red beets. However, regeneration of the pigment occurs after processing (von Elbe *et al.*, 1981). With the limited selection of red food colorants, the betalains have been the subject of much study as potential coloring agents (von Elbe *et al.*, 1974; Pasch *et al.*, 1975; von Elbe, 1975, 1986, pp. 55–56). Water is required for degradation to occur; therefore, at low water activities stability increases.

G. Enzymatic Browning

Discoloration is of concern when fresh fruits and vegetables are prepared for consumption as well as to the processor who must maintain quality of the product prior to canning, drying, or freezing processes. Phenolic compounds other than those previously discussed may be important in the color of fruits and vegetables, not because of their contribution to color but because of their role in discoloration. These compounds serve as the substrates for enzymes of the oxidoreductase group. Of particular interest in fruits and vegetables is *o*-diphenol:oxygen oxidoreductase, more commonly known as polyphenol oxidase or polyphenolase. Substrates for this enzyme are many and include diphenolic compounds such as catechin, shown below:

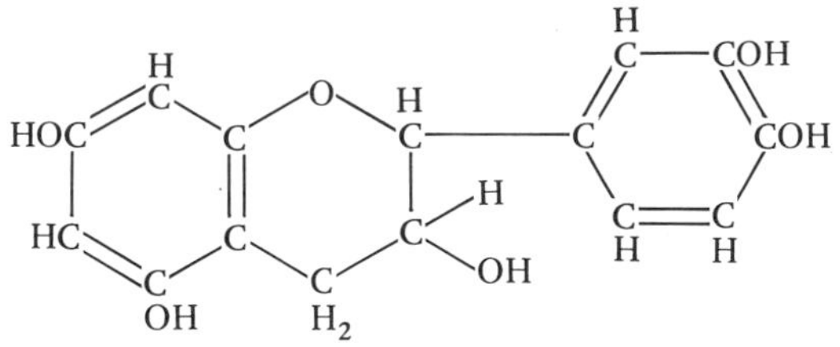

Catechin

Others include leucoanthocyanins, chlorogenic acid, caffeic acid, dicatechol, tyrosine, dihydroxyphenylalanine, and many others (Mathew and Parpia, 1971). Mushrooms, potatoes, apples, peaches, bananas, and avocados contain high concentrations of this enzyme.

1. Reaction

Several reactions are involved in formation of the yellow and dark-colored compounds associated with enzymatic browning, which occurs rapidly when the fruit or vegetable is cut or otherwise injured. In the first reaction in the series, an *o*-dihydroxyphenolic compound is oxidized in the presence of polyphenoloxidase and atmospheric oxygen to form an *o*-quinone and water.

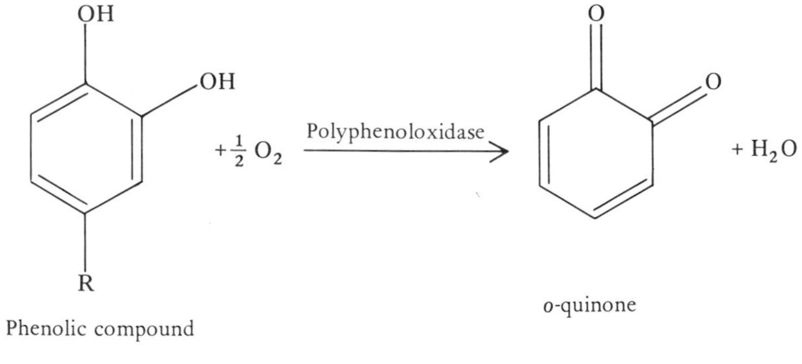

Reaction for first step in browning

The *o*-quinones are polymerized or they may complex with amino acids or proteins (Mathew and Parpia, 1971). In either nonenzymatic reaction, brown pigments are formed. Reactions with amino acids or proteins can reduce nutritive value. Reaction with casein reduces availability of lysine and digestibility of the protein (Whistler and Daniel, 1985, p. 99).

2. Control

Three components, the substrate, the enzyme, and oxygen, are necessary for the initial reaction of browning. Obviously, attention to control of the action of the enzyme and the presence of oxygen are more practical than control of the substrate. Polyphenoloxidase is denatured and therefore inactivated by a heat treatment such as blanching. However, blanching will give a cooked flavor and soft texture that may not be desirable in fruits that are to be frozen or potatoes that will be used to make potato chips. Thus fruits that will be eaten without further preparation after freezing and thawing are not blanched but fruits to be heated as for pies may be blanched before freezing.

If heating is not desirable, other conditions unfavorable to activity of the enzyme can be promoted. Sodium chloride and organic acids inhibit the enzyme system. If susceptible foods are soaked in sodium chloride solutions after peeling and slicing, enzyme activity may be inhibited. Acids such as citric, malic, or tartaric may be effective because phenyloxidase activity is greatest at a pH of 7 and decreases as the pH is lowered below 5. Level of usage is limited by sensory acceptability. For fresh-sliced potatoes, Langdon (1987) reported that dipping in baths containing citric acid, ascorbic acid, and potassium sorbate prior to packaging in vacuum packages extended shelf life to 14 days. Sulfur dioxide and bisulfites also will inhibit the enzyme system. However, their use has been prohibited in fresh fruits and vegetables, including those used on salad bars, because of the possible dangers to individuals who are sensitive to sulfites. Safety of sulfites was recently reviewed (IFT, 1986a; Taylor et al., 1986).

Exclusion of oxygen is another approach to prevention of browning. Holding vegetables in water or a salt solution after slicing and holding fruits in syrup will help to prevent browning by limiting oxygen. The desirability of reducing salt and caloric contents limits the appropriateness of using salt and sugar solutions.

Oxidation to quinones, if the previous measures are not successful, may be reversed, thus preventing polymerization to dark-colored compounds. This possibly is another facet of the protective role of sulfite. Ascorbic acid also acts as a reducing agent in prevention of browning. Erythorbic acid, an isomer of ascorbic acid, also may be effective in some systems (Sapers and Ziolkowski, 1987). Its effectiveness persists until the supply is exhausted. The use of these two chemicals on apples prior to refrigeration and freezing and on peaches prior to canning has been effective (Ponting and Jackson, 1972; Ponting et al., 1972; Luh and Phitakpol, 1972). Further discussion of these techniques is found in Chapter 13.

III. FLAVOR

The characteristic flavor of a fruit or vegetable is attributable to a mixture of many compounds, some of which are present in very small amounts. The volatile flavoring components are important; among them are organic acids, aldehydes, alcohols, and esters. Variations in ratios among the individual components of these groups are responsible for differences in flavor among various fruits and vegetables. Flavor development in fruits occurs primarily during ripening whereas flavor development in vegetables is only partial during maturation. In most vegetables flavor development occurs as cells are disrupted because enzymes and substrates are mixed.

A. Phenolic Compounds

Phenolic compounds such as catechin and leucoanthocyanins are responsible for the astringent or "puckery" taste of some foods. Whereas a slightly astringent taste is desirable in tea, coffee, cocoa, and certain fruits, an undesirable degree of astringency can be observed in certain underripe fruits such as bananas and persimmons. Time and temperature of heating during preparation of juice extracts from red currant and red raspberry fruits influence the phenolic content of the juice. As tannins (polyphenolic and phenolic compounds) increase, bitterness and astringency scores also increase (Watson, 1973). Another flavonoid compound of significance is naringin, the 7-rhamnoglucoside of naringenin, a flavone. Naringin is responsible for the bitterness of grapefruit.

B. Sugars

The amount of sugar present in a fruit or vegetable affects the perceived sweetness of that food. For example, some lemons and limes contain as little as 1% sugar whereas some oranges contain as much as 14%. Sugars present include sucrose, glucose, and fructose (Cook, 1983). Sugars are responsible in part for the superior flavor of freshly harvested vegetables. Sugars, including glucose and fructose, increase in potatoes stored at refrigerator temperatures (below 5°C) (Haard, 1985, pp. 886–888). Metabolic processes of the plant are responsible for changes in concentration of sugar both during ripening and after harvest. For example, as bananas mature, starch is hydrolyzed to sugar by the action of amylase (Mao and Kinsella, 1981). Quantitative and qualitative variations in sugar contribute to variation in taste of different fruits. During storage, sugar content may increase or decrease depending on the fruit or vegetable. Cooking may increase sugar content, as shown by Simon and Lindsay (1983) for carrots. However, canning and freezing reduced the sugar content. Sweetness was shown to be a major factor in preference determination for cooked carrots.

C. Acids

All fruits and vegetables are acidic, a factor of undoubted importance to their taste. Plants vary in the kind of acid that they contain and in the amount as reflected by pH. Acidity varies with the maturity of the plant, frequently decreasing as fruit ripens. Plants contain various organic acids, some of which are important in the intermediary metabolism of both plants and animals. Individual fruits and vegetables usually contain several acids, with one or more predominating.

Malic acid is most abundant in apples, pears, peaches, lettuce, cauliflower, green beans, and broccoli, whereas citric acid is most abundant in citrus fruits, tomatoes, leafy vegetables, strawberries, and cranberries. Many other acids, including acetic, butyric, lactic, pyruvic, fumaric, oxalic, and benzoic, are found in some fruits and vegetables at low levels.

D. Sulfur Compounds

Sulfur compounds are important contributors to the flavor of vegetables but not fruits. Flavor of two groups of vegetables—the *Allium* genus of the onion family and the *Brassica* genus of the Cruciferae family—is dependent on sulfur compounds. The typical odor of garlic is due to allicin, which is produced by the action of allinase, an enzyme of garlic, on *S*-allylcysteine sulfoxide (alliin), as shown below:

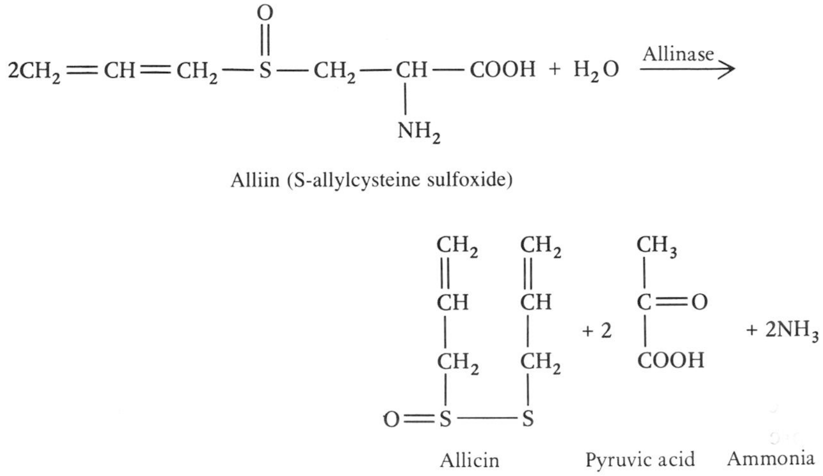

$$2CH_2 = CH = CH_2 - \overset{\overset{\displaystyle O}{\|}}{S} - CH_2 - \underset{\underset{\displaystyle NH_2}{|}}{CH} - COOH + H_2O \xrightarrow{\text{Allinase}}$$

Alliin (S-allylcysteine sulfoxide)

$$
\begin{array}{ccc}
CH_2 & CH_2 & CH_3 \\
\| & \| & | \\
CH & CH & C=O \\
| & | & | \\
CH_2 & CH_2 & COOH \\
| & | & \\
O=S\!\!-\!\!-\!\!-S & &
\end{array}
\quad + 2 \qquad\qquad\qquad + 2NH_3
$$

Allicin   Pyruvic acid   Ammonia

Reaction of alliin to form allicin, pyruvic acid, and ammonia

The odor of diallyl thiosulfinate is typical of garlic but is not unpleasant. It is an intermediate in the formation of the more volatile, unpleasant-smelling diallyl disulfide (CH= CH— CH$_2$— S—S— CH$_2$— CH= CH$_2$), garlic oil. Brodnitz and co-workers (1971) used gas chromotography to separate the volatiles of garlic oil. Mono-, di-, and trisulfides accounted for 90% of the peak areas.

Schwimmer and Weston (1961) described the production of volatile sulfur compounds from *S*-methyl-L-cysteine sulfoxide and *S*-propyl-L-cysteine sulfoxide by the action of an allinase-type enzyme. Hydrogen sulfide and *n*-propyl mercaptan (CH$_3$— CH$_2$— CH$_2$—SH) are produced from raw onions as they are crushed (Niegisch and Stahl, 1956). Onions become increasingly mild as they are boiled because of solution, vaporization, and further degradation of the sulfur compounds. The lachrymatory factor of onions has been identified by Spåre and Virtanen (1963). Propenylsulphenic acid is produced from (+)-*S*-(prop-1-enyl)-L-cysteine sulfoxide.

The characteristic flavor of raw cabbage is due to allyl isothiocyanate formed by the hydrolysis of sinigrin, a glucoside, and several related compounds (Bailey *et al.*, 1961). Sinigrin is split by the enzyme myrosinase as shown in the equation below:

$$CH_2 {=} CH {-} CH_2 {-} N {=} C \begin{matrix} {}^{OSO_3\,K} \\ {}_{S {-} C_6 H_{11} O_5} \end{matrix} \quad + \quad H_2 O \xrightarrow{\text{Myrosinase}}$$

Sinigrin

$$CH_2 {=} CH {-} CH_2 {-} N {-} C {-} S \; + C_6 H_{12}\, O_6 \; + KHSO_4$$

Allyl isothiocyanate   Glucose   Potassium
("mustard oil")   acid sulfate

Reaction of sinigrin and water

When dehydrated cabbage is reconstituted with water for slaw, it does not have the flavor of fresh raw cabbage even though it contains sinigrin. If myrosinase is added during dehydration, the flavor of fresh cabbage is restored, indicating that the precursor of the flavor component is not changed by processing, but the enzyme myrosinase is inactivated (Bailey *et al.*, 1961). A study by Schwimmer (1963) involving the use of enzymes from various sources suggested that the products produced from a common precursor differ to give the characteristic flavors. The use of an onion enzyme preparation with dehydrated cabbage gives a flavor similar to that of onions. Dimethyl disulfide, the main constituent responsible for the odor of cooked cabbage, is formed from the sulfur-containing amino acid, S-methyl-L-cysteine sulfoxide. The presence of this precursor has been shown in cabbage, turnip, cauliflower, kale, and white mustard but not in watercress or radish (Synge and Wood, 1956). The amino acid is the precursor of dimethyl disulfide, a volatile sulfur compound (Dateo *et al.*, 1957). Hydrogen sulfide also is formed when cabbage is cooked. These two volatile sulfur compounds were obtained from fresh, dehydrated, and red cabbage as well as from sauerkraut, cauliflower, broccoli (Dateo *et al.*,, 1957), and rutabaga (Hing and Weckel, 1964). When members of the cabbage family are overcooked, decomposition of the sulfur compounds continues, as indicated by the finding in one of the earliest studies on sulfur-containing vegetables that the amount of hydrogen sulfide evolved from cabbage and cauliflower increases with the cooking period (Simpson and Halliday, 1928). Maruyamu (1970) suggested that with prolonged cooking, gradual loss of pleasant volatile components occurs, allowing less pleasant sulfur components to predominate. MacLeod and MacLeod (1970) reported that younger, inner leaves of cabbage provide greater amounts of sulfur compounds than the outer, older leaves. The stronger flavor of the inner leaves is attributable to this fact. The change in taste that occurs on cooking cabbage is to be contrasted to the decreased intensity of taste that occurs on cooking onions. It furnishes additional evidence of differences in the sulfur compounds involved. Lea (1963) suggested that flavors in fresh, uncooked foods are developed because of enzymatic reactions, whereas nonenzymatic reactions are responsible for flavors in cooked foods.

IV. METHODS OF COOKING VEGETABLES

Of the many methods of vegetable cookery in common use, perhaps the most common is boiling in water. If the amount of water is less than enough to cover the vegetable, the saucepan is covered during cooking to retain steam that cooks the vegetable, but if enough water is used to cover the vegetable after it has wilted in the boiling water, it is possible to cook either with or without a cover. In "waterless" cooking, only the water clinging to the vegetable after washing or a minimum amount necessary to prevent scorching is used. A heavy pan with a tight-fitting lid usually is recommended for this method. Panning and stir frying are methods in which the finely cut vegetable is cooked with a small amount of fat in a covered pan and uncovered pan, respectively. Steaming over boiling water and baking also are used for cooking vegetables. The pressure saucepan makes it possible to cook vegetables rapidly in a small amount of water, as does microwave heating. For microwave heating of frozen vegetables, addition of water often is not necessary.

Studies on the influence of cooking method on the palatability factors—flavor, color, and texture—in general indicate the advisability of cooking vegetables in a small amount of water, which amounts to about half the weight of the vegetable, or 125–250 ml for four servings. Charles and Van Duyne (1954) reported that vegetables cooked in this proportion of water scored higher in appearance, color, and flavor than those cooked in the minimum amount of water required to prevent scorching. In a comparison of "waterless" cooking, boiling, and pressure cooking on broccoli, cabbage, cauliflower, turnips, and rutabagas, Gordon and Noble (1964) reported that flavor and color were best and ascorbic acid retention poorest with the boiling water method, whereas the pressure saucepan retained more ascorbic acid in all vegetables except turnips but with less desirable color and flavor.

When vegetables were cooked in an open pan with enough water to cover the vegetable at all times, the color of green vegetables was greener and the flavor of vegetables of the cabbage family milder than when they were cooked in steam as in a pressure saucepan, steamer, or tightly covered pan (Gordon and Noble, 1959, 1960). Whether a milder or stronger flavor is preferred is, of course, a personal matter.

Fresh broccoli boiled in a moderate amount of water (300 ml for 454 g of vegetable) was at least as satisfactory as that cooked in larger amounts of water or by methods using less water, such as steaming or pressure cooking (Sweeney et al., 1959; Gilpin et al., 1959). Results from another laboratory indicated that broccoli, Brussels sprouts, lima beans, peas, and soybeans were more palatable when cooked in a pressure saucepan, but asparagus, cauliflower, and green beans were preferred when cooked in a tightly covered pan (Van Duyne et al., 1951). Issanchou and Sauvageot (1987) compared cooking in water and steaming with and without pressure on the quality of peas and carrots. Steaming resulted in higher intensity of odor and flavor notes.

Microwave heating of vegetables also has been studied in relation to the quality of cooked vegetables. In an early study, Chapman and co-workers (1960) reported that electronically cooked fresh broccoli is slightly better than broccoli cooked by conventional boiling. Stone and Young (1985) reported that frozen green beans cooked in a microwave were greener and firmer and had stronger off-flavors than beans cooked covered in boiling water (120 ml/454 g beans).

Potatoes are a very popular item, served in several forms. Potato texture has been the subject of many studies. Panelists in a study by Leung *et al.* (1983) decided that mealy and gummy are the two most appropriate terms to describe potato texture. Mealiness is the tendency to disintegrate into small parts and gumminess is related to the cohesive nature of the potato. Microwave ovens may be used frequently for baking potatoes. Brittin and Trevino (1980) compared potatoes baked in a conventional oven and a microwave oven. A trained panel indicated that the potatoes cooked in the conventional oven were whiter, more mealy and drier, and milder in flavor (had less off-flavor) than those cooked in the microwave oven. However, a consumer panel showed no preference for potatoes cooked by either of the methods.

Sweet potatoes also have been baked in the microwave and compared to potatoes cooked by other methods (Lanier and Sistrunk, 1979). Carotenoid content was not affected by the methods used. Some differences were found in vitamin contents among the potatoes that were steamed, boiled, baked, canned, and heated in a microwave. The magnitude of the differences was probably not significant from a nutritional standpoint except for the canned potatoes, which contained less ascorbic acid, pantothenic acid, and niacin than the others. In general, the flavor and color of the microwave-heated and baked potatoes were liked more than those of the others, while the smoothness and moistness of the microwave-heated potatoes were liked less than those of the others. Losh *et al.* (1981) examined the effects of varying baking times and temperatures on the quality of baked sweet potatoes. Four oven temperatures (150, 180, 200, and 230°C) were used to bake two varieties of sweet potatoes. Mean sensory scores showed that potatoes baked at the three higher temperatures were equal in sensory quality, suggesting that any of the three could be used for the baking of sweet potatoes.

Nutritive value of vegetables also varies with cooking method. Bowman and co-workers (1975) found that microwave heating did not significantly affect the ascorbic acid content of 13 of 16 vegetables when compared to boiling. Microwave heating of frozen peas can result in greater retention of ascorbic acid than cooking conventionally in water. Peas cooked in the microwave without added water retained more ascorbic acid than did those with water added (Mabesa and Baldwin, 1979). Recently, Hudson *et al.* (1985) compared the effects of steaming, microwave heating, and boiling of frozen and fresh broccoli on ascorbic acid, thiamine, and riboflavin content. In general, steaming was least detrimental and boiling most detrimental to vitamin content, although microwave heating and steaming retained equal amounts of ascorbic acid and thiamine in frozen samples.

Effects of cook/chill systems on thiamine and ascorbic acid retention in peas and potatoes were studied by Dahl-Sawyer *et al.* (1982). They reported that conduction, convection, and microwave reheating did not differ with respect to loss of nutrients. They also reported that the greatest loss of nutrients occurred in peas with reheating and in potatoes with 24-hr chilled storage. Food service systems of food preparation may involve longer periods of hold than methods used for home preparation. The effect of a food service system on vitamin C content of 17 vegetables was studied by Carlson and Tabacchi (1988). As the vegetables passed through the steps of the system (preparation, cooking, holding, serving), ascorbic acid was lost at each step. Products purchased fresh had higher final contents than did products purchased frozen. Further information on the influence of cooking in home and food service settings may be found in reviews by Lachance and Fisher (1988, pp. 505–556) and Adams and Erdman (1988, pp. 557–605).

V. LEGUMES

Legumes are important foods in many parts of the world, providing protein, vitamins, and minerals. Many varieties of dry seed legumes, including many that have been developed for improved nutritional quality, are available for inclusion in diets around the world. Koehler *et al.* (1987) studied 36 cultivars and reported data on nutritional composition, protein quality, and sensory properties. Most of the varieties were rated above 5 on a 10-point scale for overall quality in which 10 represented excellent. In addition, flours and more refined extracts of some legumes provide functional properties in many food products. Soybeans are the most widely produced (Hartwig, 1989, p. 1). Other important varieties include peanuts, dry beans, dry peas, lentils, and chickpeas or garbanzo beans. These legumes and products derived from some of them are important in several types of products. First of all, they are important as a product consumed as harvested and processed, primarily dried. Second, they are sources of oils (see Chapter 15) and other products, such as soy sauce, peanut butter, and soy protein products. The latter include soy flours, extracts, and isolates used as ingredients in many products, soy milk, and textured soy products (see Chapter 9). An additional product produced from soy milk is tofu, which has been gaining popularity. Discussion in this chapter includes discussion of the cooking of dried legumes and production and use of tofu.

A. Processing and Cooking

The method used to cook the many varieties of dried legumes used for food must ensure softening of the seed coat and the cotyledons. Preliminary soaking shortens the cooking period. Dawson and co-workers (1952) reported that the soaking time could be reduced from overnight to 1 hr by boiling for 2 min prior to soaking. Deshpande and Cheryan (1986) reported that initial absorption is related to structure of the hilum and micropyle, which are illustrated in Fig.

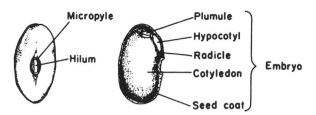

FIGURE 14.5
Structure of bean (*Phaseolus vulgaris*) seeds. *Left:* whole bean seed. *Right:* bean seed with one of the cotyledons removed. (From Hsu and Jacobson, 1974; *Home Economics Research Journal*, American Home Economics Association, Washington, DC. Reprinted with permission.)

14.5. Seed coat becomes involved after some time and increased thickness results in slower absorption of water (Deshpande and Cheryan, 1986). Time required to cook dried beans increases with storage at high temperatures (Burr *et al.*, 1968). This "hard-to-cook" phenomenon, or prolonged time needed to cook some legumes, has been the subject of many studies. Sievwright and Shipe (1986) found that storage of black beans at low temperatures and humidity (5°C, 50% relative humidity) resulted in little change in hardness and digestibility. However, other conditions of storage (30 or 40°C, 80% relative humidity) increased hardness and decreased digestibility. Soaking in a salt solution prior to cooking eliminated the effects. Similar effects of storage on textural parameters were reported for red kidney beans (Garruti and Bourne, 1985). The increase in hardness was related to a decrease in phytic acid. Some varieties of beans contain a higher than normal amount of potassium phytate, a water-soluble salt of phytic acid (Crean and Haisman, 1963). Calcium from the middle lamella complexes with phytic acid, making the calcium unavailable for the formation of insoluble salts with pectic substances. This reaction, therefore, aids in tenderization during cooking because pectin solubility contributes to cell separation during cooking (Jones and Boulter, 1983). The proportion of calcium to phytic acid is important to quality of peas (Rosenbaum *et al.*, 1966). There must be enough phytic acid to complex with calcium in the pea as well as any that might be available in the cooking water.

B. Tofu

Tofu, a bland, gelatinous, soybean curd, is a product that originated in East Asia and is prepared by adding a coagulant to soy milk to precipitate the proteins. The milk is prepared from soybeans that are soaked, pureed, and cooked (Haytowitz and Matthews, 1989, pp. 231–233). Following coagulation, the curd is pressed to remove the whey. Several coagulants may be used and influence the sensory properties of the products as well as the nutrient value. Coagulants include nigari, calcium sulfate, calcium chloride, glucono-δ-lactone (GDL), lemon juice, vinegar, and sea water. The traditional coagulant is nigari, which is derived from sea water and contains magnesium sulfate and other minerals. Calcium sulfate adds calcium and lemon juice adds ascorbic acid. Quality and nutrient content also depend on how much water is removed from the tofu. Hardness increases as more pressure is applied and water is removed (Wang *et al.*, 1983; Gandhi and Bourne, 1988) and nutrients become more concentrated (Haytowitz and Matthews, 1989, pp. 231–233). Hardness also is dependent on the coagulant used. Saio (1979) showed that tofu prepared with $CaSO_2$ was more cohesive and less fragile than that made with GDL. Wang (1984) presented a comprehensive review of factors affecting the quality of tofu.

Escueta *et al.* (1986) investigated the effects of boiling soymilk before precipitation on tofu. Heating should unfold globular proteins, facilitating curd formation. Although boiling for 30 and 60 min increased yields and altered sensory properties, acceptability did not change, suggesting that boiling for the extended period of time did not offer any advantage. The investigators indicated that tofu could be made without boiling the milk but that boiling for 12 min is appropriate to destroy the trypsin inhibitor and other antinutritional factors present in the soymilk.

It has been proposed that tofu be substituted for a portion of the egg or used as a fortifying agent to increase the nutrient content of baked products (Harrison and Konishi, 1982; Marson, 1984). Harrison and Konishi (1982) developed a tofu-with-fiber product and found that it could be incorporated into a number of products that were rated as acceptable by a group of elderly consumers. Products included cheese puffs, onion dip, casseroles, gelatin salads, potato salad, tuna salad, cakes, cookies, and fudge. Improvement of nutritional value of the products was the objective in the study. Because of its increasing use, several workers have investigated the use of other legumes (winged beans and field peas) for production of bean curds (Gebre-Egziabher and Sumner, 1983; Kantha *et al.*, 1983). The use of a mixture of legumes seems desirable.

Suggested Exercises

1. Study the characteristics of vegetables cooked by several methods. Four vegetables that represent different pigments are green beans, red cabbage, onions, and carrots. Selection of these vegetables also gives an opportunity to evaluate the effects of various methods on strong-flavored vegetables. Table I gives methods that might be compared as well as information on preparation of the vegetable. Two types of experiments may be planned with this guide. Selection of several variables from the left column and one vegetable will show effects of cooking method. Selection of one cooking method and all vegetables will show differences in vegetables. Preparation of all variations will provide an opportunity to study both aspects.

   Use 1-liter pans for all methods except the pressure saucepan. For each method, put tap water in the pan, cover, and bring to a boil. Add the vegetable immediately to avoid boiling the water away, and cover unless the vegetable is to be cooked uncovered. As soon as the water has returned to the boiling point, start timing and reduce the heat to the minimum required to keep the water boiling. Special care must be taken in some of the methods not to boil the water away and thus burn the vegetable. If

TABLE I

Methods for Cooking Vegetables[a]

Cooking method	Green beans[b]	Cabbage[c]	Carrots[d]	Onions[e]
1. Minimum water, covered pan[f]				
Volume water (ml)	70	70	70	70
Time (min)	25–35	10–15	15–20	30–40
2. Half-weight of water, covered pan				
Volume water (ml)	90	115	90	115
Time (min)	20–25	8–10	15–18	25–35
3. Water to cover, uncovered pan				
Volume water (ml)	250	400	250	375
Time (min)	20–25	8–10	15–18	25–35
4. Water to cover, covered pan				
Volume water (ml)	250	400	250	375
Time (min)	20–25	8–10	15–18	25–35
5. Water to cover, covered pan				
Volume water (ml)	250	400	250	375
Time (min)	60	60	60	60
6. Pressure saucepan				
Volume water (ml)	70	70	70	70
Time after pressure is up (min)[g]	2.5	1.5	1.5	4

[a] Weight of vegetable for each method: green beans, 180 g; cabbage, 230 g; carrots, 180 g; onions, 230 g. Amount of salt for each method: 1.5 g.

[b] Three-centimeter pieces.

[c] Shred and mix well.

[d] Four-centimeter cross-cut pieces.

[e] Quarters.

[f] Choose a heavy pan for this method.

[g] Follow manufacturer's directions for rapid cooling at the end of the cooking period.

necessary, in order to prevent burning, add a small, measured amount of boiling water to the vegetables during cooking.

Since cooking times vary with several factors, such as the maturity of the vegetable, the size of the pieces, and the conditions under which it has been held since it was harvested, the cooking times given in Table I are approximations. All vegetables should be uniformly cooked until just tender except the vegetables in method 5, which are cooked for a longer time. Record the exact cooking times for all methods.

Evaluate the color, flavor, and texture of each vegetable using an appropriate scorecard. Relate differences to the cooking methods used.

2. Select potatoes of varying cultivars and of uniform size. Bake in a conventional oven and in a microwave oven following the manufacturer's instruction. Compare the textural qualities by conducting a descriptive sensory panel. If possible, measure the TPA using a universal testing machine. For this test cores of potato of uniform height can be taken with a meat corer or a small biscuit cutter. The potatoes should be cooled to the same temperature before testing.

3. Prepare tofu using the procedure outlined by Wang (1984) or Kantha *et al.* (1983). Use a variety of coagulants ($CaCl_2$, $CaSO_4$, lemon juice, and a solution of sea salt). Determine the effect of coagulant on flavor, color, and texture. Penetrometer readings could be used to measure gel strength. For an alternative experiment, use various cultivars of beans and compare the product.

4. Prepare tofu (see exercise 3) and use as a partial egg replacement in a custard or cake.

5. Compare the texture and color of white or cream-colored vegetables cooked in water with added soda, with added acid, and without addition.

6. Add a few drops of red fruit juice to tap water with added soda, soap, or acid, or without addition. Explain the colors observed.

7. Cook a green vegetable other than spinach in water with and without soda. Compare the color of the samples and the cooking water.

8. Cook vegetables in covered and uncovered pans, in small and large amounts of water. Compare the vegetables with respect to color and determine the pH and titratable acidity of the water drained from the vegetables.

9. Compare the speed of heat penetration and quality of potatoes baked with and without aluminum foil. Speed of heat penetration can be studied by means of a thermometer or a thermocouple and temperature recorder.

10. Study the effects of various metals on pigments by adding small amounts of compounds such as ferric chloride, aluminum chloride, and stannous or stannic chloride to 5-ml samples of the vegetable or fruit extracts.

11. Compare the quality of dried beans soaked and cooked by different methods. Include distilled and tap water with and without added (0.08% solution) sodium bicarbonate (Dawson *et al.*, 1952).

12. Compare the palatability of several vegetables cooked by conventional methods and cooked in a microwave oven.

References

Adams, C. E. and Erdman, J. W., Jr. 1988. Effects of home preparation practices on nutrient content of foods. *In* "Nutritional Evaluation of Food Processing," 2nd edn., Karmas, E. and Harris, R. S. (Eds.). Van Nostrand Reinhold, New York.

Asen, S., Norris, K. H., and Stewart, R. N. 1969. Absorption spectra and color of aluminum-cyanidin-3-glucoside complexes as influenced by pH. *Phytochem.* **8**: 653–659.

Asen, S., Norris, K. H., and Stewart, R. N. 1972. Co-pigmentation of anthocyanins in plant tissues and its effect on color. *Phytochem* **11**: 1139–1144.

Bailey, S. D., Bazinet, M. L., Driscoll, J. L., and McCarthy, A. I. 1961. The volatile sulfur components of cabbage. *J. Food Sci.* **26**: 163–170.

Bartolome, L. G. and Hoff, J. E. 1972. Firming of potatoes: biochemical effects of preheating. *J. Agric. Food Chem.* **20**: 266–270.

Borchgrevink, N. C. and Charley, H. 1966. Color of cooked carrots related to carotene content. *J. Amer. Dietet. Assoc.* **49**: 116–121.

Bourne, M. C. 1976. Texture of fruits and vegetables. *In* "Rheology and Texture in Food Quality," deMan, J. M., Voisey, P. M., Rasper, V. F., and Stanley, D. W. (Eds.). Avi Publ. Co., Westport, Connecticut.

Bourne, M. C. 1986. Effect of water activity on texture profile parameters of apple flesh. *J. Texture Studies* **17**: 331–340.

Bowman, F., Berg, E. P., Chuang, A. L., Gunther, M. W., Trump, D. C., and Lorenz, K. 1975. Vegetables cooked by microwaves vs. conventional methods. Retention of reduced ascorbic acid and chlorophyll. *Microwave Energy Appl. Newsletter* **8**(3): 3–8.

Brittin, H. C. and Trevino, J. E. 1980. Acceptability of microwave and conventionally baked potatoes. *J. Food Sci.* **45**: 1425–1427.

Brodnitz, M. H., Pascale, J. V., and Derslice, L. V. 1971. Flavor components of garlic extract. *J. Agric. Food Chem.* **19**: 273–275.

Burr, H. K., Kon, S., and Morris, H. J. 1968. Cooking rates of dry beans as influenced by moisture content and temperature and time of storage. *Food Technol.* **22**: 336–338.

Carlson, B. L. and Tabacchi, M. H. 1988. Loss of vitamin C in vegetables during the food service cycle. *J. Amer. Dietet. Assoc.* **88**: 65–68.

Chang, Y. S. and Smit, C. J. B. 1973. Characteristics of pectins isolated from soft and firm fleshed peach varieties. *J. Food Sci.* **38**: 646–648.

Chapman, V. J., Putz, J. O., Gilpin, G. L., Sweeney, J. P., and Eisen, J. N. 1960. Electronic cooking of fresh and frozen broccoli. *J. Home Econ.* **52**: 161–165.

Charles, V. R. and Van Duyne, F. O. 1954. Palatability and retention of ascorbic acid of vegetables cooked in a tightly covered saucepan and in a "waterless" cooker. *J. Home Econ.* **46**: 659–662.

Clydesdale, F. M. and Francis, F. J. 1968. Chlorophyll changes in thermally processed spinach as influenced by enzyme conversion and pH adjustment. *Food Technol.* **22**: 793–796.

Clydesdale, F. M., Fleishman, D. L., and Francis, F. J. 1970. Maintenance of color in processed green vegetables. *Food Prod. Dev.* **4**(5): 127–138.

Cook, R. 1983. Quality of citrus juices as related to composition and processing practices. *Food Technol.* **37**(6): 68–71, 133.

Crean, D. E. C. and Haisman, D. R. 1963. The interaction between phytic acid and divalent cations during cooking of dried peas. *J. Sci. Food Agric.* **14**: 824–833.

Dahl-Sawyer, C. A., Jen, J. J., and Huang, P. D. 1982. Cook-chill foodservice systems with conduction, convection and microwave reheat subsystems. Nutrient retention in beef loaf, potatoes, and peas. *J. Food Sci.* **47**: 1089–1095.

Dame, C., Jr., Chichester, C. O., and Marsh, G. L. 1959. Studies of processed all-green asparagus. IV. Studies on the influence of tin on the solubility of rutin and on the concentration of rutin present in the brines of asparagus processed in glass and tin containers. *Food Research* **24**: 28–36.

Dateo, G. P., Clapp, R. C., MacKay, D. A. M., Hewitt, E. J., and Hasselstrom, T. 1957. Identification of the volatile sulfur components of cooked cabbage and the nature of the precursors in the fresh vegetable. *Food Research* **22**: 440–447.

Davis, E. A., Gordon, J., and Hutchinson, T. E. 1976. Scanning electron microscope studies on carrots: Effects of cooking on the xylem and phloem. *Home Econ. Research J.* **4**: 214–224.

Dawson, E. H., Lamb, J. C., Toepfer, E. W., and Warren, H. W. 1952. Development of rapid methods of soaking and cooking dry beans. Tech. Bull. 1051, USDA, Washington, DC.

DellaMonica, E. S. and McDowell, P. E. 1965. Comparison of beta-carotene content of dried carrots prepared by three dehydration processes. *Food Technol.* **19**: 1597–1599.

Deshpande, S. N., Klinker, W. J., Draudt, H. N., and Desrosier, N. W. 1965. Role of pectic constituents and polyvalent ions in firmness of canned tomatoes. *J. Food Sci.* **30**: 594–600.

Deshpande, S. S. and Cheryan, M. 1986. Microstructure and water uptake of phaseolus and winged beans. *J. Food Sci.* **51**: 1218–1223.

Dietz, J. M. and Gould, W. A. 1986. Effects of process stage and storage on retention of beta-carotene in tomato juice. *J. Food Sci.* **51**: 847–848.

Dreher, M. L. 1987. "Handbook of Dietary Fiber." Marcel Dekker, Inc., New York.

Eheart, M. S. and Gott, C. 1964. Conventional and microwave cooking of vegetables. *J. Amer. Dietet Assoc.* **44**: 116–119.

Eheart, M. S. and Odland, D. 1973a. Use of ammonium compounds for chlorophyll retention in frozen green vegetables. *J. Food Sci.* **38**: 202–205.

Eheart, M. S. and Odland, D. 1973b. Quality of frozen green vegetables blanched in four concentrations of ammonium bicarbonate. *J. Food Sci.* **38**: 954–958.

Escueta, E. E., Bourne, M. C., and Hood, L. F. 1986. Effect of boiling treatment of soymilk on the composition, yield, texture and sensory properties of tofu. *Can. Inst. Food Sci. Technol. J.* **19**: 53–56.

Francis, F. J. 1975. Anthocyanins as food colors. *Food Technol.* **29**(5): 52, 54.

Francis, F. J. 1985. Pigments and other colorants. *In* "Food Chemistry," 2nd edn., Fennema, O. R. (Ed.). Marcel Dekker, Inc., New York.

Gandhi, A. P. and Bourne, M. C. 1988. Effect of pressure and storage time on texture profile parameters of soybean curd (tofu). *J. Texture Studies* **19**: 137–142.

Garruti, R. D. S. and Bourne, M. C. 1985. Effect of storage conditions of dry bean seeds (*Phaseolus vulgaris* L) on texture profile parameters after cooking. *J. Food Sci.* **50**: 1067–1071.

Gebre-Egziabher, A. and Sumner, A. K. 1983. Preparation of high protein curd from field peas. *J. Food Sci.* **48**: 375–377, 388.

Gilpin, G. L., Sweeney, J. P., Chapman, V. J., and Eisen, J. N. 1959. Effect of cooking methods on broccoli. II. Palatability. *J. Amer. Dietet Assoc.* **35**: 359–363.

Goldman, M., Horev, B., and Saguy, I. 1983. Decolorization of β-carotene in model systems simulating dehydrated foods. Mechanism and kinetic principles. *J. Food Sci.* **48**: 751–754.

Gordon, J. and Noble, I. 1959. Effect of cooking method on vegetables. *J. Amer. Dietet Assoc.* **35**: 578–581.

Gordon, J. and Noble, I. 1960. Application of the paired comparison method to the study of flavor differences in cooked vegetables. *Food Research* **25**: 257–262.

Gordon, J. and Noble, I. 1964. "Waterless" vs. boiling water cooking of vegetables. *J. Amer. Dietet Assoc.* **44**: 378–382.

Gross, J. 1987. "Pigments in Fruit." Academic Press, San Diego, California.

Haard, N. F. 1985. Characteristics of edible plant tissues. *In* "Food Chemistry," 2nd edn., Fennema, O. R. (Ed.). Marcel Dekker, Inc., New York.

Halliday, E. G. and Noble, I. 1943. "Food Chemistry and Cookery." Chicago University Press, Chicago, Illinois.

Harrison, S. L. and Konishi, F. 1982. Development and evaluation of a tofu (soy curd)-with-fiber product for the elderly. *J. Nutr. Elderly* **2**: 19–29.

Hartwig, E. E. 1989. Culture and genetics of grain legumes. *In* "Legumes. Chemistry, Technology, and Human Nutrition," Matthews, R. H. (Ed.). Marcel Dekker, Inc., New York.

Haytowitz, D. B. and Matthews, R. H. 1989. Nutrient content of other legume products. *In* "Legumes: Chemistry, Technology, and Human Nutrition," Matthews, R. H. (Ed.). Marcel Dekker, Inc., New York.

Heisler, E. G., Siciliano, J., Woodward, C. F., and Porter, W. L. 1964. After cooking discoloration of potatoes: Role of organic acids. *J. Food Sci.* **29**: 555–564.

Hing, F. S. and Weckel, K. 1964. Some volatile components of cooked rutabaga. *J. Food Sci.* **29:** 149–157.

Hinton, D. M. and Pressey, R. 1974. Cellulase activity in peaches during ripening. *J. Food Sci.* **39:** 783–785.

Hsu, C. P., Deshpande, S. N., and Desrosier, N. W. 1965. Role of pectin methylesterase in firmness of canned tomatoes. *J. Food Sci.* **30:** 583–588.

Hsu, D. L. and Jacobson, M. 1974. Macrostructure and nomenclature of plant and animal food sources. *Home Econ. Research J.* **3:** 24–32.

Hudson, D. E., Dalal, A. A., and Lachance, P. A. 1985. Retention of vitamins in fresh and frozen broccoli prepared by different cooking methods. *J. Food Qual.* **8:** 45–50.

Hughes, J. C., Faulks, R. M., and Grant, A. 1975. Texture of cooked potatoes. Relationship between compressive strength, pectic substances and cell size of Redskin tubers of different maturity. *Potato Research* **18:** 495–513.

Hurt, H. D. and Crocco, S. C. 1986. Dietary fiber: Marketing implications. *Food Technol.* **40**(2): 124–126.

IFT. 1986a. Sulfites as food ingredients. A scientific status summary by the Institute of Food Technologists' Expert Panel on Food Safety and Nutrition and the Committee on Public Information. *Food Technol.* **40**(6): 47–52.

IFT. 1986b. Food colors. A scientific status summary by the Institute of Food Technologists' Expert Panel on Food Safety and Nutrition and the Committee on Public Information. *Food Technol.* **40**(7): 49–56.

Issanchou, S. and Sauvageot, F. 1987. Effects of cooking method upon flavor of carrots and peas. *J. Food Sci.* **52:** 495–496.

Jones, I. D., White, R. C., and Gibbs, E. 1963. Influence of blanching and brining treatments on the formation of chlorophyllides, pheophytins, and pheophorbides in green plant tissue. *J. Food Sci.* **28:** 437–439.

Jones, P. M. B. and Boulter, D. 1983. The causes of reduced cooking rate in *Phaseolus vulgaris* following adverse storage conditions. *J. Food Sci.* **48:** 623–626, 649.

Kantha, S. S., Hettiarachchy, N. S., and Erdman, J. W., Jr. 1983. Laboratory scale production of winged bean curd. *J. Food Sci.* **48:** 441–444, 447.

Koehler, H. H., Chang, C.-H., Scheier, G., and Burke, D. W. 1987. Nutrient composition, protein quality, and sensory properties of thirty-six cultivars of dry beans (*Phaseolus vulgaris* L.). *J. Food Sci.* **52:** 1335–1340.

Kritchevsky, D. 1985. Dietary fiber in health and disease. *In* "Food Carbohydrates," Lineback, D. R. and Inglett, G. E. (Eds.). Avi Publ. Co., Westport, Connecticut.

Lachance, P. A. and Fisher, M. C. 1988. Effects of food preparation procedures on nutrient retention with emphasis on foodservice practices. *In* "Nutritional Evaluation of Food Processing," 2nd edn., Karmas, E. and Harris, R. S. (Eds.). Van Nostrand Reinhold, New York.

Langdon, T. T. 1987. Preventing of browning in fresh prepared potatoes without the use of sulfiting agents. *Food Technol.* **41**(5): 64, 66–67.

Lanier, J. J. and Sistrunk, W. A. 1979. Influence of cooking method on quality attributes and vitamin content of sweet potatoes. *J. Food Sci.* **44:** 374–377, 380.

Lea, C. H. 1963. Some aspects of recent flavour research. *Chem. and Ind.* pp. 1406–1413.

Leung, H. K., Barron, F. H., and Davis, D. C. 1983. Textural and rheological properties of cooked potatoes. *J. Food Sci.* **48:** 1470–1474, 1496.

Losh, J. M., Phillips, J. A., Axelson, J. M., and Schulman, R. S. 1981. Sweet potato quality after baking. *J. Food Sci.* **46:** 283–286, 290.

Luh, B. S. and Phitakpol, B. 1972. Characteristics of polyphenoloxidases related to browning in cling peaches. *J. Food Sci.* **37:** 264–268.

Mabesa, L. B. and Baldwin, R. E. 1979. Ascorbic acid in peas cooked by microwaves. *J. Food Sci.* **44:** 932–933.

Mackinney, G. and Weast, C. A. 1940. Color changes in green vegetables. *Ind. Eng. Chem.* **32:** 392–395.

MacLeod, A. J. and MacLeod, G. 1970. Effects of variation in cooking methods on the flavor volatiles of cabbage. *J. Food Sci.* **35:** 744–750.

Main, G. L., Morris, J. R., and Wehunt, E. J. 1986. Effect of preprocessing treatments on the firmness and quality characteristics of whole and sliced strawberries after freezing and thermal processing. *J. Food Sci.* **51:** 391–394.

Mao, W. W. and Kinsella, J. E. 1981. Amylase activity in banana fruit: Properties and changes in activity with ripening. *J. Food Sci.* **46:** 1400–1403, 1409.

Marson, A. 1984. Tofu. *Bakers Review* July: 31. [In *FTSA Abstracts.* 1985 **17**(8): 8 G 54.]

Marty, C. and Berset, C. 1988. Degradation of trans-β-carotene produced during extrusion cooking. *J. Food. Sci.* **53:** 1880–1886.

Maruyamu, F. T. 1970. Identification of dimethyl disulfide as a major aroma component of Brassicaceous vegetables. *J. Food Sci.* **35:** 540–543.

Mathew, A. G. and Parpia, H. A. B. 1971. Food browning as a polyphenol reaction. *Adv. Food Research* **19:** 75–90.

Niegisch, W. D. and Stahl, W. H. 1956. The onion: Gaseous emanation products. *Food Research* **21:** 657–665.

Noble, A. C. 1975. Investigation of the color changes in heat concentrated tomato pulp. *J. Agric. Food Chem.* **23:** 48–49.

Odland, D. and Eheart, M. S. 1974. Ascorbic acid retention and organoleptic quality of green vegetables cooked by several techniques using ammonium bicarbonate. *Home Econ. Research J.* **2:** 241–250.

Odland, D. and Eheart, M. S. 1975. Ascorbic acid, mineral, and quality retention in frozen broccoli blanched in water, steam, and ammonia steam. *J. Food Sci.* **40:** 1004–1007.

Pasch, J. H., von Elbe, J. H., and Sell, R. J. 1975. Betalaines as colorants in dairy products. *J. Milk Food Technol.* **28:** 25–28.

Pesek, C. A. and Warthesen, J. J. 1987. Photodegradation of caarotenoids in a vegetable juice system. *J. Food Sci.* **52:** 744–746.

Ponting, J. D. and Jackson, R. 1972. Pre-freezing processing of Golden Delicious apple slices. *J. Food Sci.* **37:** 812–814.

Ponting, J. D., Jackson, R., and Watters, G. 1972. Refrigerated apple slices: Preservative effects of ascorbic acid, calcium and sulfites. *J. Food Sci.* **37:** 434–436.

Pressey, R., Hinton, D. M., and Avants, J. K. 1971. Development of polygalacturonase activity and solubilization of pectin in peaches during ripening. *J. Food Sci.* **36:** 1070–1073.

Purcell, A. E., Walter, W. M., Jr., and Thompkins, W. T. 1969. Relationship of vegetable color to physical state of the carotenes. *J. Agric. Food Chem.* **17:** 41–42.

Reeve, R. M. 1970. Relationships of histological structure to texture of fresh and processed fruits and vegetables. *J. Texture Studies* **1:** 247–284.

Reeve, R. M. 1972. Pectin and starch in preheating firming and final texture of potato products. *J. Agric. Food Chem.* **20:** 1282.

Resende, R., Francis, F. J., and Stumbo, C. R. 1969. Thermal degradation and regeneration of enzymes in green bean and spinach puree. *Food Technol.* **23:** 63–66.

Rosenbaum, T. M., Henneberry, G. O., and Baker, B. E. 1966. Constitution of leguminous seeds. VI. The cookability of field peas (Pisum sativum L.). *J. Sci. Food Agric.* **17:** 237–240.

Saio, K. 1979. Tofu—relationships between texture and fine structure. *Cereal Foods World* **24:** 342–354.

Sapers, G. M. and Ziolkowski, M. A. 1987. Comparison of erythorbic and ascorbic acids as inhibitors of enzymatic browning in apple. *J. Food Sci.* **52:** 1732–1733, 1747.

Schneeman, B. O. 1986. Dietary fiber: Physical and chemical properties, methods of analysis, and physiological effects. *Food Technol.* **40**(2): 104–110.

Schwartz, S. J. and von Elbe, J. H. 1983. Kinetics of chlorophyll degradation to pyropheophytin in vegetables. *J. Food Sci.* **48:** 1303–1306.

Schwartz, S. J., Woo, S. L., and von Elbe, J. H. 1981. High performance liquid chromatography of chlorophylls and their derivatives in fresh and processed spinach. *J. Agric. Food Chem.* **29:** 533–535.

Schwimmer, S. 1963. Alteration of the flavor of processed vegetables by enzyme preparations. *J. Food Sci.* **28:** 460–466.

Schwimmer, S. and Weston, W. J. 1961. Enzymatic development of pyruvic acid in onion as a measure of pungency. *J. Agric. Food Chem.* **9:** 301–304.

Scott, L. E., Twigg, B. A., and Bouwkamp, J. C. 1974. Color of processed sweet potatoes: Effects of can type. *J. Food Sci.* **39**: 563–564.

Shewfelt, A. L., Paynter, V. A., and Jen, J. J. 1971. Textural changes and molecular characteristics of pectin constituents in ripening peaches. *J. Food Sci.* **36**: 573–575.

Shirkhande, A. J. 1976. Anthocyanins in foods. *CRC Crit. Rev. Food Sci. Nutr.* **7**: 193–218.

Sievwright, C. A. and Shipe, W. F. 1986. Effect of storage conditions and chemical treatments on firmness, *in vitro* digestibility, condensed tannins, phytic acid and divalent cations of cooked black beans (*Phaseolus vulgaris*). *J. Food Sci.* **51**: 982–987.

Simon, P. W. and Lindsay, R. C. 1983. Effects of processing upon sensory and objective variables of carrots. *J. Amer. Soc. Hortic. Sci.* **108**: 928–931.

Simpson, J. and Halliday, E. G. 1928. The behavior of sulphur compounds in cooking vegetables. *J. Home Econ.* **20**: 121–126.

Simpson, J. and Halliday, E. G. 1941. Chemical and histological studies of the disintegration of cell-membrane materials in vegetables during cooking. *Food Research* **6**: 189–206.

Skalski, C. and Sistrunk, W. A. 1973. Factors affecting color degradation in Concord grape juice. *J. Food Sci.* **38**: 1060–1062.

Slavin, J. L. 1987. Dietary fiber: Classification, chemical analysis, and food sources. *J. Amer. Dietet. Assoc.* **87**: 1164–1171.

Spåre, C. G. and Virtanen, A. I. 1963. On the lachrymatory factor in onion (Allium cepa) vapours and its precursor. *Acta Chem. Scand.* **17**: 641–650.

Stephens, T. S. and McLemore, T. A. 1969. Preparation and storage of dehydrated carrot flakes. *Food Technol.* **23**: 1600–1602.

Sterling, C. 1968. Effect of solutes and pH on the structure and firmness of cooked carrot. *J. Food Technol.* **3**: 367–371.

Stone, M. B. and Young, C. M. 1985. Effects of cultivars, blanching techniques, and cooking methods on quality of frozen green beans as measured by physical and sensory attributes. *J. Food Qual.* **7**: 255–265.

Sweeney, J. P. 1970. Improved chlorophyll retention in green beans held on a steam table. *Food Technol.* **24**: 490–493.

Sweeney, J. P. and Martin, M. 1958. Determination of chlorophyll and pheophytin in broccoli heated by various procedures. *Food Research* **23**: 635–647.

Sweeney, J. P. and Martin, M. 1961. Stability of chlorophyll in vegetables as affected by pH. *Food Technol.* **15**: 263–266.

Sweeney, J. P., Gilpin G. L., Staley, M. G., and Martin, M. E. 1959. Effect of cooking methods on broccoli. I. Ascorbic acid and carotene. *J. Amer. Dietet. Assoc.* **35**: 354–358.

Synge, R. L. M. and Wood, J. C. 1956. (+)-(S-methyl-L-cysteine S-oxide) in cabbage. *Biochem. J.* **64**: 252–259.

Szczesniak, A. S. and Ilker, R. 1988. The meaning of textural characteristics—juiciness in plant foods. *J. Texture Studies* **19**: 61–78.

Taylor, S. L., Higley, N. A., and Bush, R. K. 1986. Sulfite in foods: Uses, analytical methods, residues, fate, exposure assessment, metabolism, toxicity, and hypertension. *Adv. Food Research* **31**: 1–76.

Twigg, B. A., Scott, L. E., and Bouwkamp, J. C. 1974. Color of processed sweet potatoes: Effect of additives. *J. Food Sci.* **39**: 565–567.

Van Buren, J. P., Hrazdina, G., and Robinson, W. B. 1974. Color of anthocyanin solutions expressed in lightness and chromaticity terms. Effect of pH and type of anthocyanin. *J. Food Sci.* **39**: 325–328.

Van Buren, J. P., Kean, W. P., and Wilkison, M. 1988. Influence of salts and pH on the firmness of cooked snap beans in relation to the properties of pectin. *J. Texture Studies* **19**: 15–25.

Van Duyne, F. O., Owen, R. F., Wolfe, J. C., and Charles, V. R. 1951. Effect of cooking vegetables in tightly covered and pressure saucepans: Retention of reduced ascorbic acid and palatability. *J. Amer. Dietet. Assoc.* **27**: 1059–1065.

von Elbe, J. H. 1975. Stability of betalaines as food colors. *Food Technol.* **29**(5): 42–43, 46.

von Elbe, J. H. 1986. Chemical changes in plant and animal pigments during food processing. *In* "Role of Chemistry in Quality of Processed Food," Fennema, O. R., Cheng, W.-H., and Lii, C.-L. (Eds.). Food and Nutrition Press, Inc., Westport, Connecticut.

von Elbe, J. H., Klement, J. T., Amundson, C. H., Cassens, R. G., and Lindsay, R. C. 1974. Evaluation of betalain pigments as sausage colorants. *J. Food Sci.* **39:** 128–132.

von Elbe, J. H., Schwartz, S. J., and Hildenbrand, B. E. 1981. Loss and regeneration of betacyanin pigments during processing of red beets. *J. Food Sci.* **46:** 1713–1715.

Wagner, J. R., Miers, J. C., Sanshuck, D. W., and Becker, R. 1969. Consistency of tomato products. 5. Differentiation of extractive and enzyme inhibitory aspects of the acidified hot break process. *Food Technol.* **23:** 247–250.

Wang, H. J. and Thomas, R. L. 1989. Direct use of apple pomace in bakery products. *J. Food Sci.* **54:** 618–620, 639.

Wang, H. L. 1984. Tofu and tempeh as potential protein sources in the western diet. *J. Amer. Oil Chemists' Soc.* **61:** 528–534.

Wang, H. L., Swain, W. F., Kwolek, W. F., and Fehr, W. R. 1983. Effect of soybean varieties on the yield and quality of tofu. *Cereal Chem.* **60:** 245–248.

Watson, E. K. 1973. Tannins in fruit extracts as affected by heat treatments. *Home Econ. Research J.* **2:** 112–118.

Weast, C. A. and Mackinney, G. 1940. Chlorophyllase. *J. Biol. Chem.* **133:** 551–558.

Whistler, R. L. and Daniel, J. R. 1985. Carbohydrates. *In* "Food Chemistry," 2nd edn., Fennema, O. R. (Ed.). Marcel Dekker, Inc., New York.

Wrolstad, R. E., Putnam, T. P., and Varseveld, G. W. 1970. Color quality of frozen strawberries: Effect of anthocyanin, pH, total acidity and ascorbic acid variability. *J. Food Sci.* **35:** 448–452.

Yang, C. S. T. and Yang, P. P. A. 1987. Effect of pH, certain chemicals and holding time-temperature on the color of lowbush blueberry puree. *J. Food Sci.* **52:** 346–347, 352.

FATS AND THEIR LIPID CONSTITUENTS

Fats are mixtures of lipids. Lipids are chemical compounds present naturally in many foods. Lipid mixtures in the form of shortenings, frying fats, and salad oils are used commonly in food preparation and are referred to here as fats. The major functions of a fat in food preparation include the following: (1) to tenderize, (2) to contribute to batter or dough aeration, (3) to serve as a heating medium, (4) to serve as a phase in an emulsion, (5) to contribute to flavor, and (6) to enhance smoothness, body, or other textural properties. The nutritional role is not covered in this book.

The subject of lipids and fats is characterized by so much interrelatedness that choosing a sequence for discussing the various aspects is not easy. Chemical structure of lipids determines both the physical properties and the chemical reactions that may occur. The relationship between the chemical properties of lipids and the physical properties of lipids and fats underlies the processing methods used for fats. The relationship between chemical and physical properties of lipids also underlies the functional properties of the fats used in food preparation. The chemical structure of lipids is discussed first because it is basic to the study of all other aspects of lipids and the fats that contain them.

I. CHEMICAL STRUCTURE OF LIPIDS

Some lipids contain only carbon, hydrogen, and oxygen; others also contain phosphorus and nitrogen; some contain sulfur. Fatty acids are important structural components of most lipids and thus constitute a starting point for a discussion of lipid structure.

A. Fatty Acids

Most naturally occurring fatty acids have straight carbon chains, even numbers of carbons, and only hydrogen atoms attached to the carbons other than the carboxyl carbon. Fatty acids differ considerably as to chain length and degree of saturation. Most fatty acids in foods have chain lengths of 4–26 carbon atoms, with 16–22 carbons predominating. Most have 0–4 double bonds, though some fish oils have more. The empirical formulas $C_nH_{2n+1}COOH$, $C_nH_{2n-1}COOH$, $C_nH_{2n-3}COOH$, $C_nH_{2n-5}COOH$, and $C_nH_{2n-7}COOH$ represent saturated, monounsaturated, diunsaturated, triunsaturated, and tetraunsaturated fatty acids, respectively. Monounsaturated, diunsaturated, triunsaturated, and tetraunsaturated fatty acids also are referred to as monoenes, dienes, trienes, and tetraenes.

Some common fatty acids with both their trivial and systematic names are given in Table I. Butyric acid is unique to butterfat, though present in rather low concentration. Lauric and myristic acids are especially abundant in palm kernel and coconut oils. Palmitic and oleic acids are major fatty acids in foods; they are widely distributed in both plant and animal lipids. Stearic acid is found in both plant and animal lipids but, except for its high concentration in cocoa butter, is more abundant in animals than in plants. The polyunsaturated fatty acids, those with at least two double bonds, are most abundant in seed oils and fish liver oils.

The locations of the double bonds in fatty acids are indicated in different ways, a common one of which can be illustrated with linoleic acid. Numbering normally begins with the carboxyl carbon, and the double bonds in linoleic acid are located between carbons 9 and 10 and between carbons 12 and 13; thus its

TABLE I
Common Fatty Acids

C:=[a]	Common name	Systematic name	Formula
4:0	Butyric	Butanoic	$CH_3(CH_2)_2COOH$
6:0	Caproic	Hexanoic	$CH_3(CH_2)_4COOH$
8:0	Caprylic	Octanoic	$CH_3(CH_2)_6COOH$
10:0	Capric	Decanoic	$CH_3(CH_2)_8COOH$
12:0	Lauric	Dodecanoic	$CH_3(CH_2)_{10}COOH$
14:0	Myristic	Tetradecanoic	$CH_3(CH_2)_{12}COOH$
16:0	Palmitic	Hexadecanoic	$CH_3(CH_2)_{14}COOH$
16:1[b]	Palmitoleic	9-Hexadecenoic	$CH_3(CH_2)_5CH{=}CH(CH_2)_7COOH$
18:0	Stearic	Octadecanoic	$CH_3(CH_2)_{16}COOH$
18:1[b]	Oleic	9-Octadecenoic	$CH_3(CH_2)_7CH{=}CH(CH_2)_7COOH$
18:2[b]	Linoleic	9,12-Octadecadienoic	$CH_3(CH_2)_4(CH{=}CHCH_2)_2(CH_2)_6COOH$
18:3[b]	Linolenic	9,12,15-Octadecatrienoic	$CH_3CH_2(CH{=}CHCH_2)_3(CH_2)_6COOH$
20:0	Arachidic	Eicosanoic	$CH_3(CH_2)_{18}COOH$
20:4[b]	Arachidonic	5,8,11,14-Eicosatetraenoic	$CH_3(CH_2)_4(CH{=}CHCH_2)_4(CH_2)_2COOH$
22:0	Behenic	Docosanoic	$CH_3(CH_2)_{20}COOH$
24:0	Lignoceric	Tetracosanoic	$CH_3(CH_2)_{22}COOH$

[a] Number of carbon atoms:number of double bonds.
[b] Cis configuration.

systematic name is 9,12-octadecadienoic acid, or octadeca-9,12-dienoic acid. Sometimes the location of the double bond in relation to the methyl (ω) end of the chain is of special interest; we recently have been hearing about the nutritional significance of ω-3 fatty acids, in which one of the double bonds is three carbons from the end of the chain.

Configuration is the arrangement of the hydrogen atoms on the double-bond carbons. Although the cis configuration is normal in naturally occurring fatty acids, the trans configuration is possible and is discussed later.

B. Glycerides

Glycerides are esters of the three-carbon alcohol glycerol and fatty acids. Glycerol, sometimes called glycerine, has three hydroxyl groups per molecule and thus is described as a polyhydric alcohol.

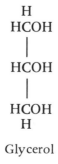

Glycerol

A glyceride, or glyceryl ester, is a monoacylglycerol, diacylglycerol, or triacylglycerol, depending on whether one, two, or three of the alcohol groups of glycerol are esterified with fatty acid(s). The terms monoglyceride, diglyceride, and triglyceride have been used for a long time and are not being easily displaced by the above preferred terms. A triglyceride may be a simple triglyceride, containing three molecules of a single fatty acid, as tristearin,

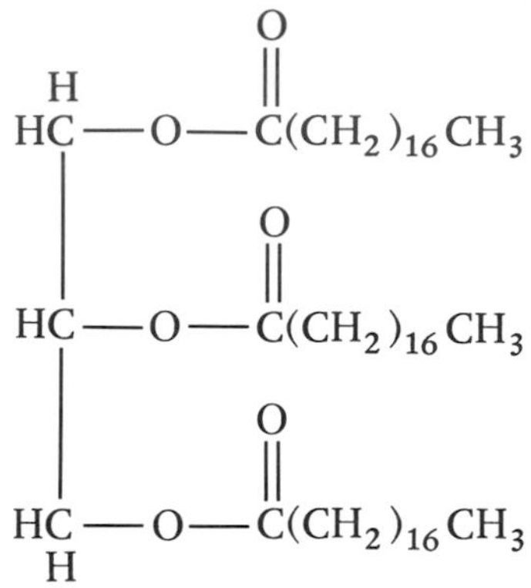

$$
\begin{array}{c}
\quad\quad\quad\quad\;\; O \\
\quad\quad\quad\quad\;\; \| \\
\text{H} \\
\text{HC}\!-\!\text{O}\!-\!\text{C}(CH_2)_{16}CH_3 \\
\; | \\
\quad\quad\quad\quad\;\; O \\
\quad\quad\quad\quad\;\; \| \\
\text{HC}\!-\!\text{O}\!-\!\text{C}(CH_2)_{16}CH_3 \\
\; | \\
\quad\quad\quad\quad\;\; O \\
\quad\quad\quad\quad\;\; \| \\
\text{HC}\!-\!\text{O}\!-\!\text{C}(CH_2)_{16}CH_3 \\
\text{H}
\end{array}
$$

Tristearin

or a mixed triglyceride, containing more than one fatty acid. Most naturally occurring triglycerides are mixed and may or may not be symmetrical. In symmetrical mixed triglycerides, the terminal fatty acids (α positions) are the same and are different from the middle fatty acid (β position). The numerals 1 and 2 also are used to designate the terminal and middle positions on the glycerol molecule. Three different fatty acids can be esterified with glycerol. Although a triglyceride molecule sometimes is represented diagrammatically with all fatty acid chains extending in the same direction, the β chain more likely extends in the opposite direction from the α chains.

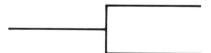

"Tuning fork" representation

Triglycerides frequently are referred to as neutral lipids because their molecules do not have free acidic or basic groups. Neutral lipids are by far the most abundant class of lipids. Even the major sources of phospholipid contain more neutral lipid than phospholipid. Natural fats consist largely of mixtures of mixed triglycerides, with traces of mono- and diglycerides and free fatty acids. Other lipid classes, such as phospholipids and glycolipids, are present in some fats.

C. Phospholipids (Phosphoglycerides)

The molecules of phospholipids contain, in addition to glycerol, two fatty acid molecules, a phosphate group, and a nitrogenous base and may be represented diagrammatically as follows:

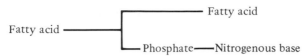

Diagrammatic representation of a phospholipid

As in a triglyceride, the fatty acids are in ester linkage with the glycerol. The phosphate is in ester linkage both with the glycerol and with the nitrogenous base. The nitrogenous base is choline

$$HO-CH_2-CH_2-\overset{+}{N}\overset{\diagup CH_3}{\underset{\diagdown CH_3}{-CH_3}}$$

Choline

in lecithin, which preferably is called phosphatidylcholine. The nitrogenous bases in the other two major phospholipids, phosphatidylethanolamine and phosphatidylserine, are ethanolamine and serine.

$$HO-CH_2-CH_2-NH_2 \qquad\qquad HO-CH_2-\underset{\underset{COOH}{|}}{CH}-NH_2$$

Ethanolamine   Serine

Sphingomyelin contains sphingosine rather than glycerol and is less abundant than the other phospholipids.

The phosphate group in a phospholipid molecule confers some acidity, but the quaternary ammonium group of choline is strongly basic. Phosphatidyl-ethanolamine and phosphatidylserine, formerly called cephalins, are acidic. The free amino group in ethanolamine or serine is not as strongly basic as is the quaternary ammonium group of choline. Because of the free carboxyl group of serine, phosphatidylserine is particularly acidic. The basic and acidic groups can be charged and phospholipids are referred to as polar lipids.

Phospholipids, like triglycerides, are variable as to their fatty acid components. In general, their fatty acids are less saturated than are those of tri-glycerides.

Phospholipids occur in egg yolk, the fat globule membrane of milk, and the intramuscular lipid of meat, as well as in plant seeds. Egg yolk and soybeans are major commercial sources. Phospholipids usually occur, along with proteins, as lipoproteins in the tissues in which they occur. Lipoproteins have structural and functional significance in the tissues that contain them and in foods into which they are incorporated.

D. Glycolipids and Sulfolipids

Glycolipids occur in small amounts in both plant and animal tissues. Like phospholipids, most glycolipids contain either glycerol or sphingosine. Those containing glycerol usually are diglycerides with a sugar unit in glycosidic linkage with glycerol at the other position. A second sugar moiety may be in glycosidic linkage with the first. The sugar units most frequently are galactose and glucose. Although present in small amounts in both plant and animal tissues, the glycosyl glycerides are more prominent in plants than in animals. Wheat galactosyl glycerides are of functional importance in yeast breads (Chapter 19).

Sulfolipids contain sulfur and occur in several forms, which are not discussed in this book. A lipid may be both a glycolipid and a sulfolipid.

More detailed discussions of the chemical structure of lipids include those of Gunstone and Norris (1983, Chapter 1), Nawar (1985, pp. 141–151), and Zapsalis and Beck (1985, pp. 415–430).

II. PHYSICAL STRUCTURE AND PROPERTIES OF FATS

The physical structure and properties of fats are discussed here before the chemical reactions of lipids in order to facilitate relating the chemical reactions to physical effects.

A. Physical Structure

The physical structure of lipids is important to the properties of shortenings, butter, and margarine, as well as such foods as peanut butter and candies that have lipid-containing coatings. The straight chains of fatty acids are able to align themselves into crystalline structure. A solid fat consists of crystals suspended in oil. The proportion of fat in the crystalline form is expressed as solid fat index (SFI). In general, the higher the SFI, the more solid is the fat at room temperature. However, the relationship between crystallinity and consistency is not that simple because a solid fat is polymorphic; it has the ability to exist in more than one crystalline form. In the β form, representing a high degree of order, crystals not only have relatively high melting points but also are relatively large and stable. A fat with a large proportion of crystals in the β form tends to be grainy. At the opposite extreme is the α-crystalline form, in which chain alignments are somewhat more random, the melting point is relatively low, and the crystals are small. A fat with a high proportion of α crystals is smooth but the crystals are unstable.

Whether a fat with small crystals remains smooth depends on the variety of molecular species present and on the processing conditions. If only a few molecular species are present, the crystals are unstable because further orientation to form larger crystals is relatively easy. A large variety of molecular species favors maintenance of small crystals but with a shift from α to β' crystals. Crystals in the β' form are much smaller than those in the β form and quite stable. Rapid cooling with agitation favors the formation of stable small crystals. A predominance of β' crystals in a solid fat is desirable from the standpoint of quality of products made with solid fats; an exception is pastry, in which the large β crystals are preferable. This brief discussion of polymorphism is an oversimplification of a complex subject.

B. Physical Properties

1. Melting Point

Melting point is the temperature at which a solid is changed to a liquid. A fat has a melting point range because of its heterogeneity. The melting point range depends to a large extent on the proportion of the fat that is present in the crystalline state. This in turn depends on several internal factors, including chain length and degree of saturation of the component fatty acids, configuration at the double bonds of unsaturated fatty acids, and variety of molecular species. Previous history of a fat also affects melting point.

The greater the ease of association between the fatty acid chains, the greater is the extent of crystallization and the higher is the temperature required to disrupt the associative forces between chains; or, in other words, the higher is the melting point of the fat. Long, straight chains of the component fatty acids contribute to high melting points of fats.

Melting point, either of free fatty acids or of glycerides, increases with increasing chain length if only chain length differs. For free saturated fatty acids, melting point increases from $-4°C$ for the 4-carbon acid to $80°C$ for the 22-carbon fatty acid (CRC, 1989–1990).

Melting point increases with increasing degree of saturation if only degree of saturation differs. The reason may not be as obvious as in the case of the chain length effect. The presence of a double bond results in a bend in the chain, as seen in the following chain segment:

$$— CH_2 — CH$$
$$\parallel$$
$$— CH_2 — CH$$

"Bent chain" segment

The chain thus is more bulky than a saturated chain and association between chains is more difficult. Other factors being equal, the larger the number of double bonds in a chain, the less extended is the chain, the more difficult is crystallization, and the less is the amount of energy required for melting once crystallization has occurred. Among the 18-carbon fatty acids, melting point of the free acids ranges from $72°C$ for stearic acid (18:0, or 18 carbons, 0 double bonds) down to $-11°C$ for linolenic acid (18:3) (CRC, 1989–1990). Most oils are liquid at room temperature because of their relatively high concentrations of unsaturated fatty acids. An exception is coconut oil, which is highly saturated but is liquid because of a preponderance of shorter chained fatty acids. Among the highly unsaturated oils, some have a higher degree of polyunsaturation than others. For example, olive oil is highly monounsaturated, and many other oils, such as soybean oil and corn oil, are highly polyunsaturated.

The third molecular characteristic that affects melting point is not a variable in fats as they naturally occur but develops as an effect of processing. Naturally occurring unsaturated fatty acids, either free or esterified, have the cis configuration at the double bond, resulting in a bend in the chain as shown previously. With the trans configuration, often produced during hydrogenation, the double bond results in a small "kink" rather than a complete turnaround:

$$— CH_2 — CH$$
$$\|$$
$$HC — CH_2 —$$

"Kinked chain" segment

The trans configuration thus permits the chain to be considerably more extended than an equally unsaturated chain having the cis configuration at the double bond(s). This results in a higher melting point of the fatty acid or glyceride than is characteristic of the corresponding lipid with the cis configuration. It is possible for a fat to be solid at room temperature and yet identical to an oil with regard to fatty acid chain lengths and extent of unsaturation. Whereas the melting point of oleic acid, cis 18:1, is 16°C, that of elaidic acid, trans 18:1, is 45°C, (CRC, 1989–1990). When trans isomers are formed, not all double bonds present are affected. The formation of trans isomers is discussed later in this chapter.

Melting point is affected also by the previous history of a fat. As discussed previously, factors such as rate of cooling and amount of agitation during cooling affect the crystallinity, and the crystallinity in turn affects the melting point of a fat. Not only does the extent of crystallinity affect melting point, but so also does the crystalline form. Even a pure sample of a simple triglyceride has different possible melting points, depending on the polymorphic state. For tristearin, these are 54.7, 64.0, and 73.3°C for the α, β', and β polymorphs, respectively (CRC, 1989–1990). The rate and extent to which the transition from α to β' to β progresses depends on the details of the cooling process. The subject is further complicated by the differing behavior of complex lipid mixtures. Shortenings differ in their β- and β'-crystallizing tendencies.

2. Solidification Temperature

Solidification temperature of a given fat is lower than the melting point; in other words, if a solid fat is melted, it must be cooled to a temperature lower than that at which it became completely melted before it will resolidify. If an oil is cooled until it solidifies, it must be warmed to a temperature higher than the temperature of solidification before it will regain fluidity.

3. Solubility

Glycerol, a viscous, odorless, somewhat sweet liquid, is extremely water soluble. The three hydroxyl groups, which constitute a large proportion of the molecule, are responsible for the water solubility. The hydrocarbon chain of a fatty acid does not have an affinity for water, whereas the carboxyl group does. Only the fatty acids with the shortest chains are truly water soluble, and solubility decreases as the chain length increases. In a triglyceride, the hydophilic portions of glycerol and fatty acids are "tied up," leaving the molecule largely hydrophobic. The molecule still has some affinity for water at the ester linkage, and again, the solubility depends on the lengths of hydrocarbon chains. As natural fats contain triglycerides having a preponderance of medium- to long-chain fatty acids, fats are insoluble in water, slightly soluble in lower alcohols, and readily soluble in the nonpolar solvents chloroform, ether, petroleum ether, benzene, and carbon tetrachloride.

Phospholipid and glycolipid molecules are more soluble in polar solvents than are triglycerides because of the hydrophilic groups they contain. These are phosphate and hydroxyl groups in phospholipids, the carboxyl group in phosphatidylserine, and the hydroxyl groups of the sugar moiety in glycolipids.

4. Density
Fats have a lower density than does water. The density of most fats is in the range of $0.90-0.92$ g/cm^3. This results in the well-known floating of unemulsified oil on the surface of water.

5. Refractive Index
The refractive index of fats is affected by several factors. It decreases with increasing temperature, increases with increasing chain length, and increases with increasing unsaturation. The significance of the temperature effect lies in the importance of temperature control in the measurement of refractive index. Refractive index is not a property that is important in the everyday use of fats, but the effect of unsaturation on refractive index provides the food industry with a quick means of estimating the extent of hydrogenation of an oil.

6. Surfactant Properties
Surfactants are discussed briefly in Chapter 6. Free fatty acids and monoglycerides have the ability to form monolayers at water–fat interfaces. This property is most easily illustrated with fatty acids. If oil containing free fatty acids is added to water without agitation, the hydrophilic carboxyl end of a fatty acid molecule is pulled into the water and the hydrocarbon chain, being hydrophobic, remains in the oil. The result is an alignment of fatty acid molecules to form a monomolecular film at the interface:

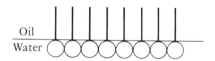

Monomolecular layer

If the mixture is agitated, the same type of film may form around oil globules, resulting in an emulsion in which each oil globule is coated by a monomolecular film:

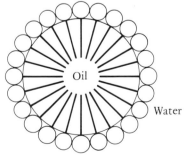

Oil globule with monomolecular layer

Monoglycerides function similarly and, therefore, are frequently added to shortenings that are likely to be used in cake batters. Their presence in shortening improves the emulsification of the shortening when batters are prepared.

III. CHEMICAL REACTIONS OF LIPIDS _____

In a discussion of the reactions to which lipids are susceptible, it is convenient to think in terms of triglycerides, though most of the reactions are applicable also to other lipids, as may be surmised from the specific molecular portions involved. Reactions of triglycerides can be classified in several ways. One basis for classification is the portion of the molecule that is involved.

Reactions involving the ester linkage:

> Hydrolysis
> Saponification
> Interesterification
> Rearrangement
> Acetylation

Reactions involving double bonds:

> Oxidation (indirectly)
> Reversion
> Hydrogenation
> Isomerization
> Halogenation

The same reactions can be classified on the basis of the general conditions under which they are most likely to occur.

Reactions occurring in commercial processing as primary reactions:

> Interesterification
> Hydrogenation
> Rearrangement
> Acetylation

Reactions occurring in commercial processing as side reactions:

> Isomerizations (cis–trans isomerization also may be the primary
> reaction)

Reactions occurring in analytical work:

> Saponification
> Halogenation

Reactions occurring in storage and use of fats and foods containing them:

   Hydrolysis
   Oxidation
   Reversion

The reactions are discussed here in the sequence of the first classification.

A. Reactions Involving the Ester Linkage

1. Hydrolysis

The ester linkage between glycerol and fatty acid is broken in hydrolysis. Complete hydrolysis of a triglyceride results in one molecule of glycerol and three molecules of fatty acid per molecule of triglyceride hydrolyzed, as illustrated with tristearin:

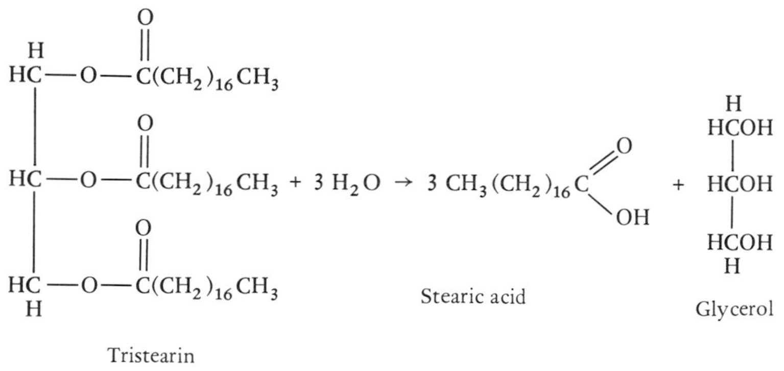

Tristearin hydrolysis

Note that the reaction is a double decomposition reaction in which water is the other reactant. One molecule of water is required for each ester linkage broken.

Lipases, enzymes that are naturally present in butterfat, nuts, whole grains, and other fatty foods that might be stored without prior heat treatment, catalyze lipid hydrolysis during storage. Most lipases preferentially split off the fatty acids in the α position on glycerol. The accumulation of free fatty acids usually results in an off-flavor, rancidity. The effect is particularly noticeable in foods containing relatively high concentrations of short-chain fatty acids, such as butter, because the lower fatty acids are volatile at room temperature. The rancid condition is a symptom that may be caused by more than one mechanism, and another mechanism is discussed later. Rancidity caused by hydrolysis is labeled hydrolytic rancidity. It should be recognized that sometimes desirable flavors result from controlled and selective enzymatic hydrolysis. Cheese, yogurt, and bread owe some of their flavor to such reactions (Nawar, 1985, p. 175).

Hydrolysis also may be catalyzed by acid in a heated system. This probably is not an important mechanism of troublesome lipid hydrolysis in food, however.

When a frying fat is used repeatedly, free fatty acids accumulate as a result of hydrolysis. In this case, heat is the catalyst. Hydrolysis is facilitated by the combination of ready availability of water (in the wet food) and the high temperatures used in frying. Determination of free fatty acid concentration by titration is sometimes used as an indication of the extent of hydrolysis.

Another reaction closely follows hydrolysis when fat is overheated. Glycerol, freed by lipid hydrolysis, is decomposd through dehydration:

$$
\begin{array}{ccc}
\text{H} & & \text{H} \\
\text{HCOH} & & \text{HC} \\
| & & \| \\
\text{HCOH} & \xrightarrow{\text{heat}} & \text{HC} \qquad +2\ H_2O \\
| & & | \\
\text{HCOH} & & \text{HC}{=}\text{O} \\
\text{H} & & \\
\text{Glycerol} & & \text{Acrolein}
\end{array}
$$

Acrolein formation

Acrolein, the decomposition product, is a volatile, irritating compound. Its formation is evidenced by the acrid fumes emanating from overheated fat. This reaction occurs also if glycerol itself is heated as a pure liquid, though this is not significant from the standpoint of food.

2. Saponification

Saponification is the formation of a metallic salt of a fatty acid; such a salt is called a soap. The reaction involves treatment of free fatty acids and/or glycerides with a base and may be considered a special case of hydrolysis when a glyceride is reacted with a base. Because only one atom of monovalent metal is taken up per fatty acid chain, regardless of chain length, the reaction provides a basis for estimating average fatty acid chain length in a fat sample. The greater the average chain length of a sample, the less sodium or potassium will a given weight of the sample take up. The amount of sodium or potassium taken up can be determined by providing a known, excessive amount of sodium hydroxide or potassium hydroxide and titrating the excess after permitting saponification to occur. Saponification is not a reaction that normally occurs in food. An exception involves excessively alkaline cakes.

3. Interesterification

Interesterification (alcoholysis, transesterification) involves the transfer of fatty acid from the original glycerol to another alcohol. The other alcohol may be free glycerol added for this purpose. In this case, the reaction sometimes is referred to as superglycerination. This is the process by which monoglycerides and diglycerides prepared commercially. Monoglycerides are useful as emulsifiers, as previously mentioned. The monoglycerides resulting from interestification can be separated from diglycerides by distillation, but it is more practical to simply add the monoglyceride and diglyceride mixture to the shortenings.

A second example of interestification is an analytical procedure. For studying the fatty acid content of lipids by gas–liquid chromatography, the fatty acids must be converted to a more readily vaporized form. This is accomplished through preparation of methyl esters by transfer of fatty acids from glycerol to methanol.

4. Rearrangement

Another reaction brought about commercially involves the use of special processing conditions so that fatty acids become randomly distributed on the glycerol molecules. To a certain extent, this may be thought of as a variant of interesterification and has become nearly as common as hydrogenation. The result of rearrangement is a greatly increased number of molecular species and, therefore, greater difficulty of crystallization and greater likelihood of the formation and maintenance of small crystals. In addition, the increased number of molecular species results in a wider melting point range and a greater degree of plasticity of the solid fat.

5. Acetylation

Acetylation is replacement of fatty acids in glycerides with acetate. The most common application probably is the acetylation of monoglycerides. Acetylated monoglycerides may be effective emulsifiers. Because their melting points can be controlled by the reaction conditions, acetylated monoglycerides are useful as ingredients in food coatings.

B. Reactions Involving Double Bonds

1. Oxidation

The type of oxidation discussed here is different from the β oxidation that occurs in lipid metabolism. The oxidation under discussion involves reaction with molecular oxygen. Because it is self-perpetuating, once started, it is referred to as autoxidation. Autoxidation actually involves a series of reactions. It begins with removal of a hydrogen atom from a carbon adjacent to a double-bond carbon, resulting in a free radical:

$$\begin{array}{ccccc} & H & H & H & H \\ -\!\!\!\! & C & - & C = C & - & C & - \\ & \bullet & & & & H \end{array}$$

Free radical

The initial removal of hydrogen requires energy. Therefore, it is influenced by storage temperature. It also is catalyzed by light and by metals, which may be present as contaminants from processing equipment. Autoxidation of lipids is catalyzed by heme iron in both raw and cooked meat. Oxidation occurs more rapidly in cooked than in raw meat, apparently because of catalysis by nonheme iron released during cooking (Rhee, 1988).

The second step is the addition of molecular oxygen to give an activated peroxide:

$$\begin{array}{ccccc} & H & H & H & H \\ -\!\!\!\! & C & - & C = C & - & C & - \\ & O & & & & H \\ & O & & & & \\ & \bullet & & & & \end{array}$$

Activated peroxide

The activated peroxide is very reactive, and relatively little energy is required for removal of a hydrogen from another carbon adjacent to a double-bond carbon (in another chain), resulting in the formation of a new free radical on the second chain and a hydroperoxide on the initial chain:

$$
\begin{array}{ccccccc}
 & H & & H & H & & H \\
- & C & - & C & = & C & - & C & - \\
 & O & & & & & & H \\
 & O & & & & & \\
 & H & & & & &
\end{array}
$$

Hydroperoxide

Hydroperoxide formation is significant from two standpoints: (1) It represents the self-perpetuating feature of autoxidation because a new free radical is formed; and (2) the hydroperoxide itself is unstable and is subject to further oxidation, a variety of rearrangements and cleavages, new reactions made possible by the cleavages, and polymerization. Among the products formed are short-chained aldehydes, ketones, acids, and hydroxyl compounds.

As with hydrolysis, an off-flavor results and is referred to as rancidity. In some foods, the off-flavor development apparently parallels the development of the breakdown product hexanal. The existence of a rancid flavor in itself does not indicate whether the fat has undergone hydrolysis or autoxidation or some of each.

Several factors affect the occurrence of lipid autoxidation.

1. Extent of unsaturation of the lipid: Polyunsaturates are oxidized before monounsaturates. Linoleic acid is oxidized 15 times and linolenic acid about 30 times as fast as oleic acid.
2. Configuration at double bonds: Trans isomers are less susceptible than are cis isomers.
3. Degree of esterification: Glycerides are less susceptible than are free fatty acids.
4. Availability of catalysts to initiate the reaction: Exposure to light is readily controlled. The catalytic activity of metal contaminants can be curtailed by the presence of chelating substances, or sequestrants. These include citrate, tartrate, phosphate, and ethylenediaminetetraacetic acid (EDTA), which form complexes with metals and thus make the metals unavailable. Citrates and EDTA in particular are frequently added to fatty foods that are to be stored.
5. Availability of oxygen: Either vacuum packaging or replacement of air with a nitrogen atmosphere has a protective effect on food to be stored.
6. The presence of antioxidants: Antioxidants apparently contribute hydrogen to stop the autoxidative process at the free radical stage; in doing so, they form stable free radicals that do not propagate further oxidation. They occur naturally along with

some foods lipids; for example, tocopherols are associated with vegetable oils. The phenolic compounds butylated hydroxyanisole (BHA), butylated hydroxytoluene (BHT), tertiary butyl-hydroquinone (TBHQ), and propyl gallate (PG) are antioxidants that are added commonly to commercially processed and manufactured foods, though their use is controversial in some cases. Dziezak (1986) and Sherwin (1978) reviewed the action of antioxidants; Dziezak also described the properties of specific antioxidants and Sherwin discussed considerations in the use of antioxidants. The phenolic antioxidants differ in their solubility and their sensitivity to heat and pH, and most frequently two antioxidants are used in combination. Chelating agents do not have a direct antioxidant effect; they enhance the protective effect of the antioxidants and thus are referred to as synergists.

7. Storage temperature: Both the initiation of autoxidation and the reactions undergone by hydroperoxides are temperature dependent.

The progress of autoxidation may be followed by measuring oxygen uptake over a period of time. In the early stages of autoxidation the oxygen uptake curve is essentially flat. The period of time represented by the flat portion of the curve is referred to as the induction period and may be quite long. During the induction period, free radicals form at a rate that depends on such factors as temperature, exposure to light, and presence of metals and of chelating agents. A sharp rise in the curve (the end of the induction period) tends to coincide with exhaustion of the supply of antioxidant. Rancidity develops primarily during the postinduction period because free radicals are not responsible for the off-flavor. In fast-food operations, where so much frying is done that addition of new fat is almost continuous, most of the frying fat at any one time probably is in the induction period with respect to oxidation. Off-flavors that develop in such fat are more likely to be hydrolytic than oxidative in origin. Accelerated storage studies, conducted routinely in the food industry for assessing shelf life of foods, are conducted at elevated temperatures to shorten the induction period and thus the time required for rancidity to develop.

Chemical tests sometimes used in the study of lipid autoxidation are the determination of peroxide number and the TBA test. Each method has limitations. Although peroxides are important as precursors of the off-flavor compounds, the increase in their concentration does not necessarily parallel the development of rancidity. The TBA test, which involves the formation of a measurable red condensation product between 2-thiobarbituric acid (TBA) and malonic dialdehyde, still is empirical in the sense that what is being measured is not what causes the off-flavor. However, because it does measure a product rather than an intermediate, it correlates quite well with flavor change in some foods, particularly meats (Tarladgis *et al.*, 1960). Recently gas–liquid and high-pressure liquid chromatographic analysis of specific volatile carbonyl compounds has provided a more direct means of studying rancidity development through autoxidation.

2. Reversion

Reversion represents a special type of oxidative degradation. Certain oils—for example, soybean and rapeseed oils—are susceptible to a relatively limited degree of oxidation involving primarily linolenic acid, which is contained in appreciable amounts by those oils that are susceptible (Sherwin, 1978). The term reversion is based on the flavor change because it occurs in oil from which a distinctive, though different, flavor has been removed in processing. The new flavor is unpleasant, described variously as grassy, beany, or fishy.

3. Hydrogenation

Treatment of highly unsaturated oils, at an elevated temperature, with hydrogen under pressure in the presence of a suitable catalyst (usually nickel) results in saturation of double-bond carbons. This chemical change in turn raises the melting point. The reaction is selective in that polyunsaturated fatty acids are hydrogenated first. Double bonds farthest from the carboxyl groups of free fatty acids, or from the ester linkages of glycerides, are saturated first. Thus if the major fatty acid in an oil is linoleic acid, as in cottonseed oil, partial hydrogenation results in a fat in which oleic acid predominates. A detailed discussion of selectivity of hydrogenation and the effects of conditions and type of catalyst is found in Coenen's review (1976).

Hydrogenation is the basis for industrial manufacture of solid shortenings and margarines from oils. Complete hydrogenation would result in excessive hardness. The process is carried just far enough to produce the physical properties desired.

An interesting example of biological hydrogenation occurs in ruminant animals. For many years it was considered impossible to produce beef and lamb having elevated levels of dienoic fatty acids, because rumen bacteria efficiently hydrogenate dienes in unprotected rations. Recent developments, however, have resulted in the production of high-linoleate beef and lamb through feeding them protected supplements (Gunstone and Norris, 1983, p. 56). Not surprisingly, freezer storage life of high-linoleate meat is relatively short because of susceptibility to lipid autoxidation (Bremner et al., 1976). Conversely, hydrogenation of a highly polyunsaturated oil reduces its susceptibility to oxidation.

4. Isomerization

Isomerization, cis–trans (geometric) and positional, occurs as a side reaction of hydrogenation. Conversion from cis to trans configuration makes a substantial contribution to the change in melting point during commercial hydrogenation. Ottenstein et al. (1977), who studied the fatty acid content of seven margarines purchased at retail stores, reported that the samples contained total 18:1 fatty acid averaging about 45% of the total fatty acid content. Nearly half of the 18:1, about 20% of the total fatty acid content, was in the trans form on the average, though the samples differed considerably. With a different catalyst and a lower hydrogen pressure, cis–trans isomerization may be largely responsible for the change in physical state. "Hard butters" appropriate for confectionary coatings may be prepared by hydrogenation of soybean oil under optimum trans conditions, resulting in a greater than 50% trans fatty acid content (Paulicka, 1976).

Positional isomerization is a migration of double bonds in fatty acid chains. Different types of isomerization actually are different reactions but tend to occur simultaneously.

5. Halogenation

The ability of a halogen to be added to double-bond carbons in unsaturated fatty acids is the basis for analytical estimation of the average extent of unsaturation of a fat. Iodine value is the number of grams of iodine absorbed by 100 g of fat and ranges from 30–40 for butterfat to more than 130 for some seed oils. Iodine value is measured less frequently than it was before the advent of gas–liquid chromatography, which provides much more specific information.

Zapsalis and Beck (1985, pp. 437–454) further discuss the major reactions of lipids.

IV. PROCESSING OF FATS

Fats do not occur free in nature but are isolated and purified from foods high in fat, such as cream, animal tissues, fruits, and seeds. Recent technological advances make possible the production of fats that are tailor made for specific uses. The variety of fats available to food manufacturers far exceeds that on the retail market; however, most consumers probably use two or three types of fat. Discussion of processing methods is limited here to principles that are related to chemical and physical properties, already covered, and to functional properties, discussed later.

A. Animal Fats

1. Butter

Cream, separated from milk by centrifugation, is pasteurized to inactivate lipase and destroy most of the organisms present. It usually is ripened by means of a mixed bacterial culture containing both lactose- and citrate-fermenting species and is churned. The churning process results in clumping of fat globules and transformation of the oil-in-water emulsion to a water-in-oil emulsion. Buttermilk is separated from the mass of fat, which is washed with water, colored for uniformity, and usually salted. The final product contains 80–81% fat. Short-chain fatty acids and diacetyl are largely responsible for the characteristic flavor of butter. Diacetyl is an oxidation product of acetylmethylcarbinol, which is produced as the first product of citrate fermentation:

$$CH_3-\overset{\overset{\displaystyle O}{\|}}{C}-\overset{\overset{\displaystyle OH}{|}}{\underset{\underset{\displaystyle H}{|}}{C}}-CH_3 \quad \xrightarrow{-2\ H} \quad CH_3-\overset{\overset{\displaystyle O}{\|}}{C}-\overset{\overset{\displaystyle O}{\|}}{C}-CH_3$$

Acetylmethylcarbinol   Diacetyl

Diacetyl formation

Consistency and texture depend on the fatty acid content of the cream and on the specific chilling and churning treatments.

2. Lard

Lard is heat rendered from fatty tissues of the hog. The fatty tissues, containing connective tissue as well as fat, are chopped and heated, with or without added water. Wet rendering is the more common procedure. The heating may be done in a steam-jacketed kettle or may consist of direct treatment with steam. Once the fat has been separated out, it may be modified in any of several ways, including bleaching, deodorization, hydrogenation, rearrangement, addition of emulsifier, and addition of antioxidant. The β-crystalline form is characteristic of lard.

Lards produced by current technology are quite different in their properties from those represented by data from early research literature. Present-day lards also are tailor made for special purposes. Whereas lard once was the major shortening used, its role today is relatively minor. Vegetable oil shortenings, which were developed as substitutes for lard in times of shortage, have become the shortenings of choice (Knightly, 1981).

3. Tallow

Several shortenings on the retail market consist of a mixture of meat fat and vegetable oil. Beef tallow, one of the fats used in such shortenings, usually is dry rendered and deodorized and may be hydrogenated. It crystallizes in the β' form (Dziezak, 1989). Like lard, it is not a major shortening substance; however, it may be a shortening component. Beef tallow is used as a frying fat or in a blend for frying; it is assumed, correctly or incorrectly, that consumers prefer the flavor of French fries prepared in fat containing beef tallow. A study involving the use of beef tallow fractions in cake is described in Chapter 21. Mutton tallow is also sometimes used, particularly in Great Britain.

4. Fish Oils

Fish oils are of interest from a nutritional standpoint because of their content of five- and six-double bond fatty acids. Although fish oils are not used as food ingredients in the United States, GRAS (generally recognized as safe) status has been requested for use of partially hydrogenated menhaden oil as a food ingredient (Dziezak, 1989).

B. Plant Fats

1. Vegetable Oils

Oilseeds are the major source of oil from plants, and their use has increased tremendously in recent years. Certain fruits, such as olives and coconuts, also are commercial sources of oil.

Seeds such as corn, cottonseed, soybean, and peanuts are cleaned and tempered and dehulled as needed, then flaked or crushed. Oil is removed by solvent extraction, application of pressure, or a combination of these methods. The flesh of olives is pressed for "virgin olive oil" and the residue and pit kernels are pressed further and/or solvent extracted for "pure olive oil." Coconut oil is pressed from dried coconut flesh. Vegetable oils undergo further processing to remove several impurities, including pigments (by adsorption), gums, free fatty acids (by light saponification), and volatile odor-causing compounds (by steam distillation under reduced pressure). Olive oil undergoes minimal processing because of the prized uniqueness of its flavor. Cottonseed oil must be freed of gossypol, a toxic polyphenolic component of cottonseed. Silicones frequently are added to frying oils to suppress foaming. Citric acid, a chelating agent, may be added for its protective effect against oxidation. The additives are used at very low levels as regulated by the FDA.

Coconut oil and oils from the oil palm tree should be mentioned because at this writing consumers are being urged to avoid foods containing them, and food manufacturers are being urged to limit their use. The basis for the negative advice is a high concentration of the C_{12} and C_{14} saturated fatty acids. Coconut oil has a 90–94% content of saturated fatty acids, with medium-chain acids predominating. Babayan (1989) wrote in support of coconut oil, pointing out that a distinction should be made between medium-chain and long-chain saturates. Two different oils are obtained from the oil palm tree, and Berger (1986) noted that a distinction is not always made between palm kernel oil, which is a seed oil, and palm oil, which is extracted from the fruit. It is the palm kernel oil that has the high concentration of medium-chain saturated fatty acids. Palm oil composition includes approximately equal concentrations of saturated and unsaturated fatty acids. Berger presented composition data and information concerning use of the oils in the food industry. It should be noted that neither palm oil nor coconut oil constitutes more than 1% of the United States fat consumption (Dziezak, 1989).

Cottonseed oil was the primary edible oil in the United States for many years. The unhydrogenated oil is used in frying. The flavor is neutral and, therefore, does not mask the flavor of foods. Cottonseed oil also is not susceptible to flavor reversion. Some oils, including cottonseed oil, are winterized to permit them to remain liquid at refrigerator temperature. Winterizing involves chilling to crystallize the saturated high-molecular-weight components, then separating the solids from the cold liquid by filtration. Winterized cottonseed oil is used in salad dressings and as a general purpose oil. Hydrogenated cottonseed oil is used in shortening and spreads. It is stable in the β'-crystalline form (Dziezak, 1989).

Soybean oil has replaced cottonseed oil as the major edible oil in the United States. When it is to be used as a frying fat, it is partially hydrogenated because its normal content of linolenic acid contributes to oxidative instability and consequent off-flavor. The concentration of linolenic acid can be reduced also by dilution with an oil that contains little linolenic acid, such as cottonseed oil. In addition, soybean oil with reduced levels of linolenic acid has been made available through plant breeding (Hammond, 1988). Large amounts of partially hydrogenated soybean oil are used also in shortenings and spreads.

Sunflower oil has become an increasingly important food oil. It is highly unsaturated but its composition is quite variable.

Most of the safflower oil that is available has an unusually high concentration of linoleic acid. Because of the high degree of polyunsaturation, it is not stable under frying conditions. However, a safflower oil with a high concentration of oleic acid is available and can be used as a cooking oil (Dziezak, 1989). Both types are used in a variety of fat and oil products.

Canola oil is relatively new in the United States. It is obtained from a genetically modified rapeseed. Whereas ordinary rapeseed oil has 20–40% erucic acid (22:1), canola oil has less than 2% (Dziezak, 1989); therefore, it is designated on food labels as low-erucic acid rapeseed (LEAR) oil. Its concentration of oleic acid is correspondingly higher than that of ordinary rapeseed oil. Canola oil is being used increasingly in oil and shortening applications. Studies involving its use in cakes are described in Chapter 21.

Liquid shortenings are used increasingly in the baking industry. They perform well because of the addition of surfactants; the kinds and concentrations of surfactants used are important. The surfactants usually are blends of monoglycerides with other surfactants such as ethoxylated sorbitan monostearate and stearyl-2-lactylates. The use of oil as shortening is advantageous from the standpoint of transportation, storage, and convenience of use in bakeries.

The residual solids after oil extraction are rather high in protein and are used in animal feeds. Interest in expanding the supply of protein for humans resulted in extensive study and considerable utilization of soybean protein in human food. For example, the defatted soy flakes can be processed into many forms, including flour (about 50% protein), soy protein concentrate (about 70% protein), or soy protein isolate (approximately 90% protein). The isolate, in turn, can be processed into a variety of textured forms such as spun fibers for use in meat analogs and granular forms for use as meat extenders, which are discussed in Chapter 9.

2. Vegetable Shortenings

Plastic solid shortenings are made by the hydrogenation of refined oils, usually mixtures of soybean and other oils; mixtures promote formation of the desired crystals in the shortenings and also extend the range of plasticity. The oils are treated with hydrogen in the presence of a nickel catalyst. The principles of hydrogenation are discussed earlier in this chapter (see Section III,B,3). When the desired extent of hydrogenation has been attained, the optional ingredients, such as preservatives and a monoglyceride–diglyceride emulsifier, are added and the shortening is chilled rapidly without agitation to 16–18°C. This treatment causes development of nuclei for crystallization. During the rapid crystallization that follows, the shortening is agitated and a large amount of air or an inert gas such as nitrogen is incorporated and finely dispersed. After packaging, plastic shortenings (and margarines for use in baked goods) are tempered for 24–72 hr at a constant temperature between 27 and 32°C. Tempering stabilizes the crystal structure so that it can withstand the temperature fluctuations encountered later (Mounts, 1989).

3. Margarines

Margarines, developed to simulate butter, most frequently are made from mixtures of vegetable oils that are hydrogenated to the desired extent. Emulsifiers, vitamin A and sometimes vitamin D, flavoring, and coloring are added to the oil. The aqueous phase, containing whey or nonfat milk solids, is cultured, pasteurized, and cooled. Salt and water-soluble preservatives, if used, are added to this phase. The two phases are emulsified under controlled conditions of temperature and agitation, forming a water-in-oil emulsion in which the fat concentration is about 80–81%.

Various modifications have resulted in proliferation of margarine types. One modification consists of hydrogenating some oil to a very hard stage and then mixing unhydrogenated oil with the hard fat. This results in the desired consistency along with a higher concentration of polyunsaturated fatty acids than is characteristic of the margarines made by hydrogenating all of the oil to the desired consistency. (Remember the selectivity of the hydrogenation reaction.) Most of the currently available products of this type are made of corn oil.

Some margarines are physically modified by whipping to incorporate nitrogen and thus increase volume. Such spreads on the market represent about a 50% increase in volume over that of conventional margarines. The products are relatively soft and easy to spread and are lower in calories only if they are substituted for a conventional spread on a volume basis. On a weight basis, they are not lower in calories. On the other hand, there are some special purpose spreads that are made with lower fat (not usually more than 50%) and higher water contents than those of butter and margarine.

Another variant on the retail market is a margarine that is similar to a conventional margarine but has a high enough concentration of unhydrogenated oil to result in a semisolid consistency even at room temperature. The cold margarine is soft enough to be easily extruded from a squeeze-type bottle. It also tends to have a somewhat lower fat concentration (70–75%) than do conventional margarines.

At least one margarine available at this writing contains buttermilk, which is added for flavor. Its fat concentration is about 75%, a little lower than that of conventional margarines. A spread that is currently available is neither margarine nor butter but a mixture of corn oil margarine and butter in 60:40 proportions.

Mounts (1989) described the margarine manufacturing process and some of the margarine types. Gunstone and Norris (1983, p. 153) described margarines of different types in relation to the combinations of oils of different iodine values used in their production.

V. FUNCTIONS OF FATS IN FOOD

The contributions of fat to food properties are related to the chemical nature of lipids and the resulting physical properties of fats. The effects of fats on food are largely textural but to a lesser extent also involve color and flavor.

A. Tenderization

One of the most important functions of fat is to tenderize baked products that otherwise might be solid masses firmly held together by strands of gluten. This function is particularly important in pastry and breads, which have little or no sugar. Fat, being insoluble in water, interferes with gluten development during mixing. Fat adsorbed on surfaces of gluten proteins interferes with hydration and thus with the development of a cohesive gluten structure. In the mixing of pastry and biscuits, intimate mixing of the fat with the other ingredients is deliberately avoided so that the fat will cause layers of dough to form. These products are likely to be flaky as well as tender. However, flakiness and tenderness are different properties and may not be entirely compatible; as flour and fat are mixed more thoroughly, the product may become more tender but less flaky. The measurement of pastry tenderness is discussed in Chapter 3.

Oils and soft solid fats, which can be spread readily through a batter or dough, are particularly effective shortening agents. However, attempts to relate shortening power to degree of lipid unsaturation, physical state, melting point, plasticity, and other properties of shortenings have not had consistent results. The importance of such characteristics has been lessened by the use of emulsifiers in shortenings used in batter products.

One of the many problems in relating fat properties to shortening power in pastry is the dependence of pastry tenderness on many factors in addition to the shortening. For example, mixing temperature and method affect pastry tenderness. Even if control of conditions is excellent and the effects of the fats on tenderness are obvious, interpretation frequently is difficult because of the heterogeneity and variability of fats and the interrelatedness of their chemical and physical properties. If an oil performs differently from a solid fat, is the effect attributable to a difference in unsaturation, in physical state, or in crystallinity? It probably is impossible to compare two fats that differ from one another in a single respect. This is not an argument against comparing fats as to shortening power but an argument for exercising caution in interpreting results, as well as for control of experimental conditions.

B. Aeration

Fat contributes to the incorporation and retention of air in the form of small bubbles distributed throughout a batter. These bubbles serve as gas cell nuclei into which carbon dioxide and steam diffuse during baking; thus they are important to grain and volume of the baked product. The role of fat in aeration may be related to its effect on batter viscosity. Finely dispersed fat enhances batter viscosity; increased viscosity, up to a point, increases gas retention. Monoglycerides added to hydrogenated shortenings increase the extent of emulsification of shortening; increased viscosity and increased batter aeration are associated with the increased emulsification.

C. Heat Transfer

Rapid growth of the fast food industry has resulted in greatly increased consumption of frying fats. Interest in quality of frying fats and of fried foods has increased accordingly. Soybean oil is the major frying fat used in the United States, though other oils and solid fats also are used. Canola oil (low-erucic acid rapeseed oil) has become a major product in Canada, where it was developed, and interest in its use is growing in the United States. Hydrogenation of canola oil reduces oxidative reactions and thus increases its frying life. The study of Warner et al. (1989) is one of many studies that have dealt with the effects of processing and/or storage conditions on sensory and oxidative stability of various frying oils.

The temperature to which a fat can be heated is limited only by its smoke point, the temperature at which degradation is sufficient to result in evolution of smoke. Smoke point is much higher than the boiling point of water; therefore, frying temperatures ranging from 175 to 195°C are used and permit much more rapid cooking than occurs in water. Surface drying and carbonyl-amine (Maillard) browning occur, and a distinctive flavor is associated with the browning. Crusts of fried foods are crisp because of the moisture loss and tender because of fat absorption. Although air can be heated to an even higher temperature than can fat, air is far less efficient as a heating medium.

Smoke points of fats currently available are beyond the temperature range used for normal frying, and problems develop only with misuse of fat. Fat degradation with consequent lowering of smoke point occurs gradually during heating, increasing both time and temperature. The availability of water affects the rate of hydrolytic degradation, as mentioned previously. Mono- and diglycerides in the fat lower smoking temperature; therefore, a fat that is ideal for a cake batter is not the best frying fat. The life of a frying fat is extended by the addition of fresh fat and by such practices as filtering out food particles between uses.

In addition to the degradation associated with smoking, frying fat gradually oxidizes, polymerizes, and increases in viscosity and in tendency to foam. Changes in sensory quality, functional properties, and nutritional value accompany the chemical changes. The chemical and physical events that occur during frying are summarized in Fig. 15.1.

Frying conditions and food composition affect the extent to which food absorbs fat during frying. The lower the temperature of the frying fat, the longer is the time required for frying and the greater is the amount of fat absorption; however, if the temperature is too high for the food being fried, surface browning is adequate before the interior is cooked. Fat absorption during the frying of flour mixtures depends on product formulation; the extent of absorption increases with increased fat and sugar concentration and is greater with a weak flour than with a strong flour. It should be noted that fat "absorption" actually includes some adsorption.

McComber and Miller (1976) reported that the total lipid content of doughnuts was affected by the type of baking powder used and by the concentration of lecithin in the formula. Bennion (1967) found the fatty acid composition of absorbed fat in fried fritter-type batter and of the frying fat to depend on the presence of baking powder in the batter; total lipid unsaturation was greater in the presence of baking powder than in its absence, but several interactions made generalization difficult. Heath et al. (1971) used different frying fats and different batter coatings in preparation of fried chicken parts. The batter was separated from the cooked chicken and each was analyzed for fatty acid content. The cooking oils not only determined the fatty acid profile in the batter samples but

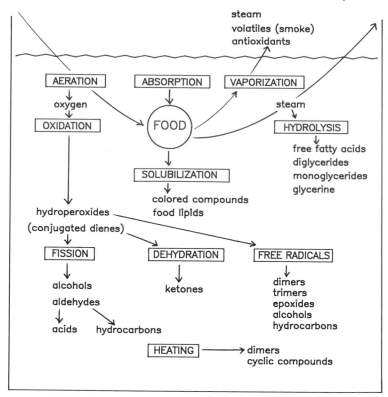

FIGURE 15.1.
Changes occurring during deep-fat frying. (From Fritsch, 1981.)

also affected the linoleic and linolenic acid contents of the tissue. When egg was omitted in the batter, the tissue content of 18:2 and 18:3 fatty acids was increased.

Stevenson *et al.* (1984) reviewed the chemical and physical changes that occur in frying fat and in food during deep frying, factors that affect the changes, methods of testing fat quality and their limitations, and implications for frying operations.

D. Contribution to Emulsion Structure

Food emulsions could not exist without fats; fats constitute one of the essential phases. Food emulsions occur naturally, as in milk, and are prepared in foods such as mayonnaise and cake batters. Except in butter and margarine, oil is the dispersed, or discontinuous, phase in food emulsions, and water is the dispersion medium, or continuous phase. Oil would not remain dispersed in an aqueous medium in the absence of a third phase, the emulsifying agent. The lipoprotein of egg yolk is a particularly effective emulsifying agent and functions in that capacity in many foods. In mayonnaise, in which the oil concentration is particularly high, dilution of egg yolk with egg white, or total substitution of egg white for yolk, results in a product that is less viscous and less stable than the product with egg yolk. Starch contributes to emulsification of fat in some foods, such as gravies and sauces. Surfactants added to shortenings function as emulsifiers in cake batters. Vegetable gums, which are used increasingly in formulated foods, function in many ways, including emulsification.

Evaporation of water from an emulsion can be disastrous. In an oil-in-water emulsion in which the amount of dispersed phase is about at its upper limit relative to the dispersion medium (water), the system is vulnerable. Gravy is a food in which the fat and water proportions vary greatly. Gravy also is a food that sometimes is subjected to prolonged heating and holding and thus to opportunity for evaporation to occur. It is not unusual, therefore, for fat to separate from a gravy during holding. If this happens to a gravy that initially appears homogeneous, the solution is to replace the lost dispersion medium or, in other words, to add water.

E. Other Contributions to Texture

Fats have additional textural effects in foods. They affect the smoothness of crystalline candies and frozen desserts through retardation of crystallization. They affect consistency of starch-thickened mixtures by influencing gelatinization. They contribute to apparent juiciness of meats. They contribute to the foam structure of whipped cream.

F. Contributions to Flavor

Fats influence flavor, whether added for that purpose or to serve another function. Often a specific fat is chosen for a specific use because of its unique flavor. Butter, bacon fat, and olive oil are examples. Fats also act as carriers or solvents for added food flavors.

VI. FAT SUBSTITUTES

Several fat substitutes have been developed or are under study as a result of consumer interest in reduction of dietary fat. They differ in the extent of potential calorie reduction.

Olestra and Simplesse have received the most publicity. Olestra is a generic name for a group of products consisting of sugar–fatty acid polyesters. Such compounds are possible because of the large number of alcohol groups on sugar molecules. Sucrose polyesters have been studied most extensively thus far. They are not digested or absorbed and, therefore, do not provide calories. They are heat stable and have many potential food applications. Those containing the higher fatty acids are possible substitutes for shortenings. Food and Drug Administration approval has been requested (in a Food Additive Petition) for their use as replacement of 35% of the fat in shortening and oil for home use and up to 75% of frying fat for commercial use.

Simplesse is egg white and/or milk protein that has been subjected to micro-particulation, a process that partially coagulates the protein under heat and shear conditions that result in minute spherical particles. The product is perceived as having a smoothness similar to that of fat. It contains protein and water in a ratio of 1:2 and provides 1.3 kcal/g (Harrigan and Breene, 1989). Its potential use is more limited than that of Olestra because its composition makes it unstable to heat. However, it lends itself to substitution in dairy products, such as cream cheese and cheese spread, ice cream, yogurt, and sour cream, as well as in salad dressings and spreads; GRAS status for its use in frozen desserts was obtained in 1990.

Additional fat substitutes mentioned in Dziezak's (1989) review are a corn maltodextrin, a tapioca dextrin, and a potato starch-based product, all of which are currently in use in certain products. Dziezak also mentioned some potential fat substitutes that are currently under study. Further developments in this area can be expected in the future.

Suggested Exercises

1. Compare the ingredient labels on a solid and a fluid margarine, preferably of the same brand, and explain the relationship between their chemical composition and their physical state.
2. Obtain conventional margarine and whipped margarine, preferably of the same brand. Determine either their density or their specific gravity (Chapter 3).
3. Compare products containing the two margarines of exercise 1 and/or those of exercise 2. In the case of the margarines of exercise 2, use them both on an equal-volume basis and on an equal-weight basis. Product formulas are in Appendix A.
4. Compare products containing a conventional margarine and a stick margarine that is more highly polyunsaturated. (How can you determine that one is more highly polyunsaturated than another?)
5. Compare cakes (basic formula in Appendix A) made with butter, with and without an added emulsifier such as a monoglyceride–diglyceride mixture.
6. Compare pastry (basic formula in Appendix A) made with shortening and with margarine substituted both on an equal-weight basis and on an equal-fat basis.
7. Compare emulsifying agents as follows:
   a. Pour 5 ml of salad oil into each of seven test tubes.
   b. Make one of the following additions to each tube:
      i. 5 ml of water
      ii. 5 ml of vinegar
      iii. 5 ml of vinegar with 1/4 tsp added paprika
      iv. 5 ml of a 1:1 aqueous dilution of egg white
      v. 5 ml of a 1:1 aqueous dilution of egg yolk
      vi. 5 ml of a gelatinized 1% starch suspension
      vii. 5 ml of a heated and cooled 1% gelatin dispersion

c. With the thumb over the end of the tube, shake each tube vigorously 50 times.

d. Record times required for beginning and completion of separation.

References

Babayan, V. K. 1989. Sense and nonsense about fats in the diet. *Food Technol.* **43**(1): 90–91, 207.

Bennion, M. 1967. Effect of batter ingredients on changes in fatty acid composition of fats used for frying. *Food Technol.* **21**: 1638–1642.

Berger, K. 1986. Palm oil products: Why and how to use them. *Food Technol.* **40**(9): 72–79.

Bremner, H. A., Ford, A. L., Macfarlane, J. J., Ratcliff, D., and Russell, N. T. 1976. Meat with high linoleic acid content: Oxidative changes during frozen storage. *J. Food Sci.* **41**: 757–761.

Coenen, J. W. E. 1976. Hydrogenation of edible oils. *J. Amer. Oil Chemists' Soc.* **53**: 382–389.

CRC. 1989–1990. "Handbook of Chemistry and Physics," 70th edn., Weast, R. C. (Ed.). CRC Press, Inc., Boca Raton, Florida.

Dziezak, J. D. 1986. Preservatives: Antioxidants. *Food Technol.* **40**(9): 94–102.

Dziezak, J. D. 1989. Fats, oils, and fat substitutes. *Food Technol.* **43**(7): 66–74.

Fritsch, C. W. 1981. Measurements of frying fat deterioration: A brief review. *J. Amer. Oil Chemists' Soc.* **58**: 272–274.

Gunstone, F. D. and Norris, F. A. 1983. "Lipids in Foods. Chemistry, Biochemistry and Technology." Pergamon Press, Oxford, England.

Hammond, E. G. 1988. Trends in fats and oils consumption and the potential effect of new technology. *Food Technol.* **42**(1): 117–120.

Harrigan, K. A. and Breene, W. M. 1989. Fat substitutes: Sucrose esters and Simplesse. *Cereal Foods World* **34**: 261–267.

Heath, J. L., Teekell, R. A., and Watts, A. B. 1971. Fatty acid composition of batter coated chicken parts. *Poultry Sci.* **50**: 219–226.

Knightly, W. H. 1981. Shortening systems: Fats, oils, and surface-active agents—present and future. *Cereal Chem.* **58**: 171–174.

McComber, D. and Miller, E. M. 1976. Differences in total lipid and fatty acid composition of doughnuts as influenced by lecithin, leavening agent, and use of frying fat. *Cereal Chem.* **53**: 101–109.

Mounts, T. L. 1989. Processing of soybean oil for food uses. *Cereal Foods World* **34**: 268–272.

Nawar, W. W. 1985. Lipids. In "Food Chemistry," 2nd edn., Fennema, O. R. (Ed.). Marcel Dekker, Inc., New York.

Ottenstein, D. M., Wittings, L. A., Walker, G., Mahadevan, V., and Pelick, N. 1977. *Trans* fatty acid content of commercial margarine samples by gas liquid chromatography on OV-275. *J. Amer. Oil Chemists' Soc.* **54**: 207–209.

Paulicka, F. R. 1976. Specialty fats. *J. Amer. Oil Chemists' Soc.* **53**: 421–424.

Rhee, K. S. 1988. Enzymic and nonenzymic catalysis of lipid oxidation in muscle foods. *Food Technol.* **42**(6): 127–132.

Sherwin, E. R. 1978. Oxidation and antioxidants in fat and oil processing. *J. Amer. Oil. Chemists' Soc.* **55**: 809–814.

Stevenson, S. G., Vaisey-Genser, M., and Eskin, N. A. M. 1984. Quality control in the use of deep frying oils. *J. Amer. Oil Chemists' Soc.* **61**: 1102–1108.

Tarladgis, B. G., Watts, B. M., Younathan, M. T., and Dugan, L., Jr. 1960. A distillation method for the quantitative determination of malonaldehyde in rancid foods. *J. Amer. Oil Chemists' Soc.* **37**: 44–48.

Warner, K., Frankel, E. N., and Mounts, T. L. 1989. Flavor and oxidative stability of soybean, sunflower and lower erucic acid rapeseed oils. *J. Amer. Oil Chemists' Soc.* **66**: 558–564.

Zapsalis, C. and Beck, R. A. 1985. "Food Chemistry and Nutritional Biochemistry." John Wiley and Sons, New York.

STARCH

The importance of starch as a dietary component has increased with recent emphasis on the benefits of decreased consumption of fat and sugar and increased consumption of complex carbohydrates. As a result, extensive efforts are being made toward improved understanding of starch and increased utilization of it in the food industry.

Starch is obtained primarily from cereal seeds and certain roots and tubers. Corn (maize) is the major commercial source of starch worldwide, but other sources, such as wheat, rice, potatoes, and cassava, are important in some countries. Wheat is the major source of starch when the starch and gluten are of particular economic importance separately, as in Australia and New Zealand (Knight and Olson, 1984, pp. 492–493). Starch has a storage function in the tissues in which it is found. The details of its isolation from plant tissue vary with the tissue and are not presented here.

The chemical nature of starch, involving the molecular structures, properties, and changes they can undergo, and the physical nature of starch, involving granular structures, properties, and changes they can undergo, need to be distinguished from, and yet related to, one another. The changes that occur in starch during food production, as well as the commercial process of starch modification and the use of modified starches are also discussed in this chapter.

FIGURE 16.1
Segment of amylose chain.

I. THE CHEMICAL AND PHYSICAL NATURE OF STARCH

A. Molecular Starch

Starch is a polysaccharide. Starch molecules of two types, amylose and amylopectin, occur together in most sources. Both amylose and amylopectin are polymers of D-glucose. Molecular weights of the amylose and amylopectin components of starch differ greatly between sources, between specific samples, and probably even between molecular species in a given sample. Averages thus are not meaningful, but amylopectin molecules are much larger than amylose molecules; molecular weight values that have been reported for amylopectin run into the millions (Hoseney, 1986, p. 39).

Each starch molecule, whether amylose or amylopectin, and regardless of size, has only one reducing end group, the group at carbon-1. Because enzymes that hydrolyze amylose and amylopectin expose reducing groups, reducing power has been used as an indication of changing molecular size during hydrolysis.

1. Amylose

The α-D-glucopyranose units in amylose are in α-1,4-glycosidic linkage (Fig. 16.1). Although amylose molecules are essentially linear, a small degree of branching exists. The side chains on those molecules that are branched are few and so long that they act similarly to unbranched molecules (Hoseney, 1986, p. 38). The degree of polymerization (number of glucose units per molecule) ranges from 500 to 2000 (Pomeranz, 1987, p. 400).

The long amylose chains readily associate with each other. In spite of the linear chemical structure, an amylose molecule has the ability to change its shape with changes in its environment. The presence of many hydroxyl groups confers a high degree of hydrogen-bonding capability; therefore, strong internal forces permit different molecular shapes to exist. A helical form apparently is common for amylose in the solid state (Whistler and Daniel, 1984, p. 161); the following is a segment of a helix:

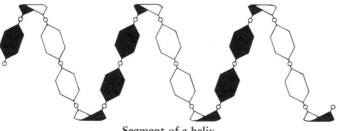

Segment of a helix

Each helix has six α-D-glucose units per turn (Whistler and Daniel, 1984, p. 161). A double helix forms when different helices pack together. The proposed double helix (French and Murphy, 1977) is represented by the following computer-drawn figure:

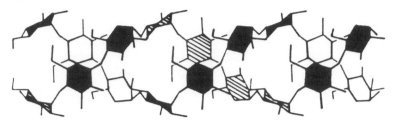

An open channel in the center of a helix permits complexing with other molecular species. For example, iodine can be visualized as a core running through the center of helices when an iodine solution is added to a starch dispersion. One molecule of iodine complexes with six glucose units in one turn of the helix. If the helix is long enough, as in amylose, the complex is blue. Fatty acids also can form complexes with long glucan chains such as amylose.

French (1979) detailed possible shapes of amylose molecules, with discussion of the available evidence. Whistler and Daniel (1984, pp. 164–167) also reviewed evidence in support of different molecular forms in water.

Amylose constitutes 20–30% of the total starch in the nonwaxy cereal starches and in potato starch. Waxy starches, such as that from waxy maize, have essentially no amylose. Some sources of relatively high-amylose starch exist. Examples are the wrinkled pea, with starch having about 66% amylose, and some corn varieties, with starch having up to 75% amylose. The high-amylose corn varieties are referred to as amylomaize. Young (1984, p. 251) presented data from several sources on amylose content of starches. Differences in amylose concentration contribute to differences in functional performance of different starches, as is discussed later in this chapter.

It is difficult to generalize about the solution properties of amylose because of the changes that occur in its isolation, the different isolated forms resulting from different isolation methods, and the dependence of solubility on the dispersion medium and temperature. Young (1984, pp. 267–268) reviewed the literature on amylose solubility.

Because of the ability of the linear amylose molecules to orient themselves with each other and associate through hydrogen bonding, it is possible to produce amylose films. Young (1984, pp. 269–274) reviewed available information on properties of amylose films produced under different conditions. Although a film from a food source might seem to be an ideal coating or packaging film for foods, the high cost of starch fractionation limits industrial application of this property of amylose. Unfractionated high-amylose starch, however, has some potential for use in films.

2. Amylopectin

The chains of amylopectin, like those of amylose, are formed from glucose units in α-1,4-glycosidic linkage; the branch points result from α-1,6-linkages (Fig. 16.2). The degree of polymerization ranges from 10^4 to 10^5, making amylopec-

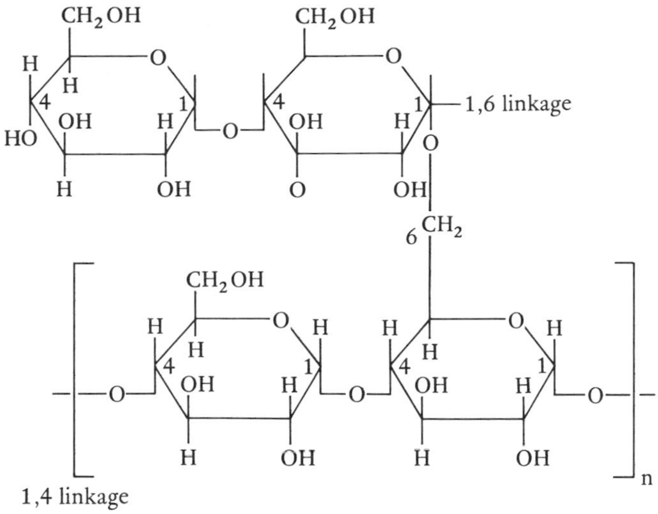

Amylopectin

FIGURE 16.2
Segment of amylopectin molecule.

tin one of the largest naturally occurring macromolecules (Manners, 1985). Its large size and highly branched structure are responsible for the high viscosity of amylopectin dispersions.

The branches in amylopectin molecules are far shorter than amylose molecules; chain lengths in amylopectin average 20–25 glucose units (Manners, 1985). Because of the shortness of the chains, the helices are too short to form a blue color with iodine; the complexes formed between amylopectin and iodine are reddish purple.

The conformation of amylopectin formerly was visualized, on the basis of limited enzymatic studies, as simply a treelike structure. As new information concerning specific enzymes and their actions was obtained and new technologies were developed, later studies involved the use of several amylases and several debranching enzymes that differ in their specificity; gel permeation chromatography of the mixtures of products produced specific information as to arrangement of chains within the macromolecules. It now seems apparent that branch points are arranged in clusters and are neither random nor regular, as proposed in early models. Several cluster models have been proposed, two of which are shown in Figs. 16.3 and 16.4. Manners (1985) discussed several models, along with some of the supporting evidence.

Three types of chains in each amylopectin molecule have long been recognized and should be described briefly in explanation of the letters that appear in Fig. 16.4. A chains are linked to B chains only through their potential reducing group (carbon-1); because each chain has only one potential reducing group, each A chain can be attached to only one other chain. B chains are linked to A chains through their carbon-6 atoms and each chain can be linked to one or more A chains; a B chain also can be linked to another B chain or to a C chain

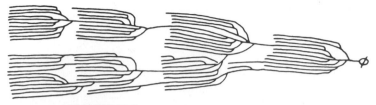

FIGURE 16.3
Cluster model for amylopectin. (From French, 1973.)

FIGURE 16.4
Cluster model for amylopectin. (From Manners and Matheson, 1981; Courtesy of Elsevier Science Publishers, Physical Sciences and Engineering Division, New York.)

through its carbon-1. The single C chain present has the only reducing end group in the macromolecule and is attached to B chains through carbon-6 atoms (Manners, 1985).

There also has been considerable interest in the relative proportions of A and B chains because they indicate the extent of multiple branching. With different analytical methods, different ratios have been obtained, with some differences among amylopectins from different sources; an A:B ratio of 1:1 to 1.5:1 probably is common (Manners, 1985).

B. Starch Granules

The physical structural unit of starch is the granule, which has a distinctive microscopic appearance for each plant source, except that sorghum starch is microscopically similar to corn starch. Both light microscopy and scanning electron microscopy (SEM) have been used in the study of granule form. Scanning electron micrographs show surface characteristics that are not apparent in optical micrographs.

A corn starch sample has a high proportion of angular granules, as well as some rounded granules (Figs. 16.5 and 16.6). Whereas the angular granules have dimpled surfaces, the rounded granules are smooth (Fitt and Snyder, 1984, p. 678). Diameters range from 5 to 25 μm (Snyder, 1984, p. 664). The hilum (original growing point according to French, 1984, p. 189) is more apparent in optical micrographs of corn starch than in those of other starches.

Wheat starch granules are of two types: relatively small, generally spherical granules and large lens-shaped granules (Fig. 16.7 and 16.8). Surfaces of different granules may be smooth, dimpled, or striated (Fitt and Snyder, 1984,

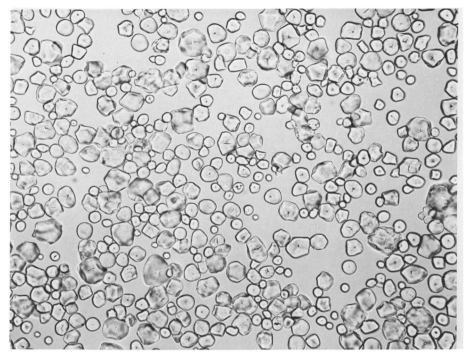

FIGURE 16.5
Corn starch, ×178.5. (From Fitt and Snyder, 1984.)

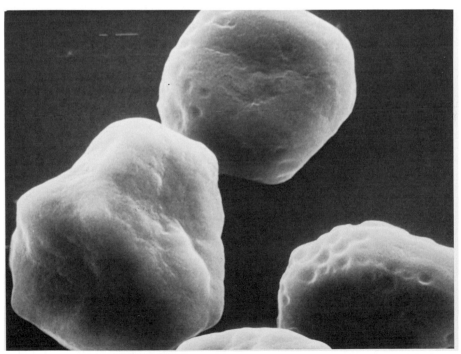

FIGURE 16.6
Corn starch, ×2550. (From Fitt and Snyder, 1984.)

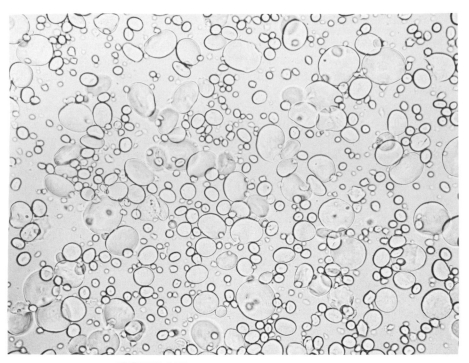

FIGURE 16.7
Wheat starch, ×178.5. (From Fitt and Snyder, 1984.)

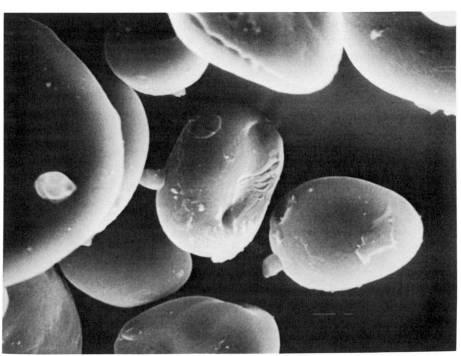

FIGURE 16.8
Wheat starch, ×2550. (From Fitt and Snyder, 1984.)

p. 686). The small granules fall largely in the range of 5–10 μm and the large granules are 25–40 μm (Hoseney, 1986, p. 43), although the size groups are not sharply defined, and some granules are of intermediate size. The hilum is not visible.

Most potato starch granules are large, oval, and particularly smooth (Figs. 16.9 and 16.10). Growth rings are more apparent in micrographs of potato starch than in those of most other starches. They appear as concentric rings, and their appearance depends on extent of hydration; they are not visible in very dry samples. French (1984, p. 208) believed growth rings to represent layers of different degrees of starch concentration that occur because of fluctuations in starch deposition during plant maturation. Granule diameters may be as great as 100 μm or more (French, 1984, p. 184).

Each starch granule contains many starch molecules. Molecular arrangement within a starch granule is at least to some extent radial, as evidenced by fibrillar fracture surfaces that, when starch granules are fractured, can be seen to extend radially rather than along growth ring boundaries (French, 1984, p. 194).

Starch granules are birefringent, indicating a high degree of internal order. Birefringence is the ability to refract light in two directions and is evidenced by distinctive polarization patterns in a polarizing microscope. The dark cross typical of native starch granules is seen in Fig. 16.11. X-Ray diffraction studies show that the molecular arrangement in a starch granule is such that there are crystalline (micellar) regions imbedded in an amorphous matrix. Loss of birefringence indicates disruption of the molecular arrangement in the crystalline areas and is used as the major criterion of gelatinization, as is discussed in IIc.

The molecular nature of the crystallinity within starch granules is yet to be fully clarified but certainly involves amylopectin in both nonwaxy and waxy starches, perhaps to an extent that differs for different starches. Amylose is responsible for the crystalline structure in amylomaize (Pomeranz, 1987, p. 399), although Boyer and Shannon (1987, p. 262) stated that the extent of crystallinity in high-amylose starches is not great. The crystalline regions are more resistant to enzymatic and chemical action and to penetration by water than are the amorphous regions in starch granules.

The amylopectin models in Figs. 16.3 and 16.4 show how branched molecules can be involved in formation of alternating crystalline and amorphous regions of the granule. The packing together of clustered branches would provide crystallinity; the branch points are considered to be in the amorphous regions. Lineback (1984) reviewed the available evidence and proposed the granule model in Fig. 16.12, which includes amylose, present in part as a helical complex with lipid in those starches that contain lipid. Figure 16.12 represents only a portion of a granule. Amylopectin chains, at least the outer ones, would be in double helices. Lineback visualized the granule surface as having protruding chains of different lengths. Whistler and Daniel (1985, p. 114) stated that the chain ends are packed tightly together and the surface resembles the bottom of a broom.

Minor components of starch granules affect their functional performance—their pasting and gelling behavior. Lipids are present both in granule interiors and on granule surfaces. Internal lipids, primarily lysophospholipids, are associated with amylose in corn and wheat starches; potato starch has very little inter-

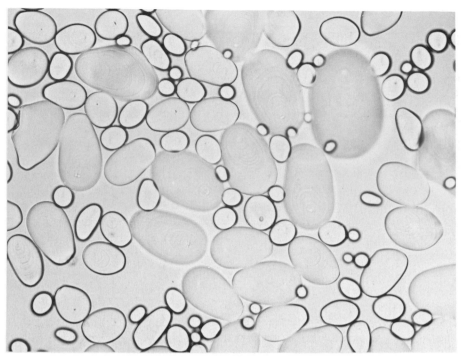

FIGURE 16.9
Potato starch, ×178.5. (From Fitt and Snyder, 1984.)

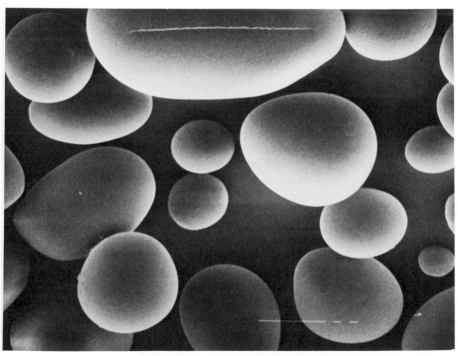

FIGURE 16.10
Potato starch, ×750. (From Fitt and Snyder, 1984.)

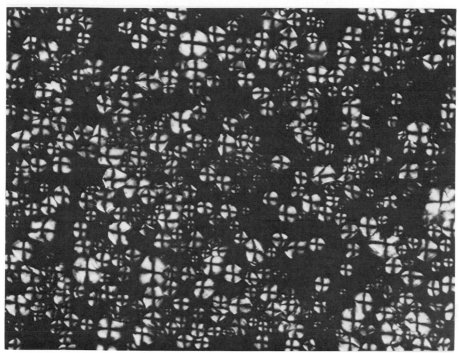

FIGURE 16.11
Corn starch photographed with polarized light to show birefringence of granules, ×357. (From Fitt and Snyder, 1984.)

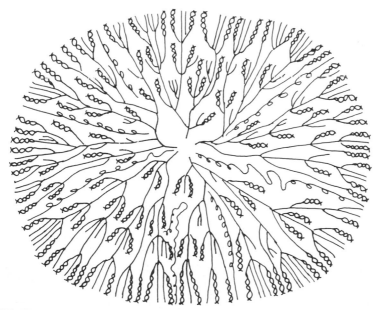

FIGURE 16.12
Schematic of the organization (structure) of a starch granule. (From Lineback, 1984.)

nal lipid. Surface lipids probably are deposited on granule surfaces from other tissue structures. Small amounts of internal and surface protein also are associated with starch granules. Potato starch has a significant amount of phosphate esterified with glucose residues of amylopectin (Galliard and Bowler, 1987, pp. 68–69).

Starch particles, as they exist in powdered starch, are agglomerates of granules. The agglomerates readily break up into a suspension of granules when the starch is combined with water. Starch is not water soluble because the granules are too large to form a solution. Molecular starch does form a colloidal dispersion, but a molecular dispersion is not attained simply by combining raw starch with water.

Starch has a relatively high density, about 1.45–1.64g/cm^3, depending on its source, its prior treatment, and the measurement method (French, 1984, pp. 219–220). It readily settles out of a suspension. Starch granules do absorb some water from an unheated suspension but the amount of swelling is limited except in mechanically damaged starch.

II. PROCESSES UNDERGONE BY STARCH

A. Enzymatic Treatment

Several enzymes act on amylose and amylopectin molecules. The α-amylases are endoenzymes, hydrolyzing α-1,4-glycosidic linkages randomly in the inner regions of molecules, including within the branch points of amylopectin. Products are of varying size and include oligosaccharides and low-molecular-weight dextrins. The dispersion shows a rapid decrease in viscosity. The β-amylases, exoenzymes, also hydrolyze α-1,4 bonds. They split amylose chains into maltose units, beginning at the nonreducing end. They free maltose from the branches of amylopectin until branch points are reached. The product, in addition to maltose, is a single large dextrin, called limit dextrin, for each amylopectin molecule attacked. Because of the accumulation of maltose, β-amylases are called saccharifying enzymes. β-Amylases are more heat sensitive than are α-amylases. Glucoamylases release glucose units from nonreducing terminals of both amylose and amylopectin. Because they are able to catalyze hydrolysis of both 1,4 and 1,6 linkages, they do not form limit dextrins (Robyt, 1984, p. 88).

Debranching enzymes attack the α-1,6 linkages in amylopectin. R-enzyme, isoamylase, amylo-1,6-glucosidase, and pullulanase are examples. They differ as to whether they act on limit dextrin or on macromolecular 1,6 bonds or both. Once a debranching enzyme has acted on branch points, β-amylase catalysis of maltose production can proceed. The literature is somewhat inconsistent as to terminology and specific action of debranching enzymes. For additional reading see Banks and Greenwood (1975, pp. 233–236) and Robyt (1984, pp. 88–89).

The various kinds of enzymes that catalyze hydrolysis of amylose and amylopectin molecules are obtained from a variety of sources—fungi, bacteria, cereals and other plants, and, in the case of α-amylase, even animal organs. Their efficiency, specificity, and optimum conditions for activity depend on the source.

A major application of enzymatic hydrolysis of starch is the production of starch syrups for use as sweeteners. The development of immobilized enzyme technology (see Chapter 6) made a large contribution to the starch syrup industry. Starch syrups are discussed in Chapter 23.

B. Starch Modification

There is a huge array of treatments that can be applied to raw starches and thus change their performance in use. Improved resistance to heat, acid, and shear stresses can be achieved. Temperatures at which thickening occurs can be altered, and specific properties of finished products can be controlled. Shelf stability of foods destined for cold storage and freezer storage can be enhanced. Starches can be treated for many kinds of modification of their functionality in starch-containing foods. Although not on the retail market, modified starches have become increasingly important in the food industry and are ingredients in many consumer products.

There are two types of instant starch. Pregelatinization is a relatively simple modification treatment that makes starches dispersible in cold water. As described by Colonna *et al.* (1987, pp. 99–100), the process involves applying either a cooked starch paste or a slurry of raw starch in water to a hot surface such as a heated drum. The process damages granules, and the dry product quickly takes up water when dispersed. Its thickening efficiency is not more than 80% of that of the corresponding untreated starch (Pomeranz, 1985, p. 73), but it is a convenient ingredient in products such as instant pudding and cake mixes. A more recently developed type of instant starch is referred to as "cold water-swelling" (CWS) starch. It differs from pregelatinized starch in having intact granules and, therefore, greater stability and better texture (Luallen, 1985). The use of CWS starches in microwavable foods was described recently (Anon., 1988), but the process of producing CWS starch was not described.

There are many chemical treatments used to modify starch; they are possible because of the many hydroxyl groups available for reaction with the chemicals used. Chemical treatment involves holding a suspension of starch in a dilute solution of the modifying reagent at a temperature that is too low to permit appreciable granule swelling but high enough to cause some disruption of internal order and permit entry of some water and reagent. The reactions between the reagent and starch molecules take place just within the granule and on the surface during the treatment (Rogols, 1986); the effects are apparent at the time of use and depend on the specific modification. The treated starch is dried for use as an ingredient.

Two major groups of chemically modified starches are cross-linked and substituted (derivatized) starches. Cross-linking involves the use of a chemical that has more than one functional group and, therefore, can react with hydroxyl groups on two different chains, forming a link between them. Polyphosphates are among the cross-linking reagents that can be used. Cross-linking does not prevent granule swelling during pasting, but it helps maintain granule integrity during heating and can prevent gelation by inhibiting loss of amylose from granules. Cross-linked starches are acid stable but not necessarily freeze–thaw stable (Luallen, 1985).

Substituted (derivatized) starches have been reacted with monofunctional reagents that have such effects as increasing the bulkiness of amylose molecules or
introducing repelling electrical charges. The openness of granule structure and
the rate of gelatinization are increased and a high paste viscosity is attained; gelation, however, is inhibited. Starch esters and starch ethers are examples. A
starch ester can be produced through treatment of starch with a reagent such as
acetic anhydride. A starch ether is formed when hydrogen molecules on hydroxyl groups are replaced by alkyl groups such as propyl groups. Substituted
starches are freeze–thaw stable but not acid stable; they also are not appropriate
in foods that require prolonged heating (Luallen, 1985).

Because both cross-linked and substituted starches exhibit inhibition of gelation, it might seem that they are functionally alike. However, differences in their
properties and in properties of starches within each group result from differences in the specific reagents used, in their concentrations, and in other conditions of the treatments. Another variable is the extent of cross-linking or of substitution. In addition, starches often are modified by more than one method,
further adding to the diversity of products. Luallen (1988) illustrated the advantage of combining different chemical treatments—cross-linking and substitution—in the production of a starch that is acid stable and freeze–thaw stable, as
well as heat stable (Fig. 16.13). The curves show the effects on viscosity of heating starch suspensions to 95°C, holding at 95°C for 30 min, and cooling to 50°C
in a Brabender amylograph; the amylograph is discussed in the section of this
chapter on gelatinization (Section II,C).

Additional chemical treatments that are applied to starches for special purposes are oxidation and acid thinning. Both oxidized starch and thin-boiling
starch can be useful in confections such as gum-type confections, in which clarity and high solids are needed.

Obtaining the Food and Drug Administration's approval for the chemical
modifications that are possible is an expensive, time-consuming process, and

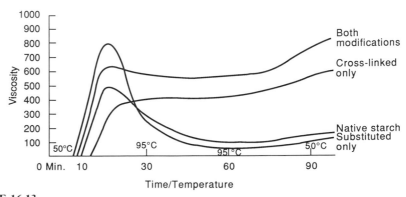

FIGURE 16.13
Separate and joint effects of cross-linking and substitution on the amylograph viscosity (in Bra-
bender units) of waxy maize starch. Starch slurries: 5% dry starch, 1% salt, pH 6.5. (From
Luallen, 1988.)

many chemicals that have potential use have not received clearance. Those that have been approved are named in the FDA Code of Federal Regulations (1988), along with conditions for use.

An important consideration in the use of modified starches in food formulation is the interaction between starch and other formula ingredients. The nature of the chemical treatments affects specific interactions and compatibility between the starch and other ingredients.

Further discussion of modified starch can be found in journal articles and book chapters by Colonna *et al.* (1987, pp. 79–114), Filer (1988), Luallen (1985), Moore *et al.* (1984, pp. 581–586), O'Dell (1979, pp. 171–181), Rogols (1986), and Rutenberg and Solarek (1984, pp. 311–388).

C. Gelatinization and Pasting

Gelatinization or partial gelatinization of starch is responsible for large changes in food properties that occur during food preparation or processing. Some terminology should be considered before discussion of the process. After considerable discussion of starch terminology and methodology at starch science and technology conferences, a committee was established to conduct a survey of starch scientists and technologists and make recommendations. In the survey, the committee obtained information about usage of certain terms and about methodology for study of certain starch phenomena; the committee (Atwell *et al.*, 1988) then proposed the following definition for gelatinization:

> *Starch gelatinization is the collapse (disruption) of molecular orders within the starch granule manifested in irreversible changes in properties such as granular swelling, native crystallite melting, loss of birefringence, and starch solubilization. The point of initial gelatinization and the range over which it occurs is governed by starch concentration, method of observation, granule type, and heterogeneities within the granule population under observation.*

The term pasting, frequently used interchangeably with gelatinization, was included in the survey and the committee proposed the following definition:

> *Pasting is the phenomenon following gelatinization in the dissolution of starch. It involves granular swelling, exudation of molecular components from the granule, and eventually, total disruption of the granules.*

The two definitions do not support interchangeable use of the terms. Whereas the terms gelatinization and pasting have often been applied to all of the changes that occur when starch is heated in water, gelatinization includes the early changes according to its proposed definition, and pasting includes later changes. It might take a while for acceptance of the definitions to be fully reflected in the literature. In the meantime, the interchangeable usage need not be troublesome if the details of these two definitions are kept in mind.

The basis for much of the current understanding of gelatinization (and pasting) was established by Meyer (1952). He pioneered in the study of many aspects of the chemical and physical nature of starch.

The completeness of gelatinization and pasting depends on the amount of

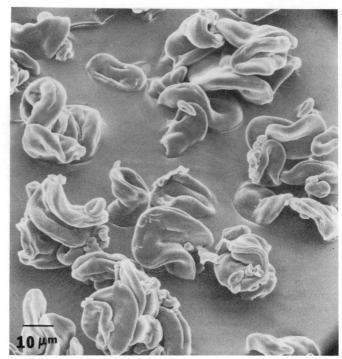

FIGURE 16.14
Wheat starch granules, heat swollen and collapsed. (From Hoseney *et al.*, 1977.)

water available and the extent of heating. As heat is applied to a starch suspension, the thermal energy permits some water to penetrate the granule surface. Water passes first into the relatively open amorphous regions. With continued heating, the energy level becomes high enough to disrupt hydrogen bonding in the crystalline areas; this effect may be observed, with the aid of a polarizing microscope, as loss of birefringence. With the entire granule structure now more "loose," water uptake proceeds readily as heating continues, resulting in rapid swelling of granules. The temperature at which rapid swelling of granules begins depends on the type of starch and on other ingredients but tends to fall in the range of 62–75°C for corn (regular and waxy) and wheat starches. It is lower for tapioca and potato starches and higher for sorghum and amylomaize starches (Pomeranz, 1987, p. 403).

As starch granules swell, their density decreases and they eventually are able to remain suspended. The increasing size of granules also results in increasing internal friction and thus contributes to increased viscosity of the suspension. Furthermore, granule surfaces become increasingly "open" with swelling so that some of the straight chains of amylose are able to leave the granules. This is the dissolution that is mentioned in the definition of pasting; the extent to which amylose is leached out of the granules is variable. Amylose that has left the gran-

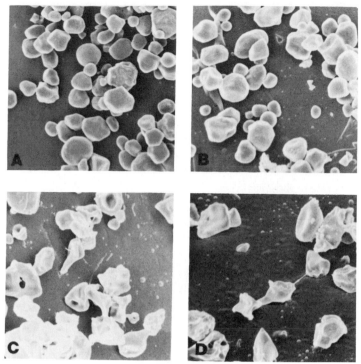

FIGURE 16.15
Scanning electron micrographs ($\times 700$) of corn starch–water dispersions at 30°C (A), 60°C (B), 70°C (C), and 80°C (D). (Courtesy of Zoe Ann Holmes.)

ules is in colloidal dispersion, and the dispersion thus is a sol in which the intact granules are in suspension. The loss of some amylose results in implosion.

Miller *et al.* (1973) applied the more general term exudate to the released granule contents. The loss of exudate results in implosion or "collapse." The collapsed granules, though still large relative to their ungelatinized state, appear somewhat shrunken and frequently appear folded and wrinkled when observed microscopically (Fig. 16.14). Whistler and Daniel (1985, p. 115) described their appearance as that of a deflated balloon. The progressive changes in the appearance of corn starch granules between suspension temperatures of 30 and 80°C are seen in Fig. 16.15.

Granule collapse need not result in thinning of the suspension because the long chains of exudate, as well as the swollen imploded granules, contribute to viscosity. In fact, Miller *et al.* (1973) observed a considerable increase in paste viscosity after maximum swelling of granules, and they attributed that further increase to the presence of exudate. In paste samples that were freeze dried, Miller *et al.* observed a filamentous network of exudate. Note that they did not say that the exudate formed a network during heating; however, network formation in the hot paste is not necessary in order for the exudate to give increased viscosity.

The temperature at which maximum viscosity is achieved depends on the starch and on other ingredients, but usually is at least 90°C. Starch-thickened mixtures are commonly heated to the boiling point to ensure maximum viscosity and disappearance of raw flavor. Miller *et al.* (1973) described the granules that remain after extensive release of exudate as very open structures that remain identifiable, even with heating to 95°C. Fragmentation, actual breaking up of granules, is not extensive during pasting unless acid is present during a long heating period or unless starch granules are relatively fragile and agitation is excessive. When fragmentation does occur, it does not involve bursting of granules but rather extensive disruption of granule structure involving breaking of glycosidic bonds by either hydrolysis or extreme shear force. The heated starch suspension is referred to as a paste.

Much of the study of starch pasting has involved controlled heating of 5–6% starch suspensions in the Brabender viscoamylograph (Fig. 16.16), in which a curve records changing viscosity as the temperature is raised 1.5°C/min to the selected temperature, usually in the range of 90–95°C. Often the paste is held at the top temperature for a specified period for observation of paste thinning. Sometimes a cooling curve then is obtained with the use of an inserted coil with cold water running through it. The cooling curve, which actually is an extension of the heating curve, is an indication of the postheating thickening referred to in the food industry as setback. During the entire procedure the suspension is stirred at a constant rate. Amylograph chart paper shows time on the horizontal axis. Because of the regular rate of temperature increase, knowing the initial temperature of the suspension makes it possible to obtain certain temperature data from the amylogram—for example, the temperature at which maximum viscosity is reached. Relative viscosity, recorded in arbitrary Brabender units, is reflected by the height of the curve. Figure 16.13 shows amylograph curves for starch suspensions that were heated to 95°C over a period of 30 min, held at 95°C for 30 min, and then cooled to 50°C over a 30-min period. Paste samples can be removed from the amylograph bowl at selected temperatures without disruption of the pasting process; a long medicine dropper can be inserted through the opening for the cooling coil. The samples then can be studied microscopically. The use of the amylograph for determination of amylolytic activity in flour is described in AACC method 22-10 (1983).

A polarizing microscope that has a hot stage is convenient for continuous observation of granules during gelatinization and for determination of the birefringence end point. Individual granules lose their birefringence within a narrow temperature range; the temperature range is broader for a starch sample because granules in a starch sample have different rates of gelatinization. Birefringence end point is the temperature at which 98% of the granules have lost their birefringence (French, 1984, p. 234); or loss of birefringence may be specified as the temperature range over which disappearance of the polarization crosses begins and is complete (Pomeranz, 1985, p. 33).

Starch is important in foods particularly as a thickening agent. Starches from different sources show different pasting behavior. Typical amylograms for some unmodified starches are shown in Fig. 16.17.

FIGURE 16.16
Brabender amylograph. (Courtesy of C. W. Brabender Instruments, Inc., South Hackensack, New Jersey.)

Waxy cereal starches have greater thickening power than do the corresponding nonwaxy starches, but their pastes are much more unstable—viscosity decreases more with holding at elevated temperatures. This is an important consideration in high-temperature processing of some canned foods. Waxy starches, as well as root and tuber starches, produce pastes described as "long" pastes. They appear to be cohesive and elastic. The viscoelasticity results from excessive swelling during cooking and consequent ease of deformation and stretching of granules (Pomeranz, 1985, p. 52). When a fatty acid or a monoglyceride is added to a root or tuber starch before heating, it complexes with linear chains, thus tightening internal structure and retarding granule swelling. A commercial application is the addition of a monoglyceride mixture in the pro-

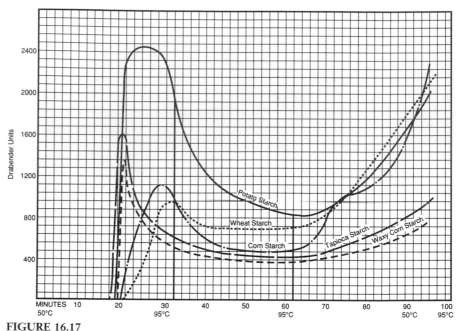

FIGURE 16.17
Amylograph viscosities of unmodified starches (10% starch, pH 6.5, distilled water). (From Moore *et al.*, 1984.)

duction of instant mashed potatoes. Lipid components do not have a similar effect on waxy starches because of their lack of amylose. Granule swelling of waxy starches can be retarded by chemical cross-linking.

The concentration of starch and the rate and extent of heating also affect pasting behavior. As starch concentration increases, paste viscosity increases. Holmes and Soeldner (1981) observed that rate of heating affected granule swelling; when they heated wheat and corn starch pastes at two different rates to the same final temperature, the rapidly heated pastes were more viscous than the more slowly heated pastes. Maximum viscosity is not attained unless the extent of heating is sufficient to cause maximum swelling of starch granules.

Added ingredients affect starch gelatinization and pasting. Sugar retards hydration of starch by competing for water. Viscosity may or may not be decreased, depending on whether the heating time is sufficient to compensate for the retardation. If the sugar concentration is excessive, compensation may not be possible because of excessive dilution of starch.

When acid is present during the heating of a starch suspension, hydrolysis of starch molecules in the granules during heating results in increased permeability of the granules, thus accelerating the gelatinization process. Ultimately extensive granule fragmentation and consequent thinning of the paste can occur.

The potential effect of acid on the thickening power of starch is recognized in recipes that call for late addition of lemon juice. However, it should be remembered that adding lemon juice after completion of heating involves addition of water as well as acid. Water added after completion of heating has little opportunity to be taken up by starch granules. In a situation such as the thickening of the cherry juice for a cherry pie, the acid is present in the juice that is thickened and there is no way to add the acid after completion of thickening. The amounts of acid used in starch-thickened mixtures need not cause undue thinning if heating is carried out rapidly. The effect of heating on the flavor of lemon juice sometimes is more apparent than a thinning effect. In commercially processed starch-thickened mixtures containing acid, starches that are modified for acid resistance are used.

When both sugar, with its retarding effect on gelatinization and pasting, and acid, with its accelerating effect, are present, they tend to counteract one another's action. However, excessive additions can result in low viscosity because of dilution of the thickening agent.

Lipids and surfactants affect pasting behavior of starch. By complexing with amylose, they retard gelatinization and decrease the release of exudate during pasting. Davis *et al.* (1986) studied the effects of different oils in model systems containing starch, water, and oil in the ratio of 1:2:0.66. They reported that safflower oil, which is highly unsaturated, delayed the loss of birefringence least.

The above discussion of gelatinization, with the exception of the reference to the study of Davis *et al.* (1986), applies to dilute starch systems. Although it is convenient to study starch gelatinization and pasting in dilute systems, most foods are limited water systems. Starch makes important contributions to structure in limited moisture systems such as bread and other baked flour mixtures. Hoseney *et al.* (1983) described the role of starch in bakery products as a water sink that "sets" the structure. Hoseney (1986, p. 56) stated that differential scanning calorimetry (DSC) is useful in the study of concentrated systems. Scanning electron micrography also is helpful, as it is for dilute systems.

There is more starch than water in bread dough or in a stiff flour–water dough. At that high starch concentration there is considerably less granule swelling and deformation by the end of the heating period than occurs in a dilute system (Fig. 16.18). The relatively low degree of pasting in bread is desirable from the standpoint of structure because intact granules are important to structure. Other ingredients in bread and other bakery foods affect gelatinization and pasting. Hoseney *et al.* (1983) stated that surfactants and fats delay pasting but not gelatinization, whereas sugars delay gelatinization but not pasting. The role of amylose in the firming of bread during cooling and the role of amylopectin in bread staling are discussed in Chapter 19.

Starch granules in angel food cake batter undergo more change during baking than do starch granules in sugar cookie or pie crust dough. The difference appears to be related more to water content than to sugar concentration. Starch granules in pie crust, which has a low water content and high fat content, change little during baking. Granules extracted from pie crust and angel food cake are shown in Fig. 16.19. The extent of starch gelatinization in various products during baking was discussed by Hoseney *et al.* (1983). The role of flour chlorination in the performance of starch in cake flour is discussed in Chapter 21.

Olkku and Rha (1978) reviewed starch gelatinization and pasting and factors affecting them. They also reviewed gelation and retrogradation.

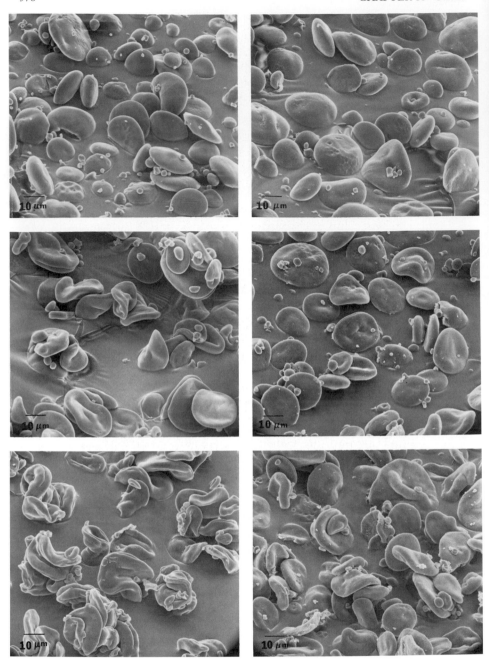

FIGURE 16.18
Starch extracted from a 10% wheat starch–water suspension (left column) and a 65% moisture
flour–water dough (right column); temperatures from top to bottom: 25, 70, and 90°C. (From
Hoseney *et al.*, 1977.)

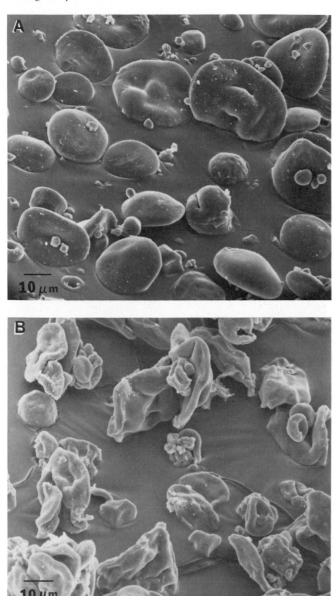

FIGURE 16.19
Starch extracted from commercially obtained baked products: A, fried pie crust; B, angel food cake. (From Hoseney *et al.*, 1977.)

D. Gelation

A gel is a solid–liquid system having a solid continuous network in which the liquid phase is entrapped. It has a distinct structural shape that is resistant to flow (Glicksman, 1982, p. 277). Gelation, the process of gel formation, is not exclusively a starch function; there are many gelling agents and they require different conditions for gelation. In the case of starch, pasting followed by cooling (without stirring) is required. If conditions favor gelation of a starch paste, extensive hydrogen bonding occurs during cooling. Probably free amylose molecules form hydrogen bonds not only with one another but also with amylopectin branches extending from swollen granules, so that the granules are a part of the solid continuous network.

Starch gels vary in their rigidity (strength). Gel strength, like paste viscosity, is affected by many factors. Starches from some sources do not gel. For example, although waxy starches and potato and tapioca starches form viscous pastes, their pastes do not form gels upon cooling. Increasing the concentration of starch increases the strength of gels formed by those starches whose pastes do gel. Extent of heating affects gelation; sufficient heating to free some amylose without excessive fragmentation of granules is important. Chemical modification can affect gelling ability, as discussed previously. Sucrose in a starch paste, by retarding granule swelling and implosion with loss of amylose, can have a negative effect on gelling ability; the extent of the effect depends on sugar concentration. Acid in a starch paste can have a negative effect on gelation through fragmentation of granules and probably also through a hydrolytic decrease in molecular size of amylose. In addition to acid concentration, the rate of heating to a given temperature influences the effect of acid, because it determines the length of time that the acid acts on the starch. As with paste viscosity, sucrose and citric acid used together tend to counteract one another's effect.

E. Retrogradation

If a starch gel is chilled or frozen, the low energy level results in further hydrogen bonding. As the starch chains associate further, the gel structure is "tightened" and water-holding capacity is decreased. Crystallinity eventually develops. The term retrogradation is applied to the phenomenon and was defined by Atwell *et al.* (1988) as follows:

> Starch retrogradation is a process which occurs when the molecules comprising gelatinized starch begin to reassociate in an ordered structure. In its initial phases, two or more starch chains may form a simple juncture point which then may develop into more extensively ordered regions. Ultimately, under favorable conditions, a crystalline order appears.

Note that, according to the definition, gelatinization—not gelation—is the necessary precursor to retrogradation. In fact, retrogradation in a starch paste that is too dilute to gel can have significant effects, though the effects often are more dramatic in a gel than in a paste. In a gravy or starch-thickened sauce, retrogradation is observed as a separation of liquid. Reheating such a mixture with stirring often reverses the separation, at least partially. A pie filling or pudding thickened with an unmodified cereal starch has an irreversible rough, spongy texture after freezing and thawing; it loses water readily. The best solution to the problem is preventing it by the use of a nongelling starch. Waxy starches represent an improvement; starches that have been cross-linked or substituted or both are even more effective.

Lineback and Rasper (1988, pp. 283–322) recently reviewed wheat starch. Much of the information is applicable to other starches as well. The subject of starch has been one of considerable change in the past two decades. Although much has been learned, much remains to be learned, and future developments—both scientific and technological—should be interesting.

Suggested Exercises

1. Compare starches from several sources, such as potato, corn, wheat, and rice. Use 6 g starch/100 ml water for pasting. Control the heating procedure. Prepare slides and observe granule appearance before and after heating. (Do not stain with iodine because it masks some granule features.) Measure line spread and penetrability or percentage sag (Appendix D).
2. Obtain several chemically modified starches for which various claims, such as acid stability, high-temperature stability, and freeze–thaw stability, have been made by their manufacturers. (Advertisements in *Food Technology* provide information about sources.) Plan and conduct experiments to test the claims.
3. Plan and conduct experiments to show the effects of various added ingredients on starch pastes and gels.
4. Consider Fig. 16.17, as well as the procedure involved in obtaining the amylograms and the conditions under which gelation occurs. Were any of the samples gelled when removed from the paste container of the amylograph?

References

AACC. 1983. "Approved Methods of the American Association of Cereal Chemists," 8th edn. AACC, St. Paul, Minnesota.

Anon. 1988. Formulating microwaveable foods with CWS starches. *Chilton's Food Engineering* **60**(5): 42, 45.

Atwell, W. A., Hood, L. F., Lineback, D. R., Varriano-Marston, E., and Zobel, H. F. 1988. The terminology and methodology associated with basic starch phenomena. *Cereal Foods World* **33**: 306–311.

Banks, W. and Greenwood, C. T. 1975. "Starch and Its Components." Edinburgh University Press, Edinburgh, Scotland.

Boyer, C. D. and Shannon, J. C. 1987. Carbohydrates of the kernel. *In* "Corn: Chemistry and Technology," Watson, S. A. and Ramstad, P. E. (Eds.). American Association of Cereal Chemists, St. Paul, Minnesota.

Colonna, P., Buleon, A., and Mercier, C. 1987. Physically modified starches. *In* "Starch: Properties and Potential," Galliard, T. (Ed.). John Wiley and Sons for the Society of Chemical Industry, Chichester, England.

Davis, E. A., Grider, J., and Gordon, J. 1986. Microstructural evaluation of model starch systems containing different types of oils. *Cereal Chem* **63**: 427–430.

FDA. 1988. Food starch—modified. *In* "Code of Federal Regulations." Title 21, Section 172.892. U.S. Govt. Printing Office, Washington, DC.

Filer, L. J., Jr. 1988. Modified food starch—an update. *J. Amer. Dietet. Assoc.* **88**: 342–344.

Fitt, L. E. and Snyder, E. M. 1984. Photomicrographs of starches. *In* "Starch: Chemistry and Technology," 2nd edn., Whistler, R. L., BeMiller, J. N., and Paschall, E. F. (Eds.). Academic Press, Orlando, Florida.

French, A. D. 1979. Allowed and preferred shapes of amylose. *Bakers Digest* **53**(1): 39–46, 54.

French, A. D. and Murphy, V. G. 1977. Computer modeling in the study of starch. *Cereal Foods World* **22**: 61–70.

French, D. 1973. Chemical and physical properties of starch. *J. Animal Sci.* **37**: 1048–1061.

French, D. 1984. Organization of starch granules. *In* "Starch: Chemistry and Technology," 2nd edn., Whistler, R. L. BeMiller, J. N., and Paschall, E. F. (Eds.). Academic Press, Orlando, Florida.

Galliard, T. and Bowler, P. 1987. Morphology and composition of starch. *In* "Starch: Properties and Potential," Galliard, T. (Ed.). John Wiley and Sons for the Society of Chemical Industry, Chichester, England.

Glicksman, M. 1982. Food applications of gums. *In* "Food Carbohydrates, "Lineback, D. R. and Inglett, G. E. (Eds.). Avi Publ. Co., Westport, Connecticut.

Holmes, Z. A. and Soeldner, A. 1981. Effect of heating rate and freezing and reheating of corn and wheat starch-water dispersions. *J. Amer. Dietet. Assoc.* **78**: 352–355.

Hoseney, R. C. 1986. "Principles of Cereal Science and Technology." American Association of Cereal Chemists, St. Paul, Minnesota.

Hoseney, R. C., Atwell, W. A., and Lineback, D. R. 1977. Scanning electron microscopy of starch isolated from baked products. *Cereal Foods World* **22**: 56–60.

Hoseney, R. C., Lineback, D. R., and Seib, P. A. 1983. Role of starch in baked foods. *Bakers Digest* **57**(4): 65–71.

Knight, J. W. and Olson, R. M. 1964. Wheat starch: Production, modification, and uses. *In* "Starch: Chemistry and Technology," 2nd edn., Whistler, R. L., BeMiller, J. N., and Paschall, E. F. (Eds.). Academic Press, Orlando, Florida.

Lineback, D. R. 1984. The starch granule—organization and properties. *Bakers Digest* **58**(2): 16–21.

Lineback, D. R. and Rasper, V. F. 1988. Wheat carbohydrates. *In* "Wheat: Chemistry and Technology," 3rd edn., vol. 1, Pomeranz, Y. (Ed.). American Association of Cereal Chemists, St. Paul, Minnesota.

Luallen, T. E. 1985. Starch as a functional ingredient. *Food Technol.* **39**(1): 59–63.

Luallen, T. E. 1988. Structure, characteristics, and uses of some typical carbohydrate food ingredients. *Cereal Foods World* **33**: 924–927.

Manners, D. J. 1985. Some aspects of the structure of starch. *Cereal Foods World* **30**: 461–467.

Manners, D. J. and Matheson, N. K. 1981. The fine structure of amylopectin. *Carbohydrate Research* **90**: 99–110.

Meyer, K. H. 1952. The past and present of starch chemistry. *Experientia* **8**: 405–420.

Miller, B. S., Derby, R. I., and Trimbo, H. B. 1973. A pictorial explanation for the increase in viscosity of a heated wheat starch-water suspension. *Cereal Chem.* **50**: 271–280.

Moore, C. O., Tuschoff, J. V., Hastings, C. W., and Schanefelt, R. V. 1984. Applications of starches in foods. *In* "Starch: Chemistry and Technology," 2nd edn., Whistler, R. L., BeMiller, J. N., and Paschall, E. F. (Eds.). Academic Press, Orlando, Florida.

O'Dell, J. 1979. The use of modified starch in the food industry. *In* "Polysaccharides in Food," Blanshard, J. M. V. and Mitchell, J. R. (Eds.). Butterworth, London, England.

Olkku, J. and Rha, C. K. 1978. Gelatinisation of starch and wheat flour starch—a review. *Food Chem.* **3**: 293–317.

Pomeranz, Y. 1985. "Functional Properties of Food Components." Academic Press, Orlando, Florida.

Pomeranz, Y. 1987. "Modern Cereal Science and Technology." VCH Publishers, New York.

Robyt, J. F. 1984. Enzymes in the hydrolysis and synthesis of starch. *In* "Starch: Chemistry and Technology," 2nd edn., Whistler, R. L., BeMiller, J. N., and Paschall, E. F. (Eds.). Academic Press, Orlando, Florida.

Rogols, S. 1986. Starch modifications: A view into the future. *Cereal Foods World* **31**: 869–874.

Rutenberg, M. W. and Solarek, D. 1984. Starch derivatives: Production and uses. *In* "Starch: Chemistry and Technology," 2nd edn., Whistler, R. L., BeMiller, J. N., and Paschall, E. F. (Eds.). Academic Press, Orlando, Florida.

Snyder, E. M. 1984. Industrial microscopy of starches. *In* "Starch: Chemistry and Technology," 2nd edn., Whistler, R. L., BeMiller, J. N., and Paschall, E. F. (Eds.). Academic Press, Orlando, Florida.

Whistler, R. L. and Daniel, J. R. 1984. Molecular structure of starch. *In* "Starch: Chemistry and Technology," 2nd edn., Whistler, R. L., BeMiller, J. N., and Paschall, E. F. (Eds.). Academic Press, Orlando, Florida.

Whistler, R. L. and Daniel, J. R. 1985. Carbohydrates. *In* "Food Chemistry," 2nd edn., Fennema, O. R. (Ed.). Marcel Dekker, Inc., New York.

Young, A. H. 1984. Fractionation of starch. *In* "Starch: Chemistry and Technology," 2nd edn., Whistler, R. L., BeMiller, J. N., and Paschall, E. F. (Eds.). Academic Press, Orlando, Florida.

FLOUR

The first section of this chapter is devoted to wheat flour and the remainder to nonwheat flours. Wheat is of continuing importance as a food crop worldwide; nonwheat sources of flour are of increasing interest as food crops.

I. WHEAT FLOUR

Flour is the major wheat product, though some wheat is used for breakfast cereals. The term flour in this part of the chapter refers specifically to white wheat flour unless indicated otherwise. The properties of flour are influenced by the raw material—wheat—and by the milling process and the treatments applied after milling.

A. Flour Production

1. Wheat

Wheat contributes about one-fifth of the calories consumed by humans (Pomeranz, 1987, p. 15). Wheat grows in many kinds of soils over the world and constitutes nearly 30% of the world's grain production (Pomeranz, 1987, pp. 14–15). In the United States, corn is the only larger field crop (Pyler, 1988, p. 300). More than two-thirds of the United States–produced wheat is processed into food (Pomeranz, 1987, p. 151).

More than 100 varieties of wheat are grown in the United States. They are classified into the following four major classes, not including the durums: hard red spring, hard red winter, soft red winter, and winter and spring white wheats (Pyler, 1988, p. 301).

Hardness refers to vitreousness or resistance of the endosperm to grinding. The starch granules are particularly tightly packed in the protein matrix, and adherence between starch and protein is particularly strong in hard wheat. Most hard wheat varieties have higher protein concentrations than do most soft wheat flours, but protein concentration does not necessarily parallel hardness. Protein concentration differs considerably within a class and even within a variety.

The greatest significance of the differences between flours of hard and soft wheats is in their baking qualities; flour made of hard wheat is especially suited to breadmaking, whereas that made of soft wheat is better for cakes, pastry, and cookies. The superiority of hard wheat flour for bread reflects the large amount and good quality of gluten it forms when mixed with water. Durum wheat varieties are both particularly hard and particularly high in protein, but they belong to a different botanical species from the common wheats and they do not make good bread flours. Durum flours are used in the making of pasta (semolina) products, for which they are well suited.

The terms red wheat and white wheat refer to the color of the kernels. The pigment that is largely responsible for the color of red wheat kernels is in the seed coat.

Spring wheat is planted in the spring and harvested in late summer, whereas winter wheat is planted in the fall so that it can develop a root system before cold weather. Winter wheat grows rapidly in the spring and is harvested in early summer; it can be grown only in areas where the root system can survive the winter.

Most of the hard wheat grown in the United States is produced in the Midwest and upper Midwest. The largest producing areas of soft wheat are in the Pacific Northwest and in Ohio and its neighboring states. Most of the durum wheat is grown in North Dakota.

Wheat quality changes with time, both long range and short range. Geneticists continually work at developing wheat varieties that combine the qualities of high yield, high disease and pest resistance, high protein content, good milling qualities, and high baking potential. The protein content of wheat can be influenced by the fertility of the soil in which it is grown. Climatic factors such as temperature, rainfall, and sunshine also affect protein content and, therefore, baking quality.

The wheat kernel (Fig. 17.1) has three parts: the bran, the endosperm, and the germ (embryo). Botanically the bran has a protective function, the endosperm is a food storage site for the developing plant, and the germ is the site of the embryonic axis, along with a small concentrated food supply for the developing seedling. The germ contains the scutellum, which produces enzymes during germination and absorbs food material from the endosperm and conducts it to the growing embryo (Pyler, 1988, p. 310).

The parts of the kernel differ in their amount and their composition. The bran, constituting 13–17% of the kernel, is made up of the layered outer pericarp and an inner portion that includes the seed coat; it has high concentrations of cellulose, hemicellulose, and minerals. The endosperm, 80–85% of the kernel, is composed of cells containing many starch granules embedded in a protein matrix. It sometimes is called starchy endosperm to distinguish it from the aleurone layer, which botanically is an outer layer of thick-walled cells of the endosperm. The aleurone layer, which is removed with the bran during milling, contains no starch or gluten protein but has reserve foods in the form of oil and nongluten protein. The germ, only 2–3% of the kernel, has a high concentration of protein, lipid, and ash; it contains most of the thiamine in the wheat kernel.

The milling process is designed to produce white flour of good baking quality

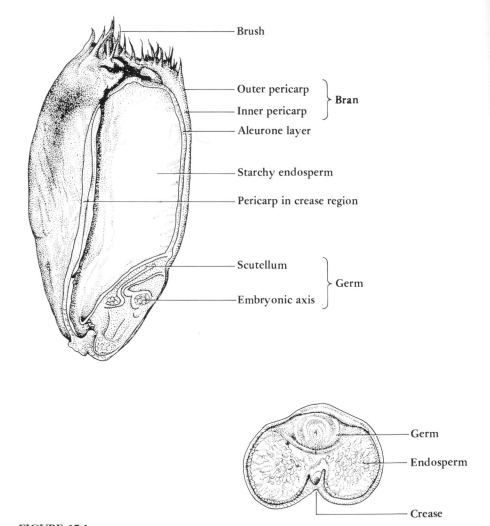

FIGURE 17.1
A kernel of wheat. Longitudinal section bisected through the crease, and lateral cross-section. (From Bradbury *et al.*, 1956.)

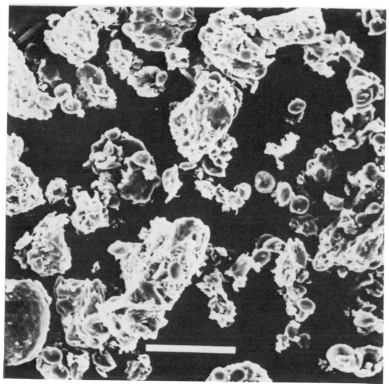

FIGURE 17.2
Scanning electron micrograph of hard wheat flour. Bar is 100 μm. (From Hoseney, 1986.)

by separating the endosperm from the bran and germ that surround it. This separation is possible because endosperm is more easily fragmented than are the bran and germ. The microscopic appearance of a hard wheat flour is shown in Fig. 17.2.

2. Milling

The process of milling is diagrammed in Fig. 17.3. Because of variations in wheats within classes, millers blend wheats of different varieties for uniformity of protein content and baking quality. Either before or after the wheat is selected and blended for the type of flour desired, it is cleaned and tempered.

Cleaning usually is a dry process involving the use of air blasts, sieves, disk separators, and scouring devices in removal of the brush hairs, dirt in the kernel crease, and foreign material such as seeds of other plants, stones, chaff, straw, and dirt.

Tempering (conditioning) is holding the grain in water for up to 24 hr, or for a shorter period if the temperature is somewhat elevated. When the grain reaches a moisture content of about 15–16%, the properties of the bran, endosperm, and germ are optimum for effective separation. The bran becomes toughened so that it does not shatter during grinding; the endosperm becomes readily

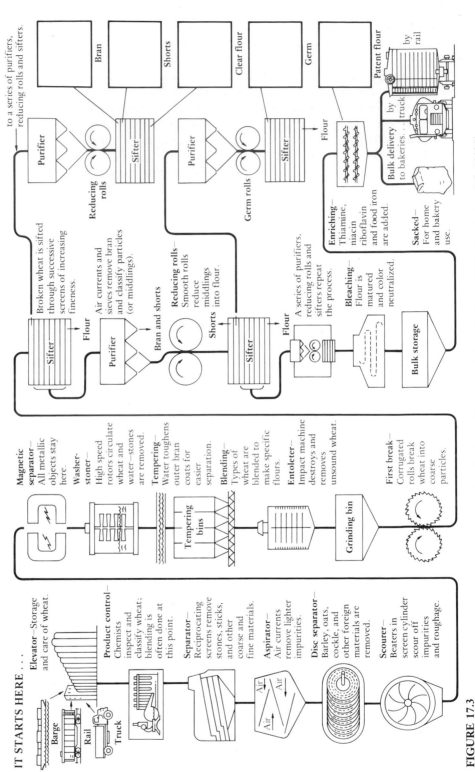

IT STARTS HERE . . .

Elevator—Storage and care of wheat.

Product control—Chemists inspect and classify wheat; blending is often done at this point.

Separator—Reciprocating screens remove stones, sticks, and other coarse and fine materials.

Aspirator—Air currents remove lighter impurities.

Disc separator—Barley, oats, cockle, and other foreign materials are removed.

Scourer—Beaters in screen cylinder scour off impurities and roughage.

Magnetic separator—All metallic objects stay here.

Washer-stoner—High speed rotors circulate wheat and water—stones are removed.

Tempering—Water toughens outer bran coats for easier separation.

Blending—Types of wheat are blended to make specific flours.

Entoleter—Impact machine destroys and removes unsound wheat.

First break—Corrugated rolls break wheat into coarse particles.

to a series of purifiers, reducing rolls and sifters.

Broken wheat is sifted through successive screens of increasing fineness.

Air currents and sieves remove bran and classify particles (or middlings).

Reducing rolls—Smooth rolls reduce middlings into flour.

A series of purifiers, reducing rolls and sifters repeat the process.

Bleaching—Flour is matured and color neutralized.

Sifter

Flour

Purifier

Bran and shorts

Shorts

Sifter

Flour

Bulk storage

Purifier

Reducing rolls

Sifter

Bran

Shorts

Germ rolls

Purifier

Clear flour

Sifter

Flour

Germ

Enriching—Thiamine, niacin riboflavin and food iron are added.

Sacked—For home and bakery use.

Bulk delivery to bakeries . . . by truck

Patent flour by rail

FIGURE 17.3

The total process of making flour. (Copyright 1981, Wheat Flour Institute, 600 Maryland Ave. SW, Suite 305W, Washington, DC, 20024.)

fractured but not to such an extent that it is pulverized too quickly; the germ becomes so pliable that it is readily flattened by rollers after it is freed from the kernel.

After tempering, the wheat goes to the five or six break rolls, which have a shearing action. The first break rolls are set rather far apart so that large particles are produced. The sheared particles are passed through sieves, progressing from the coarsest to the finest. The particles that remain on the coarsest sieve are passed to the next set of break rolls; the medium-sized particles (middlings) are primarily endosperm and go to purifiers for removal of germ and small particles of bran by air aspiration and sifting; the particles that pass the finest sieve constitute "break flour." The coarsest particles pass through successive sets of break rolls, which become increasingly finely corrugated and increasingly close together. The sifting is repeated after each break. After the third break, the germ should have been flattened and separated out. The large particles on the top sieve after the fifth or sixth break should consist primarily of bran.

Purified endosperm middlings from the different breaks pass through the reduction rolls, which are smooth and have a pulverizing action. Like the break rolls, the reduction rolls are set increasingly close together. After passing each set of reduction rolls, the reduced material is sifted to separate flour from the coarser particles that need further reduction. The mixture of bran, germ, and some endosperm that have been screened out by the time most of the endosperm has been converted to flour is designated as millfeed.

The milling process is carried out automatically and continuously, material being conveyed by such devices as elevators and tubes from one operation to the next. Each grinding and sifting step produces a certain amount of material as fine as flour, or a mill stream. A large mill may have 30–40 mill streams, each with different characteristics, which are blended to produce the grades of flour and feed desired, as indicated in Fig. 17.4. As shown in the figure, about 28 lb of feed and 72 lb of white flour can be expected from 100 lb of (cleaned) wheat. This extent of recovery of flour is referred to as a 72% rate of extraction. The chart in Fig. 17.4 actually is generic; variations are necessary because of differences in wheat and in users' specifications.

Straight flour is the result of combining all of the mill streams, that is, the entire 72 lb of flour. Frequently the mill streams from the starchier end of the mill stream spectrum (at the beginning of the reduction system) are combined to make patent flour. The remainder becomes clear flour. If only 40–60% of the streams are combined, an extra-short patent flour is produced, leaving a fancy clear. As more mill streams are combined, the patent is longer and the amount and quality of clear are reduced. Patent flours are commonly used for household and bakery needs, whereas clear flour is used in products for which its creamy to grayish color is not a diasadvantage, such as rye or whole wheat bread and pancake flour. Bread flour is a hard wheat flour having about 12% protein. All-purpose, or family, flour is made from either hard or soft wheat and has about 10.5% protein. Cake flour is a short patent soft wheat flour having about 7.5% protein. Special-purpose flours include pastry flour, which is a longer patent soft wheat flour than cake flour, and self-rising flour, to which salt and leavening agent have been added for convenient use in quickbreads.

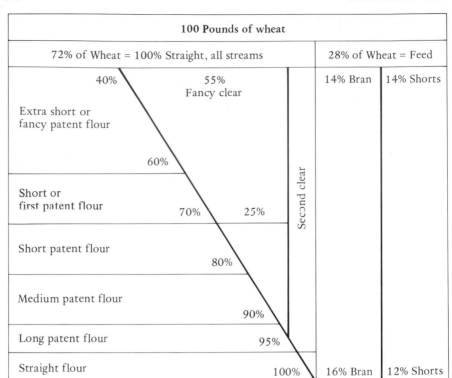

FIGURE 17.4
Chart showing the amounts of various grades of flour and feed produced from 100 lb of wheat.
(From Swanson, 1938; Courtesy of Wheat Flour Institute.)

3. Other Treatments

Air classification is sometimes used as a means of producing special-purpose flours of desired composition. It involves fractionating a flour, on the basis of particle size and density, by application of centrifugal force and air flow. The fractions differ in composition as well as in particle size and density, and the process provides the possibility of making flours that differ in their relative proportions of starch and protein and are highly specialized for different uses. Sometimes impact milling (for example, pin milling) precedes air classification. Pin milling involves forcing conventionally milled flour against pins with great force. It causes starch damage, and the process must be controlled for prevention of excessive starch damage.

Agglomeration (instantizing) involves moistening of conventionally milled flour and then drying, all with constant agitation. The particles, made sticky by the moisture, adhere to one another to form agglomerates during drying. The product is free flowing and dust free and, therefore, has advantages in conveying systems. The agglomerates are porous and, therefore, are readily dispersed without lumping, and disintegrated in water. Bulk density is relatively uniform because agglomerated flour does not pack. Miller *et al.* (1969) reviewed the properties of instantized flour and the factors that need to be considered in using it in food preparation.

Irradiation of wheat and of flour was of considerable interest in the 1950s and 1960s because relatively low doses eliminate insect infestation. There is no safety problem at the doses that are feasible, but contradictory results with respect to effects on baked flour products, along with the expense involved, resulted in considerable loss of interest. Some revival of interest seems to be apparent at the time of writing. The advantages and disadvantages of irradiation of cereal foods were reviewed by Giddings and Welt (1982), Lorenz (1975), and MacArthur and D'Appolonia (1982).

Chemical treatments of flour consist of the addition of bleaching and improving chemicals. The carotenoid pigments in flour give it a creamy to yellowish color when it is first milled. During storage for several months, its color becomes lighter because of oxidation of the pigments. Its baking quality improves also, because of oxidative effects on gluten proteins. These changes can be accelerated by the use of certain chemicals at the mill. Benzoyl peroxide is a commonly used bleaching agent. Chemicals that have both bleaching and improving actions in bread and all purpose flours include chlorine dioxide, potassium bromate, acetone peroxide, and azodicarbonamide. They differ in their rate of action and frequently are used in combination.

Cake and pastry flour are treated with chlorine. Chlorination is especially important for cake flours and is discussed further in Chapter 21.

The addition of improving chemicals is not permitted in all countries. Approved levels in the United States are specified for both mill and bakery use in the Code of Federal Regulations (FDA, 1988a,b). The approved additions of enrichment mixtures also are specified.

Bass (1988), Hoseney (1986), Pomeranz (1987), and Pyler (1988, pp. 315–327, 351–355) described the milling process and flour treatments.

B. Flour Properties

1. Composition

The average protein concentrations of different types of white wheat flour were mentioned in the preceding section. Lipid concentrations are under 2% and ash concentrations under 0.5% (12% moisture basis). Whole wheat flour has higher concentrations of protein, lipid, and ash and, therefore, less carbohydrate than do white wheat flours.

The proteins of wheat endosperm and, therefore, of white flour are of two types: gluten proteins and water-soluble proteins. The gluten proteins predominate and are responsible for the unique bread-making properties of wheat flour. Not only is gluten not a chemical entity, but its component proteins, gliadin and glutenin, are not chemical entities either. Each has several subfractions.

The gliadins and glutenins have both similarities and differences chemically. Both have high concentrations of glutamine and proline and low concentrations of lysine, though the gliadins tend to have more glutamic acid, proline, and amino acids with hydrophobic side chains and a slightly lower content of basic amino acids than do the glutenins (Làsztity, 1984, p. 37). Glutenins have much higher molecular weights than do gliadins. Gliadin molecules are compact because of intramolecular disulfide bonding; glutenin molecules are relatively extended and highly associated because of intermolecular disulfide bonding.

Gliadins are soluble in 70% alcohol; glutenins can be dispersed in alkali or in dilute acid. Both gliadins and glutenins have a high water-imbibing capacity. In combination, they take up nearly three times their weight of water. Gliadin, when hydrated separately, becomes sticky and readily extensible. Glutenin is both cohesive and elastic. When hydrated together, they form gluten, a viscoelastic network.

Sulfhydryl (—SH) groups and disulfide (—S—S—) bonds are involved in development of gluten structure. Some disulfide bonds are broken and some new ones are formed. Sulfhydryl groups contribute to the rearrangements by reducing some of the disulfide bonds. Other kinds of bonding—for example, hydrogen bonding, hydrophobic interaction, and ionic bonding—also contribute to structure development.

The water-soluble proteins of wheat endosperm are albumins and globulins. Together they constitute only 10–15% of the total protein. The major contribution of the water-soluble proteins to baking quality of flour lies in enzymes. Pyler (1988, pp. 144–178) discussed flour enzymes, as well as other enzymes that function in baking.

Amylases are discussed in Chapter 16. Their action on starch provides some sugar for yeast fermentation in bread dough. The concentration of β-amylases in flour is higher than that of α-amylase, which is in extremely low concentration, but β-amylase is inactivated at a lower temperature than is α-amylase. Neither has an appreciable effect on the molecules of intact granules at fermentation temperatures, but they can act on any damaged starch that is present in the flour.

Other flour enzymes include proteolytic enzymes which are in very low concentration in endosperm; oxidative enzymes, of which lipoxidase is the most significant because it contributes to flour bleaching; lipases, which do not affect white flour significantly because their concentrations and those of their substrate, lipids, are low in endosperm; and phytase, which hydrolyzes phytic acid, preventing its tying up divalent metallic ions such as calcium, iron, and zinc.

The carbohydrate of wheat endosperm is primarily starch, which is discussed in Chapter 16. Its contribution to bread structure is discussed in Chapter 19. Also present are hemicelluloses, which are chiefly polymers of the pentoses xylose and arabinose and of the hexoses galactose and mannose. The pentose polymers, pentosans, are of two types—water soluble and water insoluble. The water-soluble pentosans have a beneficial effect on dough consistency. Dextrins and several sugars are present in small amounts.

The lipids in wheat flour are free lipids, which are ether extractable, and bound lipids, which require polar solvents for extraction. Some of each fraction is polar lipid and some is nonpolar. The polar lipids are glycolipids and phospholipids; the nonpolar lipids are primarily triglycerides. In spite of their low concentrations, lipids affect the breadmaking performance of flour, as is discussed in Chapter 19.

Among the reviews of flour composition is that of Pyler (1988, pp. 328–350).

2. Assessment of Flour Quality

Some flour tests are conducted at the mill and at the bakery as a matter of quality control. The approved methods of the American Association of Cereal Chemists (AACC, 1983) include routine tests such as those for moisture (44-15A and 44-40), ash (08-01), nitrogen (46-13), and crude fat (30-10). The amylograph, mentioned in Chapter 16 as a means of recording relative viscosity during the heating of starch suspensions, actually was developed for determining amylolytic activity of flour. When α-amylase is present during the heating of a flour suspension, the enzyme's hydrolytic action on molecules in the starch granule results in granule breakdown and paste thinning. The details of the test are given in AACC method 22-10. A "falling number" test is another test for following amylolytic activity; it involves measurement of the time required for the stirrer in the special equipment to fall a specified distance in a hot flour paste; the time decreases with amylolytic activity.

Flour color, granularity, and hydration capacity also are properties that are frequently evaluated. Color assessment is done visually or by instrumental measurement of reflectance. Granularity—or particle size distribution—can be studied microscopically, or with standard-mesh sieves, or by sedimentation or centrifugation (AACC, 1983, method 55-10). Hydration capacity, known to affect baking performance, differs among flours and is measured by a sedimentation test (AACC, 1983, method 56-60).

Doughing behavior of flour is evaluated in the farinograph; the procedure is detailed in AACC method 54-21 (1983). Several properties are measured, including water absorption (the amount of water required with a specified amount of flour for attainment of a specified maximum dough consistency; expressed as percentage of flour weight), the time required for maximum dough consistency to develop, and stability of the dough. The curves obtained are farinograms. Figure 17.5 is a farinogram with some parameters identified. Consistency is indicated by the height of the curve; the 500 level on the chart paper has been selected arbitrarily as the standard peak level for the test. Stability time (tolerance) is found by subtraction of arrival time from departure time (see Fig. 17.5). Schiller (1984) further described the procedure. The mixograph also is a recording dough mixer, but it is for smaller samples than those studied in the farinograph. It produces a mixogram (Fig. 17.6). The mixograph method 54-40 (AACC, 1983) is less specifically defined than is the farinograph method.

Dough extensibility and resistance to extension are measured in the extensigraph (AACC, 1983, method 54-10). Measurements are made on fermented dough and are useful in studies of flour improvers.

Baking tests of flour quality have been standardized (AACC, 1983). Performance of strong flours is evaluated in bread tests (methods 10-10B and 10-11), of cake flour in angel food cake (method 10-15) and in a high-ratio cake (method 10-90), and of straight-grade soft wheat flour in cookies (method

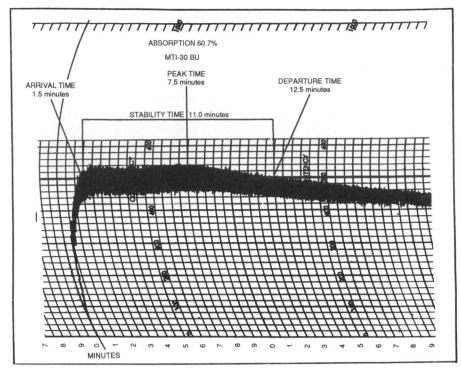

FIGURE 17.5
Farinogram. (From Schiller, 1984.)

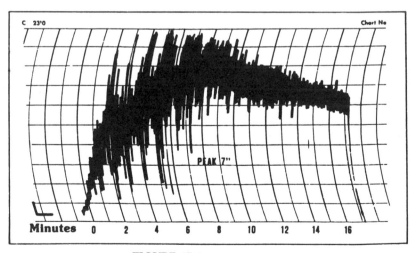

FIGURE 17.6
Mixogram. (From Schiller, 1984.)

10-50D). Volume is the major criterion of flour performance in breads and cakes. Cookie spread is the criterion of flour performance in cookies; spread is expressed as a diameter:thickness ratio, and details of the measurements are specified in the AACC method (see Chapter 3).

A useful technique for the study of flour involves fractionation of flours that differ as to baking quality into their separate components, gluten, starch, "tailings," and water solubles. The fractions then are recombined for baking tests. These are done first without interchange in order to confirm that fractionation did not alter the functional properties of the components, and then with systematic interchange in order to determine the contribution of each fraction to the quality difference between the parent flours. The procedure is time consuming and is a research method rather than a routine flour test.

II. NONWHEAT FLOURS

Research involving nonwheat flours has been stimulated tremendously by world food problems resulting primarily from expanding populations. Developing countries are much too dependent on costly imported grain; insufficient food and inadequate nutrition are widespread problems. Nonwheat flours have the potential of being economically beneficial to both developing and industrial countries and of making large nutritional contributions, particularly in developing countries. Oilseed and legume flours have higher concentrations of protein than do wheat flours. Many nonwheat flours have amino acid distributions that make their proteins complementary to wheat proteins and, in some cases, to each other.

Aside from the process of flour production and the nutritive value, especially protein quantity and quality, functional performance of nonwheat flours has been the major emphasis of research and is the aspect discussed here. Most nonwheat flours do not have any gluten-forming ability, though rye flour and triticale flour do have limited ability to form gluten.

Functional performance has been studied most extensively in bread because bread is important in diets worldwide and because negative effects of nonwheat flours are particularly apparent in bread, the structure of which is dependent on gluten. Quick breads, cookies, and cakes have been studied also. Flavor, color, grain, and volume are affected in all flour products; the effects constitute problems to different degrees. Much of the work done has dealt with means of minimizing the problems and maximizing the acceptability of products containing nonwheat flours, as well as with determining acceptable levels of specific nonwheat flours in specific products.

A. Cereal Flours

1. Rye Flour
Rye is widely distributed geographically. Its milling, though similar in principle to wheat milling, differs in some respects and was described by Rozsa (1976).

Rye flours differ in darkness of color and composition. Protein, fat, ash, and fiber concentrations increase and total carbohydrate concentration decreases with increasing darkness of the flour. The average proximate composition (12% moisture basis) is 12.1% protein, 72.8% total carbohydrate, 1.6% lipid, and 1.5% ash (calculated from Pyler, 1988, p. 379). The protein concentration may be as high as 16% in a dark rye flour. Lysine, which is the first limiting amino acid in wheat flour, is higher in rye flour; therefore, rye flour complements wheat flour nutritionally.

In spite of having a proximate composition that is similar to that of wheat flour and having greater bread-making potential than do other nonwheat flours, rye flour's performance does not measure up to that of wheat flour. Rye flour produces sticky doughs that lack cohesiveness and do not retain gas well, as compared with wheat flour doughs. The stickiness is attributed to the carbohydrate's water-soluble pentosan component, which is present in particularly high concentration in rye flour. Pentosans are polysaccharides made up primarily of the five-carbon sugars D-xylose and L-arabinose. Those pentosans that are termed "water soluble" give rye flour its high water-binding capacity. The lack of dough cohesiveness is responsible for poor gas retention. Bread made with 100% rye flour is heavy and low in volume. Most of the rye bread made in North America is made from mixtures of rye and wheat flours. Light, medium-grade, and dark rye flours can be substituted for 40, 30, and 20%, respectively, of wheat flour in bread without depressing loaf volume (Pyler, 1988, p. 387). Drews and Seibel (1976) described the rye breads of North America and Europe.

2. Triticale Flour

Triticale is a hybrid of wheat and rye. Early cross-breeding studies, mostly in the 1930s, were conducted for combining rye's ability to adapt to a wide range of growing conditions and wheat's technological advantages. In a new flurry of activity beginning in the 1950s, efforts were directed to combining rye's superior nutritive value (higher lysine concentration) with wheat's better baking quality. Triticale was perceived as a possible means of increasing and improving the food supply in developing countries. Bushuk and Larter (1980) reviewed the history of triticale's development.

Triticale has not yet reached its potential. The highest extraction rate that has thus far resulted in satisfactory triticale flour is 60–65%, which limits triticale's contribution to the world food supply. Because of the large number of possible genetic techniques available, triticale flours vary greatly in their properties. The task of finding the best rye–wheat combination is formidable. The variability in production factors probably is partly responsible for seemingly contradictory results of studies involving triticale flour in baked products.

Triticale flours have protein concentrations of 10.7–16.3%, averaging 13.4% (Pyler, 1988, p. 389). They generally have shorter dough development times than do bread flours of wheat. Triticale flours have relatively high α-amylase activity, which negatively affects water-binding capacity of the starch and thus adversely affects baking quality.

Triticale flour doughs have poorer handling properties than do wheat flour doughs, even with triticale flours that have higher protein concentrations than those of bread flours of wheat. However, handling properties are better than those of rye flour doughs. Addition of a dough conditioner such as sodium stearoyl-2-lactylate (SSL), reduction of shearing action during mixing, and reduction of fermentation time can be beneficial to bread doughs containing triticale flour.

Baked products that are not very dependent on gluten structure probably have greater potential for incorporation of triticale flour than does bread. Some of the early uses of triticale as a food ingredient were reviewed by Lorenz and Welsh (1974) and Tsen (1974). Chawla and Kapoor (1982) used triticale flour in chapatis, and Kahn and Penfield (1983) produced satisfactory snack crackers containing triticale flour.

3. Rice Flour

Little rice flour is used, though rice is a major staple food in much of the world. Milling of rice does not produce flour; it includes removal of the outer shell (hull), bran, and germ and finally polishing. Grinding of rice kernels that have been broken during milling results in rice flour. The protein concentration is only about 6.5–7% protein (Luh and Liu, 1980, p. 472). If the flour is made from waxy rice, in which the starch is almost entirely amylopectin, the starch does not retrograde at low temperatures, and waxy rice flour, therefore, has been used as a thickening agent.

Rice flour protein does not have gluten-forming properties. It can be substititued for 5% of the wheat flour in bread with little effect on bread properties; acceptable, but quite different, breads can be obtained with replacement levels up to 30% (Luh and Liu, 1980, p. 471). Luh and Liu (1980, pp. 476–480) reviewed work on rice flour in cookies.

The need for wheat-free foods in hypoallergenic diets has stimulated interest in the use of rice flour as a total replacement for wheat flour. For persons who cannot tolerate wheat, the term "acceptability" has a different meaning than it does for persons without dietary restrictions. A bread or cake that otherwise might seem to be of low quality might be preferred to no bread or cake in a hypoallergenic diet. Bean et al. (1983), Nishita et al. (1976), and Nishita and Bean (1979) studied breads and layer cakes made with rice flour; they studied several variables that affect characteristics of bread and cake containing rice flour.

Ylimaki et al. (1988) developed a bread formula with rice and potato starch in a ratio of 80:20 and added carboxymethylcellulose (CMC) to provide some structure in the absence of gluten. They were able to produce rice breads that were very similar physically to a commercial white bread used as a reference. Results of sensory evaluation were to be reported in a later paper.

A potential commercial use of rice flour is in puffed snacks. Fondevila et al. (1988) reported the development of an acceptable puffed snack containing 100% rice flour.

4. Corn (Maize) Flour

Corn, which is produced primarily for animal feed in the United States, is produced for food in some of the developing countries. Even in those countries where it is a major food, relatively little corn is used in the form of flour.

Milling of corn involves degerming of moisture-conditioned corn and dehull-

ing, both by friction, followed by partial drying, then grinding of the endosperm to differing degrees of fineness. The ground products are grits, meal, and
flour in order of increasing degree of fineness. All three products have a protein
concentration of about 7–8% protein, dry basis (Alexander, 1987, p. 356). Of
the three products, flour is the least used.

Corn flour is used commercially in dry mixes for various flour products, in
breadings, in ready-to-eat breakfast cereals, and in extruded snacks. It is a constituent of the corn–soy milk blends used in United States food aid programs
(Rooney and Serna-Saldivar, 1987, p. 401). Badi and Hoseney (1978) studied
cookies containing corn flours from different mill streams.

Instant maize flours are made by drying, milling, and sieving masa; masa is a
dough made by cooking and steeping corn in lime, washing out the pericarp and
calcium oxide, and stone grinding the residue ("nixtamal"). The masa is used for
making tortillas; tortilla-based foods have been increasingly popular in the
United States, and dried masa is available on the retail market in some sections
of the country. Because of its potential usefulness in corn-based snacks, instant
maize flour has been studied extensively (Bedolla and Rooney, 1982, 1984).

5. Oat Flour

Only a small part of the oat crop is used as food; however, the proportion increased in the late 1980s in response to a reported cholesterol-lowering effect of
oat bran. Flour constitutes a small portion of food use of oats.

Pyler (1988, p. 392) briefly described oat milling. Oats are dehulled by impact
milling; the resulting "groats," with their bran intact, are steamed, steel cut if to
be used for quick-cooking oats, and rolled to produce flakes, which are used by
the cereal industry. Grinding either groats or flakes results in flour.

Oat flour contains about 17% protein (dry weight basis) but it does not form a
gluten-type structure. McKechnie (1983) reported that it can be substituted for
as much as 30% of the wheat flour in bread. Oomah (1983) also studied the performance of oat flour in bread.

Shukla (1975) reviewed oat production, properties, milling, and processing.

6. Barley Flour

Barley is used almost entirely as feed and as a grain for brewing and for ethanol
production. However, flour can be produced; the process involves abrasive de-
hulling, resulting in "pearled" barley, then grinding, Shellenberger (1980,
p. 256) listed soups, dressings, baby foods, and specialty items as some uses for
barley flour and grits.

Some hull-less cultivars of barley have been developed, and their flours have
been studied in bread by Bhatty (1986) and by Swanson and Penfield (1988).
Bhatty found 5% replacement of wheat flour to have no serious effects on loaf
volume or appearance. Swanson and Penfield found bread containing barley
flour, whole wheat flour, and wheat bread flour in 20:30:50 proportions to be
similar to a control bread containing whole wheat flour and bread flour in 50:50
proportions except that its specific volume was 5–6% lower than that of the

control bread. McNeil *et al.* (1988) incorporated whole grain barley flour into pie pastry. An acceptable product was produced with 30% replacement of the wheat flour.

Rolled barley, similar in form to rolled oats, was studied in bread by Prentice *et al.* (1979). Although the rolled barley had a negative effect on loaf volume, bread in which rolled barley was substituted for 15% of the white bread flour was as acceptable as bread containing whole-wheat flour at a 30% replacement level. Prentice *et al.* also compared cookies containing barleymeal with oatmeal cookies. Although the texture of the barleymeal cookies was inferior to that of the oatmeal cookies, the barleymeal cookies were considered acceptable.

7. Sorghum Flour

Sorghum grain is a feed crop in industrial nations and a food crop in several developing countries. Much of the sorghum grain used as food is used in porridge; relatively little is milled into flour. Its milling is similar to that of barley. Whole grain sorghum flour is used in the making of a bread, roti, in India. The performance of sorghum flour has been studied in roti (Subramanian *et al.*, 1983), yeast bread (Perten, 1983; Olatunji *et al.*, 1982), high-ratio cakes (Glover *et al.*, 1986), and biscuits (cookies in the United States) (Olatunji *et al.*, 1982). Acceptable products were obtained with replacement levels ranging from 15% for high-ratio cakes to 45% for cookies.

B. Oilseed Flours

1. Soybean Flour

Soybeans, once produced mostly for feed and industrial products, have evolved into an important source of food. Flour, protein concentrates, and protein isolates have become important in the food industry; their protein concentrations are about 50, 70, and 90%, respectively. Circle and Smith (1978) described their production.

Soybean flours, described by Pomeranz (1985, pp. 446–447), differ in their lipid content, which ranges from 1% or less to 22%. Trypsin inhibitors are destroyed and enzymes that can cause flavor problems are inactivated by application of a moist heat treatment to the soy flakes from which the flours are produced. Protein and carbohydrate concentrations vary inversely with lipid content; on a moisture-free basis, protein concentration ranges from 59 to 47% and carbohydrate concentration from 34 to 26% with increasing fat. Ash content is relatively high, averaging more than 5% (Circle and Smith, 1978, p. 311).

Not only are soybean flours different from wheat flours in their proximate composition, but there are significant qualitative differences. Soybean proteins do not have gliadin- and glutenin-type proteins and, therefore, cannot form doughs similar to those of wheat flour. The carbohydrates do not include starch.

Pomeranz (1966) reviewed the use of soybean flour as a food ingredient. In bread, soybean flour affects water absorption, dough properties, gas retention, color, and flavor. The following three factors that can be controlled affect bread quality: coarse soybean flours are preferable to fine powders; because the extent of heating the flakes adversely affects bread, that heat treatment must be well controlled; the effect of heat treatment can be partially offset by relatively high levels of bromate. In any case, soybean flour cannot be used successfully as the only flour in bread. French (1977) and Smith and Circle (1978) also reviewed soybean flour use.

2. Cottonseed Flour

Glandless cottonseed varieties are available for flour production, but much of the cottonseed flour used is made from glanded cottonseed that is liquid-cyclone processed (LCP). The toxic pigment gossypol is removed by centrifugation in hexane. Cottonseed protein has a higher lysine concentration than does wheat protein.

Cookies containing 12 and 24% LCP cottonseed flour were rated high overall and cookies containing 36 and 48% were acceptable. Aroma was the characteristic that had decreasing scores with increasing level of substitution (Vecchionacce and Setser, 1980).

El-Minyawi and Zabik (1981) substituted LCP cottonseed flour for hard wheat flour in Egyptian balady bread. Although crispness decreased and leatheriness increased with the substitution levels up to 16%, keeping quality increased, as did protein quantity and quality. At the 8% level, the protein content was 41% higher than that of the wheat flour control and the bread was acceptable.

3. Sunflower Flour

Of the two types of sunflower, oilseed and nonoilseed, the nonoilseed type is used far more in foods. The nonoilseed varieties actually contain 21 to 33% oil, but the oilseed varieties have more than 40% and are produced primarily for their oil. Both types have high concentrations of protein.

Khan et al. (1980) compared breads in which sunflower seed flours replaced 7.5% of the wheat flour with breads containing soybean and peanut flours at the 7.5% replacement level. The flavor and color of breads containing sunflower seed flours were inferior to those of breads containing soy and peanut flours; volume was lower. D'Appolonia and MacArthur (1979) reported that roasting the sunflower meats from which flour was made improved the color, grain, and loaf volume of breads containing sunflower seed flour. The use of sunflower seed in food products was reviewed by Robertson (1975).

4. Sesame Flour

Sesame seed flour has been used for feed but only to a limited extent for food because levels of selenium can be toxic. The selenium content is not high, however, if sesame is produced in soil having low levels of selenium. The protein of sesame flour has relatively high levels of sulfur amino acids and tryptophan, making it a good complement of soy and corn proteins (Brito and Nuñez, 1982).

Problems that have been reported with the use of sesame seed flour include a coarse texture and greenish color in tortillas (Tonella et al., 1983) and low cookie spread, low top grain score, and bitterness in sugar snap cookies (Hoojjat and Zabik, 1984). In the cookies, various combinations of sesame seed flour and navy bean flour were substituted for up to 30% of the wheat flour. When the level of sesame seed flour did not exceed 15%, the cookie spread and top grain were satisfactory; the bitterness was detected with more than 10% sesame flour.

C. Legume Flours

Legumes grow worldwide and are important nutritionally because of their protein content. Whole legume seeds have such good keeping qualities when dried that they have been used chiefly in that form. However, they can be milled into flour, and legume flours can complement wheat flour nutritionally because sulfur-containing amino acids, rather than lysine, are first limiting in most legumes; their use with wheat flour also increases the protein concentration of products. Some legume seeds contain enzymes that may cause off-flavors, or they may contain antinutritional factors; roasting those seeds before milling them into flour destroys the unwanted components.

Flours made from legumes have been studied mostly in breads. In some cases the acceptable level of replacement was increased with special treatments or conditions, such as heat treatment of peanuts used for flour, defatting of winged bean flours, or the addition of SSL to formulas to increase dough strength. Satisfactory levels of replacement ranged from 5 to 15%. Effects observed with excessive replacement included those on flour absorption, dough-handling properties, loaf volume, crumb grain, flavor, and color.

Legume flours that have been studied in bread include faba bean flour (D'Appolonia, 1977; Hsu et al., 1980; Lorenz et al., 1979), mung bean and navy bean flours (D'Appolonia, 1977), lentil flours (D'Appolonia, 1977; Hsu et al., 1980), pinto bean flour (D'Appolonia, 1977; Staley and Pelaez, 1984), yellow pea flour (Hsu et al., 1980; Repetsky and Klein, 1982), cowpea flour (Mustafa et al., 1986; Okaka and Potter, 1977), lupine flour (Ballester et al., 1984; Campos and El-Dash, 1978), peanut flour (Khan et al., 1978), and winged bean flour (Kailasapathy and MacNeil, 1985; Okezie and Dobo, 1980). (Note: The soybean also is a legume and its inclusion with oilseed flours rather than here is arbitrary.)

Some studies have dealt with legume flours in quick breads. A replacement level of 35% was satisfactory for navy bean flour in pumpkin bread (Dryer et al., 1982) and in pumpkin spice muffins (Cady et al., 1987). In a plain quickbread, Raidl and Klein (1983) found defatted soybean flour to perform satisfactorily with up to 15% replacement of wheat flour, but yellow field pea flour gave both an off-taste and aftertaste, even at the 5% level.

In cookies, the reported satisfactory levels of replacement of wheat flour with legume flour have ranged from 10 to 30%. Characteristics affected have been dough-handling properties, cookie softness, cookie spread, top grain (of sugar cookies), color, and flavor. The legume flours studied have included defatted peanut flour (Beuchat, 1977; McWatters, 1978), defatted field pea flour (McWatters, 1978), navy bean flour (Cady et al., 1987; Dreher and Patek, 1984), and pinto bean flour (Staley and Pelaez, 1984).

Staley and Pelaez (1984) found substitution of pinto bean flour for all-purpose flour to be satisfactory in certain cakes. El-Samahy *et al.* (1980) produced acceptable cakes from mixes in which defatted peanut flour was substituted for 10% of the wheat flour.

D. Root Flours

Root flours have been studied less than other types, possibly because their protein concentrations are lower than those of seed flours. Studies involving root and tuber flours tend to be aimed at utilization of native foods.

Cassava and yam flours were studied in bread by Ciacco and D'Appolonia (1978). They found replacement at a 15% level to result in satisfactory bread, especially when SSL was added. Sweet potato flour at replacement levels up to 21% did not have serious deleterious effects on yeast-leavened doughnuts (Collins and Aziz, 1982). El-Samahy *et al.* (1980) produced acceptable cakes with sweet potato flour used at replacement levels up to 30%. Ciacco and D'Appolonia (1977) reviewed earlier studies involving root and tuber starches.

Suggested Exercises

1. Compare cake flour, all-purpose flour, and bread flour as to:
   a. rate of settling from a suspension (3.2 g flour in 50 ml water) in a stoppered 100-ml graduated cylinder (control mixing; read volume of sediment in bottom of cylinder after 5 min)
   b. microscopic appearance
   c. amount of water required to give dough of a certain consistency
   d. elasticity of dough, as measured by area covered by a weighed amount of dough rolled on an oiled surface to a controlled thickness and permitted to stand for 1 min
2. Substitute different types of wheat flour in baked products (basic formulas in Appendix A.).
3. Use a nonwheat flour in a yeast bread, quick bread, cookie, or cake (basic formulas in Appendix A.). There are many possible variables, including the following: the kind of nonwheat flour, the product in which used, replacement of flour at a given level versus addition of the nonwheat flour, and the level of replacement.

References

AACC. 1983. "Approved Methods of the American Association of Cereal Chemists," 8th edn. AACC, St. Paul, Minnesota.

Alexander, R J. 1987. Corn dry milling: Processes, products, and applications. *In* "Corn: Chemistry and Technology," Watson, S. A. and Ramstad, P. E. (Eds.). American Association of Cereal Chemists, St. Paul, Minnesota.

Badi, S. M. and Hoseney, R. C. 1978. Corn flour: Use in cookies. *Cereal Chem.* **55**: 495–504.

Ballester, D., Zacarias, I., Garcia E., and Yañez, E. 1984. Baking studies and nutritional value of bread supplemented with full-fat sweet lupine flour (L. albus cv. Multolupa). *J. Food Sci.* **49**: 14–16, 27.

Bass, E. J. 1988. Wheat flour milling. *In* "Wheat: Chemistry and Technology," Vol. 2, Pomeranz, Y. (Ed.). American Association of Cereal Chemists, St. Paul, Minnesota.

Bean, M. M., Elliston-Hoops, E. A., and Nishita, K. D. 1983. Rice flour treatment for cake-baking applications. *Cereal Chem.* **60:** 445–449.

Bedolla, S. and Rooney, L. W. 1982. Cooking maize for masa production. *Cereal Foods World* **27:** 219–221.

Bedolla, S. and Rooney, L. W. 1984. Characteristics of U.S. and Mexican instant maize flours for tortilla and snack preparation. *Cereal Foods World* **29:** 732–735.

Beuchat, L. R. 1977. Modification of cookie-baking properties of peanut flour by enzymatic and chemical hydrolysis. *Cereal Chem.* **54:** 405–414.

Bhatty, R. S. 1986. Physiochemical and functional (breadmaking) properties of hull-less barley fractions. *Cereal Chem.* **63:** 31–35.

Bradbury, D., Cull, I. M., and MacMasters, M. M. 1956. Structure of the mature wheat kernel. I. Gross anatomy and relationships of parts. *Cereal Chem.* **33:** 329–342.

Brito, O. J. and Nuñez, N. 1982. Evaluation of sesame flour as a complementary protein source for combinations with soy and corn flours. *J. Food Sci.* **47:** 457–460.

Bushuk, W. and Larter, E. N. 1980. Triticale: Production, chemistry, and technology. *Adv. Cer. Sci. and Technol.* **3:** 115–157.

Cady, N. D., Carter, A. E., Kayne, B. E., Zabik, M. E., and Uebersax, M. A. 1987. Navy bean flour substitution in a master mix used for muffins and cookies. *Cereal Chem.* **64:** 193–195.

Campos, J. E. and El-Dash, A. A. 1978. Effect of addition of full fat sweet lupine flour on rheological properties of dough and baking quality of bread. *Cereal Chem.* **55:** 619–627.

Chawla, V. K. and Kapoor, A. C. 1982. Chemical composition and protein quality of wheat-triticale chapatis. *J. Food Sci.* **47:** 2015–2017.

Ciacco, C. F. and D'Appolonia, B. L. 1977. Functional properties of composite flours containing tuber flour or starch. *Bakers Digest* **51**(5): 46–50, 141.

Ciacco, C. F. and D'Appolonia, B. L. 1978. Baking studies with cassava and yam. II. Rheological and baking studies of tuber-wheat flour blends. *Cereal Chem.* **55:** 423–435.

Circle, S. J. and Smith, A. K. 1978. Processing soy flours, protein concentrates, and protein isolates. *In* "Soybeans: Chemistry and Technology," Vol. 1, rev., Smith, A. K. and Circle, S. J. (Eds.). Avi Publ. Co., Westport, Connecticut.

Collins, J. L. and Aziz, N. A. A. 1982. Sweet potato as an ingredient of yeast-raised doughnuts. *J. Food Sci.* **47:** 1133–1139.

D'Appolonia, B. L. 1977. Rheological and baking studies of legume-wheat flour blends. *Cereal Chem.* **54:** 53–63.

D'Appolonia, B. L. and MacArthur, L. A. 1979. Utilization of sunflower flour derived from untreated and roasted sunflower meats in bread baking. *Bakers Digest* **53**(1): 32–36.

Dreher, M. L. and Patek, J. W. 1984. Effects of supplementation of short bread cookies with roasted navy bean flour and high protein flour. *J. Food Sci.* **49:** 922–924.

Drews, E. and Seibel, W. 1976. Bread-baking and other uses around the world. *In* "Rye: Production, Chemistry, and Technology," Bushuk, W. (Ed.). American Association of Cereal Chemists, St. Paul, Minnesota.

Dryer, S. B., Phillips, S. G., Powell, T. S., Uebersax, M. A., and Zabik, M. E. 1982. Dry roasted navy bean flour incorporation in a quick bread. *Cereal Chem.* **59:** 319–320.

El-Minyawi, M. A. and Zabik, M. E. 1981. Cottonseed flour's functionality in Egyptian baladi bread. *Cereal Chem.* **58:** 413–417.

El-Samahy, S. K., Morad, M. M., Seleha, H., and Abdel-Baki, M. M. 1980. Cake-mix supplementation with soybean, sweet potato or peanut flours. II. Effect on cake quality. *Bakers Digest* **54**(5): 32-33, 36.

FDA. 1988a. Bakery products. *In* "Code of Federal Regulations," Title 21. Section 136. U.S. Govt. Printing Office, Washington, DC.

FDA. 1988b. Cereal flours and related products. *In* "Code of Federal Regulations," Title 21, Section 137. U.S. Govt. Printing Office, Washington, DC.

Fondevila, M. P., Liuzzo, J. A., and Rao, R. M. 1988. Development and characterization of a snack food product using broken rice flour. *J. Food Sci.* **53:** 488–490.

French, F. 1977. Bakery uses of soy products. *Bakers Digest* **51**(5): 98–103.

Giddings, G. G. and Welt, M. A. 1982. Radiation preservation of food. *Cereal Foods World* **27:** 17–20.

Glover, J. M., Walker, C. E., and Mattern, P. J. 1986. Functionality of sorghum flour components in a high ratio cake. *J. Food Sci.* **51:** 1280–1283, 1292.

Hoojjat, P. and Zabik, M. E. 1984. Sugar-snap cookies prepared with wheat-navy bean-sesame seed flour blends. *Cereal Chem.* **61:** 41–44.

Hoseney, R. C. 1986. "Cereal Science and Technology." American Association of Cereal Chemists, St. Paul, Minnesota.

Hsu, D., Leung, H. K., Finney, P. L., and Morad, M. M. 1980. Effect of germination on nutritive value and baking properties of dry peas, lentils, and faba beans. *J. Food Sci.* **45:** 87–92.

Kahn, C. B. and Penfield, M. P. 1983. Snack crackers containing whole-grain triticale flour: Crispness, taste, and acceptability. *J. Food Sci.* **48:** 266–267.

Kailasapathy, K. and MacNeil, J. H. 1985. Baking studies with winged bean (Psophocarpus tetragonolobus L.DC) flour-wheat flour blends. *J. Food Sci.* **50:** 1672–1675.

Khan, M. N., Mulsow, D., Rhee, K. C., and Rooney, L. W. 1978. Bread making properties of peanut flour produced by direct solvent extraction procedure. *J. Food Sci.* **43:** 1334–1335.

Khan, M. N., Wan, P., Rooney, L. W., and Lusas, E. W. 1980. Sunflower flour: A potential bread ingredient. *Cereal Foods World* **25:** 402–404.

Làsztity, R. 1984. "The Chemistry of Cereal Proteins." CRC Press, Inc., Boca Raton, Florida.

Lorenz, K. 1975. Irradiation of cereal grains and cereal grain products. *CRC Crit. Rev. Food Sci. Nutr.* **6:** 317–382.

Lorenz, K. and Welsh, J. R. 1974. Food product utilization of Colorado-grown triticales. *In* "Triticale: First Man-Made Cereal," Tsen, C. C. (Ed.). American Association of Cereal Chemists, St. Paul, Minnesota.

Lorenz, K., Dilsaver, W., and Wolt, M. 1979. Faba bean flour and protein concentrate in baked goods and in pasta products. *Bakers Digest* **53**(3): 39–45, 51.

Luh, B. S. and Liu, Y.-K. 1980. Rice flours in baking. *In* "Rice: Production and Utilization," Luh, B. S. (Ed.). Avi Publ. Co., Westport, Connecticut.

MacArthur, L. A. and D'Appolonia, B. L. 1982. Microwave and gamma radiation of wheat. *Cereal Foods World* **27:** 58–60.

McKechnie, R. 1983. Oat products in bakery foods. *Cereal Foods World* **28:** 635–637.

McNeil, M. A., Penfield, M. P., and Swanson, R. B. 1988. Sensory evaluation of pastry containing barley flour. *Tennessee Farm and Home Science* **146:** 4–7.

McWatters, K. H. 1978. Cookie baking properties of defatted peanut, soybean, and field pea flours. *Cereal Chem.* **55:** 853–863.

Miller, B. S., Trimbo, H. B., and Derby, R. I. 1969. Instantized flour—physical properties. *Bakers Digest* **43**(6): 49–51, 66.

Mustafa, A. I., Al-Wessali, M. S., Al-Basha, O. M., and Al-Amir, R. H. 1986. Utilization of cowpea flour and protein isolate in bakery products. *Cereal Foods World* **31:** 756–759.

Nishita, K. D. and Bean, M. M. 1979. Physicochemical properties of rice in relation to rice bread. *Cereal Chem.* **56:** 185–189.

Nishita, K. D., Roberts, R. L., and Bean, M. M. 1976. Development of a yeast-leavened rice-bread formula. *Cereal Chem.* **53:** 626–635.

Okaka, J. C. and Potter, N. N. 1977. Functional and storage properties of cowpea powder-wheat flour blends in breadmaking. *J. Food Sci.* **42:** 828–833.

Okezie, B. O. and Dobo, S. B. 1980. Rheological characteristics of winged bean composite flours. *Bakers Digest* **54**(1): 35–41.

Olatunji, O., Akinrele, I. A., Edwards, C. C., and Koleoso, O. A. 1982. Sorghum and millet processing and uses in Nigeria. *Cereal Foods World* **27:** 277–280.

Oomah, B. D. 1983. Baking and related properties of wheat-oat composite flours. *Cereal Chem.* **60:** 220–225.

Perten, H. 1983. Practical experience in processing and use of millet and sorghum in Senegal and Sudan. *Cereal Foods World* **28:** 680–683.

Pomeranz, Y. 1966. Soy flour in breadmaking. *Bakers Digest* **40**(3): 44–48, 78.

Pomeranz, Y. 1985. "Functional Properties of Food Components." Academic Press, Orlando, Florida.

Pomeranz, Y. 1987. "Modern Cereal Science and Technology." VCH Publishers, Pullman, Washington.

Prentice, N., Burger, W. C., and D'Appolonia, B. L. 1979. Rolled high-lysine barley in breakfast cereal, cookies, and bread. *Cereal Chem.* **56**: 413–416.

Pyler, E. J. 1988. "Baking Science and Technology," 3rd edn., vol. 1. Sosland Publishing Co., Merriam, Kansas.

Raidl, M. A. and Klein, B. P. 1983. Effects of soy or field pea flour substitution on physical and sensory characteristics of chemically leavened quick breads. *Cereal Chem.* **60**: 367–370.

Repetsky, J. A. and Klein, B. P. 1982. Partial replacement of wheat flour with yellow field pea flour in white pan bread. *J. Food Sci.* **47**: 326–327, 329.

Robertson, J. A. 1975. Use of sunflower seed in food products. *CRC Crit. Rev. Food Sci. Nutr.* **6**: 201–240.

Rooney, L. W. and Serna-Saldivar, S. O. 1987. Food uses of whole corn and dry-milled fractions. *In* "Corn: Chemistry and Technology," Watson, S. A. and Ramstad, P. E. (Eds.). American Association of Cereal Chemists, St. Paul, Minnesota.

Rozsa, T. A. 1976. Rye milling. *In* "Rye: Production, Chemistry, and Technology," Bushuk, W. (Ed.). American Association of Cereal Chemists, St. Paul, Minnesota.

Schiller, G. W. 1984. Bakery flour specifications. *Cereal Foods World* **29**: 647–651.

Shellenberger, J. A. 1980. Advances in milling technology. *Adv. Cer. Sci. and Technol.* **3**: 227–269.

Shukla, T. P. 1975. Chemistry of oats: Protein foods and other industrial products. *CRC Crit. Rev. Food Sci. Nutr.* **6**: 383–431.

Smith, A. K. and Circle, S. J. 1978. Protein products as food ingredients. *In* "Soybeans: Chemistry and Technology," Vol. 1, rev., Smith, A. K. and Circle, S. J. (Eds.). Avi Publ. Co., Westport, Connecticut.

Staley, L. L. and Pelaez, J. 1984. Full flavor pinto bean flour—a versatile ingredient. *Cereal Foods World* **29**: 411–414.

Subramanian, V., Jambunathan, R., and Rao, N. S. 1983. Textural properties of sorghum dough. *J. Food Sci.* **48**: 1650–1653, 1673.

Swanson, C. O. 1938. "Wheat Flour Quality." Burgess Publ. Co., Edina, Minnesota.

Swanson, R. B. and Penfield, M. P. 1988. Barley flour level and salt level selection for a whole-grain bread formula. *J. Food Sci.* **53**: 896–901.

Tonella, M. L., Sanchez, M., and Salazar, M. G. 1983. Physical, chemical, nutritional and sensory properties of corn-based fortified food products. *J. Food Sci.* **48**: 1637–1643.

Tsen, C. C. 1974. Bakery products from triticale flour. *In* "Triticale: First Man-Made Cereal," Tsen, C. C. (Ed.). American Association of Cereal Chemists, St. Paul, Minnesota.

Vecchionacce, L. M. and Setser, C. S. 1980. Quality of sugar cookies fortified with liquid cyclone processed cottonseed flour with stabilizing agents. *Cereal Chem.* **57**: 303–306.

Ylimaki, G., Hawrysh, Z. J., Hardin, R. T., and Thomson, A. B. R. 1988. Application of response surface methodology to the development of rice flour yeast breads: Objective measurements. *J. Food Sci.* **53**: 1800–1805.

LEAVENING AGENTS

Batter and dough products are leavened by water vapor, air, carbon dioxide, and sometimes ammonia. The gases are distributed as small bubbles in batters and doughs, and the fineness of their dispersion is responsible for the grain of the baked products.

I. LEAVENING GASES

Vapor is formed from water during baking of batters and doughs. Air is incorporated as preformed gas cells in beaten egg white and also in other ways, such as by creaming the shortening and by beating a viscous batter. Carbon dioxide is formed in foods by decomposition of a bicarbonate or by the chemical reaction of sodium bicarbonate with an acid, or by the biological action of microorganisms.

The leavening gases seldom, if ever, act singly. Although water vapor is important in leavening pastry, popovers, and cream puffs, it may be assisted by air, especially in pastry. Air usually is thought of as the principal leavening agent in foods containing beaten egg whites, such as angel food cakes; however, Barmore showed, in a classic study (1936), that water vapor actually is responsible for the major part of the expansion of angel food cakes during baking. In foods leavened by carbon dioxide, whether formed by chemical or biological action, air and water vapor also are present during baking, because some air is incorporated

FIGURE 18.1
Cakes containing oil and leavened by (left) carbon dioxide, air, and water vapor; (middle) air and water vapor; (right) water vapor (air-evacuated batter). (From Hood and Lowe, 1948.)

into batters and doughs during mixing and vapor is formed from water during baking. However, the important function that carbon dioxide plays in leavening some of these products is demonstrated easily by omission of the baking powder or yeast. A heavy product usually results, with a few exceptions such as certain cakes that normally contain a relatively small proportion of baking powder along with a large proportion of egg.

The roles of the different leavening gases were demonstrated in two other classic studies, that of Hood and Lowe (1948) and that of Carlin (1944). Hood and Lowe studied shortened cakes mixed by a modified conventional method. When the formula contained the normal baking powder, the leavening gases were carbon dioxide, water vapor, and air. When the baking powder was omitted, the leavening gases were water vapor and air. When the batter containing no baking powder was evacuated with a vacuum pump, the only leavening gas was water vapor. Cakes made by these three treatments are shown in Fig. 18.1, and the proportion of the total volume increase (relative to batter volume) contributed by each of the three gases is shown in Fig. 18.2. The major increase in cake volume was produced by carbon dioxide followed by water vapor. Air was responsible for only a small proportion of the total increase in volume but was important because the effectiveness of water vapor as a leavening gas depended on the presence of air in the batter. Carlin (1944) demonstrated the importance of air distribution to carbon dioxide's effectiveness; carbon dioxide did not form new gas cells but diffused into and expanded the existing air cells, now called gas cell nuclei.

Gas cell nuclei are the small gas cells that form during mixing. They consist largely of the air that is incorporated and dispersed, but the rapid evolution of some of the carbon dioxide at room temperature also may make a contribution. Gas cell nuclei are important to the ultimate grain of the product because, as mentioned above, carbon dioxide and water vapor formed during baking diffuse into the gas cell nuclei and expand them rather than forming new cells.

II. CHEMICAL LEAVENING AGENTS _____

Chemical leavening agents are used primarily as formulated baking powders in noncommercial food preparation. Commercial bakers might formulate their

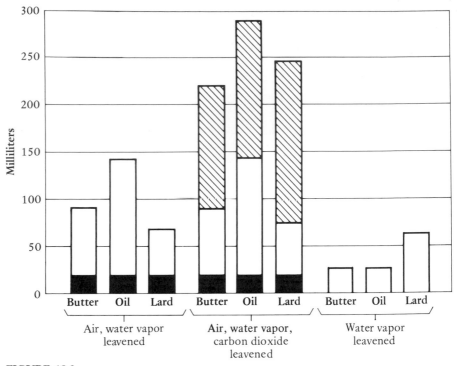

FIGURE 18.2
Cakes made from butter, oil, and hydrogenated lard; leavened by (1) air and water vapor, (2) carbon dioxide, air, and water vapor, and (3) water vapor. The total volume increase over the initial volume of the batter and the proportion of the increase attributed to carbon dioxide (hatched), air (solid), and water vapor (open) are shown. (From Hood and Lowe, 1948.)

own baking powders to meet special needs, or they might add the leavening ingredients separately in commercial baking and in the production of mixes; the principles of leavening action are the same in these different situations.

A. Chemical Nature of Leavening Action

Carbon dioxide can be produced by heat decomposition of certain compounds. For example, the heating of dissolved ammonium bicarbonate results in volatile products. The absence of a residual

$$NH_4HCO_3 \rightarrow NH_3 + CO_2 + H_2O$$

Decomposition of ammonium bicarbonate

salt is advantageous, but the ammonia imparts a detectable taste unless it is able to escape completely. Therefore, the one important use of ammonium bicarbonate is in commercially baked cookies that have a large surface relative to their mass and are baked at high temperatures.

Sodium bicarbonate does not require an acid for release of carbon dioxide, as indicated by the following reaction that occurs with heating:

$$2 \, NaHCO_3 \;\rightarrow\; CO_2 \;+\; Na_2CO_3 \;+\; H_2O$$

Decomposition of sodium bicarbonate

However, the sodium carbonate produced has an unpleasant taste and its alkalinity has other undesirable effects, for example, on crumb color. Sodium bicarbonate, therefore, is not used alone and ideally is not used in excess when used with an acid.

Chemical leavening nearly always involves carbon dioxide production through the reaction in water of sodium bicarbonate with an acid. The reaction can be expressed as follows:

$$HX \;+\; NaHCO_3 \;\rightarrow\; CO_2 \;+\; H_2O \;+\; NaX$$

Reaction of acid and sodium bicarbonate

The yield of carbon dioxide is greater than that obtained by decomposition of sodium bicarbonate. The acid may be that of a separate acidic ingredient such as buttermilk. The acidic ingredient is combined with sodium bicarbonate in baking powders.

B. Baking Powders

All baking powders on the retail market are formulated to yield at least 12% (required by law), usually more nearly 14%, carbon dioxide when water is added and heat is applied. Acid salts, being more stable than acids as such, are combined in proper proportions with sodium bicarbonate; corn starch is added for the purpose of standardization. The corn starch also has a protective effect against atmospheric moisture.

Baking powders differ from one another as to their specific acidic constituents. They frequently are classified accordingly as sulfate–phosphate (SAS–phosphate) or phosphate types. Tartrate baking powder formerly was available but disappeared from the retail market because of its cost; phosphate baking powders apparently are becoming rare on the retail market. The proportions of sodium bicarbonate and one or more of the various acid constituents to be used in formulation of a baking powder are determined experimentally by titration, rather than stoichiometrically, because in some cases the exact reactions are not known. The results of such titrations are expressed as neutralizing value (NV), the number of parts by weight of sodium bicarbonate that 100 parts of the leavening acid will neutralize. For example, the NV of sodium aluminum sulfate (SAS) and of sodium aluminum phosphate is 100, and that of monocalcium phosphate monohydrate is 80 (LaBaw, 1982); therefore, monocalcium phosphate monohydrate would need to be used in a higher concentration than would sodium aluminum sulfate.

Baking powders differ from one another as to the speed with which carbon dioxide is formed during mixing and baking. This difference is a function of the solubility of the acidic constituents. The dough rate of reaction test, as de-

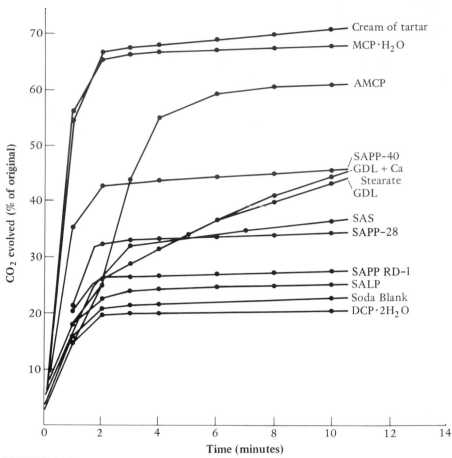

FIGURE 18.3
Dough rate of reaction curves for various leavening acids at 27°C with 3 min of stirring.
MCP·H_2O, Monocalcium phosphate monohydrate; AMCP, anhydrous monocalcium phosphate;
SAPP, sodium acid pyrophosphate; GDL, glucono-δ-lactone; SAS, sodium aluminum sulfate;
SALP, sodium aluminum phosphate; DCP·$2H_2$O, dicalcium phosphate dihydrate. (Reprinted from
Kichline and Conn, 1970, with the permission of Van Nostrand Reinhold, New York.)

scribed by Barackman (1931), is used for predicting the rate of carbon dioxide
release during mixing and holding unheated. The results of the test with several
leavening acids, as reported by Kichline and Conn (1970), are seen in Fig. 18.3.
The extremely high rates for cream of tartar and monocalcium phosphate
monohydrate (MCP·H_2O) are obvious and show why these acid salts are not
used singly. When a tartrate baking powder was made, it contained both tartaric
acid and potassium bitartrate.

   At the opposite extreme is the reaction rate of dicalcium phosphate dihydrate (DCP·$2H_2$O), which reacts with sodium bicarbonate very little at mixing
temperature and, in fact, reacts very little at all until a late stage of baking. If
used with another acid that can carry the major burden of the leavening task,
DCP·$2H_2$O can be useful in the baking industry by helping prevent dipping in
the center of cakes.

Sulfate–phosphate baking powders are readily available. Monocalcium phosphate monohydrate, the usual phosphate in a sulfate-phosphate baking powder, reacts with sodium bicarbonate during mixing, to provide nucleating gas. Sodium aluminum sulfate (SAS) reacts relatively little during mixing. Most of its reaction with sodium bicarbonate occurs during baking, providing carbon dioxide to diffuse into and expand the gas cell nuclei. A baking powder containing these two salts often is called a double-acting baking powder. Hoseney (1986, p. 250) described a newer double-acting baking powder containing sodium aluminum phosphate (SALP) along with monocalcium phosphate monohydrate and sodium bicarbonate. A phosphate baking powder, often called a single-acting baking powder, most likely contains a monocalcium phosphate that is coated to delay its dissolving.

The goal for leavening action during baking is release of most of the available carbon dioxide before firming of the product, but not too rapid release. If release of gas is not complete before the structure is firm, volume is relatively low. In a product such as a cake, bulges and cracks may develop in the center portion after the outer portions are firm. If release of gas is too rapid, the fluid or soft batter or dough permits coalescence of bubbles and the result is either a coarse structure or excessive loss of leavening gas (Kichline and Conn, 1970).

Encapsulated leavening ingredients are available to bakers. Microencapsulation is a technique of growing importance. The pharmaceutical industry's timed release capsules have been in use for many years, and the microencapsulation technique lends itself to the improvement of many food ingredients, including enrichment nutrients, flavoring and coloring agents, acidulants, and leavening agents. The application of an inert coating to minute gaseous, liquid, or solid particles can be used to change physical properties, permit delayed release, and protect from air, moisture, and chemical reactants. Dziezak (1988) reviewed the methods and materials in use, some criteria for making decisions concerning encapsulation, and some specific ingredient applications. Obviously the protection of leavening agent components from premature reaction contributes to storage stability, not only of the leavening agents themselves but also of mixes containing them.

The amount of baking powder in a formula affects the properties of the product. The effects on cake quality of varying the level of sulfate–phosphate baking powder are shown in Fig. 18.4. The volume of the cakes first increases and then decreases as baking powder level is increased. Too little baking powder, because of insufficient carbon dioxide to expand cells properly, results in a fine grain and compact crumb. Too much baking powder causes overextension and breaking of cell walls. The result is an open structure with large cells just below the crust and a compact cell structure near the bottom of the cake when a moderate excess of baking powder is used, and a fallen cake when a great excess is used. The crumb is harsh and crumbly with an excess of baking powder. Maximum palatability does not necessarily coincide with maximum volume.

A wide variety of leavening acids is used in commercial baking and in the commercial production of mixes. Different acids are used for different products in commercial baking. Differences among them involve NV, granularity, solubility, reaction rate at room temperature, stability in refrigerated and frozen doughs, response to heat, and final pH of the baked product. Pyler (1988b,

FIGURE 18.4
Effects of increasing amounts of baking powder on the volume and cell structure of cakes. The amounts of SAS-phosphate baking powder per cup of flour are $\frac{1}{2}$, 1, $1\frac{1}{2}$, 2, and $2\frac{1}{2}$ tsp in cakes 1, 2, 3, 4, and 5. Cake 2 is considered ideal. (Courtesy of General Foods Corporation and Calumet © Baking Powder.)

p. 929) listed eight leavening acids with their chemical formulas, NV, and final pH of the baked products containing them. LaBaw (1982) described several leavening acids that are used commercially; he also discussed the principles of leavening and the production of baking powders. Reiman (1983) discussed various phosphates in some detail and described their use in prepared mixes and refrigerated and frozen doughs.

Glucono-δ-lactone (GDL) is one of the compounds listed by Pyler (1988b, p. 929). It has the following formula:

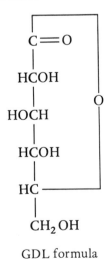

GDL formula

A lactone is an inner ester; the ester linkage in GDL is between the alcohol group on carbon-5 and the carboxyl group of gluconic acid. The ester linkage is hydrolyzed and the gluconic acid formed reacts with sodium bicarbonate when heat is applied. Because of stability at low temperatures, GDL is used in a few commercial products in spite of its slight off-taste and its relatively high cost. Its NV is only 50 (Pyler, 1988b, p. 929).

The successful interchange of sweet and sour milk in flour mixtures is related to leavening agents. If a product contains no buttermilk or other acidic ingredient, the leavening agent in its formula is baking powder. If buttermilk is substituted for the sweet milk, some sodium bicarbonate is needed to neutralize the lactic acid; however, total substitution of sodium bicarbonate for the baking powder, either on a weight basis or on a volume basis, would result in excessive alkalinity and excessive leavening. The substitution procedure described here is only a rough guide because of several variables involved, but in practice it works quite well. The acid in 1 c of moderately sour milk is neutralized by 1/2 tsp (about 2 g) of sodium bicarbonate, and the above combination of sour milk and sodium bicarbonate provides leavening equivalent to about 2 tsp (6–8 g) of baking powder. Thus if a formula calls for 3/4 cup of milk and $2\frac{1}{2}$ tsp of baking powder, the appropriate amount of sodium bicarbonate with the use of 3/4 cup of buttermilk is 3/8 tsp (1.5 g). Leavening equivalent to that of $1\frac{1}{2}$ tsp of baking powder would be taken care of and 1 tsp (3–4 g) of baking powder would need to be used along with the buttermilk and sodium bicarbonate.

III. BIOLOGICAL LEAVENING SYSTEMS (YEAST FERMENTATION)

Many breads containing yeast not only are leavened by yeast-produced carbon dioxide but also owe their distinctive flavor to the metabolic products of yeast. Other breads are yeast leavened (and flavored) and also have a sour flavor that results from bacterial production of acid.

A. Yeast Alone

Bakers' yeasts are available in different forms. The form that has the greatest gassing power is compressed yeast. It contains about 71% moisture. Because of the high moisture content, it is unstable at elevated temperatures, but under refrigeration it is fairly stable for several weeks.

Bakers also use bulk yeast, which is essentially the same product as compressed yeast but in particulate form. Like compressed yeast, it requires refrigeration; however, it may be more stable because its physical form permits more rapid cooling (Pyler, 1988a, p. 205).

Active dry yeast (ADY) is prepared from special strains of yeast that are unusually stable to drying. The moisture level is only 4–8% (Pyler, 1988a, p. 206); therefore, it can be held for several weeks at room temperature and for several months under refrigeration. Packaging ADY under an inert atmosphere further extends its storage life. The high stability of ADY as compared with that of compressed yeast is gained at the expense of some gassing power. For best results, it needs to be rehydrated at a temperature of about 40°C rather than at lower temperatures.

Instant active dry yeast (IADY) is a relatively recent development. It is finely granulated and does not need to be rehydrated; in fact, Bruinsma and Finney (1981) reported a loss of some viability if it was rehydrated. It has a moisture content of 4–8% (Pyler, 1988a, p. 209) and has a long shelf life; Bruinsma and Finney (1981) observed that vacuum-packed samples stored under refrigeration were stable for at least 22 months. A disadvantage of IADY is its sensitivity to some bread ingredients and some other variables in bread production. Its proof time is longer than that of compressed yeast (Jackel, 1983) and it is more expensive.

Although compressed yeast is still available on the retail market, the active dry forms have gained tremendous popularity. Their longer storage life probably is responsible.

The yeast species that is responsible for carbon dioxide production by any of the available forms is *Saccharomyces cerevisiae*, of which there are different strains. Yeast is capable of either aerobic or anaerobic fermentation, but fermentation in bread dough is anaerobic after the early rapid consumption of oxygen by yeast and bacteria (Hoseney, 1986, p. 219). The overall reaction below

$$C_6H_{12}O_6 \rightarrow 2C_2H_5OH + 2CO_2$$

Overall yeast fermentation reaction

does not show the complex, step-wise nature of yeast fermentation but is descriptive of the total process.

Yeast activity is affected by available substrate. Sugars are available from several sources. Flour normally contains sugars totaling about 1.5% on the dry weight basis. Glucose is readily fermented. Sucrose usually is added to yeast doughs; it is hydrolyzed rapidly to glucose and fructose by yeast invertase at the cell membrane or just inside. Some maltose is formed through the action of flour amylases on starch. Maltose is transported into the yeast cell, where it is hydrolyzed into glucose. The rate of utilization of maltose depends on the supply of glucose and sucrose; the rate is low in the presence of glucose and sucrose and more rapid if the supply of glucose and sucrose is depleted.

Yeast activity is affected also by pH and temperature. The optimum pH for fermentation by *Saccharomyces cerevisiae*, 4.5–5.5, is attained readily in a fermenting dough as some of the carbon dioxide formed becomes dissolved in water. A temperature of approximately 35°C is optimum (DuBois, 1984).

Sucrose, which provides a substrate for yeast, also can have a retarding effect on fermentation if used in excess. The effect probably is osmotic, resulting in dehydration of yeast cells and consequent interference with their metabolic activity. Sweet doughs need levels of yeast sufficient to overcome the fermentation-retarding effect of their relatively high levels of sugar.

Sodium chloride also has a retarding effect on yeast fermentation as a result of osmotic pressure. Although used in smaller amounts than sugar, sodium chloride on a weight basis has about 12 times the osmotic effect of sucrose and 6 times the osmotic effect of glucose because of its low molecular weight and its formation of two ions.

Sucrose and sodium chloride can be balanced against each other in formulation for control of fermentation rate so that it is neither too rapid nor too slow; for example, a relatively low level of salt can be used to balance the retarding effect of the high level of sucrose in a sweet dough. An alternative to reducing the sodium chloride level in a sweet dough is increasing the yeast concentration. On the other hand, it might be necessary to reduce the level of yeast in a low-sugar bread made without sodium chloride, as for a low-sodium diet.

Pyler (1988a, pp. 204–212) and Sanderson (1985) described commercial yeast products, and Pyler (1988a, pp. 195–202) reviewed the biochemistry of yeast fermentation. Reviews of the fermentation by *Saccharomyces cerevisiae* are those of Magoffin and Hoseney (1974) and Pomper (1969).

B. Yeast in the Presence of Bacteria

Several breads are products of both yeast and bacterial fermentation. The yeasts produce carbon dioxide for leavening and the bacteria produce the acids that contribute a distinctive sour flavor.

Rye bread is leavened by several *Saccharomyces* species, including *Saccharomyces cerevisiae;* it is soured by acid-forming bacteria, *Lactobacilli* (Lorenz, 1983). Some of the *Lactobacillus* species produce not only lactic acid, but also some acetic acid. The proportion of the lactic acid should be high (Pomeranz, 1987, p. 240). The starter can be made from rye flour with reliance on air-borne and flour-contained microorganisms. However, commercial starters are available and their use contributes to product uniformity. With regular baking of rye bread, the starter can be carried over from one batch to the next.

San Francisco sourdough French bread is leavened by another *Saccharomyces*, *S. exiguus.* The San Francisco sourdough bread system, described by Sugihara *et al.* (1970), is particularly acidic as a result of the activity of a bacterium that is present along with *S. exiguus* in the natural San Francisco sourdough cultures. The bacterium, for which Kline and Sugihara (1971) suggested the name *Lactobacillus sanfrancisco*, does not compete with the yeast; in fact, the organisms show great compatibility. *Lactobacillus sanfrancisco* produces rather large quantities of both lactic acid and acetic acid, and *S. exiguus* thrives at low pH levels.

Salt-rising bread represents yet another biologically leavened product. The starter culture that is used carries both "wild yeast" cells and flavor-producing bacteria. This combination sounds similar to that in the San Francisco sourdough French bread, but the organisms are quite different (Matz, 1972). As with rye bread, commercial cultures are commonly used.

Suggested Exercises

1. Compare the reaction rates of different baking powders, including the low-sodium baking powder in exercise 4 below, the tartrate baking powder in exercise 2, and a phosphate baking powder, if one is available. Quickly mix 4 g of each baking powder with 15 ml of a 1:1 dilution of egg white in a 100-ml mixing cylinder. Record the volume of each dispersion every minute for 5 min. Set the cylinders in a pan of water that has been brought to

a boil and removed from the heat. Again record the volume of each dispersion every minute for 5 min.

2. Formulate a tartrate baking powder with cream of tartar, sodium bicarbonate, and corn starch in the proportions 2:1:1 by volume. Prepare pancake
   batters, one containing a commercial double-acting baking powder, one a
   commercial single-acting baking powder if one is available, and the third
   the laboratory-formulated tartrate baking powder. Immediately pour 100
   ml of each batter into a 250-ml graduated cylinder. Over a period of an
   hour, plot percentage increase in volume against holding time for each
   batter. At the end of the hour measure the viscosity or do linespread tests
   on the three batters. Explain the results and their significance. (*Note:* Use
   Conn, 1981, as a reference.)

3. Study the effects of various levels of one or more types of baking powder on
   the quality of biscuits, muffins, or cakes (basic formulas in Appendix A).

4. The following baking powder formula is recommended for persons on a
   low-sodium diet (when dietary modifications permit additional potassium
   within limits):

Potassium bicarbonate (g)	19.9
Corn starch (g)	14.0
Tartaric acid (g)	3.8
Potassium bitartrate (g)	28.0

   The formula (KDDA, 1973), which can be prepared by a pharmacist, is
   accompanied by a statement that $1\frac{1}{2}$ times as much low sodium as "regular"
   baking powder should be used. Test the accuracy of the statement.

5. The chocolate cake formulas in Appendix A are equivalent as to total leavening, and their batters should be of approximately neutral pH. Prepare
   these cakes and three additional ones: one with buttermilk and baking powder, one with sweet milk and an amount of sodium bicarbonate that is
   equivalent in leavening power to the 5 g of baking powder, and one with
   sweet milk and 5 g of sodium bicarbonate (i.e., excessive leavening as well
   as excessive alkalinity). Compare the batters as to pH and color, and compare the cakes as to color, grain, and flavor.

6. Make yeast doughs (basic formula in Appendix A) with various combinations of three levels each of yeast, sugar, and salt (moderate, lower, and
   higher). Plan the combinations so that valid comparisons can be made.
   (Two doughs cannot be compared without benefit of statistics if they differ
   in the amounts of two different ingredients, but a single combination of
   levels of the three ingredients can be involved in several comparisons.)

References

Barackman, R. A. 1931. Chemical leavening agents and their characteristic action in doughs. *Ce-
reaI Chem.* **8**: 423–433.

Barmore, M. A. 1936. The influence of various factors, including altitude, in the production of
angel food cake. Tech. Bull. 15, Colorado Agric. Exp. Stn., Fort Collins, Colorado.

Bruinsma, B. L. and Finney, K. F. 1981. Functional (bread-making) properties of a new dry yeast.
Cereal Chem. **58**: 477–480.

Carlin, G. T. 1944. A microscopic study of the behavior of fats in cake batters. *Cereal Chem.* **21:** 189–199.

Conn, J. F. 1981. Chemical leavening systems in flour products. *Cereal Foods World* **26:** 119–123.

DuBois, D. 1984. What is fermentation? It's essential to bread quality. *Bakers Digest* **58**(1): 11–14.

Dziezak, J. D. 1988. Microencapsulation and encapsulated ingredients. *Food Technol.* **42**(4): 136–140, 142–148, 151.

Hood, M. P. and Lowe, B. 1948. Air, water vapor, and carbon dioxide as leavening gases in cakes made with different types of fats. *Cereal Chem.* **25:** 244–254.

Hoseney, R. C. 1986. "Cereal Science and Technology." American Association of Cereal Chemists, St. Paul, Minnesota.

Jackel, S. S. 1983. Leavening is basic to baking. *Bakers Digest* **57**(5): 38–42.

KDDA, 1973. "The Knoxville Area Diet Manual." Knoxville District Dietetic Assoc., Knoxville, Tennessee.

Kichline, T. P. and Conn, T. F. 1970. Some fundamental aspects of leavening agents. *Bakers Digest* **44**(4): 36–40.

Kline, L. and Sugihara, T. F. 1971. Microorganisms of the San Francisco sour dough bread process. II. Isolation and characterization of undescribed bacterial species responsible for the souring activity. *Appl. Microbiol.* **21:** 459–465.

LaBaw, G. D. 1982. Chemical leavening agents and their use in bakery products. *Bakers Digest* **56**(1): 16–21.

Lorenz, K. 1983. Sourdough processes—methodology and biochemistry. *Bakers Digest* **57**(4): 41–45.

Magoffin, C. D. and Hoseney, R. C. 1974. A review of fermentation. *Bakers Digest* **48**(6): 22–27.

Matz, S. A. 1972. Formulations and procedures for yeast-leavened bakery foods. *In* "Bakery Technology and Engineering," 2nd edn., Matz, S. A. (Ed.). Avi Publ. Co., Westport, Connecticut.

Pomeranz, Y. 1987. "Modern Cereal Science and Technology." VCH Publishers, New York.

Pomper, S. 1969. Biochemistry of yeast fermentation. *Bakers Digest* **43**(2): 32–38.

Pyler, E. J. 1988a. "Baking Science and Technology," 3rd edn., vol. 1. Sosland Publishing Co., Merriam, Kansas.

Pyler, E. J. 1988b. "Baking Science and Technology," 3rd edn., vol. 2. Sosland Publishing Co., Merriam, Kansas

Reiman, H. M. 1983. Chemical leavening systems. *Bakers Digest* **57**(4): 37–40.

Sanderson, G. W. 1985. Yeast products for the baking industry. *Cereal Foods World* **30:** 770–775.

Sugihara, T. F., Kline, L., and McCready, L. B. 1970. Nature of the San Francisco sour dough French bread process. II. Microbiological aspects. *Bakers Digest* **44**(2): 51–57.

YEAST BREADS

A large variety of commercially baked breads is available on the retail market. In addition, interest in home production of bread has increased in recent years. The first part of this chapter deals with white wheat bread (white pan bread in the baking industry); variety breads are discussed later in the chapter.

I. WHITE WHEAT BREAD

A. Ingredients

The major ingredients of bread are discussed individually in other chapters. In this chapter, they are discussed in relation to their functional roles in bread. Although bread can be produced with only flour, water, and yeast, salt is considered to be a fourth essential ingredient in a normal bread (Hoseney, 1986, p. 205). Sugar and shortening are frequent additions, and several other ingredients are commonly added in commercial baking. Pyler (1988a, Chapters 7–13) discussed the ingredients of bread and other flour mixtures.

1. Flour

Yeast breads are particularly dependent on flour. The individual gluten proteins of flour interact during mixing with water to provide a framework of gluten that is stretched during fermentation and coagulates during baking to help form the structure of the loaf. The starch, which is embedded in the protein network along with other dough constituents, also is essential to structure. Strong flours from hard wheats are desirable for bread making. They have a high protein content; they have a high absorption (high water requirement for doughing); they form dough with a cohesive, elastic gluten structure, resulting in dough with good handling properties; and they produce bread with large volume and good crumb characteristics. In commercial baking, vital gluten, a dry form of gluten, often is added when a flour does not have the desired strength.

a. Role of Flour Proteins and Lipids. The gluten proteins, gliadin and glutenin, are discussed in Chapter 17. Their disulfide (—S—S—) bonds are affected by mechanical stress, and disulfide–sulfhydryl relationships are responsive to the action of reducing and oxidizing agents.

Sulfur groups are involved in more than one type of change during the mixing of flour and water. Some of the existing disulfide bonds are broken through the applied mechanical energy and through the presence of reducing groups, for example the sulfhydryl (—SH) of cysteine, one of the amino acids of flour protein. New disulfide bonds form at other sites. When oxidizing agents, such as bromates and peroxides, are present, as is common in commercial baking, they participate in the exchange reactions by reacting with some of the sulfhydryl groups. These changes involving covalent bonds, along with the effects of noncovalent bonds, result in the development of gluten structure. Properly developed gluten makes the dough easy to handle, either by hand or by machine, and it permits a large amount of expansion and gas retention during fermentation and the early stages of baking.

The contributions of ionic bonding, hydrogen bonding, and hydrophobic interaction to dough structure are not always recognized. Some ionic bonding occurs because some of the amino acid residues of the gluten proteins carry charges. Hydrogen bonding is important because of the presence of many hydroxyl, carboxyl, carbonyl, and amide groups. Hydrophobic interaction involves nonpolar side chains of amino acid residues. Although these three types of bonds are weak as compared with covalent bonds, the total effect of noncovalent bonds is considerable.

Flour fractionation studies are mentioned in Chapter 17. Interchange of gliadin and glutenin fractions in flours of good and poor baking quality has produced evidence that gliadin proteins are responsible for the loaf volume potential and glutenin proteins for the mixing requirement of bread flours. Mixing requirement is the mixing time and intensity (speed) required for dough development.

Flour lipids, in spite of their low concentration, also are involved in dough development. The polar galactosylglycerides are bound simultaneously to gliadins and glutenins (Fig. 19.1). The hydrophilic portion of the glycolipids is

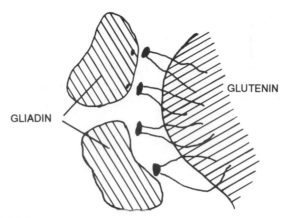

FIGURE 19.1
Binding of galactosyl glycerides to gliadin and glutenin. (From Chung *et al.*, 1978.)

bound to gliadin through hydrogen bonding, and the hydrophobic portion is bound to glutenin by hydrophobic interaction.

Pomeranz (1988) discussed the functionality of flour proteins and lipids, as well as of other flour components.

b. Contribution of Starch.   Starch granules become closely associated with gluten during dough mixing. Gelatinization of starch during baking contributes to structure of the baked loaf. If the starch has been damaged extensively during milling, it absorbs water at room temperature, thus competing with gluten proteins for available water. The action of flour amylases on damaged starch during dough fermentation provides sugar for yeast, and sugar beyond that utilized by yeast is available for carbonyl-amine browning during baking. An excessive amount of damaged starch has a deleterious effect on bread quality.

Replacement of wheat starch with corn starch in a flour reconstituted from fractions resulted in inferior baking properties in studies by D'Appolonia and Gilles (1971), Hoseney *et al.* (1971), and Sandstedt (1961). Hoseney *et al.* (1971) also used rye, barley, milo, rice, oat, and potato starches, as well as starches from several hard and soft wheat varieties, in reconstituted flours in which gluten and water solubles were constant. Of the nonwheat starches, only rye and barley starches came close to performing as well as wheat starches in baking tests. Of the wheat starches, only that from durum wheat performed poorly.

2. Liquid

Water obviously is essential for hydrating flour proteins during mixing, and the optimal level of water is important to dough properties. The optimal level of water varies directly with protein quantity and quality. This relationship is particularly important commercially because yield of bread increases with increasing water level. Water also dissolves sugar and salt and serves as a dispersion medium for yeast cells. Water is required for the starch gelatinization that occurs during baking. Milk sometimes is used as the liquid in noncommercial bread making, though bakers use water, often with whey and sometimes with nonfat dry milk. If fluid milk is used, it must be scalded to prevent a deleterious effect on dough consistency and loaf volume. Volpe and Zabik (1975) reported a study that implicated a dialyzable proteose–peptone component as the volume-depressing protein. Milk components contribute to bread flavor and to crust color.

3. Yeast

The primary function of yeast, as described in Chapter 18, is to produce carbon dioxide for leavening. Yeast cells contain not only the enzymes required for carbon dioxide production from glucose but also those required to make substrate available from flour. Maltase and sucrase are present and hydrolyze, respectively, maltose, which is formed through amylolytic activity, and sucrose, which frequently is added to yeast doughs. Another function of yeast is its effect on gluten during fermentation; it decreases the viscous flow properties of the dough and increases the elasticity. Hoseney (1986, pp. 220–221) noted that the effect, which appears to be an oxidizing effect, is brought about directly by the yeast, not by its metabolic products. Yeast activity also makes an important contribution to bread flavor. As indicated in Chapter 18, the rate of yeast activity is affected by temperature and by the concentrations of sugar and salt present. The level of yeast used needs to be reduced slightly at high altitudes and increased in dough that is to be frozen.

4. Salt

The use of sodium chloride at a level of 2% of the flour weight is common. It has a desirable effect on flavor. As discussed in Chapter 18, salt retards yeast activity through an osmotic effect on yeast cells. When salt is omitted completely, as in a low-sodium bread, reduction of the level of yeast is helpful in keeping the rate of fermentation under control. Sodium chloride also has a strengthening effect on dough, probably through an effect on the charged groups of the gluten proteins (Hoseney, 1986, p. 205).

Interest in reduction of dietary sodium has resulted in studies of bread with less sodium chloride than the usual 2%. Somewhat promising results have been obtained with simply reduced levels (Guy, 1985), with substitution of mixtures of sodium chloride and potassium chloride (Wyatt and Ronan, 1982), and with substitution of mixtures of sodium chloride and either magnesium acetate or magnesium chloride (Salovaara, 1982). Dwyer (1982) discussed the many factors that must be considered in sodium reduction in bakery products.

5. Sugar

The concentration of added sucrose varies from none to about 8% or a little higher. When no sucrose is added, fermentation is slow initially, because sugar for yeast substrate must be formed by amylolytic activity. If sugar content exceeds 10% of the weight of the flour, fermentation is retarded by the osmotic effect of the dissolved solute on the yeast cells. Excessive sugar also may interfere with gluten development by competing with gluten proteins for water. At its higher levels, sugar contributes to flavor. Also, when more sugar is added than is used in fermentation, the residual reducing sugars formed by sucrase are available for participation in carbonyl-amine browning during baking.

6. Shortening

Up to a point (about 3% of flour weight), added shortening results in increased loaf volume, improved grain, and reduced rate of crumb firming. The effect of shortening is seen in Fig. 19.2 (1 vs. 3, and 2 vs. 4). The retarding effect of shortening on the "setting" of dough during baking apparently is at least partly re-

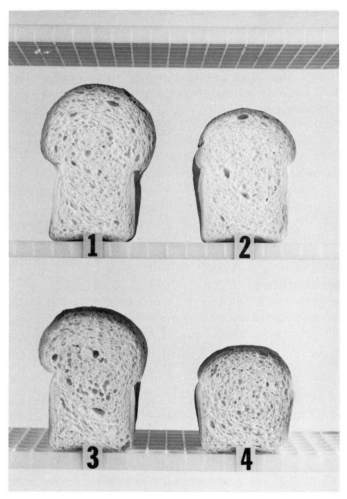

FIGURE 19.2
Loaf volumes of bread made from 100 g of flour in an optimized formula and with optimum mixing (4¾ min) except as otherwise indicated: (1) 998 cm³; (2) undermixed (mixed 2 min), 748 cm³; (3) no shortening, 875 cm³; and (4) undermixed (mixed 2 min), no shortening, 575 cm³. (Courtesy of K. F. Finney.)

sponsible for the increased loaf volume (Junge and Hoseney, 1981). The effects of shortening on loaf volume and crumb grain depend on the quality of the flour proteins and on the flour lipids present. The beneficial effects of shortening are greatest with good-quality flour containing an adequate concentration of galactolipids.

7. Other Ingredients

Commercial breads frequently contain mold inhibitors, dough improvers, yeast foods, and surfactants. The common mold inhibitors are the propionates. Dough improvers are oxidizing agents, which are mentioned in Chapter 17 as possible flour additives at the mill. Yeast foods are salts, usually ammonium salts, that accelerate yeast activity; these may be combined with oxidizing agents and buffers.

Surfactants (emulsifiers) have a crumb-softening effect in bread and enhance shelf life. The crumb-softening effect of surfactants results from their complexing with starch (Pyler, 1988a, p. 477). The antistaling effect is discussed later in this chapter. Sodium-stearoyl-2-lactylate (SSL), propylene glycol fatty acid esters, and monoglycerides and their derivatives are examples of surfactants used in bread. Some surfactants—SSL, for example—are useful in strengthening doughs, and thus they may be especially helpful in doughs containing some nonwheat flour. The dough-strengthening effect involves interaction between surfactants and gluten protein (Pyler, 1988a, p. 477). The common chemical characteristic of surfactants is their having both hydrophilic and lipophilic portions in their molecules. The relative proportions of the two moieties are expressed numerically with ratios (HLB) ranging from 0 to 20; HLB (hydrophilic–lipophilic balance) differs among surfactants. Pyler (1988a, pp. 473–487) discussed the surfactants used in baking.

B. Mixing Methods

In order for bread ingredients to perform their functions, they must be mixed to the extent required for development of a viscoelastic dough. The straight-dough and sponge methods are basic in making bread.

1. Straight Dough

The straight-dough method, in which all the ingredients are mixed together at one time and allowed to rise, is used for most noncommercial and some commercial bread making. The order in which ingredients are added during mixing is of little importance if care is taken that fresh fluid milk, if used, is scalded and that yeast is not destroyed by excessive heat. Shortening can be melted or not as desired. Because the flour's absorption (water uptake) cannot always be predicted, it is usually more convenient to add flour to liquid than liquid to flour. Flour should be added to give a soft, rather sticky dough. Dough temperature should be about 28°C if compressed yeast is used but a few degrees higher if active dry yeast is used (Chapter 18). The temperature of the liquid required to produce dough at the desired temperature depends on the temperatures of the room and the ingredients. Mixing and kneading are done as a single-machine operation in bakeries, whereas bread made at home is mixed and kneaded by hand unless a food processor or a mixer with a sufficiently powerful motor and a dough hook attachment is available. This type of equipment is useful in experimental work. When hand kneading is necessary, it should be done with a light, folding motion in which the fingers do the folding and the heels of the hands do the pushing of the folded dough. After each pushing motion, the dough is given a one-quarter turn. Only a light film of flour is used on the board to avoid making the dough too stiff.

As kneading progresses, the tendency for the dough to stick decreases, and the dough becomes smooth and elastic. When a dough has been kneaded sufficiently, small blisters can be seen on the surface. The time required for kneading varies with the type of motions used, with the size of the dough mass, and with the flour, being longer for stronger than for weaker flours. Dough that has not been kneaded enough does not hold the gas well. The bread from such a dough has a coarse, irregular crumb, small volume, and an uneven break along one side of the loaf. Insufficient kneading probably is common in noncommercial bread. Indeed, it seems difficult to overknead by hand dough made from strong flours by the straight-dough method, but it is possible with weak flours or with the sponge method, in which the gluten of part of the flour has been softened by fermentation. In a bakery where the entire process of mixing and kneading is done by machine, it would be easy to overmix without strict control. Overmixed dough results in bread with heavy cell walls and poor volume. The effect of undermixing is seen in Fig. 19.2 (2 vs 1, and 4 vs 3). The mixed dough is subjected to bulk fermentation.

The straight-dough method can be modified to omit kneading. This has been called the no-knead or batter method. The dough can be dropped directly into the pans, before or after rising, or chilled and shaped with the hands. It is always allowed to rise in the pan before baking. Sweet rolls and coffee cakes are made more often without keading than is bread because the best results are obtained with this method when a soft, rich dough is used. The method results in a more open grain than is obtained with kneading.

It is sometimes convenient to hold dough in a refrigerator and to bake portions of it at intervals. A roll dough made by a standard or a no-knead method is put into a refrigerator immediately after mixing. Care is taken to prevent crust formation, and the dough is punched periodically, as some rising occurs even with refrigeration. Because the dough is cold when taken out for use, the rising time after shaping is longer than with freshly made dough. The storage time for such doughs is less than a week. A dough that is to be frozen might be kept cool during mixing and then frozen immediately.

2. Sponge

The sponge-dough method of processing is used widely by commercial bakers and occasionally in noncommercial baking. The dough requires more handling and a longer total fermentation time but is more tolerant of variations, especially of fermentation time and temperature, than is a straight dough. The flavor of bread made by the two procedures is slightly different because of the longer fermentation time and, therefore, greater flavor development with the sponge method.

The sponge is made by mixing yeast, liquid, and about half to two-thirds of the flour. The amount of yeast may be less than that for the straight dough. Fermentation of the sponge is a long process; the rate can be controlled by variations in temperature and in the sugar and salt additions. In rising, the sponge becomes light and frothy and sometimes falls back because viscosity is too low for gas retention. The method derives its name from the spongelike appearance of this mixture. After the sponge has risen, the remaining ingredients are added and the dough is handled like a straight dough, with a bulk fermentation, scaling, molding, and proofing. The rising time of the dough is rather short because of the long sponge fermentation. The sponge method is not desirable with a weak flour because of the long total fermentation period.

3. Modifications

A continuous-process mixing method that was popular in the baking industry several years ago might be considered a modification of the sponge method. The ingredients, one of which is a prefermented sponge, are metered into the mixer. The mixed ingredients are passed on to the dough developer and from there to the divider and extruder. The extruded dough is ready for fermentation in the pans (proofing). Bulk fermentation is bypassed because of the addition of the preferment. Flavor of the product is bland, and crumb characteristics are poor; therefore, the continuous process now is seldom used for pan bread. Pomeranz (1987, pp. 236–238) compared the continuous and conventional processes.

The Chorleywood no-time process is a continuous process that involves the use of relatively high levels of yeast and the replacement of bulk fermentation by extensive mixing with fast-acting oxidizing agents. It can result in excellent bread and is widely accepted in more than 30 countries (Kulp, 1988, p. 401).

The liquid ferment process is a widely accepted modification of the sponge-dough method. It differs in having a lower flour:water ratio in its preferment than does the conventional sponge-dough process.

C. Dough Structure

The nature of the glutenin macromolecule apparently is basic to the structure of the gluten network and to certain of its properties. Ewart (1972) proposed that the glutenin subunits are linked into long strings, or concatenations in which each subunit is joined to the next by two disulfide bonds. He considered the concatenations to be partially coiled and also entangled with one another, with cross-links of various types at the points of entanglement. Later, Ewart (1977) modified his hypothesis to decrease the emphasis on entanglement, and in 1979 he concluded that there is only one disulfide between adjacent chains. Coiled portions of the chains are considered important for viscoelasticity. During the stress of mixing, the coils permit stretching of the chains because of the ease of breaking secondary bonds; then with removal of stress, the coils reform (Fig. 19.3). As indicated in the figure, it now is believed that the subunits are joined head to tail (Shewry *et al.*, 1984).

Reducing agents, whether in the flour or added, decrease dough structure by breaking down concatenations. If an excessive amount of breaking of disulfide bonds occurs during mixing, as a result of either excessive reducing agents or overmixing, the glutenin chains are reduced in length so that the glutenin behaves more like a viscous liquid than an elastic solid. The uniform, long fibers formed by the glutenins of bread flour are seen in Fig. 19.4. They are formed by interaction between glutenin concatenations. When dough is first moistened, glutenin polymers are randomly oriented to one another. During mixing, secondary bonds are broken, and glutenin polymers become aligned with one another with formation of new secondary bonds between them (Schofield, 1986, p. 17).

In a dough, early reducing action by sulfhydryl groups keeps the dough fluid long enough to permit formation of the important new disulfide bonds (Bushuk, 1984). Later, removal of sulfhydryl groups (by slow-acting oxidants) is desirable; their removal apparently is more important than further formation of new disulfide bonds. Lásztity (1984, p. 53) stated that the number of disulfide bonds is less important than their locations.

The above discussion of the nature of glutenin does not include the interaction between glutenin and gliadin to form gluten. Gliadin is viewed as a plasticizer; gliadin molecules interact with each other and with glutenin molecules, thus limiting the association between glutenin polymers (Ewart, 1979). If the concentration of gliadin is too low, the dough is too cohesive; if it is too high, the dough is too weak. The gluten in developed dough is viewed as a continuous network of fibrils covered with a protein membrane (Bushuk, 1984). The starch granules are embedded in the fibrillar network. Fig. 19.5 is an SEM micrograph of mixed dough.

Also observed in the dough microstructure are discrete masses of fat and yeast cells rather uniformly distributed through the spaces between sheets of gluten.

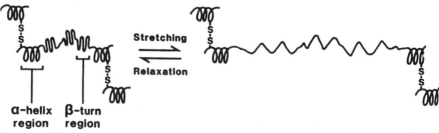

α-helix   β-turn
region   region

FIGURE 19.3
Schematic representation of a polypeptide subunit of glutenin within a linear concatenation. The subunits are joined head to tail via disulfide (—S—S—) bonds to form polymers with molecular weights of up to several million. (From Schofield, 1986.)

FIGURE 19.4
Scanning electron micrograph of purified glutenin of the bread wheat variety Manitou. (From
Orth *et al.*, 1973.)

Gas cell nuclei, representing air incorporated during mixing, also are distributed
throughout the spaces between sheets of gluten and become enlarged during
fermentation as a result of diffusion of carbon dioxide into them.

D. Fermentation

Doughs mixed by the straight-dough or sponge process are subject to bulk fermentation. The carbon dioxide formed is dissolved in the aqueous phase of the
dough until the water is saturated. After the water becomes saturated, additional
carbon dioxide produced diffuses into the gas cell nuclei, increasing the internal
pressure and expanding the cells. Bulk dough is allowed to rise at a temperature
of 27–35°C, depending on the type of yeast used (Chapter 18). The desired
dough temperature can be achieved by adjusting the water temperature, taking
into account the flour temperature, the room temperature, and the temperature

The gluten content appears to be inversely related to the staling rate of bread. Its role probably is largely indirect. Gluten affects specific volume, which in turn influences the amount of starch present in a given volume of the crumb—the greater the specific volume of the loaf, the lower is the concentration of starch on a volume basis. Kim and D'Appolonia (1977) reported that the effect of increased flour protein involves dilution of starch by gluten.

Lipids, like proteins, have been credited with retarding staling. The binding of lipids to gluten proteins during doughing results in increased volume, however, and again the effect may be partially indirect. Added shortening also forms complexes with amylose, as do surfactants, especially monoglycerides, which frequently are added to commercial breads for their crumb-softening effect. Possibly some complexing occurs also between surfactant chains and amylopectin chains that extend from starch granules. Starch chains that are complexed with lipid or surfactant are not free to participate in starch–starch association.

Moisture levels in bread have been of interest because of several early findings: Increasing water level in bread dough retards firming (however, the maximum water content of white pan bread in the United States is 38%, as specified by the FDA, 1988); complete prevention of moisture loss from a loaf of bread does not prevent staling; moisture gradients exist in the loaf; and migration of moisture from the crumb to the crust occurs during storage. Moisture transfers from starch to gluten and from gluten to starch have been studied and discussed extensively, but results have not been conclusive.

Storage temperature affects the rate of staling. Lowering the storage temperature above the freezing point increases the staling rate. Staling is rapid at refrigerator temperature but very slow at or below the freezing point.

Complete understanding of the process of bread staling awaits further research findings. Among the reviews of staling are those of Kulp and Ponte (1981) and Pyler (1988b, Chapter 20).

II. VARIETY BREADS

Several groups of breads might be considered variety breads. These include specialty white breads, breads containing whole-wheat flour, breads containing nonwheat flours, special dietary breads, and flat breads. There is some overlap among these categories.

A. Specialty White Breads

1. Hearth Breads
French, Italian, and Vienna breads are characterized by a large amount of crisp crust. The crustiness is achieved with the use of formulas containing little or no fat or sugar and with baking directly on a hearth or on a baking sheet. A steam-filled atmosphere in the oven also contributes to the crustiness. Because of the lack of support, the level of water in the dough is lower than that of a pan bread dough so that the dough will hold its shape. Differences among French, Italian, and Vienna breads are not great and are not described with consistency.

Sourdough French bread is a French bread that has a sour flavor achieved with the use of acid-forming bacteria, or sometimes by the addition of a commercial preparation of lactic and acetic acids. San Francisco sourdough bread is a unique version that is leavened by a special yeast species, *Saccharomyces exiguus*, and soured by the lactic and acetic acid-forming bacterial species, *Lactobacillus sanfrancisco* (Kline and Sugihara, 1971). Although starter cultures are available commercially, production of the bread requires conditions that exist in entirety only in San Francisco.

2. Sweet Dough Breads

Sweet dough breads are included here, although they are not universally considered to be variety breads. They not only contain relatively high levels of sugar and fat, but they usually also contain milk and eggs. When the sugar level is especially high, the amount of yeast needs to be increased. The addition of egg and a relatively large amount of fat necessitates reduction of the amount of liquid for the dough to hold its shape. Sugar, fat, and egg interfere with gluten development; therefore, relatively short fermentations are used in compensation for low structural strength of the dough. The protein and lactose in milk contribute to crust browning.

Some sweet dough products are flaky because fat is rolled in and dough is folded. A Danish roll is an example of a flaky sweet dough product.

3. Salt-Rising Bread

Like sourdough bread, salt-rising bread is made from dough that undergoes both yeast and bacterial fermentation. However, wild yeast cultures are used for leavening and the bacterial cultures produce a unique flavor that is not sour. As with sourdoughs, starter cultures are available to bakers.

4. Fruit Breads

Raisin bread is the major example of a fruit bread. A strong flour is needed for support of the added fruit. A relatively high level of yeast compensates for the fermentation-inhibiting effect of the acid contained in the fruit. The dough is a sweet dough.

5. English Muffins

A true English muffin is a yeast-fermented product that has a bland, slightly acidic flavor. However, sourdough muffins have become available, as well as raisin muffins and those containing whole wheat flour. The dough is a relatively soft dough and the product is flat. The desired shape requires a relatively high level of water, but a very soft dough does not perform well in an automated system; therefore, ice water is used in formulation and the dough is chilled. The high water content also contributes to leavening, along with yeast. On the other hand, it makes the addition of propionates for mold inhibition during storage especially important.

English muffin dough is cooked on one side in grill cups, then turned out onto griddle plates for cooking on the other side. During the cooking phase in the cups, the shape is established as the heating dough softens and flows to the edges of the cups and then becomes firm. An English muffin should be tough and chewy before it is toasted for use and crunchy after it is toasted. The grain is coarse.

Pfefer (1976) described the production of English muffins.

B. Breads Containing Whole Wheat Flour

Breads containing whole wheat flour alone or with white flour in varying proportions are termed wheat breads (although white bread also is made from a wheat flour). The bran and germ in whole wheat flour dilute the gluten proteins, resulting in a weaker gluten structure than that obtained with bread flour. With 100% whole wheat flour, the volume is low and the crumb is compact. Obviously the color and flavor also are affected by substitution of whole wheat flour for bread flour. As the proportion of whole wheat flour is decreased, the loaf volume increases, the density of the crumb decreases, the color becomes lighter, and the flavor becomes blander. Other forms of wheat, such as rolled and cracked wheat, sometimes are added, primarily for their textural effects.

C. Breads Containing Nonwheat Flours

Nonwheat flours are described in Chapter 17, along with information concerning replacement levels that have been found to give acceptable products. The effects of substituting nonwheat flours are summarized earlier in this chapter.

Rye flour probably is the nonwheat flour that has been used most extensively. It does have a small amount of gluten-type protein, but it also has a particularly high concentration of water-soluble pentosans, which dilute the flour protein. In the United States bread seldom is made with 100% rye flour.

There are many types of rye bread. They differ in crumb darkness; differences in darkness result from differences in the flour used or in the level of substitution, except that caramel color sometimes is used for particularly dark rye breads. Rye breads differ also in flavor; they are sourdough breads and the degree of sourness varies. The source of the sourness differs also. Bacterial starters produce acid in the dough. Sometimes mixtures of lactic and acetic acids are added directly; the ratios of the two acids vary. Although some rye breads are baked in pans, many are baked on hearths. Pumpernickel is a rye bread that contains 100% rye meal, (coarsely ground whole rye). It is dark and compact.

Multigrain breads have increased in popularity. They generally have a coarse grain and a compact crumb, as well as a rather dark color. Commercially prepared grain mixtures are available to bakers.

D. Special Dietary Breads

1. High-Protein Breads

The addition of soy flour to a bread flour thus far has been the most satisfactory method of increasing the protein concentration in bread. Soy flour also increases the quality of the bread protein; because lysine is the limiting amino acid in wheat flour but not in soy flour, the soy protein complements the wheat protein. Pomeranz (1987, p. 362) stated that a rich formula, a high-protein wheat flour, and optimum oxidant levels are desirable, and the addition of synthetic glycolipids increases the levels of supplementation that are possible. High-protein breads are not among the most well-accepted variety breads.

2. High-Fiber Breads

The addition of fiber to bread has been studied for some time and has involved various grain and legume fibers and cellulose, singly and in combination. No standards have been established for fiber levels in high-fiber breads in the United States; levels have ranged from 15 to 25% of the flour weight.

Several adjustments are beneficial to the quality of high-fiber breads. Because the fibrous material dilutes the gluten proteins, volume is decreased unless compensation is made; addition of vital wheat gluten solves that problem. The use of a strong flour is important also. In addition to the dilution effect, the fiber tends to cut gluten strands; the addition of SSL (sodium stearoyl-2-lactylate) is helpful. A high level of added water is necessary, because the fiber has a high absorption and interferes with hydration of gluten proteins unless sufficient water is added to hydrate both the fiber and the gluten. The high level of water makes the product especially susceptible to mold growth unless an adequate amount of propionate is added. Care must be taken to avoid undermixing, because the point of full dough development is hard to judge in the presence of added fiber.

Lorenz (1976) studied the feasibility of using triticale bran in high-fiber breads. Triticale bran of two degrees of coarseness was used at replacement levels of 0, 5, 10, and 15% of the wheat flour in bread made by the straight-dough procedure. The 10 and 15% levels of triticale bran resulted in increased absorption, increased mixing time, reduced proof time, and a color difference in the baked breads. Use of the fine bran resulted in a much better grain than did use of the coarse bran, but the fineness of the bran did not affect loaf volume. The total sensory scores for the breads made with fine triticale bran at the 5 and 10% replacement levels were similar to those for the all-wheat control, and the quality of bread at the 15% replacement level was considered good. The samples containing 10 and 15% fine bran retained softness during storage to a greater extent than did the controls.

Pomeranz et al. (1977) compared the effects of wheat bran, oat hulls, and seven commercial celluloses on bread when used at levels of up to 15% replacement of wheat flour. The fibers differed in their effects on water absorption and mixing time. The level of substitution had a greater effect on loaf volume than did the type of fiber. At the high fiber levels, volume reduction was considerably greater than was expected simply on the basis of gluten dilution. According to Pomeranz and co-workers (1977), the results suggested that gas retention was affected. Microscopic study showed that crumb structure of the baked loaves was disrupted by fiber. Although oat hulls were found unsuitable, acceptable breads were obtained with wheat bran and microcrystalline celluloses at a replacement level of about 7%.

The incorporation of oat bran into human diets has been of special interest recently because of its suggested cholesterol-lowering effect. Krishnan et al. (1987) substituted commercial oat bran of three particle size ranges for 10 and 15% of the wheat flour in bread. The flour–bran blends had higher absorptions and longer dough development times than did the wheat flour control. (Farinograph) absorption increased with increased level of substitution and with decreased particle size of the bran. Dough stability decreased at the 15% level of substitution with all brans and also at the 10% level with the small-particle bran. Loaf volume decreased with increased level of substitution and with decreased particle size; the use of potassium bromate decreased that effect. Grain was coarser in the oat bran breads than in the wheat bread, but with the large-particle bran at the 10% level of substitution and the highest bromate level used, the grain was considered good. Volume under those conditions was equal to that of the control. A sensory panel preferred the 15% large-bran bread and the 10% intermediate-bran bread. The bread containing 15% small bran received the lowest scores for all characteristics.

Dubois (1978) discussed the factors that must be considered in the production of high-fiber breads. Krishnan *et al.* (1987) pointed out that the effectiveness of any nutritional supplementation depends on the acceptance of the ingredient by the population, the existence of the technology, and similarity of the supplemented product to the one being improved.

E. Flat Breads

Flat breads have been used since ancient times in the Mediterranean area and in Europe, but they have become popular in the United States only in recent years. They are dense, consisting mostly of crust. They usually, but not always, are round, and they vary considerably in size. Their formulas are simple; most contain little or no sugar, salt, or fat. Some flat breads are made entirely from white wheat flour; others are made from other flours, such as whole wheat, rye, sorghum, potato, and barley or from mixtures of flours. Some are fermented; some are not. Some flat breads are cooked on griddles or other hot flat surfaces; some are cooked in ovens; a few are steamed; a few are fried. Several of the flat breads that are most popular in the United States are described here.

1. Egyptian Balady Bread
Balady bread contains a high-extraction (85–100%) flour, with water, salt, and yeast. It is baked on a very hot hearth (about 500°C) for only about a minute. The puffing that occurs during baking results more from steam formation and expansion than from carbon dioxide; the top crust separates from the lower crust, forming a pocket that becomes apparent when the bread is halved. The pocket frequently is filled with food, such as various bean or vegetable mixtures, when the bread is eaten.

2. Arabic Bread
Arabic bread actually is a group of leavened flat pocket breads. White Arabic bread (shamy, mafrood, pita) is made from 72% extraction flour and brown Arabic bread from 90–95% extraction flour. Airborne yeasts and bacteria are the traditional fermenting agents; bakers' yeast may be used, but the yeasty flavor is not the preferred flavor. The dough is dry as compared with the balady bread

dough, and steam, therefore, contributes less to the dough expansion and separation of layers than it does in balady bread. Arabic bread is baked at about 450°C.

The following three breads are not yeast breads. They are included here because they are flat breads.

3. Chapati (Roti)

Chapati, or roti, is an unleavened flat bread that is commonly used in India and some other Asian countries. Coarse flour from wheat is used when available; if it is not, flour from sorghum, millet, or maize might be used. The dough is very soft and is cooked on a heated plate for 2–5 min. Chapati dough puffs during cooking, but the use of a nonwheat flour, especially millet, reduces the amount of puffing.

4. Crisp Flat Breads

Crisp flat breads are common to the Scandinavian countries; they are a major export of Sweden. Wheat flour or rye flour or a blend of the two is used without added leavening. The bread can be brown and it can be very light. The production of the light crisp flat bread that is most popular worldwide is automated. Thickness varies among the products. Norwegian lefse is almost paper thin; the rye crisps of Norway, Sweden, and Finland are a little thicker.

5. Tortillas

Tortillas are a staple food in Latin American countries and are used increasingly in the United States because of the increased consumption of Mexican foods. Traditional tortillas are made from corn flour, although (wheat) flour tortillas have become increasingly popular in recent years. Wheat flour dough is easier to handle than is corn flour dough, because it is more cohesive. Its cohesiveness makes larger tortillas possible; larger tortillas are convenient for foods, such as burritos, in which tortillas serve as wraps. Corn tortillas, too, have certain advantages. They are more flavorful, and they constitute a good source of calcium in a diet in which they are a staple item; the origin of the calcium is the lime in which the corn is steeped before it is ground. Baked tortillas are used for foods in which the tortillas serve as wraps; baked tortillas are fried for foods in which they need to be crisp, such as tacos and nachos.

Gisslen (1985, pp. 44–60) provided formulas and instructions for several variety breads, as did Miller and Dickerson (1981). Pyler (1988b) discussed variety breads other than flat breads. Faridi (1988) described a large variety of flat breads. Pomeranz (1987) described breads of countries around the world. The extensive bibliographies with the latter three references provide many sources of additional information.

Suggested Exercises

1. Compare different types of wheat flour in basic yeast doughs (basic formula in Appendix A). For class work with yeast dough, time can be saved with the use of a microwave oven during fermentation. The following fermenta-

tion and baking procedure, involving two 450-g portions of bread dough, can be used:

   a. Heat 840 ml distilled water to 85°C on high power.
   b. Put the two portions of dough (in oiled glass loaf pans, 22 × 11 × 6 cm), covered with waxed paper, into the oven with the water.
   c. Proof under the following conditions:
      i. 2 min at 30% power (136 W)
     ii. 5 min at 10% power (53 W)
    iii. 10 min with no power
   d. Sheet the dough to 1.5 cm thickness, shape the loaves, replace in loaf pans, and again cover with waxed paper.
   e. Proof under the following conditions:
      i. 5 min at 10% power
     ii. 10 min with no power
   f. Bake at 218°C for 20 min.

2. Compare different levels of yeast, salt, sugar, and/or shortening in yeast bread.
3. Compare one or more nonwheat flours with wheat flour in bread.
4. Substitute at least two types of fiber for some wheat flour in yeast bread, or substitute one type of fiber at several levels.
5. Develop a formula and procedure for making starch bread (a bread containing no flour).

References

Baker, A. E., Dibben, R. A., and Ponte, J. G., Jr. 1987. Comparison of bread firmness measurements by four instruments. *Cereal Foods World* **32**: 486–489.

Bloksma, A. H. 1986. Rheological aspects of structural changes during baking. *In* "Chemistry and Physics of Baking," Blanshard, J. M. V., Frazier, P. J., and Galliard, T. (Eds.). Royal Society of Chemistry, London, England.

Bushuk, W. 1984. Functionality of wheat proteins in dough. *Cereal Foods World* **29**: 162–164.

Chung, O. K., Pomeranz, Y., and Finney, K. F. 1978. Wheat flour lipids in breadmaking. *Cereal Chem.* **55**: 598–618.

D'Appolonia, B. L. and Gilles, K. A. 1971. Effect of various starches in baking. *Cereal Chem.* **48**: 625–636.

Dubois, D. K. 1978. The practical application of fiber materials in bread production. *Bakers Digest* **52**(2): 30–33.

Dwyer, R. 1982. Sodium in bakery products. *Bakers Digest* **56**(6): 10–11.

Ewart, J. A. D. 1972. Recent research and dough visco-elasticity. *Bakers Digest* **46**(4): 22–28.

Ewart, J. A. D. 1977. Re-examination of the linear glutenin hypothesis. *J. Sci. Food Agric.* **28**: 191–199.

Ewart, J. A. D. 1979. Glutenin structure. *J. Sci. Food Agric.* **30**: 482–492.

Faridi, H. 1988. Flat breads. *In* "Wheat: Chemistry and Technology," 3rd edn., vol. II, Pomeranz, Y. (Ed.). American Association of Cereal Chemists, St. Paul, Minnesota.

FDA. 1988. Bakery products. *In* "Code of Federal Regulations," Title 21, Section 136.110. U.S. Govt. Printing Office, Washington, DC.

Gisslen, W. 1985. "Professional Baking." John Wiley and Sons, New York.

Guy, E. J. 1985. Effect of sodium chloride levels on sponge doughs and breads. *Cereal Foods World* **30**: 644–648.

Hoseney, R. C. 1986. "Cereal Science and Technology." American Association of Cereal Chemists, St. Paul, Minnesota.

Hoseney, R. C., Finney, K. F., Pomeranz, Y., and Shogren, M. D. 1971. Functional (breadmaking) and biochemical properties of wheat flour components. VIII. Starch. *Cereal Chem.* **48:** 191–201.

Junge, R. C. and Hoseney, R. C. 1981. A mechanism by which shortening and certain surfactants improve loaf volume in bread. *Cereal Chem.* **58:** 408–412.

Kim, S. K. and D'Appolonia, B. L. 1977. The role of wheat flour constituents in bread staling. *Bakers Digest* **51**(1): 38–44.

Kline, L. and Sugihara, T. F. 1971. Microorganisms of the San Francisco sour dough bread process. II. Isolation and characterization of undescribed bacterial species responsible for the souring activity. *Appl. Microbiol.* **21:** 459–465.

Krishnan, P. G., Chang, K. C., and Brown, G. 1987. Effect of commercial oat bran on the characteristics and composition of bread. *Cereal Chem.* **64:** 55–58.

Kulp, K. 1988. Bread industry and processes. *In* "Wheat: Chemistry and Technology," 3rd edn., vol. I, Pomeranz, Y. (Ed.). American Association of Cereal Chemists, St. Paul, Minnesota.

Kulp, K. and Ponte, J. G., Jr. 1981. Staling of white pan bread: Fundamental causes. *CRC Crit. Rev. Food Sci. Nutr.* **15:** 1–48.

Lásztity, R. 1984. "The Chemistry of Cereal Proteins." CRC Press, Inc., Boca Raton, Florida.

Lorenz, K. 1976. Triticale bran in fiber breads. *Bakers Digest* **50**(6): 27 30, 52.

Marston, P. E. and Wannan, T. L. 1976. Bread baking—the transformation from dough to bread. *Bakers Digest* **50**(4): 24–28, 49.

Miller, B. S. and Dickerson, J. 1981. Commercial formulations and home recipes. *In* "Variety Breads in the United States," Miller, B. S. (Ed.). American Association of Cereal Chemists, St. Paul, Minnesota.

Orth, R. A., Dronzek, B. L., and Bushuk, W. 1973. Studies of glutenin. IV. Microscopic structure and its relations to breadmaking quality. *Cereal Chem.* **50:** 688–696.

Pfefer, D. N. 1976. Formulation and production of English muffins. *Bakers Digest* **50**(2): 32–33, 36–37.

Pomeranz, Y. 1987. "Modern Cereal Science and Technology." VCH Publishers, New York.

Pomeranz, Y. 1988. Composition and functionality of wheat flour components. *In* "Wheat: Chemistry and Technology," 3rd edn., vol. II, Pomeranz, Y. (Ed.). American Association of Cereal Chemists, St. Paul, Minnesota.

Pomeranz, Y., Shogren, M. D., Finney, K. F., and Bechtel, D. B. 1977. Fiber in breadmaking—effects on functional properties. *Cereal Chem.* **54:** 25–41.

Pyler, E. J. 1988a. "Baking Science and Technology," 3rd edn., vol. 1. Sosland Publishing Co., Merriam, Kansas.

Pyler, E. J. 1988b. "Baking Science and Technology," 3rd edn., vol. 2. Sosland Publishing Co., Merriam, Kansas.

Salovaara, H. 1982. Sensory limitations to replacement of sodium with potassium and magnesium in bread. *Cereal Chem.* **59:** 427 430.

Sandstedt, R. M. 1961. The function of starch in the baking of bread. *Bakers Digest* **35**(3): 36–44.

Schoch, T. J. 1965. Starch in bakery products. *Bakers Digest* **39**(2): 48–57.

Schofield, J. D. 1986. Flour proteins: Structure and functionality in baked products. *In* "Chemistry and Physics of Baking," Blanshard, J. M. V., Frazier, P. J., and Galliard, T. (Eds.). Royal Society of Chemistry, London, England.

Shewry, P. R., Field, J. M., Faulks, A. J., Parmar, S., Miflin, B. J., Dietler, M. D., Lew, E. J.-L., and Kasarda, D. D. 1984. The purification and N-terminal amino acid sequence analysis of the high molecular weight gluten polypeptides of wheat. *Biochim. Biophys. Acta* **788:** 23–34.

Volpe, T. and Zabik, M. E. 1975. A whey protein contributing to loaf volume depression. *Cereal Chem.* **52:** 188–197.

Wyatt, C. J. and Ronan, K. 1982. Evaluation of potassium chloride as a salt substitute in bread. *J. Food Sci.* **47:** 672–673.

QUICK BREADS, EXTRUDED FOODS, AND PASTA

Quick breads derive their generic name from the speed of their leavening action. Less development of gluten is desired in quick breads than in yeast breads, and the function of flour in some cases is overshadowed by those of other ingredients. Chemical leavening agents and other basic ingredients of quick breads are discussed in earlier chapters. Soft wheat all-purpose flour is used for most quick breads, especially in the southeastern part of the United States, where self-rising flour also is popular. The volume of research dealing with quick breads is small as compared with that on yeast breads, and only a few quick breads are considered in this chapter.

Extruded foods are produced by the commercial process of extrusion. The principles and some applications are discussed in this chapter. Pasta is discussed separately, although it constitutes one group of extruded foods

I. QUICK BREADS

A. Muffins

Muffins contain flour, liquid, egg, sugar, salt, shortening, and baking powder. The liquid, usually milk, represents about half the amount of flour on a volume basis, and it approximately equals the amount of flour on a weight basis. The amounts of egg, sugar, and shortening are low in a basic muffin formula as compared with the proportions in a cake; therefore, muffin batter cannot tolerate the amount of mixing to which cake batters are subjected. The mixing method that is referred to as the muffin method was developed for the purpose of permitting minimal mixing. It involves stirring a mixture of milk, egg, and oil or melted shortening into the combined dry ingredients. A satisfactory alternative is cutting the fat rather finely into the dry ingredients and then adding milk and egg mixture. In either case, mixing the ingredients of a basic muffin should be just sufficient after the liquid is added to dampen the dry ingredients; the batter should not be smooth.

Muffins from properly mixed batter rise evenly and well during the early part of the baking period because the batter contains many gas cells in readily extensible gluten. They are large and symmetrical, without peaks, and their crust is pebbled rather than rough or smooth. They have a fairly even, open grain, without tunnels. They are tender. Overmixing decreases tenderness and produces a muffin with a crust that is light in color, smooth, and peaked, with tunnels that go toward the peak, and with a fine grain between the tunnels. Properly mixed and overmixed muffins are shown in Fig. 20.1.

The volume of muffins becomes larger with moderate overmixing and smaller with extensive overmixing. The peaks and tunnels characteristic of overmixed muffins are probably the result of overdevelopment of gluten and loss of carbon dioxide during overmixing. The batter changes from a lumpy mixture that does not hold together to one that is smooth and falls in strands from the spoon, indicating excessive gluten development. Both the strong gluten and the decreased amount of carbon dioxide prevent the muffin from rising normally during the early part of the baking period. After a crust has formed, however, sufficient steam pressure develops internally to push through the batter, forming tunnels in the crumb and a peak at the soft center of the crust. If some of the dough is pushed through the original crust, a knob is formed. The peak usually is at the point where the spoon left the batter as it was dropped into the pan, because the gluten strands follow the spoon. More tunnels are formed if an excessively high oven temperature causes a crust to form early in the baking period. Anything that interferes with gluten overdevelopment, such as extra sugar and shortening or the substitution of cornmeal or a nonwheat flour for part of the flour, decreases tunnel formation. Commercial muffin mixes usually have relatively high proportions of sugar to make them more nearly foolproof. They tend to have a more cakelike crumb than do basic muffins. Relatively rich mixtures frequently are used also in large-quantity food production because of the relative difficulty of avoiding overmixing with mechanical equipment.

Agitation should be minimized as muffin batter is being put into the pans. The use of an ice cream dipper has been suggested for this purpose and, in any case, the batter should be placed rather than dropped into the pans. In experimental work, the portions are weighed; an ice cream dipper is helpful in achieving the desired weight with a minimum amount of agitation.

With substitution of flours other than white wheat flours in muffins and with various additions, such as fruits and nuts, a huge variety of products is possible. Flour substitutions are not usually total unless hypoallergenic foods are being developed.

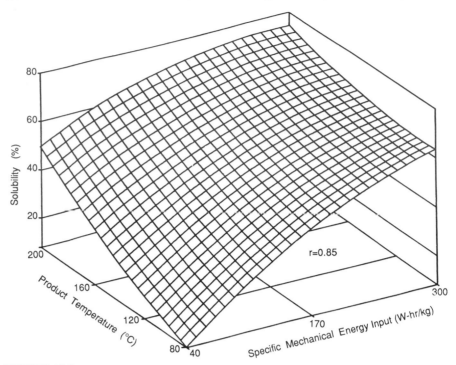

FIGURE 20.2
Influence of product temperature and specific mechanical energy input on solubility of extruded wheat starch. (From Meuser *et al.*, 1987.)

sion increases. The cohesiveness and mechanical strength of the product are determined by the extent of protein denaturation. A high degree of denaturation results in a low extent of cohesiveness. Factors that decrease the expansion tend to increase the mechanical strength of extruded products. Increased protein content also has a strengthening effect.

Product shape, color, and flavor are subject to variation. Although the product shape is affected primarily by the design of the die, the presence of vegetable gums can influence the shape. Food colors that might be added differ in their heat stability and their interactions with other ingredients, but some can withstand the processing conditions. Color coating of the extruded food is a possibility when color retention during extrusion is a problem. Many flavor compounds cannot withstand the high temperatures attained in extruders. Microencapsulated flavors are more stable than traditional flavors, and external application of flavoring agents is even better. Sometimes a combination of internal flavoring and external application might be used.

Study of the effects of extrusion variables can be complex because the properties of extruded products are affected by a large number of parameters. Meuser *et al.* (1987) developed a systems analysis approach for predicting the results of using specified extrusion conditions or for determining the extrusion conditions

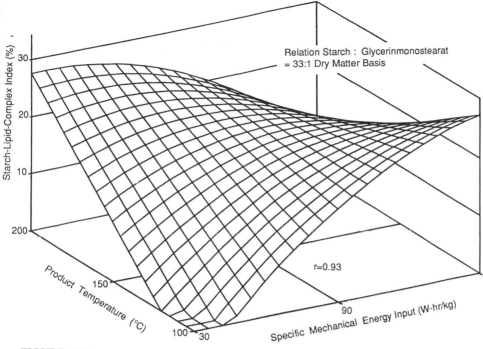

FIGURE 20.3
Influence of product temperature and specific mechanical energy input on the formation of a
complex of wheat starch and lipid (stearyl monoglyceride). (From Meuser *et al.*, 1987.)

needed for obtaining a desired product property. Computer-drawn response
surfaces, such as Figs. 20.2 and 20.3, illustrate the results of varying some conditions. Figure 20.2 is a relatively simple response surface showing increased solubility of extruded starch with both increased mechanical energy input and increased product temperature. Figure 20.3 shows the effects of the same variables
on the extent of starch–lipid complex formation during extrusion cooking. The
surface in this example illustrates an interaction between the two variables. (Formulate a statement concerning the effect of temperature at an energy input of 30
W-hr/kg; then formulate a statement concerning the effect of temperature at an
energy input of 150 W-hr/kg. Does the effect of temperature depend on the
energy input? Formulate a statement concerning the effect of energy input at
100°C; then formulate a statement concerning the effect of energy input at
200°C. Does the effect of energy input depend on the temperature?)

Extrusion cooking is used in the production of an ever-increasing array of
snack foods, including some that are fried after extrusion. Expanded ready-to-
eat cereals from various grains are available in many physical forms and flavors.
Instant cereals to be eaten hot constitute another application of extrusion. As
mentioned previously, pasta can be formed by cold extrusion; instant or quick-
cooking pasta and noodles can be made by extrusion cooking, as can crackers

and wafers. Instant soup mixes and beverage mixes are powdered products that can result from extrusion cooking. Coextrusion, which became possible with the development of twin-screw extruders, permits the production of interesting items, including filled snacks, as well as multicolored and multiflavored products (Berset, 1989).

Guy (1986) compared extrusion cooking and conventional baking. The mechanisms for expansion differ in that the gas cell nuclei in extruded foods are formed by spontaneous nucleation by steam during cooking, whereas they are occluded during mixing of a batter or dough for conventional baking. The starch granules in extrusion-cooked foods lose their structure and are compacted into a continuous phase; the starch granules in conventional oven-baked products are pasted to varying degrees, depending on the water level, but they generally retain their identity. Extruded products have a crunchier texture than is possible in an oven-baked product.

Phillips (1989) described the molecular changes produced in proteins by extrusion cooking. Denaturation increases protein digestibility and occurs more readily than does protein degradation. Carbonyl-amine (Maillard) browning is the main cause of loss of nutritional quality, but Phillips pointed out that non-enzymatic browning occurs mostly in high-carbohydrate foods in which protein does not make a major nutritional contribution.

Harper (1989) and Linko *et al.* (1981) reviewed the extrusion process and applications.

III. PASTA

Pasta products (also known as alimentary pastes, macaroni products, and semo-lina pastes) constitute a special class of extruded foods. They usually are cold extruded rather than extrusion cooked. The extruded products then are dried. Although their ingredients vary somewhat, pasta products differ from one another primarily in physical form, having a multitude of shapes and sizes, in addition to hollow tubes (macaroni), and solid rods (spaghetti).

Pasta usually is made from durum wheat, which is very hard but is not a good bread wheat; the protein in durum wheat differs qualitatively from that of a good bread wheat. Durum is the wheat of choice for pasta, however. Durum wheat is milled into several classes of endosperm products, one of which is semolina (durum middlings). Sometimes farina (middlings of bread wheat) is used for pasta, but semolina is preferred. The desirable properties that it confers on pasta include a yellow color, resistance to falling apart during cooking, firmness of the cooked product, and lack of stickiness.

Noodles sometimes are considered to be a type of pasta, although they are made from flour rather than from the coarser middlings. In the United States they must contain egg. Some noodles are extruded and some are sheeted and cut. Some are precooked.

Hoseney (1986) described the production and properties of pasta and noodles.

Suggested Exercises

1. Vary the proportion of sugar in muffins (basic formula in Appendix A).
2. Substitute nonwheat flours and/or sources of fiber at different levels for wheat flour in muffins, quick loaf breads, and biscuits (basic formulas in Appendix A).
3. Vary the amount of manipulation of muffin batter and biscuit dough.

References

Berset, C. 1989. Color. *In* "Extrusion Cooking," Mercier, C., Linko, P., and Harper, J. M. (Eds.). American Association of Cereal Chemists, St. Paul, Minnesota.

Bowman, F., Dilsaver, W., and Lorenz, K. 1973. Rationale for baking wheat-, gluten-, egg- and milk-free products. *Bakers Digest* **47**(2): 15–22.

Chen, H., Rubenthaler, G. L., Leung, H.K., and Baranowski, J. D. 1988. Chemical, physical, and baking properties of apple fiber compared with wheat and oat bran. *Cereal Chem.* **65**: 244–247.

Dryer, S. B., Phillips, S. G., Powell, T. S., Uebersax, M. A., and Zabik, M. E. 1982. Dry roasted navy bean flour incorporation in a quick bread. *Cereal Chem.* **59**: 319–320.

Eidet, I. E., Newman, R.K., Gras, P. W., and Lund, R.E. 1984. Making quick breads with barley distillers' dried grain flour. *Bakers Digest* **58**(5): 14–17.

Fulton, L.and Davis, C. 1978. Baked products for the fat-controlled, low-cholesterol diet. *J. Amer. Dietet. Assoc.* **73**: 261–269.

Guy, R. C. E. 1986. Extrusion cooking versus conventional baking. *In* "Chemistry and Physics of Baking," Blanshard, J. M. V., Frazier, P. J., and Galliard, T. (Eds.). Royal Society of Chemistry, London, England.

Harper, J. M. 1989. Food extruders and their applications. *In* "Extrusion Cooking," Mercier, C., Linko, P., and Harper, J. M. (Eds.). American Association of Cereal Chemists, St.Paul, Minnesota.

Hoseney, R. C. 1986. "Cereal Science and Technology." American Association of Cereal Chemists, St. Paul, Minnesota.

Linko, P., Colonna, P., and Mercier, C. 1981. High-temperature, short-time extrusion cooking. *Adv. Cer. Sci. and Technol.* **4**: 145–235.

Meuser, F., Pfaller, W., and Van Lengerich, B. 1987. Technological aspects regarding specific changes to the characteristic properties of extrudates by HTST extrusion cooking. *In* "Extrusion Technology for the Food Industry," O'Connor, C. (Ed.). Elsevier Applied Science Publ., London, England.

Oring, K. E. and Gehringer, J. 1989. Development of a rice flour muffin recipe. *J. Amer. Dietet. Assoc.* **89**: 261–262.

Phillips, R. D. 1989. Effect of extrusion cooking on the nutritional quality of plant proteins. *In* "Protein Quality and the Effects of Processing," Phillips, R. D. and Finley, J. W. (Eds.). Marcel Dekker, Inc., New York.

Raidl, M. A. and Klein, B. P. 1983. Effects of soy or field pea flour substitution on physical and sensory characteristics of chemically leavened quick breads. *Cereal Chem.* **60**: 367–370.

SHORTENED CAKES

Much of the research literature on cakes deals with commercial products but has general application. Only shortened cakes are discussed in this chapter. Angel food and sponge cakes, which do not contain shortening or chemical leavening agents, are included in Chapter 7.

I. INGREDIENTS

The major ingredients of plain shortened cakes are similar to those of muffins, though the proportions are quite different. Egg, shortening, and sugar levels all are higher in cakes than in muffins. Vanilla and other flavoring ingredients that are not used in muffins are common cake ingredients.

Balance of cake ingredients is of long-recognized importance, but what constitutes balance undergoes change as new ingredients are developed and old ones are modified. This is particularly true in the baking industry but also affects home baking. For example, the addition of surfactants to shortenings is an ingredient modification that affects cake formulation in the baking industry; the most likely effect on home baking is improved performance of the available shortenings and of cake mixes.

The rules for cake formulation are based largely on balance between tenderizing ingredients, sugar and fat, and structural ingredients, flour and egg. For many years the weight of sugar in cakes could not exceed the weight of flour. With improvements in cake flour and cake shortenings, it became possible to use larger amounts of sugar and liquid. Today the high-ratio (high-sugar) cake formulas, in which the weight of sugar exceeds that of flour, are used almost universally. In addition to their greater richness, high-ratio cakes have greater moistness and longer shelf life than do those from less rich formulas. The sugar:flour ratio can vary over a wide range, with 140:100 being the maximum and 125:100 average for white and yellow cakes. The proportions of other tenderizing and structural ingredients should be established so that the weight of shortening does not exceed the weight of eggs. Increasingly, with further improvements in shortenings, the weight of shortening has been less than that of egg. In white cakes, the proportion of egg white might well be higher than that of whole egg in yellow cake, because the protein concentration in egg white is lower than that in whole egg.

Balance between liquids and the batter constituents that have an affinity for water is the basis for the rule that the weight of the liquids, including fluid milk, water, and eggs, should equal or slightly exceed the weight of the sugar; otherwise the high sugar concentration interferes with hydration of proteins and gelatinization of starch.

Further information concerning cake formulation is found in Pyler's discussion (1988, pp. 981–987). Gisslen (1985, pp. 199–200) also discussed the balancing of formulas. Both Pyler and Gisslen provided some suggestions for adjustments to be made when special ingredients such as cocoa, liquid sugar, and fruit are added. Pyler (1988, pp. 1001–1002) also listed recommendations for formula adjustments at high altitudes. The need for changes begins at 2500–3000 ft; the changes include increased levels of the structural ingredients, flour and egg, and a decreased proportion of leavening agent. Often the water is increased because of the rapid rate of evaporation at high altitudes.

The formulas in Appendix A are presented in metric units. In the research literature, formula ingredients are sometimes presented in formula percentages (totaling 100%); ingredients in formulas for flour mixtures are customarily presented in percentages of flour weight (Bakers percent). These methods of presenting formulas are discussed in Chapter 1 and illustrated in Table I of that chapter.

The ingredients of cakes are discussed individually in other chapters. They are considered here only in relation to their respective roles in cakes.

A. Flour

Cake flour, a chlorinated short patent soft wheat flour, is used in cakes. As mentioned in Chapter 17, the chlorine treatment improves performance in cakes. Cakes made with chlorinated flour have greater volume and finer grain than do those made with unchlorinated flour. Effects of chlorination on several flour constituents have been noted. Tsen *et al.* (1971) found the effects of chlorine on flour proteins to include breaking of inter- and intramolecular hydrogen bonds and breaking of some peptide bonds. The changes in flour proteins, resulting in increased dispersibility, were accompanied by improved baking quality up to a point; overchlorination caused further changes in the flour proteins but deterioration of baking quality. They theorized that the changes in proteins might be responsible for the improving effect of chlorine, but they did not establish a cause-and-effect relationship. Gaines and Donelson (1982) and Donelson *et al.* (1984) presented evidence that the beneficial effects of flour chlorination represent effects on flour lipids.

Other workers have related cake quality improvement with flour chlorination to effects on starch. Huang *et al.* (1982a,b) obtained evidence of depolymerization and oxidation of starch during flour chlorination and increased swelling capacity of starch granules during baking. Although most flour constituents react with chlorine, and flour fractionation and reconstitution studies have shown all of the flour fractions to contribute to cake quality, Hoseney (1986, p. 271) stated that such studies have shown that the effect of chlorination on starch is responsible for the improved baking quality. Apparent discrepancies in the results obtained by different workers might be explained by variables such as cake formula (Johnson and Hoseney, 1979) and aging of flour (Clements and Donelson, 1982).

The flour that is used in box cake mixes is ground with the shortening before its incorporation into the mix. This "finishing" treatment contributes to incorporation, distribution, and retention of air when the mix is used (Hoseney *et al.*, 1988, p. 440).

Soft wheat all-purpose flour, which is available in some areas of the United States, performs quite similarly to cake flour. Hard wheat all-purpose flour can be used in cakes if substituted on a weight basis or if used in smaller volume than soft wheat flours (1 c minus 2 tbsp of hard wheat all-purpose flour replacing 1 c of a soft wheat flour). Even with appropriate substitution, cakes made from hard wheat all-purpose flour have less even grain, less velvety crumb, and less compressibility than cakes made from soft wheat flours. Although flour is an important structural ingredient of cakes, the flour proteins are dispersed as discrete particles in cake batter, rather than forming a continuous structure as in yeast dough. Thus starch plays a larger role in cake structure than do flour proteins. Its gelatinization causes increased batter viscosity that prevents collapse during baking.

B. Liquid

The liquid in cakes serves as a solvent for sugar, salt, and the chemical leavening agent. It hydrates flour proteins and gelatinizes starch. The extent of starch gelatinization is greater in cake than in bread. Milk, the usual liquid in cakes, contains carbonyl-amine reactants, thus contributing to crust browning. As with bread, relatively high levels of liquid contribute to keeping quality.

C. Leavening

As in bread doughs, the carbon dioxide that is formed in cake batters diffuses into gas cell nuclei that consist of small air bubbles incorporated during mixing. Expansion of these gas cells during baking results in the volume increase and consequent cake lightness. Leavening systems in cakes are similar to those in biscuits and muffins except that lower levels of leavening agent can be used in cakes. Cake batters are less cohesive and thus offer less resistance to expansion than do quick bread batters and doughs. If an acidic ingredient such as buttermilk replaces fresh milk in the formula, sodium bicarbonate is substituted for a portion of the baking powder, as described in Chapter 18. The leavening acids available to commercial bakers differ in their buffering capacity; because pH affects crumb color, bakers' choice of leavening acid is important.

D. Egg

The readily coagulable proteins of egg contribute to structure of the baked cake, though the rather high lipid concentration in egg yolk must be considered a tenderizing factor. Egg also emulsifies added fat. The role of egg in emulsification of shortening is attributable to the lipoproteins of the yolk; both high-density and low-density lipoproteins have been implicated as emulsifiers. The effects of egg on color, flavor, and nutritive value are obvious.

E. Sugar

Sugar obviously affects flavor. It also contributes to moisture retention of the baked cake because of its hygroscopicity. In addition, sugar has a tenderizing effect through its dilution of flour proteins and delaying of starch gelatinization. The sugar that ordinarily is used in cakes is sucrose, which is nonreducing; therefore, the contribution to browning is minimal.

F. Shortening and Surfactants

Shortening tenderizes by coating protein and starch particles, thereby disrupting continuity of their structure. Shortening also enhances apparent moistness, fineness and uniformity of grain, and keeping quality. If either margarine or butter is used, flavor also is imparted; hydrogenated shortening ordinarily is used in cake, however. The effectiveness of shortenings has been increased in recent years by the addition of surfactants, which facilitate emulsification and air incorporation and dispersion in batters. Shortenings on the retail market usually contain added monoglycerides, either in combination with diglycerides or in distilled form. Shortenings for commercial baking are likely to contain additional surfactants, such as propylene glycol–fatty acid esters, glycerol–lactic acid esters, and sorbitol–fatty acid esters. With the surfactants that are currently available, the importance of the nature of the fat source is minimized. A major development, made possible by the use of emulsifiers, is extensive use of liquid shortenings in the baking industry, where their fluidity is advantageous from the standpoint of bulk handling operations such as pumping and metering. When liquid shortenings are used, the shortening level usually is reduced somewhat.

G. Other Ingredients

Many of the additional ingredients that might be in cakes are flavoring materials such as vanilla, spices, and synthetic flavors, which are used in small amounts. These ingredients have little effect other than that for which they are added. Other ingredients that also are added primarily for flavor affect other properties. For example, cocoa, chocolate, fruit juices, apple sauce, other forms of fruit, and molasses lower batter pH. Unless adjustments are made, texture and other properties in turn are affected. With the addition of cocoa or chocolate, the sugar:flour ratio may be as high as 180:100. Shortening is reduced because of the fat, especially in chocolate, and liquid may be increased because of the starch, especially in cocoa. In the baking industry, small amounts of hydrophilic colloids, such as carboxymethyl cellulose and some other gums, are used to control batter viscosity and stability.

Pyler (1988, pp. 980–989) discussed cake ingredients.

II. MIXING

The conventional (multistage, creaming) method or one of its modifications is a time-honored and successful method of mixing cakes. Currently it is used much more frequently in commercial baking than in noncommercial baking. In commercial baking, mixing is done by machine. When mixing is done by hand, plastic shortening is creamed and sugar is added gradually while creaming is continued. Shortening contributes to entrapment of air during the creaming stage. Either whole egg or yolk is beaten into the creamed mixture. Gradual addition of ingredients is important in hand mixing to permit proper incorporation of air without excessive exertion, but it is not essential with an electric mixer, with which the shortening and sugar, with or without eggs, can be beaten at one time. Flour, sifted with baking powder and salt, is added alternately with milk. If the eggs have been separated or if white cake is being made, beaten egg whites are folded into the batter after the other ingredients have been added.

In a modification of the conventional method, the conventional–meringue method, about half of the sugar is beaten into the egg whites before they are added to the batter. In another modification, the conventional–sponge method, about half of the sugar is beaten into whole eggs until the mixture is foamy and stiff; then the mixture is folded into the batter after all other ingredients have been added. Both of these modified methods have proved successful, especially with soft shortenings. They make creaming easier, especially if the amount of shortening is small, because less sugar is added to the shortening; these methods also incorporate additional air in the form of a meringue or sponge.

The single-stage (quick-mix, one-bowl, dump) method is especially adapted to electric mixers but also is satisfactory with hand mixing. All of the ingredients, including plastic shortening at room temperature, are combined in one or two steps. If a second step is used, part of the liquid and usually the eggs are added after some of the beating has been done. Leavening may be added at this time instead of with the dry ingredients. The method is particularly successful with high-ratio cakes and has received general acceptance, undoubtedly because of a combination of convenience and cake quality.

In the pastry-blend method, the shortening and flour are creamed together first. The remaining steps include blending a mixture of sugar, baking powder, and half of the milk into the flour–shortening mixture and finally adding the egg and remaining milk. The baking powder can be sifted with the flour if desired. The flour–batter method sometimes used in commercial baking also begins with creaming of shortening and flour. The egg, however, is beaten with the sugar; that mixture is combined with the creamed mixture, after which the milk is added gradually.

In the muffin method, egg, milk, and melted or liquid shortening are combined, then beaten with the dry ingredients into a smooth batter. This rapid method was probably more popular before cake mixes were available than it is now. Although muffin cakes are acceptable, especially when warm, they are usually inferior in eating and keeping qualities to cakes made by other methods. A modification of the muffin method, in which egg whites with or without part of the sugar are added after the other ingredients are combined, is especially successful if the shortening is oil. When sugar is beaten into the egg whites, this modification is the muffin–meringue method.

The choice of a method for mixing a cake depends on convenience, on whether an electric mixer is available, and on the type of shortening to be used. With many formulas, any one of several methods used correctly would give excellent results. Hunter et al. (1950) compared the single-stage, conventional, and pastry-blend methods, using formulas containing three levels of sugar (100, 125, and 140% of the weight of the flour). The formulas were balanced by the use of increased amounts of shortening, egg, and milk with the higher levels of sugar. The appearance of the cakes is shown in Fig. 21.1. The authors reported that, although these methods of mixing produced no marked differences in palatability, the pastry-blend method appeared to be adapted to a wider range of conditions, such as formula, kind of shortening, and temperature, than the other methods investigated.

It usually is recommended that ingredients be at room temperature when cake is mixed unless the day is very warm. However, the unfavorable effects on cake quality commonly attributed to high and low ingredient temperatures were not evident in a study by Hunter et al. (1950), in which ingredients at 8, 22, and 30°C were used for cakes made with three different shortenings and mixed by two methods. In commercial baking, mixing times are relatively long and there is a potential for a considerable temperature increase in the batter during mix-

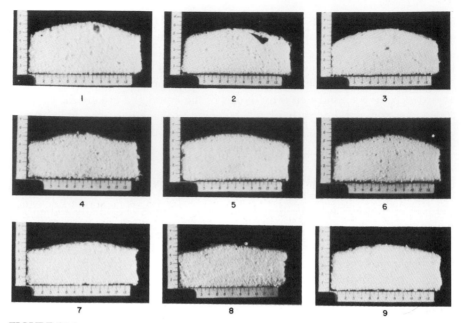

FIGURE 21.1
Cakes prepared by three methods of mixing from three formulas having different levels of sugar.
Cakes are arranged by number as follows:

	Method of mixing		
Sugar level (%)	Conventional	Pastry blend	One bowl
100	1	2	3
125	4	5	6
140	7	8	9

(From Hunter *et al.* 1950.)

ing. Pyler (1988, pp. 996–997), writing about commercial baking, stated that
the temperature attained by the batter is important because it affects viscosity,
which in turn affects batter aeration and stability. Because sugar also reduces
viscosity, a somewhat lower mixing temperature might be used for high-sugar
cakes than for other cakes.

The continuous phase in a cake batter is aqueous. In this aqueous phase, some
batter constituents, such as sugar, salt, and leavening salts, are dissolved. Other
constituents are colloidally dispersed (proteins) or suspended (starch granules,
fat globules, and gas cells). Ideally both the shortening and the incorporated air
are finely distributed. Emulsifiers have a direct role in the dispersal of fat. The
increased viscosity resulting from increased dispersal of fat undoubtedly contributes to fine dispersion and retention of incorporated air. It should be noted
that Shepherd and Yoell (1976, p. 225) stated that air bubbles are located in fat
particles in most cake batters having solid shortening; Hoseney (1986, p. 286)

stated that with the creaming method, air bubbles are held in the shortening until the fat melts during baking, whereas air is in aqueous dispersion in a single-stage batter.

Pyler (1988, pp. 989–996) described mixing methods and equipment used in the baking industry.

III. BAKING

Fat melts and batter viscosity decreases in the early stages of baking. Leavening gases are formed and diffuse into the gas cell nuclei; new gas cells are not formed. The batter must remain sufficiently viscous to keep starch suspended and to minimize coalescence of gas bubbles; excessive coalescence causes loss of gas because large bubbles tend to rise to the surface and escape. In the later stages of baking, flour and egg protein coagulate and the starch granules take up a large amount of water and swell, thus increasing batter viscosity and contributing to the eventual firming. The sugar concentration controls the temperature at which starch gelatinizes (Ngo and Taranto, 1986); the reason for cake failure in the presence of excessive sugar is the inability of starch to gelatinize. The internal temperature reached during baking of a cake is only slightly above 100°C (Hoseney *et al.*, 1988, p. 437).

Because the batter receives heat primarily from the pan, the batter at the bottom and sides of the pan expands first. The top center of the cake becomes firm last (Fig. 21.2); dipping in the top center occurs if the starch in that portion is not able to gelatinize because of either excessive sugar or insufficient water.

IV. CAKE QUALITY

Cake quality is assessed by measurement of volume (or determination of a volume index), compressibility, and breaking strength and by sensory evaluation. Objective measurements and sensory evaluation are discussed in Chapters 3 and 4, respectively. In addition to having a high specific volume, a high-quality cake is symmetrical. A loaf cake has a rounded top, but a layer cake needs to be flat for stacking. The crumb of a high-quality cake has a fine grain, with small cells

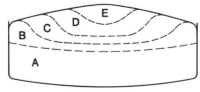

FIGURE 21.2
Diagram of the expansion of cake batter in a pan in successive stages of baking. (From Hoseney *et al.*, 1988.)

and thin cell walls. It is moist but not soggy and tender but not crumbly. It has a velvety mouth feel. Many variations in ingredients and in baking conditions have been found to affect cake quality.

A. Effects of Ingredients

1. Flour

The effect of chlorination has been discussed. It is of such importance that cake flour is routinely chlorinated. Also important is the fineness of granulation. A short patent soft wheat flour has smaller particles than do other flours. In addition, particle size in cake flour is further reduced by pin milling, which improves its performance in cake (Pyler, 1988, p. 912) if starch damage is not too extensive.

Many nonwheat flours have been studied in relation to their effects on cake quality. Some studies have involved partial replacement of wheat flour for improvement of nutritive value or increased utilization of a particular source of flour, or both. Other studies have involved total replacement of wheat flour in the development of hypoallergenic cakes.

Glover *et al.* (1986) replaced up to 50% of the wheat flour in high-ratio cakes with sorghum flour. Volume and texture were affected negatively, and a fractionation/reconstitution study showed starch to be responsible. At 5, 10, and 15% replacement levels the cakes were acceptable, though different from the control. Substitution of dextrose for sucrose resulted in improvement of volume and texture, apparently because of a lowering of the particularly high gelatinization temperature of sorghum starch.

El-Samahy *et al.* (1980) incorporated peanut, soybean, and sweet potato flours in cake mixes. They produced acceptable cakes with mixes in which defatted peanut flour, soybean flour, and sweet potato flour were substituted for 10, 15, and 30%, respectively, of the wheat flour.

Nonwheat flours were used by Bowman *et al.* (1973) in the development of hypoallergenic cakes. Chocolate cake was made successfully from a formula that included a mixture of rice and rye flours. The formula did contain milk and egg. Orange cup cakes containing rice flour and no milk or eggs also were acceptable as to appearance and texture and had a pleasing flavor. There was not sufficient structural strength for the batter to be baked as whole cakes. Plain rice flour pound cakes were similar to conventional pound cake except in texture. All three products were considered acceptable.

Rice flour has continued to be used in hypoallergenic foods. Bean *et al.* (1983) made layer cakes with 100% rice flour. The formula contained no milk or egg. They found that hydrating the rice flour before mixing the ingredients improved cake quality.

Perez and Juliano (1988) prepared flours from 10 rice varieties that differed in their amylose content and gelatinization temperature (GT). They made cakes with the formula of Bean *et al.* (1983), hydrating the rice flour with formula water. All of the cakes made with flours having intermediate GT collapsed during baking. When the level of sucrose was reduced, the extent of collapse was reduced but volume was poor and the crumb was hard. Of the cakes made with flours having low GT, all had good crumb quality; cake height increased with increased amylose content.

2. Liquid

Both the amount and the kind of liquid affect cake properties. Too little liquid can result in a peaked center and/or cracked crust because of excessive batter viscosity (Gisslen, 1985, p. 204); as mentioned previously, it also can cause dipping because of insufficient swelling of starch in the top center of the cake. Too much liquid can result in a heavy cake with low volume (Gisslen, 1985, p. 204) because of the viscosity-lowering effect and consequent decrease in retention of incorporated air.

The type of liquid used in a cake can have effects that depend on the formula. For example, the effect of using buttermilk depends not only on total leavening equivalency but also on the resulting pH. With balance between buttermilk and sodium bicarbonate along with equivalent total leavening, the effects of the substitution of buttermilk may be slight. With excessive acidity or excessive alkalinity, flavor, crumb color, crust browning, cake grain, and volume may be affected. Flavor ranges from somewhat sour with excessive acidity to bitter or soapy with excessive alkalinity. Crumb color is relatively dark in excessively alkaline plain cake and relatively red and dark in excessively alkaline chocolate cake. Crust browning occurs increasingly as alkalinity increases; this represents the effect of pH on carbonyl-amine browning. Cake grain, within limits, tends to increase in fineness with decreased pH and to become coarse, with thick cell walls, with increased pH. Volume tends to be relatively low with excessive acidity and to increase with alkalinity up to a point beyond which volume decreases. Flavor and color effects are noted before changes in grain and volume become drastic. Assuming total leavening equivalency, insufficient sodium bicarbonate to neutralize the acid is preferable to excessive sodium bicarbonate when buttermilk is used. A study in which differences in cake characteristics were observed when various combinations of fresh whole milk, buttermilk, baking powder, and sodium bicarbonate were used was that of Briant *et al.* (1954). Crumb pH ranged from 5.45 to 6.60.

3. Leavening Agent

Obviously too little leavening results in poor volume and a heavy cake. Excessive leavening up to a point increases volume but results in a crumb that is coarse and irregular as well as crumbly (Gisslen, 1985, p. 204).

4. Egg

Too little egg can result in a coarse or irregular crumb. Too much egg has a toughening effect.

5. Sugar

A low sugar concentration results in toughness and possibly a tunneled crumb. An excessive sugar concentration results in a dense, heavy crumb because of interference with structure development and consequent inability to retain leavening gases; the cake is crumbly (Gisslen, 1985, p. 204).

A moderate increase in sugar without other changes is possible with some formulas. More often, however, higher levels of sugar are accompanied by increased levels of total liquid, egg, and fat. The liquid and egg are increased for reasons mentioned in the discussion of formulation earlier in the chapter. The reason for increasing the level of shortening is related to batter aeration. Without increased fat, batter viscosity is likely to be relatively low as a result of the other changes, and retention of air is correspondingly reduced. Emulsification of added fat, on the other hand, results in increased viscosity and thus greater batter aeration as compared with that obtained when the sugar and liquid levels are increased but the shortening level is not.

In the previously mentioned study of Hunter *et al.* (1950), cakes made with the higher levels of sugar were more tender, moist, and velvety than those made with the lowest level of sugar. Although the judges seemed to prefer the flavor of cakes that were low in sugar, their total scores favored the cakes that were high in sugar.

Considerable recent research has dealt with the use of sugars other than crystalline sucrose and of nonsugar sweeteners. Cardello *et al.* (1979) compared the sweetness of cakes containing sucrose and fructose in equal amounts. The sensory panel did not perceive a difference in sweetness, perhaps because of the high sugar concentration in cake.

Honey, when substituted for part of the sugar in a cake, contributes to moisture retention during storage. This is because honey contains fructose (levulose), which is extremely hygroscopic. The reducing sugars of honey contribute to crumb browning; the effect on browning might be excessive unless the batter pH is lowered in the late stage of baking through incorporation of a slow-release leavening acid in the formula. That solution applies to a commercial baking situation. At the consumer level, the use of honey might need to be restricted to products in which the crumb is brown anyway. If the use of honey in a white cake batter is desired, carbonyl-amine browning might be controlled by slight acidification of the batter, as by addition of cream of tartar, if one is willing to accept the flavor effect and the possibly reduced volume.

McCullough *et al.* (1986) substituted high-fructose corn syrup (HFCS) for sucrose at replacement levels of 50 and 75%. The replacement resulted in decreased volume, increased perception of moistness, and increased browning. They considered the effects on sensory quality to be slight.

Total replacement of sucrose with HFCS in layer cake results in reduced volume, excessive tenderness, and darkness of both crumb and crust unless the formula is adjusted. The adjustments include reduction of shortening to 10% of the flour weight, substitution of oil for plastic fat, and use of a composite emulsifier and a slow-acting leavening acid (Pyler, 1988).

Sorbitol and polydextrose have been studied as partial substitutes for sucrose in low-calorie foods. Sorbitol is a hexahydric alcohol with a sweet taste. It does not reduce calories but is so slowly absorbed from the intestine that high blood glucose levels do not develop. The reduction in calories must come from reduction of other ingredients. Kamel and Rasper (1988) used sorbitol as replacement for 30% of the sucrose in layer cake containing no shortening. According to Friedman (1978), the use of sorbitol as replacement for 15% of the sucrose in cakes results in increased moistness and softness of crumb. He also stated that sorbitol can totally replace sucrose in cakes and cake mixes; the products are acceptable, though somewhat less sweet than those containing sucrose.

Polydextrose, a randomly cross-linked polymer of glucose, is of interest because of its low caloric value, about 1 kcal/g. It has humectant properties. It provides bulk, or body, rather than sweetness. Torres and Thomas (1981) stated that it can be used as total replacement for sucrose in dietetic baked products, including cakes; because it is only slightly sweet, it must be used with an appropriate noncaloric sweetener. Kamel and Rasper (1988) found replacement of sucrose with polydextrose at a higher level than 30% to have adverse effects on both physical and sensory properties of high-ratio cakes containing no shortening. Freeman (1982) presented formulas for cakes containing shortening with polydextrose and sugar in almost equal amounts.

Rasper and Kamel (1989) studied emulsifier/oil systems in reduced calorie cakes. They used several emulsifier blends with canola and corn oils. Sucrose was replaced with different levels of sorbitol and of polydextrose. With a relatively high level of the best performing emulsifier blend, cakes were comparable to the controls containing plastic shortening, when the oil level was as low as 17% and the sucrose level as low as 60% of the shortening and sucrose levels in the control.

Kamel and Washnuik (1983) were successful in eliminating shortening in yellow layer cakes, while reducing egg concentration by 50% and sugar by 10%. The reduced calorie products were moist and light and had a good taste and acceptability. The use of five times the recommended level of an emulsifier blend made satisfactory products possible.

Hess and Setser (1983) made cakes that contained shortening but no sucrose. To a basic formula containing bulked (with spray-dried maltodextrin) aspartame they added additional aspartame (nonbulked) or fructose. Each additional sweetener was used at two levels. Added fructose gave sweeter cakes than did added aspartame; the sweetest cakes were those with the higher level of fructose. Added fructose in most cases gave greater tenderness and cell uniformity and higher overall eating quality than did added aspartame.

Neville and Setser (1986), with the use of response surface methodology, developed high-ratio reduced calorie layer cakes with optimized formulations. Fructose, the intense sweeteners saccharin and aspartame, and the bulking agent polydextrose replaced the sucrose.

6. Shortening and Surfactants

Both the amount and kind of shortening affect cake quality. Increasing the concentration of shortening results in increased tenderness, fineness of grain, and apparent moistness. A low level results in toughness. Shortenings have undergone considerable change in recent years, and with the emulsifiers that are available today, the source of the fat is not as important as it once was; some unusual fats and oils can be used as shortening in cakes.

Edible oils have been imported in increasing quantities for use in the food industry. At the same time a large amount of beef tallow has been produced in the United States, and only a small percentage of it has been used in food. The possibility of fractionating beef tallow and substituting the different fractions for different types of fat used in the manufacture of foods has been studied, and there has been some industrial application of the technology. Bundy *et al.* (1981) developed a fractionation procedure that does not require the use of organic solvent and investigated the potential of solid and liquid fractions as shortening agents in cake. The solid and liquid fractions were used with and without an emulsifier blend in white layer cakes. The liquid fraction did not perform well in the absence of emulsifier; the solid fraction gave somewhat better results, and both fractions performed well with the emulsifier (Fig. 21.3). The investigators concluded that beef tallow has potential as an economical substitute for vegetable shortenings.

Canola oil was developed in Canada and is of commercial importance there. Its source is a genetically modified rapeseed, which was developed over a period of many years. The erucic acid (22:1) content is very low and the oleic acid (18:1) content is high, as compared with the concentrations in traditional rapeseed oil. This difference is significant from a nutritional standpoint because of recent evidence of the efficacy of oleic acid in lowering serum cholesterol while maintaining high-density lipoprotein concentrations. The use of canola oil in margarines has been limited by polymorphic instability; the small β' crystals shift to large β crystals after hydrogenation (see Chapter 15). Various methods of increasing crystal stability have been studied; mixing products that have been hydrogenated to different extents is among the methods that have shown some promise.

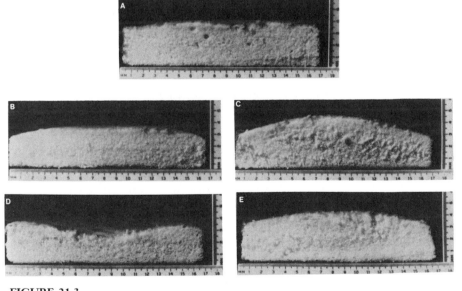

FIGURE 21.3
Cross-sections of cakes made with (A) a commercial shortening; (B) a 25°C solid tallow fraction; (C) a 25°C solid tallow fraction plus emulsifier; (D) a 35°C olein tallow fraction; and (E) a 35°C olein tallow fraction plus emulsifier. (From Bundy *et al.*, 1981.)

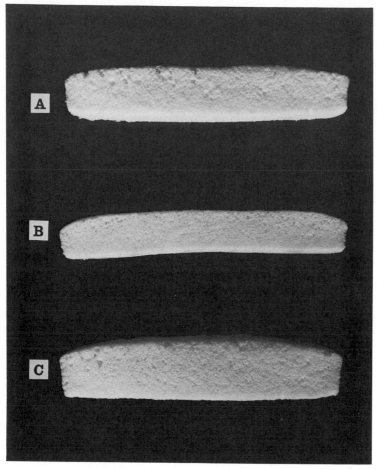

FIGURE 21.4
Effect of surfactant system on canola oil cakes. All cakes contained 52.5% fat or oil and 137%
water (percentage flour weight). (A) Hydrogenated, emulsified shortening; (B) canola oil; (C) canola oil with 9.5% emulsifier system (hydrated form of mono- and diglycerides, polysorbate 60, and
stearoyl-lactylate). (From Vaisey-Genser and Ylimaki, 1989.)

Vaisey-Genser and Ylimaki (1989) described canola oil and shortening and
discussed their use in baked products, including cakes. In spite of the presence of
β crystals, the shortening apparently has good creaming power; the routine use
of emulsifiers in shortenings facilitates excellent performance in quick-mix
methods. The increasing use of liquid shortenings prompted Vaisey-Genser *et
al.* (1987) to study white layer cakes in which canola oil was used with an emulsifier blend with excellent results (Fig. 21.4). They found that needed adjustments
with the use of canola oil included an increased amount of water and of emulsifier at the lowest oil level used.

Ebeler *et al.* (1986) compared white layer cakes containing no emulsifier, sucrose–fatty acid esters, and a commercial mono/diglyceride emulsifier. The cakes containing sucrose esters were inferior to those containing mono/di-glycerides; they had lower batter-specific gravity and higher cake volume than did those with no emulsifier but collapsed upon cooling.

The importance of surfactants in cakes is well established. Because of special needs resulting from various substitutions and replacements of ingredients, blends of surfactants are used increasingly. Along with monoglycerides, a hydro-philic surfactant, such as polysorbate 60, and the strengthener sodium or cal-cium stearoyl lactylate might be used. A hydrophilic surfactant is particularly dispersible in cake batter; the strengthening effect of a stearoyl lactylate, com-monly used in bread, can be useful in cakes in which formulation differs radi-cally from that of standard cakes.

Knightly (1988) described surfactants used in baked foods and their specific uses.

7. Other Ingredients

Whey and other dairy products have been used as functional additives to cakes. Dried whey is a nutritious and readily available by-product of cheesemaking. As indicated in Chapter 8, it is available in such large quantities that it is a potential environmental pollutant. Both sweet whey and acid whey are available because of differences in cheese-making procedures. Whey solids contribute to flavor, tenderness, crumb softness, browning, and shelf life of cakes; fragility is a poten-tial problem. Guy (1982) reported a study in which he compared sweet whey solids (SWS) and nonfat dry milk (NFDM) in both rich and lean yellow layer cakes. Evaluation of the cakes included special attention to fragility. Replace-ment of the NFDM in the rich formula with SWS resulted in decreased cake height and increased fragility. Replacement of the NFDM in the lean formula with SWS resulted in cakes that were comparable in height and fragility, as well as in sensory properties, to rich-formula cakes containing NFDM.

Pearce *et al.* (1984) incorporated NFDM into lean cake batters containing no fat or egg (Kissell, 1959). The addition of NFDM did not affect either the spe-cific gravity of the batter or the cross-sectional area of the cake. There was evi-dence, however, that the NFDM destabilized the foam; the cross-sections had large, nonuniform cells over most of their area and compact crumb at the outer edges. Other evidence indicated that NFDM caused inhibition of starch granule swelling, probably because of water binding by milk proteins and lactose. (*Note:* The Kissell formula is so lean that effects of variables are exaggerated.)

Whey protein concentrate, which is 75% protein, has been used as partial re-placement for egg in some baked products, including cakes. However, its use requires extensive formula rebalancing, especially if it is substituted for whole egg (Cocup and Sanderson, 1987).

Salt is a minor ingredient in cakes from a quantitative standpoint, but concern about hypertension prompts interest in reducing dietary sodium wherever pos-sible. Guy (1986) studied the effects of eliminating salt from the formulas of high -ratio white, yellow, spice, and devil's food cakes. Omitting the salt resulted in slightly decreased cake volume index and compressibility. In the absence of

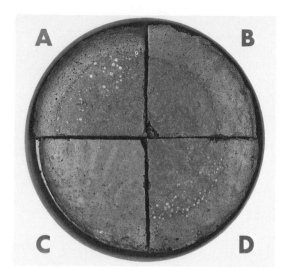

FIGURE 21.5
Effect of sugar and salt on devil's food cake crust appearance. (A) 0% salt, 150% sugar; (B) 4% salt, 150% sugar; (C) 0% salt, 120% sugar; (D) 2% salt, 150% sugar. (From Guy, 1986; photograph by A. Rivenburg.)

salt, some spottiness occurred on the crust surface of devil's food cake, which had the highest concentration of sugar; its basic formula also had the highest concentration of salt, 4% of the flour weight, as compared with 2.5–3.1% in the basic formulas for the other cakes. Evidence indicated that the spots consisted of crystallized sugar; development of spots was prevented when a fine grade of sugar was used, when the sugar level was reduced (Fig. 21.5), or when part of the sugar was dissolved in formula water. When the water level in white cake was reduced to two levels below the 99% (flour weight basis) in the basic formula, dipping occurred in the cakes containing salt but not in the salt-free cakes (Fig. 21.6). The salt-free cake containing 89% water was similar to that containing salt and 99% water. This water–salt interaction was not observed with the other cakes. In spite of some effects on baking quality, omission of salt did not impair sensory qualities of the cakes, as indicated by panelists' scores.

Current interest in food fiber from a health standpoint can be expected to result in attempts to add fiber to as many foods as possible. In recent research, various sources of fiber have been tested particularly in yeast breads and quick breads. Some earlier studies involved the partial replacement of wheat flour in cakes with wheat bran and middlings (Brockmole and Zabik, 1976; Rajchel *et al.*, 1975; Springsteen *et al.*, 1977) and with microcrystalline cellulose (Brys and Zabik, 1976). When replacement levels above 40% were used, cakes tended to be gummy, and excessive negative effects on volume and on sensory scores were noted. With levels ranging from 12% to 40% for different fiber sources, cakes generally were at least acceptable, in spite of some sensory effects, especially on color and flavor.

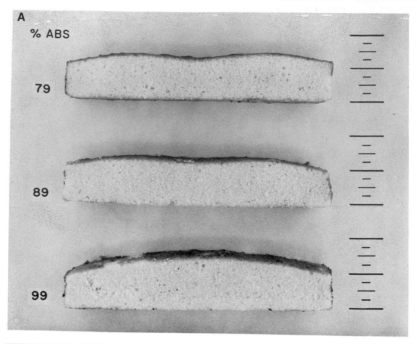

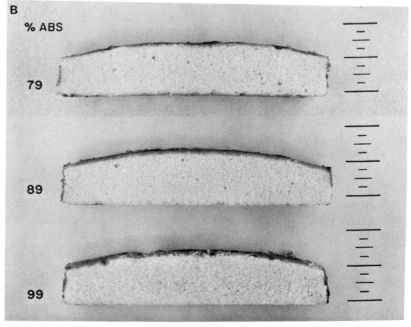

FIGURE 21.6

Effect of water level percentage (percentage absorption, or % ABS) on the profile of white layer cakes containing (A) 2.5% salt and (B) no salt. (From Guy, 1986; photograph by A. Riverburg.)

B. Effects of Baking Conditions

In early studies of baking conditions, Charley (1950, 1952) studied the effects of baking pan material, size, and shape. Baking was more rapid in dark or dull-finished pans than in shiny pans. Cakes baked in faster baking pans had greater volume and better crumb quality, though they tended to be peaked. Cakes baked in shallow pans tended to be larger, more tender, less brown, and flatter topped than those baked in deeper pans. Those baked in round and square pans of the same depth were similar.

Oven temperature is an important factor affecting speed of heat penetration. If the temperature is too low, coagulation of proteins and gelatinization of starch are slow, and gas is lost from the batter. As gas is lost from the cells, the remaining cells enlarge, their walls thicken, volume is reduced, and the cake may settle in the middle. If the oven temperature is too high, a crust forms on the cake before it has risen fully. The soft batter in the middle of the cake then rises and forms a hump or even a crack. These effects were noted in a study in which oven temperatures of 149, 163, 191, and 218°C were used with a high-ratio cake formula (Jooste and Mackey, 1952). Each increase of temperature within this range resulted in increased palatability and, in the presence of an emulsifier, increased volume. The cakes are shown in Fig. 21.7.

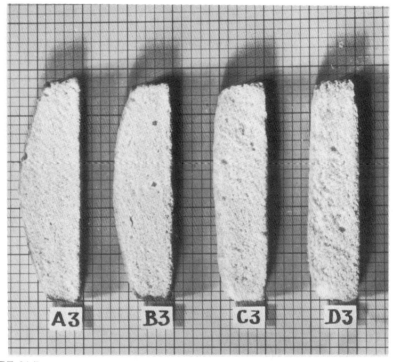

FIGURE 21.7
Cakes made with butter containing 6% glyceryl monostearate and baked at four temperatures. Cake A3 was baked at 218°C, B3 at 191°C, C3 at 163°C, and D3 at 149°C. (From Jooste and Mackey, 1952.)

Odland and Davis (1982) baked several products, including yellow cake, in preheated and nonpreheated standard gas ovens. They also compared cakes baked in four different kinds of gas and electric ovens. With all types of oven, energy was saved with elimination of preheating. Products for all methods of baking were similar, except that shear value of cakes was higher with no preheating than with preheating; a difference in tenderness was not detected by the sensory panel, however, and it was concluded that preheating is not necessary.

The increased use of microwave ovens has provided impetus to the development of formulas and procedures for microwave cooking of an ever-widening variety of foods. Some of the research reported has dealt with cakes, and cake mixes developed for microwave baking are on the retail market.

Hill and Reagan (1982) compared yellow butter cakes baked in a microwave oven, with and without a turntable, with cakes baked in a conventional oven. The formula was adapted from a microwave guidebook. The cakes baked in a microwave oven were baked at 75% power, and those that were not baked on a turntable were rotated 180° at the midpoint of the cooking period. Conventionally baked cakes received the highest sensory scores and had the lowest shear values. Of the microwave-baked cakes, those baked on a carousel received the higher scores for flavor and appearance and had the lower shear values (greater tenderness). The moisture loss during baking was similar for cakes baked under the three sets of conditions. Although microwave baking had adverse effects, the microwave-baked cakes were considered satisfactory.

Because microwave-baked cakes have pale crusts, microwave/convection (MW/C) ovens have been tested for cakes. Stinson (1986a) compared devil's food and yellow cakes baked in a MW/C oven with cakes baked in a conventional oven. The cakes were prepared from commercial mixes and were baked in shiny aluminum pans. The MW/C oven was used at a temperature of 163°C and at 10% microwave power. The conventional oven was operated at 177°C. The cakes baked in the MW/C oven had drier crumb and moister crusts than did the conventionally baked cakes. Color and symmetry were somewhat affected by the oven used. Tenderness, as indicated by penetrability values and sensory scores, was not affected; neither were shrinkage and uniformity. When two layers were baked simultaneously in the MW/C oven, they were superior to single layers baked in the MW/C oven and comparable in quality to the conventionally baked cakes.

Stinson (1986b) also studied the effects of different MW/C oven conditions and of pan type on the quality of devil's food cake prepared from a mix. Starting the baking in a nonpreheated oven resulted in slightly sunken cakes; using a preheated oven did not. Cake layers baked two at a time were superior to cake layers baked singly. Whereas cakes baked in shiny or dull aluminum pans were flat, those baked in shiny or dull microware pans were peaked. (*Note:* whereas metal pans normally are not used in microwave ovens, Stinson stated that aluminum, as well as microware, pans are recommended for combination cooking by the manufacturer of the MW/C oven used.) Slight differences among the cakes in specific qualities were observed but all of the cakes were considered acceptable.

Suggested Exercises

1. Note that the muffin method and the single-stage (one-bowl) method done in a single step are the same except that plastic fat is melted for the former. Compare cakes made with the same formula (Appendix A) and with the same fat, in plastic form at room temperature and melted.

2. Study the effects of under- and overmixing on cake made by the single-stage method.

3. Compare cakes made from a standard high-ratio formula and mixed by several different methods.

4. Compare cakes made from flours of different protein contents. Also substitute wheat starch for all of the cake flour in one batter. If the cake made with starch is unsatisfactory, try increasing the proportion of egg in a cake containing starch rather than flour.

5. Compare cakes made from different types of flour to which starch has been added as needed to equalize protein concentration. Wheat starch is preferable for this purpose. If it is not available, use corn starch.

6. Compare cakes mixed by the single-stage method and containing butter, with and without added emulsifier. Distilled monoglycerides, added to the butter at a level of about 2% of the butter weight, perform well.

7. Compare cakes to which dried whey has been added at different levels.

8. Use nonfat dried milk to increase the concentration of protein in cake. Determine the maximum level of addition that is feasible for a given set of conditions as to formula and mixing method.

9. Compare cakes containing bran substituted at different levels for flour—for example, 10, 20, and 30% replacement by weight.

10. Develop a satisfactory formula and procedure for increasing the level of polyunsaturated fatty acids in cake through the use of a vegetable oil that is highly polyunsaturated.

11. Compare cakes containing several nonwheat flours at a single replacement level or one nonwheat flour at several replacement levels.

12. Compare gingerbreads with combinations of sweet milk, buttermilk, baking powder, and sodium bicarbonate such as to produce a range of pH values, with equivalent total leavening.

13. In Fig. 21.5 several comparisons are possible. What is shown by A vs C? A vs B? Two comparisons (B vs C and D vs C) seem to confound the variables because two factors are varied at once. Why are these reasonable comparisons, however?

References

Bean, M. M., Elliston-Hoops, E. A., and Nishita, K. D. 1983. Rice flour treatment for cake-baking applications. *Cereal Chem.* **60**: 445–449.

Bowman, F., Dilsaver, W., and Lorenz, K. 1973. Rationale for baking wheat-, gluten-, egg-, and milk-free products. *Bakers Digest* **47**(2): 15–22.

Briant, A. M., Weaver, L. L., and Skodvin, H. E. 1954. Quality and thiamine retention in plain and chocolate cakes and in gingerbread. Mem. 332, Cornell Univ. Agric. Exp. Stn., Ithaca, New York.

Brockmole, C. L. and Zabik, M. E. 1976. Wheat bran and middlings in white layer cakes. *J. Food Sci.* **41**: 357–360.

Brys, K. D. and Zabik, M. E. 1976. Microcrystalline cellulose replacement in cakes and biscuits. *J. Amer. Dietet. Assoc.* **69**: 50–55.

Bundy, K. T., Zabik, M. E., and Gray, J. I. 1981. Edible beef tallow substitution in white layer cakes. *Cereal Chem.* **58**: 213–216.

Cardello, A. V., Hunt, D., and Mann, B. 1979. Relative sweetness of fructose in model solutions, lemon beverages and white cake. *J. Food Sci.* **44**: 748–751.

Charley, H. 1950. Effect of baking pan material on heat penetration during baking and on quality of cakes made with fat. *Food Research* **15**: 155–168.

Charley, H. 1952. Effects of the size and shape of the baking pan on the quality of shortened cakes. *J. Home Econ.* **44**: 115–118.

Clements, R. L. and Donelson, J. R. 1982. Role of free flour lipids in batter expansion in layer cakes. I. Effects of "aging." *Cereal Chem.* **59**: 121–124.

Cocup, R. O. and Sanderson, W. B. 1987. Functionality of dairy ingredients in bakery products. *Food Technol.* **41**(10): 86–90.

Donelson, J. R., Yamazaki, W. T., and Kissell, L. T. 1984. Functionality in white layer cake of lipids from untreated and chlorinated patent flours. II. Flour fraction interchange studies. *Cereal Chem.* **61**: 88–91.

Ebeler, S. E., Breyer, L. M., and Walker, C. E. 1986. White layer cake batter emulsion characteristics: Effects of sucrose ester emulsifiers. *J. Food Sci.* **51**: 1276–1279.

El-Samahy, S. K., Morad, M. M., Seleha, H., and Abdel-Baki, M. M. 1980. Cake-mix supplementation with soybean, sweet potato or peanut flours. II. Effect on cake quality. *Bakers Digest* **54**(5): 32–33, 36.

Freeman, T. M. 1982. Polydextrose for reduced calorie foods. *Cereal Foods World* **27**: 515–518.

Friedman, T. M. 1978. Sorbitol in bakery products. *Bakers Digest* **52**(6): 10–13, 48.

Gaines, C. S. and Donelson, J. R. 1982. Contribution of chlorinated flour fractions to cake crumb stickiness. *Cereal Chem.* **59**: 378–380.

Gisslen, W. 1985. "Professional Baking." John Wiley and Sons, New York.

Glover, J. M., Walker, C. E., and Mattern, P. J. 1986. Functionality of sorghum flour components in a high ratio cake. *J. Food Sci.* **51**: 1280–1283, 1292.

Guy, E. J. 1982. Evaluation of sweet whey solids in yellow layer cakes with special emphasis on fragility. *Bakers Digest* **56**(4): 8–12.

Guy, E. J. 1986. Effect of salt removal on the baking quality and hedonic ratings of white, yellow, spice, and devil's food cakes. *Cereal Foods World* **31**: 890–895.

Hess, D. A. and Setser, C. S. 1983. Alternative systems for sweetening layer cakes using aspartame with and without fructose. *Cereal Chem.* **60**: 337–341.

Hill, M. and Reagan, S.P. 1982. Effect of microwave and conventional baking on yellow cakes. *J. Amer. Dietet. Assoc.* **80**: 52–55.

Hoseney, R. C. 1986. "Cereal Science and Technology." American Association of Cereal Chemists, St. Paul, Minnesota.

Hoseney, R. C., Wade, P., and Finley, J. W. 1988. Soft wheat products. *In* "Wheat: Chemistry and Technology," 3rd edn., vol. II, Pomeranz, Y. (Ed.). American Association of Cereal Chemists, St. Paul, Minnesota.

Huang, G., Finn, J. W., and Varriano-Marston, E. 1982a. Flour chlorination. I. Chlorine location and quantitation in air-classified fractions and physicochemical effects on starch. *Cereal Chem.* **59**: 496–500.

Huang, G., Finn, J. W., and Varriano-Marston, E. 1982b. Flour chlorination. II. Effects on water-binding. *Cereal Chem.* **59**: 500–506.

Hunter, M. B., Briant, A. M., and Personius, C. J. 1950. Cake quality and batter structure. Bull. 860, Cornell Univ. Agric. Exp. Stn., Ithaca, New York.

Johnson, A. C. and Hoseney, R. C. 1979. Chlorine treatment of cake flours. II. Effect of certain ingredients in the cake formula. *Cereal Chem.* **56**: 336–338.

Jooste, M. E. and Mackey, A. O. 1952. Cake structure and palatability as affected by emulsifying agents and baking temperatures. *Food Research* **17**: 185–196.

Kamel, B. S. and Rasper, V. F. 1988. Effects of emulsifiers, sorbitol, polydextrose, and crystalline cellulose on the texture of reduced-calorie cakes. *J. Texture Studies* **19**: 307–320.

Kamel, B. S. and Washnuik, S. 1983. Composition and sensory quality of shortening-free yellow layer cakes. *Cereal Foods World* **28**: 732–733.

Kissell, L. T. 1959. A lean-formula cake method for varietal evaluation and research. *Cereal Chem.* **36**: 168–175.

Knightly, W. H. 1988. Surfactants in baked foods: Current practice and future trends. *Cereal Foods World* **33**: 405–412.

McCullough, M. A. P., Johnson, J. M., and Phillips, J. A. 1986. High fructose corn syrup replacement for sucrose in shortened cakes. *J. Food Sci.* **51**: 536–537.

Neville, N. E. and Setser, C. S. 1986. Textural optimization of reduced-calorie layer cakes using response surface methodology. *Cereal Foods World* **31**: 744–749.

Ngo, W. H. and Taranto, M. V. 1986. Effect of sucrose level on the rheological properties of cake batters. *Cereal Foods World* **31**: 317–322.

Odland, D. and Davis, C. 1982. Products cooked in preheated versus non-preheated ovens. *J. Amer. Dietet. Assoc.* **81**: 135–145.

Pearce, L. E., Davis, E. A., and Gordon, J. 1984. Thermal properties and structural characteristics of model cake batters containing nonfat dry milk. *Cereal Chem.* **61**: 549–554.

Perez, C. M. and Juliano, B. O. 1988. Varietal differences in quality characteristics of rice layer cakes and fermented cakes. *Cereal Chem.* **65**: 40–43.

Pyler, E. J. 1988. "Baking Science and Technology," 3rd edn., vol. II. Sosland Publishing Co., Merriam, Kansas.

Rajchel, C. L., Zabik, M. E., and Everson, E. 1975. Wheat bran and middlings. A source of dietary fiber in banana, chocolate, nut and spice cakes. *Bakers Digest* **49**(3): 27–30.

Rasper, V. F. and Kamel, B. S. 1989. Emulsifier/oil systems for reduced calorie cakes. *J. Amer. Oil Chemists' Soc.* **66**: 537–542.

Shepherd, I. S. and Yoell, R. W. 1976. Cake emulsions. *In* "Food Emulsions," S. Friberg. (Ed.). Marcel Dekker, Inc. New York.

Springsteen, E., Zabik, M. E., and Shafer, M. A. M. 1977. Note on layer cakes containing 30 to 70% wheat bran. *Cereal Chem.* **54**: 193–198.

Stinson, C. T. 1986a. A quality comparison of devil's food and yellow cakes baked in a microwave/convection versus a conventional oven. *J. Food Sci.* **51**: 1578–1579.

Stinson, C. T. 1986b. Effects of microwave/convection baking and pan characteristics on cake quality. *J. Food Sci.* **51**: 1580–1582.

Torres, A. and Thomas, R. D. 1981. Polydextrose and its applications in foods. *Food Technol.* **35**(7): 44–49.

Tsen, C. C., Kulp, K., and Daly, C. J. 1971. Effects of chlorine on flour proteins, dough properties, and cake quality. *Cereal Chem.* **48**: 247–254.

Vaisey-Genser, M. and Ylimaki, G. 1989. Baking with canola oil products. *Cereal Foods World* **34**: 246–255.

Vaisey-Genser, M., Ylimaki, G., and Johnston, B. 1987. The selection of levels of canola oil, water, and an emulsifier system in cake formulations by response-surface methodology. *Cereal Chem.* **64**: 50–54.

PASTRY AND COOKIES

The term pastry includes a wide variety of products made from doughs containing medium to large amounts of fat. Some pastries, such as Danish pastry, are made with a medium amount of fat and leavened with yeast. Other pastries do not contain an added leavening agent. Puff pastry contains a large amount of fat, part of which is rolled into the dough. The dough is folded and rolled many times, resulting in a multilayered product. Formation of steam during baking results in puffing, with separation of layers. Only the plain pastry ordinarily used for pies is considered in this chapter.

Cookies (biscuits in the United Kingdom) usually, though not always, contain a chemical agent; the concentration is usually lower in cookies than in cakes. Cookie dough consistency ranges from stiff for rolled cookies to fairly soft for drop cookies. As is mentioned in Chapter 17, rolled sugar cookies are the usual test product for evaluation of straight-grade soft wheat flours. Cookie spread and surface cracking are the criteria of baking quality. Cookies also are used frequently for assessment of the baking quality of nonwheat flours and of the effects of potential additives, such as various fiber sources.

Pie pastry is not made in the home to the extent that it once was because of the availability of ready-to-bake pie crusts. Similarly, noncommercial baking of cookies probably has decreased with the increased variety of cookies available. Both types of product are important in the food industry, however, and the food science student should have some understanding of them.

I. PASTRY

A pastry formula is simple, containing flour, salt, shortening, and water; the shortening content is relatively high and the water content very low. The mixing of the ingredients and the handling of the dough are extremely important to the properties of the product.

A. Ingredients

In noncommercial baking, all-purpose flours are widely and successfully used to produce pastries that are both tender and flaky; the soft wheat all-purpose flour frequently used in the South and flour of lower protein content give more tender pastry than that obtained with hard wheat flour. In commercial baking, pastry flour, a long-patent soft wheat flour, is more commonly used. To produce pastries of equal tenderness, less fat is required with pastry or soft wheat all-purpose flour than with hard wheat all-purpose flour. In an early study (Denton et al., 1933), the breaking strengths of pastries made from different flours were as follows: cake or pastry flours, 347–399 g; all-purpose flours, 565–751 g; bread flours, above 950 g.

McNeil et al. (1988) reported the effects of replacing up to 40% of the wheat flour in pastry with whole grain barley flour. The pastries containing barley flour were darker in color than the controls. Pastry containing barley flour at the 40% level was less flaky, more tender, and less acceptable in flavor and appearance than the control pastry, and its overall acceptability was lower. However, pastries containing whole grain barley flour at the 20% and 30% replacement levels did not differ significantly from the controls in overall acceptability or any sensory characteristic except flakiness.

Various shortenings are discussed in Chapter 15. Fats that are the softest while being incorporated into a dough seem to make the most tender pastry. Butter and margarine, largely because of their water content, produce less tender pastry than does hydrogenated vegetable shortening when the fats are used on an equal-weight basis. Oil usually produces more tender pastry than do plastic fats when the conventional mixing method is used, but the pastry may be crumbly and greasy, especially if the oil is distributed uniformly throughout the flour. If the oil is cut in so that flour–fat clumps are large, a pastry that is both tender and flaky can be obtained.

As would be expected, the tenderness of pastry increases with the amount of shortening in the formula, and an excessive amount makes pastry too tender to handle easily. In many of the research formulas, about 40 g of fat is used for

100 g of all-purpose flour. This proportion is equivalent to about 1/4 c of hydrogenated shortening per cup of flour. With this relatively lean formula, differences in tenderness attributable to the type of shortening and to other variables are shown more clearly than with the higher proportion of fat in the formulas that are frequently used in noncommercial baking.

The amount of water in pastry is important because pastry dough that is too dry is crumbly and difficult to roll, whereas an excessive amount of water results in tough pastry. Striking results were obtained by Swartz (1943), who made pastries with 100 g of pastry flour, 44 g of fat, and 25, 32.5, or 40 g of water. When the ingredients were at 21–24°C and blended at low speed in a mixer for 30 sec after addition of the water, breaking strengths for pastries with these three amounts of water were 394, 683, and 1035 g, respectively.

B. Mixing Methods and Conditions

The conventional method of mixing pastry involves cutting the solid fat into the flour and salt and gradually adding just enough water to make the dough manageable for rolling. Several variations in the method of combining the ingredients of pastry are possible. In one procedure, half of the shortening is cut into the flour until the mixture is as fine as corn meal and the remainder is cut in until the large particles are the size of peas. In another, the water is mixed with part of the flour, the fat is cut into the remaining flour until it is the size of peas, and the two mixtures are combined. In a modified puff pastry method, about 2 tbsp of the flour–fat mixture are removed before the addition of water and later are sprinkled over the pastry sheet. The sheet is rolled up like a jelly roll, then cut and rolled out for the pie pan. For the hot-water method, boiling water and fat are beaten together and the dry ingredients are added last.

The properties of tenderness and flakiness can occur together or separately. Ideally a pastry has both properties. Thorough mixing of fat and flour usually results in more tender pastry than does minimal mixing. Pastry usually is more tender when the ingredients and dough are warm than when they are cold. During mixing, soft fat coats the flour particles, whereas solid fat is broken into discrete particles. Room temperature and temperature of ingredients, therefore, should be controlled during experimental work with pastry.

Flakiness of the baked pastry is best achieved by cutting cold fat into the flour as rather large particles and by using cold water and minimal mixing to keep the fat in discrete particles. Thin horizontal layers are evident in flaky pastry. The thorough blending of flour and fat obtained when pastry is made by the hot-water method or with oil in the conventional method usually produces pastry that is mealy rather than flaky, though it is possible to produce flaky pastry with intimate mixing of fat and flour.

Lengthening the mixing time after water is added to the dough usually toughens pastry because of excessive gluten development. In a study of pastry made with lard, the toughening effect of added mixing with water was much more pronounced when the fat and water were at refrigerator temperature than when they were at room temperature (Swartz, 1943). This might reflect the greater plasticity of fat at the higher temperature.

C. Baking

Pastry is baked at a high temperature and is crisp because the low water concentration greatly inhibits starch gelatinization. The high concentration of fat also contributes to crispness.

II. COOKIES

Cookie formulations are diverse, but generally the proportion of water in a cookie dough is low, and sugar and fat concentrations are high. The sugar competes with both starch and gluten for the small amount of water that is present. The baked cookie most often is tender–crisp because little or no starch gelatinization occurs, and the sugar and fat interfere with gluten development. When a formula contains a higher proportion of water, the product is softer and more cakelike because of starch gelatinization.

The rolled snap type of cookie is used in test baking. Those of us who remember the gingersnaps of our childhood did not realize how much there was to notice about them in addition to their gingery flavor. Snap cookies get their name from the fact that they break with a snap. Snap is a different characteristic from hardness, and it requires a day or two to develop.

As mentioned previously, snap cookies also are expected to have cracked surfaces; the patterns of cracking are both important and variable. Cookie spread, also mentioned previously, is an important characteristic and is reported as the width/thickness ratio (W/T) (see Chapter 3). The plain sugar-snap cookie apparently is used more often in test baking than is the gingersnap.

A. Mixing and Baking

During the creaming of shortening and sugar, air is incorporated and forms gas cell nuclei. With the addition of the other ingredients and further mixing, about half of the sugar dissolves. Although gluten proteins are hydrated, little development occurs because of the relatively large amounts of sugar and shortening and the kind of flour used.

In the oven the dough first becomes softer as the shortening melts. Additional sugar dissolves, further contributing to fluidity and, therefore, to cookie spread. Carbon dioxide from the chemical leavener diffuses into the gas cell nuclei formed during mixing. With continued heating, steam also forms and contributes to dough expansion, though in most cookies its contribution is small as compared with that of carbon dioxide from the leavening agent. The flour proteins, which are discontinuous in the dough, expand and form a continuous structure; the horizontal spreading stops when the viscosity becomes too great to permit flow (Doescher *et al.*, 1987a). In those cookies that develop a cracked surface pattern, sucrose crystallizes at the surface during baking, freeing water to evaporate. The drying surface then cracks with continuing expansion of the dough. Hoseney (1986, pp. 257–261) described the changes that occur in cookie dough during baking.

The surface cracking described above can occur only in intermediate-moisture (6–8%) cookies. Many commercial cookies are baked to water concentrations of 3–4%, at which there is too little water to dissolve the sucrose and the sucrose, therefore, exists in the amorphous glassy state.

Immediately after baking and cooling, a snap cookie can be bent. Hoseney (1986, pp. 260–261) related the development of snap to continuing recrystallization of sucrose. Sugar syrup is responsible for the cookie's flexibility, which is lost as sucrose crystallizes out of solution.

B. Factors Affecting Cookie Spread and Surface Grain

1. Flour

The high degree of cookie spread that is considered desirable in the baking industry is best achieved with a soft flour other than cake flour, which is not used because cookie spread is reduced by chlorination (Kulp and Olewnik, 1989). The softer flour takes up less water than does a stronger flour, leaving more

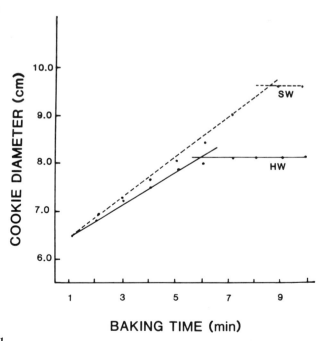

FIGURE 22.1
Changes in cookie diameter during baking of cookies made with hard wheat flour (HW) and soft wheat flour (SW). (From Abboud *et al.*, 1985.)

BAKING
TIME
(MIN)

COOKIE QUALITY

POOR   EXCELLENT

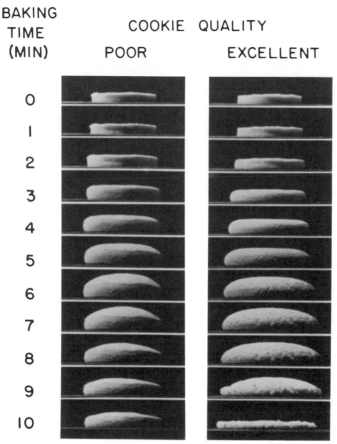

0
1
2
3
4
5
6
7
8
9
10

FIGURE 22.2
Time-lapse photographs of cookies made from an excellent and a poor-quality cookie flour. (From Yamazaki and Lord, 1971.)

water to dissolve sugar; the sugar syrup thus formed makes the dough more slack and permits more spreading before the dough becomes too viscous to flow. That point is reached at a higher temperature with soft wheat flour than with hard wheat flour (Doescher *et al.*, 1987a). Tanilli (1976) reported that weak, low-protein flour gave cookies with W/T values ranging from 8.5 to 10.0, as compared with 6.5–8.0 for strong, high-protein flours. The effects of two types of flour on the rate and extent of cookie spread are seen in Figs. 22.1 and 22.2. Starch does not contribute to the increased viscosity during baking because it is not gelatinized; therefore, differences between flours in their effects on spread are not related to starch. Gaines (1985) studied a large number of soft wheat cultivars and found the spread of sugar-snap cookies to be correlated negatively with protein content.

Gaines and Donelson (1985) made cookies with straight grade and whole wheat flours from 69 soft wheat cultivars. Across the cultivars there was a significant correlation between cookie diameters of straight grade and those of whole wheat flours; therefore, diameter of cookies made with straight grade flour can predict effects of whole wheat flour from the same wheat on cookie diameter. Spread was greater for cookies made with whole wheat flour than for those made with straight grade flour.

Several nonwheat flours have been used successfully as partial replacements for wheat flour. McWatters (1978) reported that soybean, peanut, and field pea flours gave spread ratios and top grains similar to those of wheat flour controls at levels of 10, 20, and 30%, respectively; however, the use of field pea flour at the 30% level resulted in a beany flavor. McWatters was able to produce soy cookies with normal spread and top grain with as much as 30% replacement when dough consistency was reduced by addition of water.

Beuchat (1977) replaced wheat flour with untreated and hydrolyzed peanut flours at levels up to 25% in sugar cookies. Untreated peanut flour decreased the spread and had a deleterious effect on top grain; the hydrolyzed flour did not.

Oomah (1983) compared wheat–oat flour blends containing up to 25% oat flour in cookies. Of the two oat flours used, one had no significant effect on cookie spread, even at the 25% level. The other resulted in increased spread with increased level of substitution.

Olatunji et al. (1982) prepared sorghum and millet flours and tested them in biscuits (cookies). The composite flours also contained 5% soy flour. At levels up to 45% of the total flour weight, with both sorghum and millet flours, cookie spread was at least as great as that of the controls, and cookie quality was highly acceptable.

Hoojjat and Zabik (1984) substituted various combinations of navy bean and sesame seed flours for 20 and 30% of the wheat flour in sugar-snap cookies. Both cookie spread and top grain scores decreased with increased percentage of navy bean and/or sesame flour.

Sunflower protein isolate in sugar-snap cookies, even at a replacement level as low as 5%, caused reduced cookie spread and poor surface grain (Claughton and Pearce, 1989). The addition of 1–2% lecithin to the formula improved the spread and surface grain. The highest acceptable level of replacement (15%) when lecithin was present was dictated by sensory qualities.

2. Shortening

Abboud et al. (1985) reported that cookie spread was not affected by an increase in shortening from 30 to 35%; however, the fineness of the top grain was increased. When the shortening level was decreased to 25%, the top grain was coarse. They also varied the kind of shortening. Whether the shortening was a nonemulsified shortening, a shortening containing monoglyceride, a nonemulsified oil, or the nonemulsified shortening melted and cooled at room temperature did not affect cookie spread when a noncreaming method was used.

3. Sucrose and Other Sweeteners

Vetter et al. (1984) reported greater cookie spread with 60% sugar than with 50 or 40% (flour basis). The particle size also had an effect: Three sugars (sucrose)

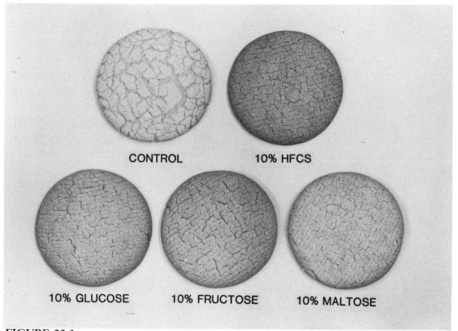

FIGURE 22.3
Effect of small amounts of corn syrup and various sugars on the cracking pattern of cookies.
HFCS, High-fructose corn syrup. (From Hoseney, 1986.)

of increasing particle size range resulted in cookies having average W/T values
of 9.50, 8.84, and 7.82.

Abboud *et al.* (1985) reported increased cookie spread with increased sucrose
concentration from 50 to 70% of flour weight. When the quantity of sugar was
decreased from 60 to 50%, cookie diameter was not affected but the top grain
became finer.

When Doescher *et al.* (1987b) used sucrose, both dissolved and undissolved,
the rate of expansion, rate of water loss, and cookie diameter increased with increased sucrose whether the sucrose was dissolved or not. When they used granular sucrose, and sucrose, glucose, and fructose syrups, the spread was greatest
with granular sucrose and least with fructose syrup. When other sugars are only
partially substituted for sucrose, they interfere with crystallization and decrease
surface cracking (Fig. 22.3).

4. Other Ingredients

Although emulsifiers are not used routinely in cookies, Rusch (1981) stated that
they are capable of increasing cookie spread. Of the emulsifiers that are commercially available, SSL has a particularly marked effect.

Gaines and Finney (1989) tested 17 commercial enzymes in sugar-snap
cookies. Papain most effectively increased cookie spread and improved top

grain. With its addition they were able to produce cookies from two hard wheat flours, obtaining products that were as large as control cookies made from soft wheat flour.

5. Mixing Method

Vetter *et al.* (1984) reported that the all-in mixing method gave cookies with greater spread than did the creaming method. They attributed the difference to the coating of sugar particles with fat and consequent retardation of sugar dissolution with the creaming method.

C. Effect of Type of Sweetener on Sensory Characteristics

Redlinger and Setser (1987) studied the perceived sweetness of shortbread-type cookies containing various sweeteners (sucrose, fructose, aspartame, acesulfame K, sodium saccharin, and calcium cyclamate). Approximate equisweet concentrations had been determined in preliminary work. They also varied the addition of flavor (none added, vanilla, lemon).

Initial, maximum, and residual sweetness (about 30 sec after swallowing), as well as nature and intensity of nonsweet aftertaste were evaluated for unbaked and baked cookies. The perception of sucrose in baked cookies was different from that of all the other sweeteners; although saccharin gave a perception of sweetness in baked cookies similar to that of sucrose, it had a rather intense nonsweet aftertaste. The greatest difference in profiles between unbaked and baked cookies was for aspartame, which lost some perceived sweetness and gained considerable nonsweet aftertaste during baking. The added flavorings did not affect perceptions of sweetness or nonsweet aftertaste in the baked cookies.

Hardy *et al.* (1979) compared sucrose and fructose in sugar cookies. Cookies containing sucrose were rated higher for relative sweetness, flavor, texture, and overall acceptance than were those containing fructose. The cookies containing fructose were chewy, whereas those containing sucrose were crisp. The cookies containing fructose also browned more than did those containing sucrose.

Cookie ingredients and cookie production were discussed by Hoseney *et al.* (1988, pp. 425–436) and Pyler (1988, pp. 1012–1021). Pyler included descriptions of several types of bakery cookies: deposit (drop), wire cut (dough extruded and sliced by a wire), cutting machine (dough sheeted, then cut or stamped), and rotary molded (dough forced into biscuit-shaped dies for shaping, then removed). In commercial baking the entire process is mechanized and continuous.

In reading the research literature on cookies, one notices that researchers have not always specified the type of cookie made, although they sometimes provide the formula when they do not identify the cookie type. When there is no identification, either by name or by formula, it seems likely that the product is a sugar-snap cookie. Also apparent is the much greater frequency with which cookie spread is reported than is surface grain. Perhaps surface grain desirability can be assumed to parallel cookie spread in those cookies that develop a cracked surface pattern. Perhaps, too, the nature of the surface is considered in evaluation of appearance in sensory testing; for some studies, sensory evaluation is not reported.

Suggested Exercises

1. Compare pastries made from scratch with pastry made from retail commercial mixes and from ready-to-bake dough. Compare the products as to quality, cost, and time required to produce them. [*Note:* Pastry is subject to considerable manipulation, and control (discussed in Chapter 1, Section IV) is important. The use of a mixer method contributes to control; one such method is provided with the pastry formula in Appendix A.]
2. Compare pastries made with oil and solid shortening.
3. Compare oil pastries in which oil is cut in finely and coarsely with a pastry blender.
4. Compare pastries made with soft wheat and hard wheat flours.
5. Compare pastries mixed by different methods.
6. Vary ingredient temperatures in making pastry.
7. Study the ingredient lists on the packages of some cookies that you know to lack surface cracking and some that you know to have surface cracking. Can you explain the difference in surface grain?
8. Compare cookies (formula in Appendix A) made with soft wheat and hard wheat flours.
9. Partially replace wheat flour with nonwheat flours in cookies.
10. Vary the sugar concentration and/or the type of sugar in cookies.
11. Compare cookies made by the creaming method with cookies made by the all-in method.

References

Abboud, A. M., Rubenthaler, G. L., and Hoseney, R. C. 1985. Effect of fat and sugar in sugar-snap cookies and evaluation of tests to measure cookie flour quality. *Cereal Chem.* **62**: 124–129.

Beuchat, L. R. 1977. Modification of cookie-baking properties of peanut flour by enzymatic and chemical hydrolysis. *Cereal Chem.* **54**: 405–414.

Claughton, S. M. and Pearce, R. J. 1989. Protein enrichment of sugar-snap cookies with sunflower protein isolate. *J. Food Sci.* **54**: 354–356.

Denton, M. C., Gordon, B., and Sperry, R. 1933. Study of tenderness in pastries made from flours of varying strengths. *Cereal Chem.* **10**: 156–160.

Doescher, L. C., Hoseney, R. C., and Milliken, G. A. 1987a. A mechanism for cookie dough setting. *Cereal Chem.* **64**: 158–163.

Doescher, L. C., Hoseney, R. C., Milliken, G. A., and Rubenthaler, G.L. 1987b. Effects of sugars and flours on cookie spread evaluated by time-lapse photography. *Cereal Chem.* **64**: 163–167.

Gaines, C. S. 1985. Associations among soft wheat flour particle size, protein content, chlorine response, kernel hardness, milling quality, white layer cake volume, and sugar-snap cookie spread. *Cereal Chem.* **62**: 290–292.

Gaines, C. S. and Donelson, J. R. 1985. Evaluating cookie spread potential of whole wheat flours from soft wheat cultivars. *Cereal Chem.* **62**: 134–136.

Gaines, C. S. and Finney, P. L. 1989. Effects of selected commercial enzymes on cookie spread and cookie dough consistency. *Cereal Chem.* **66**: 73–78.

Hardy, S. L., Brennand, C. P., and Wyse, B. W. 1979. Fructose: Comparison with sucrose as sweetener in four products. *J. Amer. Dietet. Assoc.* **74**: 41–46.

Hoojjat, P. and Zabik, M. E. 1984. Sugar-snap cookies prepared with wheat-navy bean-sesame seed flour blends. *Cereal Chem.* **61**: 41–44.

Hoseney, R. C. 1986. "Cereal Science and Technology." American Association of Cereal Chemists, St. Paul, Minnesota.

Hoseney, R. C., Wade, P., and Finley, J. 1988. Soft wheat products. *In* "Wheat: Chemistry and Technology," 3d edn., vol. II, Pomeranz, Y. (Ed.). American Association of Cereal Chemists, St. Paul, Minnesota.

Kulp, K. and Olewnik, C. 1989. Functionality of protein components of soft wheat flour in cookie applications. *In* "Protein Quality and the Effects of Processing," Phillips, R. D. and Finley, J. W. (Eds.). Marcel Dekker, Inc., New York.

McNeil, M. A., Penfield, M. P., and Swanson, R. B. 1988. Sensory evaluation of pastry containing barley flour. *Tennessee Farm and Home Science* Issue 146: 4–7.

McWatters, K. H. 1978. Cookie baking properties of defatted peanut, soybean, and field pea flours. *Cereal Chem.* **55**: 853–863.

Olatunji, O., Akinrele, I. A., Edwards, C. C., and Koleoso, O. A. 1982. Sorghum and millet processing and uses in Nigeria. *Cereal Foods World* **27**: 277–280.

Oomah, B. D. 1983. Baking and related properties of wheat-oat composite flours. *Cereal Chem.* **60**: 220–225.

Pyler, E. J. 1988. "Baking Science and Technology," 3rd edn., vol. II. Sosland Publishing Co., Merriam, Kansas.

Redlinger, P. A. and Setser, C. S. 1987. Sensory quality of selected sweeteners: Unbaked and baked flour doughs. *J. Food Science* **52**: 1391–1393, 1413.

Rusch, D. T. 1981. Emulsifiers: Uses in cereal and bakery foods. *Cereal Foods World* **26**: 111–115.

Swartz, V. 1943. Effect of certain variables in technique on the breaking strength of lard pastry wafers. *Cereal Chem.* **20**: 121–126.

Tanilli, V. H. 1976. Characteristics of wheat and flour for cookie and cracker production. *Cereal Foods World* **21**: 624–628, 644.

Vetter, J. L., Bright, H., Utt, M., and McMaster, G. 1984. Cookie formulating. *Bakers Digest* **58**:(4): 6–9.

Yamazaki, W. T. and Lord, D. D. 1971. Soft wheat products. *In* "Wheat: Chemistry and Technology," 2nd edn., Pomeranz, Y. (Ed.). American Association of Cereal Chemists, St. Paul, Minnesota.

SUGARS AND ALTERNATIVE SWEETENERS

Sugars are mentioned frequently in this book because of their contributions to the properties of many types of food. The chemical and physical nature and the functions of sugars in foods are summarized in this chapter. Both crystalline sugars and sugar syrups are discussed. Alternative sweeteners are considered also.

I. SUGARS

A. Chemical Nature

The monosaccharides have the smallest carbohydrate molecules. The monosaccharides of particular interest in food science are the hexoses (glucose, fructose, and galactose), though the pentoses (arabinose, xylose, and ribose) also are of interest. Glucose and galactose are aldohexoses and fructose is a ketohexose. Their formulas are represented in the Haworth cyclic structures shown below:

α-D-glucopyranose   α-D -fructofuranose

A discussion of the significance of the α (versus β) and D (versus L) notations can be found in any biochemistry textbook. The Haworth structures for the sugars are unrealistic in that all of the atoms in the ring are represented as being in a single plane; actually different arrangements in space are possible among sugars, and a chair conformation is most stable for the pyranose sugars. A specific example is the following representation of α-D-glucopyranose:

Chair structure, α-D -glucopyranose

Similarly, the ring in fructose is not planar either, but the conformation is different because furanoses are five-membered rings. Zapsalis and Beck (1985, p. 326) stated that the envelope and twist conformations probably both exist in a fructose solution. Conformation (molecular shape) apparently depends on configuration (arrangements of atoms and groups about individual carbon atoms).

The carbonyl groups (or potential carbonyl groups) in sugar molecules are important with regard to the chemical reactions in which sugars participate. Note in the formulas presented above that the C-1 of glucose is the aldehydic carbon and the C-2 of fructose is the keto carbon. These are reducing groups, both of which are "tied up" when glucose and fructose combine (in glycosidic linkage) to form the disaccharide sucrose:

Sucrose (α-D-glucose, β-D-fructose)

The sucrose molecule, therefore, does not have any reducing ability. A disaccharide is defined as a carbohydrate that yields two monosaccharide units when hydrolyzed. When the disaccharide sucrose is hydrolyzed to glucose and fructose, two reducing groups are freed for each glycosidic bond broken. In maltose, two glucose molecules are joined through the C-1 of one and the C-4 of the other; in other words, one reducing group is tied up and the other remains free. This disaccharide, therefore, is a reducing sugar, but hydrolysis of maltose results in a mixture with yet greater reducing power than that of the intact molecule. The lactose molecule consists of glucose and galactose linked through the C-4 of glucose and the C-1 of galactose; thus the situation with regard to reducing ability is the same as that for maltose.

Reduction of carbonyl groups of reducing sugars produces sugar alcohols, which are somewhat sweet and are discussed later in this chapter. The sugar alcohols sorbitol, mannitol, xylitol, and maltitol are produced by the reduction of glucose, fructose, xylose, and maltose, respectively. Oxidation of aldehyde groups produces acids, for example gluconic acid from glucose. Oxidation of the alcohol group at the C-6 of a hexose produces a uronic acid, for example glucuronic acid from glucose.

Hydrolysis of disaccharides is an important reaction in foods. From a chemical standpoint, the significance lies not in the increased potential for participation in oxidation–reduction reactions (though this is useful analytically), but in the freeing of carbonyl groups that are potential reactants in browning (Chapter 6). In addition, because sugars differ in properties such as sweetness and solubility, the properties of a reaction mixture are different from those of the disaccharide that is hydrolyzed. One such property change accompanying hydrolysis is responsible for some common terminology. The hydrolysis specifically of sucrose is called inversion because the mixture of products has a different rotatory power from that of sucrose. Sucrose in solution rotates a plane of polarized light to the right; it is dextrorotatory. Fructose, one of the products of sucrose hydrolysis, is so strongly levorotatory that the mixture of products is levorotatory. Because the direction of rotation by the solution is inverted, the enzyme that catalyzes the reaction is called invertase, and the mixture of glucose and fructose resulting from the reaction is called invert sugar.

Several examples of enzymatic catalysis of sugar hydrolysis are mentioned in other chapters. These examples include the hydrolysis of maltose and sucrose by yeast enzymes in bread making (Chapter 19); the treatment of milk with lactase to reduce development of grittiness in ice cream, resulting from low solubility of lactose (Chapter 8); and the treatment of milk with lactase to permit the use of milk by individuals who are intolerant of lactose. The role of enzymatic hydrolysis in production of sugar syrups is noted later in this chapter.

Hydrolysis also can be catalyzed by a combination of heat and dilute acid. Its occurrence and significance in food processing depend on the conditions of processing, such as time, temperature, and pH. The significance of this method of hydrolysis in relation to sugar in candy is discussed in connection with crystallization (Chapter 24).

Caramelization of sugars is discussed briefly in Chapter 6. Like carbonyl-amine browning, caramelization is a series of reactions rather than a single reaction. It should be noted that the browning that occurs when caramels are made is primarily carbonyl-amine browning rather than caramelization.

The hydroxyl groups on sugar molecules permit their reacting with other compounds through either glycosidic or ester linkages. Anthocyanins, discussed in Chapter 14, are examples of compounds in which sugars are in glycosidic linkage with nonsugar moieties. Sinigrin, a flavor component also mentioned in Chapter 14, is another example of a glycoside. Sucrose esters, used as surfactants, are examples of sugar in ester linkage with another compound.

Sugars polymerize in nature to form polysaccharides that have functional importance in foods. These include cellulose (Chapter 14) and starch (Chapter 16). Pectic substances (Chapter 14) are polymers of sugar derivatives.

Further information concerning the chemical nature of sugars can be found in any food chemistry book.

B. Physical Properties

The physical properties of sugars result from their large number of hydroxyl groups and consequent hydrogen-bonding ability. Sugar molecules form hydrogen bonds with water, with each other, and with other polar compounds.

Solubility in water differs among sugars. Because sugars are similar with respect to their content of hydroxyl groups, it would seem that their differences in solubility must be related to configuration and/or conformation. At room temperature, fructose is most soluble, followed by sucrose, glucose, maltose, and lactose. Although fructose is the most soluble sugar and lactose the least soluble over a broad temperature range, the solubility rankings of sucrose, glucose, and maltose vary somewhat with the temperature at which solubility is measured.

The practical significance of differences in solubility lies in the appropriateness of different sugars in food applications involving high sugar concentrations. Lactose, for example, has too low a solubility to be used successfully as a major food ingredient. As a component of milk, it is in dilute solution.

Hygroscopicity differs among sugars as does solubility. Fructose is particularly hygroscopic. Because sugars differ in their hygroscopicity, their uses differ. For example, lactose and maltose are useful in powders that need to remain dry and free flowing, and invert sugar is useful in products that need to retain moisture. Honey, which contains a rather high proportion of fructose, frequently serves as a humectant in home-baked foods because of the ability of its fructose component to absorb moisture from the atmosphere.

Crystallizability is a physical property that varies inversely with solubility. Crystallization is discussed in the next chapter.

Sweetness obviously is an important property of sugars. Ranking sugars as to sweetness is a quite different matter from ranking them as to solubility because sweetness is a sensory quality. Its perception depends on many conditions of the test, including the concentration at which sugars are compared, the temperature of the solutions, and the viscosity of the carrier medium. Regardless of the conditions, fructose usually is ranked as the sweetest sugar and lactose as the least sweet sugar. The actual values assigned to designate sweetness relative to that of sucrose depend on the conditions.

Sweetness of sugars apparently varies with hydrogen bonding. Perception of sweetness depends on hydrogen bonding between glycol groups on the sugar molecules and receptor sites on the tongue. If adjacent intramolecular hydroxyl groups that have the potential of eliciting the sweet taste response are involved in hydrogen bonding with one another, their ability to elicit the response is restricted. Differences in configuration and conformation among different sugars would result in differences in the distances between groups that have hydrogen bonding potential and, therefore, differences in sweetness of the sugars.

Shallenberger and Birch (1975, pp. 50–71) discussed the physical properties of sugars in more detail.

C. Functions in Foods

The functions of sugars in food are discussed in various chapters and will only be listed here. Sugars participate in nonenzymatic browning reactions. They modify the properties of egg, starch, and gelatin gels. They dehydrate pectin micelles to permit pectin gels to form. They stabilize egg white foams. They serve as substrate for yeast fermentation in bread dough and for lactic acid fermentation in cultured buttermilk and some other dairy products. They have a large role in the aeration of batters and a tenderizing function in batter and dough products. Sugars contribute to flavor through nonenzymatic browning, and they contribute sweetness to many foods. They crystallize to form the structure of crystalline candies. Crystals fail to form in noncrystalline candies because of a high sugar concentration, a high concentration of interfering substances, or a combination of these factors. In hard candies, the moisture level is so low (sugar concentration so high) that the mixture becomes rigid quickly and the sugar molecules do not have an opportunity to become oriented into crystals. Interfering substances function through mechanisms that are discussed in connection with crystallization.

D. Sources

Sucrose occurs in many plants, but the commercial sources are sugar cane and sugar beets. The extracted and refined sucrose is available in several forms, including granulated sugars of different degrees of fineness, achieved through control of the crystallization process and appropriate sieving (Pyler, 1988, p. 415). The other forms of sucrose are adequately described in elementary textbooks.

Dextrose (glucose) is obtained through hydrolysis of starch. Because a reducing group is exposed with every glycosidic bond broken, the reducing power of the reaction mixture increases as hydrolysis proceeds. The extent of hydrolysis is expressed as dextrose equivalent (DE), the percentage of the total dry substance represented by reducing sugar content calculated as dextrose. Dextrose is a common ingredient in commercially produced foods. Because it is a reducing sugar, it promotes browning of baked products. Its relatively high osmotic pressure enhances retardation of bacterial growth.

Fructose has gained in importance as a food ingredient since the technology for converting glucose to fructose was developed. The conversion involves treatment with the enzyme glucose isomerase. Some research dealing with the use of fructose in baked flour mixtures is mentioned in Chapters 21 and 22.

Lactose is crystallized from whey. Pyler (1988, pp. 439–440) summarized some applications in different foods.

Sugar syrups frequently are used along with crystalline sugar. Honeys differ in the ratio of fructose to glucose, which may range from about 1 : 1 to 2 : 1. The properties of products containing honey are affected by the hygroscopicity and browning properties of fructose. The effects can be beneficial or not, depending on the product and on the level of substitution.

Corn syrups, sometimes called starch syrups, have increased in commercial importance because of increased cost of sucrose and developments in the technology of producing syrups from corn starch. Corn syrups contain dextrose, maltose, higher sugars, and dextrins; they differ in the extent of conversion. The corn syrups that have been available for many years are produced by acid hydrolysis or enzyme hydrolysis of corn starch. With either method, a range in the extent of hydrolysis is possible, resulting in products ranging from about 12 DE to about 68 DE. Products of particularly high DE are obtained with enzyme conversion. Because of the shift toward dextrose, the higher the extent of conversion, the greater are the sweetness and humectant properties of the syrup.

High-fructose corn syrups (HFCS) represent a more recent technological development. The use of glucose isomerase in the production of fructose from glucose is mentioned above. Immobilized enzyme technology, described briefly in Chapter 6, is applied to the process. Glucose isomerase is used also in the conversion of corn syrup to HFCS. Fructose concentration in the product can be as high as 90% (dry basis). Bakers use a HFCS with about 42% fructose. Its composition is similar to that of invert sugar syrup, which also is used in commercial food production.

Pomeranz (1985, pp. 122–135) and Strickler (1982, Chapter 2) described corn sweeteners and their use in food. Hobbs (1986) and Horn (1981) described corn syrups and their properties, and Saussele *et al.* (1983) described high-fructose corn syrups and their use in baked products.

II. ALTERNATIVE SWEETENERS

Alternative sweeteners have been of interest for many years, but consumers' increasing calorie consciousness has stimulated development and testing of additional products. Some sweeteners are noncaloric compounds that are intensely sweet and can be used in small amounts. Because sweetness is not the only function of sugar, adjustments must be made in formulation. Others do provide calories but either provide fewer calories for a given degree of sweetness or are preferred by some because of other nutritional benefits.

A. Nonnutritive Sweeteners

Saccharin has been in use for most of this century. It is 200–300 times as sweet as sucrose but has a bitter aftertaste, especially at higher concentrations. It is quite stable over a wide temperature range. Saccharin was on the food additives GRAS (generally recognized as safe) list of 1958 but became suspect in 1977.

When the FDA proposed a ban because of possible carcinogenicity, public pressure on Congress prevented implementation, and the ban has been delayed to 1992. In the meantime, evidence substantiating a health risk to humans when saccharin is used in normal dietary amounts has not appeared (IFT, 1989).

The cyclamates are about 30 times as sweet as sucrose and are heat stable (IFT, 1986). They were well received because of their similarity to sucrose in taste. However, after years of use in soft drinks and other products, they were removed from the market in 1970 because of evidence of possible carcinogenicity. By that time, a market for diet beverages had been established, providing incentive for development and testing of additional sweeteners. At the time of writing, cyclamates might be about to make a comeback because of additional scientific evidence (Anon., 1989).

Acesulfame-K was approved by the FDA in 1988 for use in chewing gum, beverages, and certain foods (IFT, 1989). It is about 200 times as sweet as sucrose and leaves no bitter aftertaste. It is stable in the ranges of time, temperature, and pH used in its processing and storage. Acesulfame-K shows promise for use in combination with certain other alternative sweeteners, with which it is synergistic. Acesulfame-K and sorbitol complement each other. Sorbitol provides the body and texturizing properties, and acesulfame-K improves the flavor.

Polydextrose is not a sweetener but a bulking agent used with nonnutritive sweeteners. It contributes textural effects that normally are provided by sucrose. Its caloric contribution is only 1 kcal/g; therefore, its use with a noncaloric sweetener does permit reduction of caloric value. Maltodextrin, a concentrated solution or powder of low DE (less than 20) glycosides, also is a bulking agent.

B. Nutritive Sweeteners

Aspartame is widely used as a sweetener. It is the methyl ester of the dipeptide aspartylphenylalanine. Although it provides as many calories as does sucrose on a weight basis, its intense sweetness permits calorie reduction in products. It is about 180–200 times as sweet as sucrose and is one of the more satisfactory alternative sweeteners from the standpoint of taste. The perception of sweetness tends to linger, which sometimes is desirable; otherwise the level can be reduced or aspartame can be combined with other sweeteners or with substances that counteract some of the sweetness. The potency of aspartame relative to that of sucrose varies with concentration, temperature, pH, and other flavors present. It is used satisfactorily and synergistically with other sweeteners, including sucrose (Homler, 1984).

Dry aspartame and dry products containing it are storage stable. The stability of aspartame in solution decreases with increased temperature. It is most stable in the pH range 3–5. The chemical changes that can occur involve hydrolysis of the ester linkage and formation of diketopiperazine; aspartic acid and phenylalanine can ultimately form from hydrolysis of peptide linkage between the two amino acids. The diketopiperazine and the amino acids have no sweetening power; therefore, the product loses sweetness. The potential for the occurrence of these changes poses no problem in the products in which aspartame is used. The major application is in carbonated beverages (Homler, 1984). Aspartame also is used as a free-flowing sugar substitute and as a sweetener in dry beverage mixes, fruit beverages, chewing gum, breath mints, puddings, and dessert toppings (IFT, 1989).

Baked products were not among the products for which approval for use of aspartame was sought and received. Aspartame is not stable during the prolonged heating involved in production of baked goods. However, approval has been requested for an encapsulated form of aspartame, which is not released until the end of the baking period. Pszczola (1988) described its performance in toppings, frostings, and fillings. For the flour products, bulking agents would need to be added. Cake mixes containing aspartame with bulking agents probably will be available to consumers.

Sugar alcohols (polyhydric alcohols, polyols) that are used as alternative sweeteners include xylitol, sorbitol, and mannitol. They are naturally occurring substances. Igoe (1989, p. 181) summarized the relative sweetness of some sugar alcohols, as well as some sugars and artificial sweeteners. Xylitol is similar to sucrose in sweetness, and the other sugar alcohols are considerably less sweet. They are useful, however, because they do contribute some sweetness and can be used along with nonnutritive sweeteners; they provide the bulking and other properties that nonnutritive sweeteners cannot contribute. Sugar alcohols do not participate in carbonyl-amine browning reactions (Dwivedi, 1986). On a weight basis, sugar alcohols provide as many calories as does sucrose. However, sorbitol is absorbed more slowly from the intestine and, therefore, is useful in special diets. Another advantage of sugar alcohols over sucrose is their not causing dental caries (IFT, 1986). In fact, Pepper and Olinger (1988) stated that xylitol may be caries inhibiting. Sugar alcohols present some problems of tolerance, causing diarrhea when consumed in excess. Disaccharide alcohols, such as maltitol, are tolerated in larger amounts than are monosaccharide alcohols (IFT, 1988).

Sorbitol and mannitol are listed as permitted sweeteners in the United States. Maximum levels are specified for chewing gum, hard candies, and cough drops. Sorbitol is very hygroscopic, mannitol only moderately so (Dwivedi, 1986). Xylitol also is approved. It is used in chewing gum, tableted products such as mints, hard candies, certain candy coatings, and chocolate. It does not have the bodying effect of sucrose. Pepper and Olinger (1988) described the adjustments in ingredients and procedures that are required when xylitol is used in various products.

Maltitol has no history of use in the United States, though it has been used in Japan for many years. A request for GRAS status was made in 1986 (IFT, 1988).

Many potential sweeteners are currently under study. Two that are expected to be approved by 1991 are sucralose and alitame (Best, 1989). Sucralose is a selectively chlorinated sucrose derivative with sweetness about 400–800 times as great as that of sucrose (IFT, 1988). Approval was requested for its use in 14 food categories, including baked goods and baking mixes.

A point that is made frequently in the literature on alternative sweeteners is that there is not an ideal sweetener. Perhaps there never will be, but the existing sweeteners, especially when used in combination, are meeting current needs fairly well. The sweet substances presently under study, as well as those that are not yet known, possibly will eventually extend the satisfactory use of alternative sweeteners. As with other food additives, new products will be slow in reaching the consumer because the time required for the necessary safety testing is long.

Suggested Exercises

1. Make caramels from a modified formula containing no corn syrup. (Formula: 200 g sucrose; 14 g margarine; 127 ml milk; 110 ml half-and-half. Stir constantly while heating to a boiling point of 120°C.) Repeat with glucose (dextrose). Note the color of the glucose mixture at a boiling point of 120°C but continue heating to a boiling point of 138°C. Explain the results. What would happen if an attempt were made to make caramels or any other candy with lactose?
2. Test the ability of individuals to perceive a difference between sweetened tea or another sugar-containing product and its counterpart containing a sugar substitute (see exercise 1 in Chapter 4).

References

Anon. 1989. Showcase: Sweeteners. *Prepared Foods* **158**(4): 82.

Best, D. 1989. High-intensity sweeteners lead low-calorie stampede. *Prepared Foods* **158**(3): 97–98.

Dwivedi, B. K. 1986. Polyalcohols: sorbitol, mannitol, maltitol, and hydrogenated starch hydrolysates. *In* "Alternative Sweeteners," Nabors, L. O. and Gelardi, R. C. (Eds.). Marcel Dekker, Inc., New York.

Hobbs, L. 1986. Corn syrups. *Cereal Foods World* **31**: 852–858.

Homler, B. E. 1984. Properties and stability of aspartame. *Food Technol.* **38**(7): 50–55.

Horn, H. E. 1981. Corn sweeteners: Functional properties. *Cereal Foods World* **26**: 219–223.

IFT. 1986. Sweeteners: Nutritive and non-nutritive. *Food Technol.* **40**(8): 195–206.

IFT. 1988. Future ingredients. *Food Technol.* **42**(1): 60–64.

IFT. 1989. Low-calorie foods. *Food Technol.* **43**(4): 113–125.

Igoe, R. S. 1989. "Dictionary of Food Ingredients," 2nd edn. Van Nostrand Reinhold, New York.

Pepper, T. and Olinger, P. M. 1988. Xylitol in sugar-free confections. *Food Technol.* **42**(10): 98–101, 104, 106.

Pomeranz, Y. 1985. "Functional Properties of Food Components." Academic Press, Orlando, Florida.

Pszczola, D. E. 1988. Applications of aspartame in baking. *Food Technol.* **42**(1): 56, 58.

Pyler, E. J. 1988. "Baking Science and Technology," 3rd edn., vol. 1. Sosland Publishing Co., Merriam, Kansas.

Saussele, H., Jr., Ziegler, H. F., and Weideman, J. H. 1983. High fructose corn syrups for bakery applications. *Bakers Digest* **57**(4): 26–28.

Shallenberger, R. S. and Birch, G. G. 1975. "Sugar and Chemistry." Avi Publ. Co., Westport, Connecticut.

Strickler, A. J. 1982. Corn syrup selections in food applications. *In* "Food Carbohydrates," Lineback, D. R. and Inglett, G. E. (Eds.). Avi Publ. Co., Westport, Connecticut.

Zapsalis, C. and Beck, R. A. 1985. "Food Chemistry and Nutritional Biochemistry." John Wiley and Sons, New York.

CRYSTALLIZATION

Crystallization is important in the development of structure of certain foods. The structure of many candies depends on crystallization of sugar, and the structure of frozen desserts depends on crystallization of water. In each case, crystallization occurs from a sugar solution. Formation of sugar crystals in a crystalline candy requires supersaturation. As sugar crystallizes out of solution, the solution becomes decreasingly concentrated because of removal of solute. Eventually the solution is no longer supersaturated and crystallization cannot continue. During formation of ice crystals in a frozen dessert, on the other hand, the sugar solution becomes increasingly concentrated because solvent is removed from the solution. Eventually the solution is so concentrated that its freezing point is too low for further freezing under the existing conditions.

I. CRYSTALLINE CANDY

Different kinds of crystalline candy are discussed in numerous elementary textbooks. The discussion here is confined to the underlying principles.

A. Structure

In a crystalline candy, many small crystals of sugar are suspended in a small amount of a concentrated sugar solution. Other ingredients, such as acid, are in the solution. Still others, such as fat, may be adsorbed on crystal surfaces.

B. Effects of Procedures

The relatively large crystals of free-flowing sugar are dissolved. The solution is concentrated by boiling. The concentrated solution is supersaturated by cooling, and crystallization occurs during beating of the supersaturated solution. The resulting product is sufficiently cohesive and firm to hold its shape and contains crystals so small as not to be apparent.

1. Concentration

Two points must be remembered: (1) As water is boiled out of an aqueous solution, the solution becomes increasingly concentrated. (2) For a given solute, a given boiling point represents a specific concentration (Chapter 6). The sugar concentration achieved during boiling determines the amount of sugar that ultimately will crystallize and, therefore, the firmness of the product. Underconcentration results in a soft product, whereas overconcentration results in a hard product. If the concentration step is not done properly, succeeding steps cannot compensate. It is important, therefore, that the thermometer be calibrated with boiling water and that it be read carefully at the end-point. The bulb of the thermometer should be covered with the boiling syrup. If the boiling point exceeds the desired temperature, it is better to add some water and bring the solution back to the proper concentration than to gamble. Unless the mixture contains milk solids, which are subject to scorching, there is no need to stir once the sugar is dissolved.

The rate of heating is not critical unless an acidic ingredient is present, but it should be similar for all samples being compared in an experiment. Heating periods of about 15–20 min are satisfactory for solutions containing a cup of sugar. If an acidic ingredient, such as cream of tartar, is present, the rate of heating is extremely important because inversion of sucrose occurs. Inversion, which occurs during heating, affects the ability of the sucrose to crystallize in a later step. The extent of inversion depends on the concentration of the acidic ingredient and on the length of time that it has an opportunity to act. A small amount of cream of tartar can have as great an effect in a solution heated slowly to the final boiling point as a large amount of cream of tartar in a solution heated rapidly to the same boiling point.

2. Cooling

The solution should be undisturbed during cooling; otherwise, premature crystallization is likely to occur, resulting in a grainy product. Although the solution becomes supersaturated when cooling is begun, the extent of supersaturation (which is important to the crystallization process) increases as cooling proceeds. However, a factor in opposition to the increasing supersaturation is the increasing viscosity that also occurs during cooling, as is discussed below.

3. Beating

Crystallization is initiated when beating is begun. If beating is begun when the solution is quite hot, crystallization is rapid, and the product is grainy (Fig. 24.1). If beating is delayed until the solution has cooled to room temperature, beating is difficult, crystallization is slow, and the product is smooth. The pre-

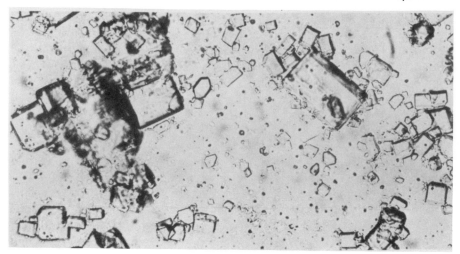

FIGURE 24.1
Sugar crystals in fondant made from sucrose and water and beaten while hot (103°C). (Courtesy of Vida Wentz.)

ceding two statements at first seem to contradict theory because if the extent of supersaturation increases with cooling, one might expect crystallization to occur more readily in the cooler, more highly supersaturated solution. Viscosity also is involved, however. In crystallization, the molecules align themselves into definite patterns to form crystals, each containing many oriented molecules. The molecules can move into their proper positions more readily in a syrup of relatively low viscosity than in one of high viscosity. A moderate extent of cooling provides for adequate development of supersaturation, in a properly concentrated syrup, without excessive viscosity.

Once initiated by beating, crystallization continues spontaneously. However, beating is continued for the purpose of keeping the crystals small and the product smooth. If crystallization occurs very rapidly, as in a hot syrup, even vigorous beating may not result in a smooth product. On the other hand, cooling to room temperature would be rather wasteful of time and energy. A temperature of 40–50°C for the beginning of beating favors both reasonably rapid crystallization and the formation of small crystals (Fig. 24.2).

C. Effects of Ingredients

In practice one would be likely to have additional constituents present for the sake of palatability. Other constituents may have either chemical or physical effects on crystallization.

1. Chemical Effects

As stated above, the chemical effect of cream of tartar is inversion of sucrose, or hydrolysis to a mixture of glucose and fructose. The reaction varies as to extent, depending on the heating conditions. The effect is retardation of crystallization. Crystallization is less rapid from a solution of mixed sugars than from a solution of a single sugar. The different kinds of molecules must sort themselves out dur-

ing crystal formation. Crystals tend to be small because of the slower crystallization (Fig. 24.3 as compared with Fig. 24.2). It is possible for hydrolysis, by producing too much invert sugar, to interfere with crystallization sufficiently to
produce a sticky or runny product. (*Note:* Interference does not necessarily
mean prevention.)

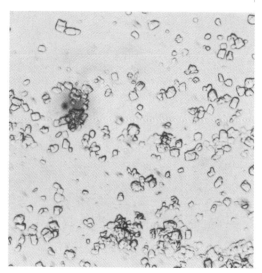

FIGURE 24.2
Sugar crystals in fondant made from sucrose and water and cooled to 40°C before beatng. (Courtesy of Vida Wentz.)

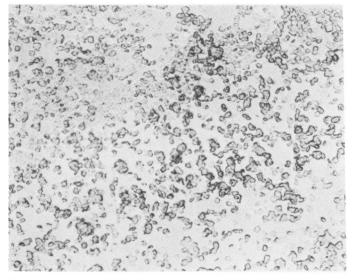

FIGURE 24.3
Sugar crystals in fondant made from sucrose, water, and cream of tartar and cooled to 40°C before
beating. (Courtesy of Vida Wentz.)

A factor that has not been taken into account in the above discussion is the effect of hydrolysis of sucrose on the boiling point of the solution. After hydrolysis, the same weight of sugar consists of more molecules than originally; therefore, the effect on boiling point is greater and a specified boiling point is reached at a lower sugar concentration than if no hydrolysis occurs. Some of the retardation of crystallization probably results from underconcentration. If the extent of hydrolysis were known, a correction in the final boiling point could be made; however, that is not the case. Although the only compensation would involve guesswork, this factor at least should be recognized.

2. Physical Effects

Fat is an example of a substance that has a physical effect on crystallization. It tends to be adsorbed on surfaces of crystals as they form and thus interfere with growth of large crystals. Protein, such as that in milk solids, acts similarly to fat. These substances exert their effect during the beating stage. Addition of corn syrup or honey also retards crystallization and thus promotes smoothness. With either of these substances, the effect is similar to that of the cream of tartar except that added sugars, rather than sugars produced chemically, have the retarding effect.

II. FROZEN DESSERTS

Frozen desserts include ice cream, for which federal and state regulations require specified minimum levels of fat and milk solids; French ice cream, which contains some egg; ice milk, which contains too little fat to qualify as ice cream; mellorine; frozen yogurt; milk sherbet; water ice; and reduced calorie frozen desserts. The discussion below pertains primarily to commercially frozen ice cream but the information has rather general application.

A. Structure

An unfrozen ice cream mix is an oil-in-water emulsion. The aqueous phase in which the fat globules are dispersed also serves as a dispersion medium for proteins in colloidal dispersion and sugars and salts in true solution. Commercial ice cream mixes are homogenized and the fat globules, therefore, are relatively small. Their surfaces are coated with a stabilizing film composed primarily of casein (Thomas, 1981).

During freezing, the characteristic structure of ice cream results from the freezing of water, with formation of crystals. Air is incorporated so that the frozen mixture is a foam. The air cells not only make the ice cream light and soft but also contribute to smoothness by mechanically hindering the formation of ice crystals. The fat globules become solidified and to a large extent they form clusters. Ideally all of the fat agglomerates, but none churns out as discernible particles. The aqueous phase never completely freezes because as ice crystals form, making the solution increasingly concentrated, the freezing point of the

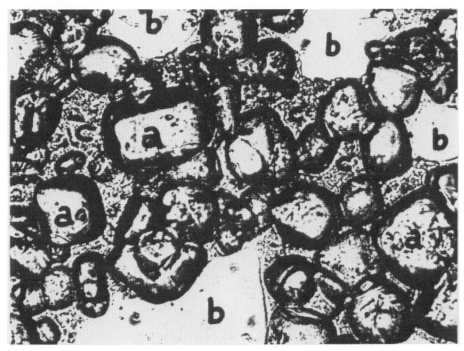

FIGURE 24.4
The internal structure of ice cream. (a) Ice crystals, average size 45–55 μm; (b) air cells, average size 110–185 μm; (c) unfrozen material. Average distance between ice crystals or between ice crystals and air cells, 6–8 μm. Average distance between air cells, 100–150 μm. (From Arbuckle, 1986.)

remaining solution decreases until the limit of the freezing capability is reached. The structure is seen in the light and electron micrographs in Figs. 24.4 and 24.5, respectively. Berger (1976) discussed ice cream emulsions in detail.

B. Effects of Procedures

The temperature at which an unfrozen mix is stored affects its stability. Coalescence of fat globules accelerates as the storage temperature is increased. The temperature to which the mix is frozen affects firmness. Decreasing temperature results in increased firmness, partly as a result of further freezing of water and partly as a result of increased solidification of fat. Butterfat is so heterogeneous that fat crystallization occurs over a wide temperature range (Berger, 1976, p. 158).

The incorporation of air with agitation of the ice cream mix during freezing results in an increase in volume, called overrun. The desirable amount of overrun depends on the specific mixture. A mixture with a high total solids content can support a higher overrun than can one with a low total solids content (Arbuckle, 1986, p. 257). The producers of commercial ice cream must meet federal

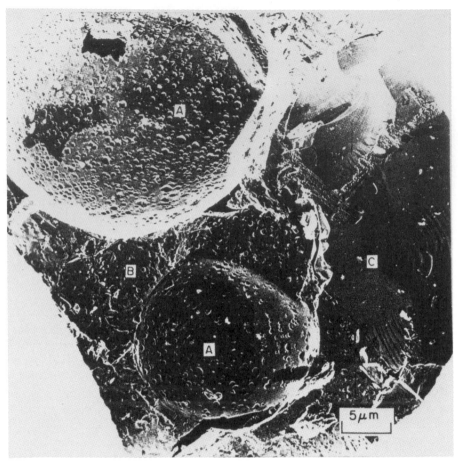

FIGURE 24.5
Structural elements in ice cream: (A) Air cells coated by fat globules; (B) continuous phase showing fat globules; and (C) ice crystal. (Reprinted from Berger, 1976.)

and state regulations, not only concerning fat content of ice cream but also concerning factors such as weight per gallon, which depends on overrun. Arbuckle (1986, Chapter 10) described overrun calculations for certain situations, as well as total solids calculations and methods for calculation of freezing point.

Agitation also brings about some coalescence of fat globules, which is desirable to a certain extent because it contributes to body and stiffness. Agitation tends to keep the clumps from becoming too large. However, if the clumps of fat are too large to begin with or form while viscosity of the mix is low, the result is poor aeration and a coarse-textured product.

Freshly made ice cream is rather soft. Arbuckle (1986, p. 233) stated that 33–67% of the water is frozen during the initial freezing phase. Firmness increases with further freezing during the hardening phase in which the mixture is not agitated. An additional 23–57% of the water may freeze during the hardening phase (Arbuckle, 1986, p. 233). Hardening involves growth of existing crystals rather than formation of new ones; therefore, the development of a fine foam and fine crystal structure during initial freezing is important to the smoothness of the ultimate product. Quick hardening helps keep the crystals small.

Temperature fluctuation during continued storage of the frozen and hardened ice cream results in migration of water vapor from smaller ice crystals to larger ones. The result is decreased smoothness. Minimization of temperature fluctuation, therefore, is desirable (Fig. 24.6).

C. Effects of Ingredients

Small ice crystals, not too close together, result in a smooth texture in ice cream; therefore, factors that control crystal size contribute to smoothness. Such factors include concentrations of nonfat milk solids and of fat that are sufficient to interfere with crystal growth. Sugar lowers the freezing point of the water, resulting in greater distances between crystals because of a smaller extent of freezing at a given temperature. Other solids, both fat and nonfat, interfere mechanically with crystallization.

Stabilizers (gums) are common ingredients of commercial frozen desserts. They are hydrocolloids. By binding large amounts of water and increasing viscosity, they inhibit crystallization. They also probably immobilize water in a gel structure. Gums not only retard crystal growth but they have some protective effect against temperature fluctuations that occur during storage. Water freed from crystals during temperature increases is held by stabilizers rather than added to existing crystals during temperature decreases. Many different gums have been used over the years. Carboxymethyl cellulose is used frequently in combination with other gums. Gelatin is added to ice cream occasionally.

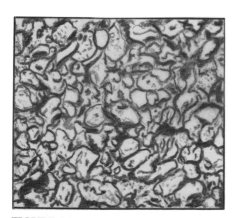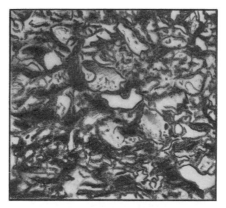

FIGURE 24.6
Micrographs of the same ice cream stored under different conditions. *Left:* Stored for 1 day at a constant low temperature. *Right:* Stored for 5 days at widely fluctuating temperatures. (From Cole, 1932.)

Corn syrups are used extensively as sweetening agents. In addition to their sweetening effect, they contribute to smoothness and to total solids content. Corn sweeteners, as liquid syrup or as dried corn syrup, are used in blends with sucrose. Arbuckle (1986, p. 75) stated that corn syrup solids usually constitute no more than 25–35% of the total sweeteners in commercial ice cream. Pearson and Ennis (1979) compared sensory properties of ice creams containing high-fructose corn syrup and corn syrups of different dextrose equivalents (DE) with properties of a control containing sucrose. Most of the combinations were preferred over the control. Mouth feel, rather than sweetness, dictated preference.

Dextrose also is used extensively in frozen desserts. The level of its use is limited to about 25% of the total sugar because, on an equal-weight basis, it lowers freezing point more than does sucrose (Arbuckle, 1986, p. 74).

High lactose concentrations promote the development of sandiness resulting from lactose crystallization. It is interesting that lactose crystallization is more likely to occur at temperatures above −18°C than at lower temperatures, possibly because of greater mobility of the lactose molecules at the higher temperatures. The availability of low-lactose milk products has made it possible to produce a high-solids ice cream that does not become gritty during storage (Arbuckle, 1986, p. 66).

Total solids content may have a profound effect on body of ice cream. Low concentrations of solids result in excessive overrun and fluffiness, hence, poor body. Very high concentrations of solids, on the other hand, can cause gumminess. Fat, fruits, chocolate, and corn syrup solids tend to depress overrun; whey solids, egg yolks, emulsifiers, and certain stabilizers enhance overrun (Arbuckle, 1986, p. 258).

Added emulsifiers contribute to desirable dryness and stiffness of ice cream but not by stabilizing the emulsion. In fact, they have a destabilizing effect on the fat emulsion during freezing. Properly used, emulsifiers result in a large number of small air cells, small ice crystals, and optimally clumped fat. The product is smooth and well bodied (Thomas, 1981). Sucrose–fatty acid esters are among the newer emulsifiers. Those with relatively high HLB (hydrophilic–lipophilic balance) were found by Buck *et al.* (1986) to promote a desirable amount of clumping and a good product.

As with other food products, there has been interest in reducing calorie content of frozen desserts. Substitution of aspartame for sucrose is one means of reducing calories. As discussed in Chapter 23, a bulking agent is needed. Corn syrup can be used as a bulking agent, and it also contributes to sweetness. Goff and Jordan (1984) used polydextrose/sorbitol combinations as bulking agents in a frozen dairy dessert with a reduced fat content and with aspartame as sweetener. The product with the highest proportion of polydextrose had an off-flavor, but the products with the other combinations were acceptable. Goff and Jordan (1984) also studied various combinations of aspartame, polydextrose, corn syrup solids (CSS), and microcrystalline cellulose (MCC). Either CSS or MCC increased the perception of smoothness and the acceptability of the frozen dairy dessert. Homler *et al.* (1987) presented a suggested formulation for a frozen dessert with reduced fat and substitution of aspartame for sucrose. Best (1989) stated that a mixture of polydextrose K, sorbitol, and maltodextrin is a good sugar substitute in frozen desserts.

The most effective means of reducing calories in dairy products is reduction of their fat content. The smooth mouth feel normally provided by fat must be supplied by a low-calorie texturizer such as a gum (Best, 1989).

In a mellorine product, milk fat is replaced with an animal or vegetable fat, most commonly a cottonseed–soybean oil blend, though Arbuckle (1986, p. 281) stated that coconut oil is most satisfactory. The fat level may or may not be reduced. Additional stabilizer and emulsifier might be used for a soft-serve mellorine containing oil. Nonfat milk solids and emulsifier might be reduced and sweetener increased for a hardened mellorine (Arbuckle, 1986, p. 281).

Berger (1976) discussed ice cream emulsions in detail.

Suggested Exercises

1. Compare the properties of two fondants containing 0.8 g cream of tartar/100 g sucrose, one heated as rapidly as possible to 114°C and the other heated very slowly to 114°C. Explain the results.
2. Compare ice creams with varying levels of added whey solids.
3. Develop ice cream or ice milk formulas for special dietary needs, for example: reinforced with extra milk solids; made with lactase-treated milk; made with a highly polyunsaturated oil (beaten into nonfat dry milk that has been reconstituted to triple strength); sweetened with various combinations of sucrose, corn syrup, sorbitol, and aspartame.

References

Arbuckle, W. S. 1986. "Ice Cream," 4th edn. Avi Publ. Co., Westport, Connecticut.

Berger, K. G. 1976. Ice cream. *In* "Food Emulsions," Friberg, K. G. (Ed.). Marcel Dekker, Inc., New York.

Best, D. 1989. High-intensity sweeteners lead low-calorie stampede. *Prepared Foods* **158**(3): 97–98.

Buck, J. S., Walker, C. E., and Pierce, M. M. 1986. Evaluation of sucrose esters in ice cream. *J. Food Sci.* **51**: 488–493.

Cole, W. C. 1932. A microscopic study of ice cream texture. *J. Dairy Sci.* **15**: 421–427.

Goff, D. H. and Jordan, W. K. 1984. Aspartame and polydextrose in a calorie-reduced frozen dairy dessert. *J. Food Sci.* **49**: 306–307.

Homler, B., Kedo, A., and Shazer, W. R. 1987. FDA approves four new aspartame uses. *Food Technol.* **41**(7): 41–42, 44.

Pearson, A. M. and Ennis, D. M. 1979. Sensory properties of high fructose corn syrup ice cream formulations. *J. Food Sci.* **44**: 810–812.

Thomas E. L. 1981. Structure and properties of ice cream emulsions. *Food Technol.* **35**(1): 41–48.

BASIC FORMULAS
AND PROCEDURES

Baking Powder Biscuits

Formula

All-purpose flour (g)	110.0
Baking powder (SAS–phosphate) (g)	5.4
Salt (g)	3.0
Fat (g)	25.0
Milk (ml)	~81.0

Procedure

1. Sift the flour, baking powder, and salt into a mixing bowl. Set oven at 218°C.
2. Cut the fat into the dry ingredients using a pastry blender. Continue cutting until no fat particles are larger than peas.
3. Pour all the milk into the flour–fat mixture at one time. Mix vigorously with a fork until the dough is stiff, cutting through

the center of the dough with the fork several times. Total number of mixing strokes should be about 30.

4. Lightly flour a rolling pin and bread board or pastry cloth. Turn the dough out onto the board and knead, with or without folding (be consistent), for 15 strokes.
5. Roll the dough until it is 1–1.5 cm thick. Use rolling guides. Cut with a biscuit cutter about 5 cm in diameter, flouring the cutter each time it is used.
6. Place biscuits on an unoiled baking sheet, using a spatula. Leave 1–1.5 cm of space between biscuits.
7. Bake at 218°C until the biscuits are golden brown (about 12 min).

Cake, Angel Food

Formula

Cake flour (g)	31.0
Sugar (g)	85.0
Egg white (g)	81.0
Salt (g)	0.4
Cream of tartar (g)	1.2

Procedure

1. Set oven at 177°C. Wash a pan approximately 19 × 9.5 × 6 cm thoroughly with detergent to remove any traces of fat.
2. Sift the flour with one-quarter of the sugar.
3. Beat the egg white until foamy. Add the salt and cream of tartar.
4. Beat until egg white is stiff but not shiny.
5. Add remaining sugar in three portions, beating enough to blend (about 10 sec) between additions.
6. Sift one-quarter of the flour–sugar mixture over the meringue and fold with a wire whip just enough to blend. Repeat until all is used.
7. Pour into ungreased pan. Bake at 177°C for about 30 min.
8. Invert on a wire rack until cake reaches room temperature before removing it from the pan. If the cake comes over the top of the pan, suspend the inverted pan between two custard cups.
9. Cut with a serrated knife.

Cake, Chocolate

Formula	*With buttermilk*	*With sweet milk*
Cake flour (g)	100.0	100.0
Baking powder (SAS–phosphate) (g)	2.2	5.0
Sodium bicarbonate (g)	0.7	—
Salt (g)	1.0	1.0

Cocoa (g)	18.0	18.0
Sugar (g)	150.0	150.0
Shortening (g)	66.0	66.0
Egg (g)	100.0	100.0
Buttermilk (ml)	68.0	—
Sweet milk (ml)	—	68.0
Vanilla (ml)	2.0	2.0

Procedure
1. Prepare pan by oiling and fitting the bottom with waxed paper.
2. Set oven at 185°C.
3. Put all of the ingredients into the bowl of an electric mixer.
4. Blend at low speed, then mix at medium speed 2½ min, stopping twice to clean the sides of the bowl with a rubber scraper.
5. Transfer 300 g of batter to a pan approx 19.9 × 9 × 6 cm. Save the remainder of the batter for display and for pH determination.
6. Bake the cake at 185°C until it springs back under the pressure of a finger, about 30 min.
7. Cool 10 min on a rack before removing from the pan.

Cake, Plain Shortened

Formula
Cake flour (g)	78.0
Baking powder (SAS–phosphate) (g)	2.9
Salt (g)	1.5
Vanilla (ml)	1.2
Milk (ml)	79.0
Shortening (hydrogenated) (g)	41.0
Sugar (g)	100.0
Egg (g)	36.0

Procedures

Regardless of the mixing method to be used: prepare the pan (19 × 9 × 5 cm, if available; otherwise use about 0.375 g of batter/ml capacity rather than the specified 300 g) by oiling the fitting the bottom with waxed paper; set oven at 190°C; sift together the flour, baking powder, and salt; add vanilla to the milk. Mix by one of the following methods; then weigh 300 g of batter into a loaf pan and bake at 190°C until the cake springs back under the pressure of a finger, about 30 min. Cool at least 10 min on a rack before removing from the pan.

Conventional Method (Mixer)
1. Cream shortening 1 min; add sugar gradually for 1 min; continue creaming for 90 sec, using medium-high speed.

2. Add egg gradually for 1 min; continue creaming for 1 min. Stop the mixer and scrape sides of bowl thoroughly to the bottom.
3. Add one-third of the flour mixture and one-third of the milk and beat for 45 sec, using low speed. Repeat twice. Using medium-high speed, blend batter for 45 sec.

Conventional Method (Hand)

1. Cream fat in a mixing bowl with a wooden spoon or a rubber scraper until it is plastic and very light. Add sugar, about a tablespoon at a time, beating after each addition until mixture looks fluffy.
2. Add egg in three portions, beating well after each addition. Beat for 1 min after the last addition of egg.
3. Add a heaping tablespoon of the sifted flour mixture; stir until flour is dampened, and beat 40 strokes.
4. Add one-third of the remaining flour mixture and one-third of the milk. Stir until flour is dampened and beat 40 strokes. Repeat twice. Beat an additional 50 strokes after all ingredients have been added.

Conventional–Sponge Method (Mixer)

1. Cream the fat for 1 min, using medium-high speed; add half the sugar gradually for 1 min; continue creaming for 90 sec. Stop the mixer and scrape sides of bowl thoroughly to the bottom.
2. With a rotary beater, beat egg until it forms soft peaks. Add remaining half of the sugar gradually, beating well after each addition. Continue beating until mixture is thick and looks like meringue. If a hand-operated beater is used, the total number of turns should be at least 250.
3. Add one-third of the flour mixture and one-third of the milk to the fat and sugar and beat for 45 sec, using low speed. Repeat twice. Stop the mixer and scrape and beat for 45 sec, using low speed.
4. Add the egg–sugar mixture and fold with a wire whisk until thoroughly combined.

Conventional–Sponge Method (Hand)

1. Cream the fat in a mixing bowl with a wooden spoon or a rubber scraper until it is very plastic and light. Add half the sugar, about a tablespoon at a time, beating after each addition until the mixture looks fluffy.
2. With a rotary beater, beat egg until it forms soft peaks. Add remaining half of the sugar gradually, beating well after each addition. Continue beating until mixture is thick and looks like meringue. If a hand-operated beater is used, the total number of turns should be at least 250.

3. To the fat–sugar mixture, add a heaping tablespoon of the sifted flour mixture, stir until the flour is dampened, and beat 40 strokes.
4. Add one-third of the remaining flour mixture and one-third of the milk. Stir until flour is dampened and beat 40 strokes. Repeat twice.
5. Add the egg–sugar mixture and fold until well combined.

Conventional–Meringue Method

Proceed as for conventional–sponge method, except add the egg yolk to the fat and half the sugar. Make a meringue with the egg white and the remainder of the sugar.

Single-Stage (One-Bowl) Method (Mixer)

1. To the flour, baking powder and salt, add the sugar, fat (room temperature), and about 50 ml of the milk.
2. Beat at low speed for 1 min and at medium speed for 1 min. Stop the mixer and scrape the bowl from the bottom.
3. Add egg and remaining milk. Beat at medium speed 2 min.

Single-Stage (One-Bowl) Method (Hand)

1. To the flour, baking powder and salt, add the sugar, fat (room temperature) and about 50 ml of the milk.
2. Stir with a wooden spoon or rubber scraper until flour is dampened, then beat hard for 2 min. Do not count time taken out to scrape the bowl and spoon or to rest during the 2 min of mixing.
3. Add egg and remaining milk; beat hard for 2 min.

Pastry-Blend Method (Mixer)

1. Beat flour mixture and fat (room temperature) at medium speed for 1 min and at medium-high speed for 2 min. Stop the mixer and scrape.
2. Using medium speed, gradually add a mixture of the sugar and half of the milk in 2 min; continue beating for 1 min. Stop the mixer and scrape.
3. Add a mixture of the egg and the remainder of the milk gradually while beating at low speed for 1 min. Stop the mixer and scrape. Beat at medium speed for 2 min.

Muffin Method (Hand)

1. Mix the sugar with the other dry ingredients.
2. Beat egg until foamy; add milk, vanilla, and fat that has been heated just enough to melt it; mix well. Use the mixture immediately for step 3.
3. Make a well in the center of the dry ingredients. Add the liquid ingredients; stir vigorously with a rubber scraper until flour has been dampened, and then beat vigorously for 1 min.

Cookies, Sugar

Formula

Flour (g)	112
Shortening (g)	32
Sugar (g)	65
Salt (g)	1
Baking powder (g)	3
Water (ml)	~9

Procedure

1. Mix by the conventional cake method.
2. Roll between guide strips to a thickness of about 6 mm.
3. Cut cookies and bake 10 min at 204°C.

Custard

Formula

Milk (ml)	250
Egg (g)	50
Sugar (g)	25
Salt (g)	0.34
Vanilla (ml)	1.2

Procedure

Stirred Custard

1. Scald milk over boiling water or in a microwave oven before measuring.
2. Beat egg slightly; add sugar and salt. Add milk very slowly at first, then more rapidly, stirring constantly.
3. Cook over hot water that is kept at simmering temperature (about 90°C). Stir constantly. Continue cooking until the mixture forms a coating on a metal spoon. Add vanilla.
4. Remove from heat at once and quickly pour into a small labeled container. Allow the custard to cool to 50°C before testing.

Baked Custard

1. Set an oven at 177°C. Scald milk over boiling water or in a microwave oven before measuring. Boil additional water.
2. Beat egg slightly; add sugar and salt. Add milk very slowly at first, then more rapidly, stirring constantly. Add vanilla.
3. Pour into matched, labeled custard cups to the top line (about ⅞ full). Place cups in a large, shallow baking pan, set pan on oven rack, and pour in boiling water until it reaches the level of the custard mixture.
4. Bake at 177°C until a knife inserted in one of the custards comes out clean. (Be consistent in testing technique.) Record baking time.

5. Remove from water at once with tongs. Cover with foil. Refrigerate overnight, if possible, or for 30 min if testing must be done on the day of preparation.

Fondant

Formula

Sugar (g)	200
Distilled water (ml)	118

Procedure

1. Determine the boiling point reading in water for each thermometer that is to be used. The final boiling point of the solutions in the experiment is 114°C; thus the end-point is indicated by the thermometer reading that is 14°C higher than that when the thermometer bulb is immersed in actively boiling water.
2. Put the ingredients in a 600-ml beaker or a pan having a diameter at the base of not more than 10 cm. Stir slightly but not enough to distribute sugar on the sides of the beaker or pan.
3. Place the beaker or pan on a small burner. Suspend the thermometer from a ring stand in such a way that its bulb is completely immersed in the center of the sugar solution. Covering the pan with a piece of aluminum foil pressed against the thermometer and sides of the pan for about 3 min permits steam to dissolve sugar crystals on the sides of the container.
4. Try to regulate the heat so that the total cooking time is about 15 min. Remove the aluminum foil after the first 3 min of boiling. Cook to a boiling point of 114°C. Remember to correct the reading if necessary.
5. Transfer quickly to a 15-cm casserole in which a calibrated thermometer has been placed. Do not scrape the syrup into the casserole or permit it to drip. Do not move the thermometer in the solution or otherwise agitate the syrup during cooling. Allow to cool to 50°C. Note whether crystals form on the surface while the syrup is cooling.
6. Record the time and beat rapidly and vigorously with a wooden spoon until the mixture loses its gloss and becomes stiff. This might happen suddenly. Record time.
7. Immediately work the fondant vigorously with the hand for 60 sec.
8. Store in a tightly covered container until microscopic examination is carried out.
9. Prepare a slide for microscopic observation by putting a drop of turpentine and a tiny grain of candy from the center of the ball on a slide. Mix with a toothpick. Add a cover slip and slide it back and forth to further disperse crystals. Observe under a magnification of about ×200. Move the slide about until a good,

uncrowded field is found. Sketch crystals from samples representing the different treatments.

Gingerbread

Formula

Flour (g)	112.0
Baking powder (g)	1.5
Baking soda (g)	1.5
Salt (g)	1.0
Ginger (g)	2.5
Molasses (g)	164.0
Buttermilk (g)	61
Shortening (g)	25
Egg (g)	24

Procedure

1. Mix by the single-stage cake method.
2. Weigh 300 g of batter into pan prepared as for cakes.
3. Bake at 190°C until the cake springs back under the pressure of a finger, about 30 min.

Ice Cream

Formula

Whipping cream (ml or g)	236
Milk (ml or g)	236
Sugar (g)	62
Vanilla (ml)	5

Procedure

1. Scald freezer cans and dashers before use.
2. Combine the ingredients. Stir to dissolve the sugar.
3. Pour the mixture into a freezer can of premeasured inside depth; measure the distance from the top of the can down to the surface of the unfrozen mixture. Calculate the depth of the mix by difference.
4. Pack the space between the freezer can and the outer container with an 8:1 ice:salt mixture. Freeze according to the freezer instructions.
5. Remove the dasher, smooth the surface of the frozen mixture, and measure from the top of the can down to the surface of the frozen mixture. Obtain the new depth by difference.
6. Calculate percentage overrun: 100(final depth − initial depth/ initial depth). Although the percentage increase in depth is calculated, overrun represents percentage increase in volume because of the relationship between depth and volume of a cylinder.

Muffins

Formula

All-purpose flour (g)	110.0
Sugar (g)	12.0
Baking powder (SAS–phosphate) (g)	5.4
Salt (g)	3.0
Egg (g)	24.0
Milk (ml)	125.0
Oil (g)	21.0

Procedure

1. Oil six muffin cups. Set oven at 218°C.
2. Sift the flour, sugar, baking powder, and salt into a mixing bowl.
3. Beat the egg slightly; add milk and oil and mix well.
4. Make a well in the dry ingredients. Add the liquid ingredients and stir immediately with a rubber scraper for 16 strokes, or until the dry ingredients are just dampened. The batter should look lumpy at the end of the mixing period.
5. Transfer the batter to the oiled muffin cups with as little agitation as possible. Each cup should be a little more than half-full.
6. Bake the muffins at 218°C until they are golden brown, about 20 min. Remove from the pans immediately.

Pancakes

Formula

Milk (ml or g)	122.0
Egg (g)	24.0
Fat (melted) (g)	12.5
Flour (g)	84.0
Baking powder (g)	3.0
Salt (g)	1.0

Procedure

1. Mix by the muffin method; beat as for a cake rather than as for a muffin.
2. Cook on a griddle without added fat; use a cooking spray, if necessary.

Pastry

Formula

All-purpose flour (g)	41
Salt (g)	1
Shortening (hydrogenated) (g)	18
Water (ml)	13

Procedure

Conventional Method
1. Have all ingredients at room temperature. Set oven at 218°C.
2. Put flour, salt, and fat in a mixing bowl having a diameter of about 15 cm. Cut in the fat with a pastry blender until no particles are larger than peas.
3. Sprinkle water on the flour–fat mixture. Then cut in with a pastry blender. After every five strokes, remove dough from blender with a rubber scraper. Count strokes, using only enough to make a dough that will form a ball when pressed together. A total of 25 strokes usually is enough. If necessary, add more water. Record the total amount used.
4. Put ball of dough on a piece of waxed paper large enough to roll it on and flatten to a thickness of about 2.5 cm with the hands. Lay a guide strip, ~3 mm thick, on each side of the dough. Cover with another piece of waxed paper and roll the dough to the thickness of the guides. Keep the dough in a rectangular shape while rolling.
5. Peel off the upper paper and invert the pastry on a baking sheet. Remove the other piece of paper.
6. Cut the wafers 4.5 × 9 cm but do not separate them from each other or remove the extra dough around the edges. Prick the wafers uniformly with a meat tenderizer.
7. Bake at 218°C until the edges of the dough are light brown. Cool before making measurements.

Mixer Method (McNeil *et al.*, 1988, Chapter 22)
1. Have all ingredients at room temperature.
2. Sift flour and salt together.
3. Cut shortening into the flour mixture for 1.5 min, using a low mixer speed.
4. Add water, and mix 45 sec. (Some variations may require shorter mixing times.)
5. Place dough on waxed paper, pat it into a ball, and roll it out between metal rolling guides to a thickness of 3 mm.
6. Invert the dough onto a baking sheet, prick with a meat tenderizer, and score into wafers of uniform size.
7. Bake at 218°C for 18 min.

Quick Loaf Bread

Formula

Margarine (softened) (g)	56
Sugar (g)	75
Egg (g)	48
Buttermilk (g)	120
Flour (g)	116
Baking soda (g)	1

| Baking powder (g) | 0.8 |
| Salt (g) | 2.7 |

Procedure
1. Mix by the conventional cake method.
2. Bake at 177°C in a pan prepared as for a cake.
3. Cool 10 min on a rack before removing from pan.

Yeast Bread

Formula

Yeast (active dry) (g)	4
Water (g or ml)	118
Sugar (g)	13
Salt (g)	4
Flour (hard wheat all-purpose) (g)	200
Shortening (g)	8

Procedure
1. Weigh all ingredients except the water. Adjust the water temperature to 45°C just before weighing or measuring and use immediately.
2. Add the yeast to part of the water.
3. Put sugar, salt, and remaining water into the mixing bowl.
4. Add the yeast dispersion and about one-quarter of the flour; mix well.
5. Cut the shortening into about one-quarter of the flour with a pastry blender. Add to batter.
6. Stir while adding enough of the remaining flour to make a ball of dough that is irregular in shape, rough and dull in appearance, and somewhat sticky to handle. Since it is important to know the total amount of flour in the dough, flour to be used in kneading and shaping the dough is taken from the weighed amount that was not needed for mixing. If the original 200 g of flour are not enough to make a dough of the right consistency, weigh additional flour for use in the dough and on the board.
7. Turn the dough out on a lightly floured board and knead 100 strokes.
8. Roll the ball of dough in an oiled bowl to lightly oil its surface. Record the time and allow to rise in a moist warm place, such as in a cabinet or unheated oven containing a bowl of boiling water or on a dish drainer over but not touching hot water in a sink, or on a wire rack over a large bowl of hot water. (Cover with a towel.)[1]
9. When the dough has risen until a light finger impression remains in the dough, record the time and gently punch the dough down, turning the edges toward the center. Weigh 300 g.

10. With the hands, flatten the dough on a lightly floured board and shape into a rectangle. Roll from end to end. Pinch to seal the seam and the ends, and place, seam down, in an oiled pan about 19 × 9 × 6 cm.
11. Allow to rise as in step 8. Record the time required for proofing.[1]
12. Bake at 218°C for 20–30 min, until the bread sounds hollow when removed from the pan and knocked on the bottom with the knuckles.
13. Remove bread from pan and cool to room temperature on a wire rack before carefully cutting through the center with a bread knife.
14. Weigh the flour not needed for mixing, kneading, and shaping. Subtract from the amount originally weighed to find the amount in the bread.

[1] Considerable time can be saved with the use of the microwave fermentation procedure described in Exercise 1 of Chapter 19.

CONVERSION TABLE FOR OVEN TEMPERATURES

Degrees	
Fahrenheit	Celsius
150	66
175	79
200	93
225	107
250	121
275	135
300	149
325	163
350	177
375	190
400	204
425	218
450	232
475	246
500	260
525	274
550	288

SOURCES OF EQUIPMENT

APPENDIX C

Type of equipment	Sources
Amylograph, Brabender (3)[a]	C. W. Brabender Instruments, 50 E. Wesley Street, South Hackensack, New Jersey 07606
Color difference meter, Hunter (4)	Hunterlab, 11491 Sunset Hills Road, Reston, Virginia 22090
Compressimeter, Baker (2)	Watkins Corp. P.O. Box 1445, W. Caldwell, New Jersey 07006
Consistometer, Bostwick (1)	Central Scientific, Cenco Center, 2600 Kostner Avenue, Chicago, Illinois 60623
Farinograph (3)	C. W. Brabender (see Amylograph)
Hydraulic press, Harco (2)	Harco Industries, 10802 N. 21st Street, Phoenix, Arizona 85029
Loaf volumeter (1)	National Manufacturing, 544 "J" Street, Lincoln, Nebraska 68502
Pressure tester, Magness Taylor (1)	D. Ballauf Co., 619 H Street NW, Washington, DC 20001
Shear, Warner–Bratzler (2)	G-R Manufacturing Co., 1317 Collins Lane, Manhattan, Kansas 66502
Shortometer, Bailey (2)	Computer Controlled Machines, 1700 Cannon Road, Northfield, Minnesota 55057
Universal penetrometer (2)	Scientific supply companies
Universal testing machines, Instron (4)	Instron Corp., 2500 Washington Street, Canton, Massachusetts 02021
Texture test system (4)	Food Technology Corp., 12300 Parklawn Drive, Rockville, Maryland 20852
Rheo-meter, Sun Scientific Co., Ltd. (3)	Surimi, Inc., Fisherman's Commerce Bldg., 4039 21st Avenue W, Suite 303, Seattle, Washington 98199
Viscometer, Brookfield (2)	Brookfield Engineering Laboratories, Inc., 240 Cushing Street, Stoughton, Massachusetts 02072

[a] Price range: 1, less than $500; 2, $500–$5000; 3, $5000–$9999; 4, $10,000 or more.

D

IMPROVISED TESTS

LINE SPREAD

The line-spread test is used to measure the consistency of foods in terms of the distance that they spread on a flat surface in a given period of time. It is suitable for foods such as white sauce, soft custard, applesauce, starch puddings, cake batters, and cream filling.

A hollow cylinder having a diameter of 5 to 8 cm is needed for this test and can be made by removing the handle from a biscuit cutter, by cutting both ends from a cookie cutter, or by cutting heavy copper tubing or plastic tubing into cylinders approximately 7 cm high. It also is possible to purchase plumbing fittings that are suitable for line-spread rings.

In addition, a diagram of concentric circles drawn 1 cm apart, the smallest having a diameter equal to the inside edge of the linespread ring, is needed. The smallest ring has no number, but the second one is numbered 1, and others are numbered consecutively. Alternatively, the circles other than the inner one may be omitted and four lines drawn at right angles to one another outward from the circle. The numbers representing centimeters are placed along these lines. Readings of spread are facilitated by marking lines into millimeters as well as centimeters.

To conduct the test, place a flat glass plate or large pie plate over the diagram. Check for evenness with a spirit level. Place the cylinder directly over the smallest circle and fill with the food to be tested. Level off with a spatula. Remove any food that falls onto the plate. Suspend a thermometer into the test material.

When the material reaches the desired temperature, lift the cylinder and allow the food to spread for exactly 2 min. Quickly take readings (at four equally spaced axes) on the limit of the spread of the substance. The line-spread value is found by averaging the four readings. The value represents the distance in centimeters that the material spreads in 2 min.

PERCENTAGE SAG

Percentage sag is a measure of gel strength. Weak gels will have higher values than will strong gels.

Insert the probe of a vernier caliper into the center of the gel and measure the height in centimeters. Remove the probe. Loosen the gel from the top of the cup with a spatula and turn out onto a flat plate. Insert the probe of a vernier caliper through the center of the gel 30 sec after it is removed from the cup. Measure the height as before. Calculate percentage sag as follows:

$$\text{Percentage sag} = \left(\frac{\text{Height in container} - \text{height out of container}}{\text{Height in container}} \right) \times 100$$

SYNERESIS OF GELS

Invert the gel on a fine wire screen supported on a funnel or into a funnel covered with nylon mesh or cheesecloth and allow to drain for 1 hr. Measure the volume of the liquid in a graduated cylinder.

FOAM STABILITY

Support a funnel (about 125 mm in diameter) with a funnel support or support stand. Place a 25-ml graduated cylinder under the funnel. Place glass wool over the stem but do not stuff it into the stem. Transfer the prepared foam to the funnel, using a plate scraper. Cover with a watch glass or plastic film. Record the volume of liquid draining from the foam every 10 min for 1 hr.

Make a graph of the results. Place drainage time on the horizontal axis and volume of liquid drained from the foam on the vertical axis. All of the variables from one experiment can be recorded on one graph if properly labeled. If a microcomputer with graphic capabilities is available the graph could be prepared with it.

SPECIFIC GRAVITY

Specific gravity is found by dividing the weight of a material by the weight of an equal volume of water. The result, being a ratio, carries no units. For the test, a small container, such as a crystallizing dish, having a capacity of about 50 ml or a diameter of about 5 cm and a smooth even rim is needed. Weigh the container (dry) to the nearest 0.1 g. Partially fill the container with boiled distilled water that is at room temperature. Place on a balance and add more water until the container is completely full. Fullness should be judged at eye level. Weigh to the nearest 0.1 g.

Fill the dry cup with the material to be tested in such a way that no air pockets are left. If egg white foam or creamed shortening and sugar are being tested, fill the cup half-full and tap on a folded towel 12 times, rotating frequently. Then add excess batter and again tap 12 times. Remove the excess batter with a spatula, starting in the center. Wipe the outside of the container and weigh to the nearest 0.1 g. Calculate the specific gravity as follows:

$$\text{Specific gravity} = \left(\frac{\text{Weight of filled container} - \text{weight of container}}{\text{Weight of water-filled container} - \text{weight of container}} \right)$$

VOLUME BY SEED DISPLACEMENT

The volume of baked products can be measured with an improvised seed displacement test. For the test, a sturdy wooden or metal box or straight-sided can, larger than the products to be measured, is needed. The container must have a flat, open top. In addition, a vernier caliper or ruler, a shallow box or pan, a 1- or 2-liter graduated cylinder, and light seed such as rapeseed (obtain from feed store) are needed.

To find the volume of the empty box or can, put it into the shallow pan or box and pour seed into it until it overflows. Always pour in the same way and from the same height to avoid variations in packing. Avoid shaking the box. Level the seed by passing a ruler across the top of the box once. Using a graduated cylinder, measure the volume of seed held by the box. Avoid shaking.

The product should not be too soft when the determination is made. Cake is more easily handled if it is about 24 hr old. If desired, it can be dusted lightly with corn starch so that the seeds will be less likely to stick during the determination. Place the product in the box or can, pour seed on it until the box overflows, and level the seed as before. Measure the volume of the seed.

Find the volume of the product by subtracting the volume of the seed that was around the product from the volume of the empty container.

Find the specific volume of the product by dividing its volume by its weight. If each product in a series is made from the same weight of dough or batter, the volumes can be compared directly.

An alternative method can be used to determine volume if the sides of the pan extend well above the top of the product baked in it, and is especially useful for delicate cakes such as sponge or angel food. The cake is cooled in the pan and its top is dusted with cornstarch. Seeds are added and leveled as in the previously described method. The volume of the cake is found by subtracting the volume of the seeds held by the pan containing the product from the volume of the empty pan.

INDEX TO VOLUME _____

In this method, the area of a slice of a baked product as found with a polar planimeter is used as an index to the volume of the product. This index furnishes a valuable comparison of products that are similar as to shape. Products should be made from equal amounts of batter or dough and baked in identical pans if a comparison of the values is to be meaningful.

For the test a compensating polar planimeter measuring in square centimeters and a flat level surface that will not be damaged by the needle point of the pole arm or by tape are needed. A large board such as a bread board is appropriate.

Baked products of uniform age should be used for testing. Cut two or more slices 1.5–2.5 cm thick from representative parts of the product. A cutting box should be used to ensure uniformity. Place the slice on a large piece of paper and draw around it with a sharp, soft lead pencil. Be careful not to press the product out of shape and to hold the pencil at the same angle for the entire sample and for all samples to be compared.

Mark each tracing at one point. This becomes the starting and finishing point for the measurement.

Place the paper on the flat surface and tack or tape in place.

Assemble the planimeter as indicated in the directions. Place the assembled planimeter on the paper with the tracer arm to your right and perpendicular to the front of the table.

Test the placement of the planimeter by quickly moving the tracer point around the area to be traced. The planimeter should be placed so that the tracer arm moves freely and stays on the paper. The angle between the two arms must stay between 15 and 186°. Reposition the planimeter if the conditions are not met.

Prior to using the planimeter for the first time, practice reading it. The first digit of the reading is the number on the dial at which the pointer is standing or which it has just passed. The pointer usually will fall between two numbers. The smaller of the two numbers is recorded. The second and third digits are read from the measuring wheel. Observe the line that is opposite or has just been passed by the zero line on the vernier scale. The second digit is the number, or major division on the measuring wheel, and the third is found from the single divisions on the measuring wheel. The last digit is the graduation on the vernier that makes a straight line with the wheel graduation. The value of the units depends on the calibration of the instrument.

It may be advantageous to practice finding the area of a known square or rectangle. Place the tracer point on the starting line. Record the reading. Follow the outline of the figure carefully by moving the tracer point in a clockwise direction. Avoid any counterclockwise movement. If the tracer point leaves the line slightly, the error can be compensated for by moving the tracer arm an equivalent amount away from the line in the other direction. Do not use a guide in tracing straight lines. When the tracer has returned to the starting point, make a second reading. The difference between the two readings, multiplied by the appropriate factor (0.1 for instrument calibrated in centimeters), is the area of the figure. The area found for the known square or rectangle should agree with the calculated area within 2%.

Measure the area of the outlined baked product as described. Repeat the measurement at least once. The areas found should agree within 2%. Average the readings.

TABLE FOR SENSORY DIFFERENCE TESTS [a]

	Paired difference and duo–trio				Triangle				Paired preference		
	Probability levels				Probability levels				Probability levels		
(n)	0.05	0.01	0.001	(n)	0.05	0.01	0.001	(n)	0.05	0.01	0.001
5	—	—	—	5	4	5	—	—	—	—	—
6	—	—	—	6	5	6	—	—	—	—	—
7	7	7	—	7	5	6	7	7	7	—	—
8	7	8	—	8	6	7	8	8	8	8	—
9	8	9	—	9	6	7	8	9	8	9	—
10	9	10	10	10	7	8	9	10	9	10	—
11	9	10	11	11	7	8	10	11	10	11	11
12	10	11	12	12	8	9	10	12	10	11	12
13	10	12	13	13	8	9	11	13	11	12	13
14	11	12	13	14	9	10	11	14	12	13	14
15	12	13	14	15	9	10	12	15	12	13	14
16	12	14	15	16	9	11	12	16	13	14	15
17	13	14	16	17	10	11	13	17	13	15	16
18	13	15	17	18	10	12	13	18	14	15	17
19	14	15	17	19	11	12	14	19	15	16	17
20	15	16	18	20	11	13	14	20	15	17	18
21	15	16	18	21	11	13	15	21	16	17	19
22	16	17	19	22	12	14	15	22	17	18	19
23	16	18	20	23	12	14	16	23	17	19	20
24	17	19	20	24	13	15	16	24	18	19	21
25	18	19	21	25	13	15	17	25	18	20	21
26	18	20	22	26	14	15	17	26	19	20	22
27	19	20	22	27	14	16	18	27	20	21	23
28	19	21	23	28	15	16	18	28	20	22	23
29	20	22	24	29	15	17	19	29	21	22	24
30	20	22	24	30	15	17	19	30	21	23	25
31	21	23	25	31	16	18	20	31	22	24	25
32	22	24	26	32	16	18	20	32	23	24	26
33	22	24	26	33	17	18	21	33	23	25	27
34	23	25	27	34	17	19	21	34	24	25	27

Paired difference and duo–trio				Triangle				Paired preference			
Probability levels					Probability levels					Probability levels		
(*n*)	0.05	0.01	0.001	(*n*)	0.05	0.01	0.001	(*n*)	0.05	0.01	0.001
35	23	25	27	35	17	19	22	35	24	26	28
36	24	26	28	36	18	20	22	36	25	27	29
37	24	26	29	37	18	20	22	37	25	27	29
38	25	27	29	38	19	21	23	38	26	28	30
39	26	28	30	39	19	21	23	39	27	28	31
40	26	28	30	40	19	21	24	40	27	29	31
41	27	29	31	41	20	22	24	41	28	30	32
42	27	29	32	42	20	22	25	42	28	30	32
43	28	30	32	43	20	23	25	43	29	31	33
44	28	31	33	44	21	23	26	44	29	32	34
45	29	31	34	45	21	24	26	45	30	32	34
46	30	32	34	46	22	24	27	46	31	33	35
47	30	32	35	47	22	24	27	47	31	33	36
48	31	33	36	48	22	25	27	48	32	34	36
49	31	34	36	49	23	25	28	49	32	34	37
50	32	34	37	50	23	26	28	50	33	35	37
60	37	40	43	60	27	30	33	60	39	41	44
70	43	46	49	70	31	34	37	70	44	47	50
80	48	51	55	80	35	38	41	80	50	52	56
90	54	57	61	90	38	42	45	90	55	58	61
100	59	63	66	100	42	45	49	100	61	64	67

[a]Number of correct or concurring responses needed to establish a significant difference for paired difference and duo–trio (one tailed) tests; triangle tests (one tailed); and paired preference tests (two tailed).

INDEX

FOOD SCIENCE AND TECHNOLOGY
A Series of Monographs

Maynard A. Amerine, Rose Marie Pangborn, and Edward B. Roessler, PRINCIPLES OF SENSORY EVALUATION OF FOOD. 1965.

Martin Glicksman, GUM TECHNOLOGY IN THE FOOD INDUSTRY. 1970.

L. A. Goldblatt, AFLATOXIN. 1970.

Maynard A. Joslyn, METHODS IN FOOD ANALYSIS, second edition. 1970.

A. C. Hulme (ed.), THE BIOCHEMISTRY OF FRUITS AND THEIR PRODUCTS. Volume 1—1970. Volume 2—1971.

G. Ohloff and A. F. Thomas, GUSTATION AND OLFACTION. 1971.

C. R. Stumbo, THERMOBACTERIOLOGY IN FOOD PROCESSING, second edition. 1973.

Irvin E. Liener (ed.), TOXIC CONSTITUENTS OF ANIMAL FOODSTUFFS. 1974.

Aaron M. Altschul (ed.), NEW PROTEIN FOODS: Volume 1, TECHNOLOGY, PART A —1974. Volume 2, TECHNOLOGY, PART B—1976. Volume 3, ANIMAL PROTEIN SUPPLIES, PART A—1978. Volume 4, ANIMAL PROTEIN SUPPLIES, PART B— 1981. Volume 5, SEED STORAGE PROTEINS—1985.

S. A. Goldblith, L. Rey, and W. W. Rothmayr, FREEZE DRYING AND ADVANCED FOOD TECHNOLOGY. 1975.

R. B. Duckworth (ed.), WATER RELATIONS OF FOOD. 1975.

Gerald Reed (ed.), ENZYMES IN FOOD PROCESSING, second edition. 1975.

A. G. Ward and A. Courts (eds.), THE SCIENCE AND TECHNOLOGY OF GELATIN, 1976.

John A. Troller and J. H. B. Christian, WATER ACTIVITY AND FOOD. 1978.

A. E. Bender, FOOD PROCESSING AND NUTRITION. 1978.

D. R. Osborne and P. Voogt, THE ANALYSIS OF NUTRIENTS IN FOODS. 1978.

Marcel Loncin and R. L. Merson, FOOD ENGINEERING: PRINCIPLES AND SELECTED APPLICATIONS. 1979.

Hans Riemann and Frank L. Bryan (eds.), FOOD-BORNE INFECTIONS AND INTOXICATIONS, second edition. 1979.

N. A. Michael Eskin, PLANT PIGMENTS, FLAVORS AND TEXTURES: THE CHEMISTRY AND BIOCHEMISTRY OF SELECTED COMPOUNDS. 1979.

J. G. Vaughan (ed.), FOOD MICROSCOPY. 1979.

J. R. A. Pollock (ed.), BREWING SCIENCE, Volume 1—1979. Volume 2—1980.

Irvin E. Liener (ed.), TOXIC CONSTITUENTS OF PLANT FOODSTUFFS, second edition. 1980.

J. Christopher Bauernfeind (ed.), CAROTENOIDS AS COLORANTS AND VITAMIN A PRECURSORS: TECHNOLOGICAL AND NUTRITIONAL APPLICATIONS. 1981.

Pericles Markakis (ed.), ANTHOCYANINS AS FOOD COLORS. 1982.

Vernal S. Packard, HUMAN MILK AND INFANT FORMULA. 1982.

George F. Stewart and Maynard A. Amerine, INTRODUCTION TO FOOD SCIENCE AND TECHNOLOGY, second edition. 1982.

Malcolm C. Bourne, FOOD TEXTURE AND VISCOSITY: CONCEPT AND MEASUREMENT. 1982.

R. Macrae (ed.), HPLC IN FOOD ANALYSIS. 1982.

Héctor A. Iglesias and Jorge Chirife, HANDBOOK OF FOOD ISOTHERMS: WATER SORPTION PARAMETERS FOR FOOD AND FOOD COMPONENTS. 1982.

John A. Troller, SANITATION IN FOOD PROCESSING. 1983.

Colin Dennis (ed.), POST-HARVEST PATHOLOGY OF FRUITS AND VEGETABLES. 1983.

P. J. Barnes (ed.), LIPIDS IN CEREAL TECHNOLOGY. 1983.

George Charalambous (ed.), ANALYSIS OF FOODS AND BEVERAGES: MODERN TECHNIQUES. 1984.

David Pimentel and Carl W. Hall, FOOD AND ENERGY RESOURCES. 1984.

Joe M. Regenstein and Carrie E. Regenstein, FOOD PROTEIN CHEMISTRY: AN INTRODUCTION FOR FOOD SCIENTISTS. 1984.

R. Paul Singh and Dennis R. Heldman, INTRODUCTION TO FOOD ENGINEERING. 1984.

Maximo C. Gacula, Jr., and Jagbir Singh, STATISTICAL METHODS IN FOOD AND CONSUMER RESEARCH, 1984.

S. M. Herschdoerfer (ed.), QUALITY CONTROL IN THE FOOD INDUSTRY, second edition. Volume 1—1984. Volume 2 (first edition)—1968. Volume 3 (first edition)—1972.

Y. Pomeranz, FUNCTIONAL PROPERTIES OF FOOD COMPONENTS. 1985.

Herbert Stone and Joel L. Sidel, SENSORY EVALUATION PRACTICES. 1985.

Fergus M. Clydesdale and Kathryn L. Wiemer (eds.), IRON FORTIFICATION OF FOODS. 1985.

John I. Pitt and Ailsa D. Hocking, FUNGI AND FOOD SPOILAGE. 1985.

Robert V. Decareau, MICROWAVES IN THE FOOD PROCESSING INDUSTRY. 1985.

S. M. Herschdoerfer (ed.), QUALITY CONTROL IN THE FOOD INDUSTRY, second edition. Volume 2—1985. Volume 3—1986. Volume 4—1987.

F. E. Cunningham and N. A. Cox (eds.), MICROBIOLOGY OF POULTRY MEAT PRODUCTS. 1986.

Walter M. Urbain, FOOD IRRADIATION. 1986.

Peter J. Bechtel, MUSCLE AS FOOD. 1986.

H. W.-S. Chan, AUTOXIDATION OF UNSATURATED LIPIDS. 1986.

Chester O. McCorkle, Jr., ECONOMICS OF FOOD PROCESSING IN THE UNITED STATES. 1987.

Jethro Jagtiani, Harvey T. Chan, Jr., and William S. Sakai, TROPICAL FRUIT PROCESSING. 1987.

Solms et al.: FOOD ACCEPTANCE AND NUTRITION. 1987.

R. Macrae, HPLC IN FOOD ANALYSIS, Second Edition, 1988.

Pearson and Young: MUSCLE AND MEAT BIOCHEMISTRY, 1989.